LabVIEW™
Student Edition

LabVIEW™
Student Edition

Robert H. Bishop

Marquette University

Upper Saddle River   Boston   Columbus   San Francisco   New York
Indianapolis   London   Toronto   Sydney   Singapore   Tokyo   Montreal
Dubai   Madrid   Hong Kong   Mexico City   Munich   Paris   Amsterdam   Cape Town

Vice President and Editorial Director, ECS: *Marcia J. Horton*
Acquisitions Editor: *Norrin Dias*
Editorial Assistant: *Sandra Rodriguez*
Senior Managing Editor: *Scott Disanno*
Procurement Manager: *Mary Fischer*
Senior Specialist, Program Planning & Support: *Maura Zaldivar-Garcia*
Product Marketing Manager: *Bram van Kempen*
Field Marketing Manager: *Demetrius Hall*
Marketing Assistant: *Jon Bryant*
Cover Design: *National Instruments*
Composition: *Jouve India*

Library of Congress Cataloging-in-Publication Data on File.

10 9 8 7 6 5 4 3 2

ISBN-13: 978-0-13-401133-2
ISBN-10:   0-13-401133-3

*To my parents, W. Robert Bishop
and Anna Maria DiPietro Bishop*

CONTENTS

2   Virtual Instruments   57

PREFACE

Learning with LabVIEW™ is the textbook that accompanies the *LabVIEW Student Edition* from National Instruments, Inc. This textbook, as well as the LabVIEW software, has undergone a significant revision from the previous edition. *Learning with LabVIEW* teaches basic programming concepts in a graphical environment and relates them to real-world applications in academia and industry. Understanding and using the intuitive and powerful LabVIEW software is easier than ever before. As you read through the book and work through the examples, we hope you will agree that this book is more of a personal tour guide than a software manual.

The LabVIEW graphical development environment was built specifically for applications in engineering and science, with built-in functionality designed to reduce development time for design and simulation in signal processing, control, communcations, electronics and more. The *LabVIEW Student Edition* delivers all the capabilities of the full version of LabVIEW, widely considered the industry standard for design, test, measurement, automation, and control applications. With LabVIEW, students can design graphical programming solutions to their homework problems and laboratory experiments—an ideal tool for science and engineering applications—that is also fun to use! The *LabVIEW Student Edition* affords students the opportunity for self-paced learning and independent project development.

The goal of this book is to help students learn to use LabVIEW on their own. With that goal in mind, this book is very art-intensive with over 400 figures in all. That means that there are numerous screen captures in each section taken from a typical LabVIEW session. The figures contain additional labels and pointers added to the LabVIEW screen captures to help students understand what they are seeing on their computer screens as they follow along in the book.

The most effective way to use *Learning with LabVIEW* is to have a concurrent LabVIEW session in progress on your computer and to follow along with the steps in the book. A directory of virtual instruments has been developed by the author exclusively for use by students using *Learning with LabVIEW* and is available on **www.pearsonhighered.com/bishop**. These virtual instruments complement the material in the book. In most situations, the students are asked to develop the virtual instrument themselves following instructions given in the book, and then compare their solutions with the solutions provided by the author to obtain immediate feedback. In other cases, students are asked to run a specified virtual instrument as a way to demonstrate an important LabVIEW concept.

GAINING PRACTICAL EXPERIENCE AND SOLVING REAL-WORLD PROBLEMS

With higher education emphasizing hands-on laboratory experience, many educational institutions have improved their laboratory facilities in order to increase student exposure to practical problems. College graduates are gaining vital experience in acquiring and analyzing data, constructing computer-based simulations of physical systems, and multipurpose computer programming. LabVIEW offers a powerful, efficient, and easy-to-use development environment, allowing educators to teach their students a wide range of topics with just one open, industry-standard tool. It can also transform the way engineers, scientists, and students around the world design, prototype, and deploy cutting-edge technology. Customers and students at more than 25,000 companies and schools are using LabVIEW and modular hardware from National Instruments to simplify technology development and increase productivity. From testing next-generation gaming systems to creating breakthrough medical devices, the resulting innovative technologies are impacting millions of people worldwide.

The cover of this edition of *Learning with LabVIEW* shows ten interesting application areas that use LabVIEW in the solution process.

1. Marine Mammals
2. Cycling Speed
3. NASA Microshutters
4. Cancer Lab
5. Roman Amphitheater
6. Nucor Steel Plant
7. La Selva Rainforest
8. ARV Drone
9. Tumor Treatment
10. Autonomous Vehicle

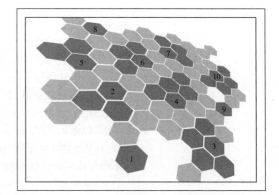

1. Marine Mammals

LabVIEW was used to develop a reliable data acquisition system that can collect and analyze ultrasonic signals produced by killer whales to perform live audio monitoring and movement tracking to study how the marine mammals use echolocation in their natural habitat.

2. **Cycling Speed**

   Using LabVIEW, NI hardware, and off-the-shelf components, a unique wind tunnel was created to conduct continuous loops of development and testing for cyclists and their bikes by evaluating and optimizing their riding positions.

3. **NASA Microshutters**

   The challenge of synchronizing the motion of a magnet moving more than 1 meter per second with the opening and closing of tens of thousands of tiny microelectromechanical system microshutters was addressed for application on the James Webb space telescope to enable the next stepping-stone toward understanding the universe.

4. **Cancer Lab**

   LabVIEW helps advance a unique and more accurate form of cancer treatment where accelerated charged particles can precisely target deep-seated cancer cells opening new perspectives for patients with respect to surgery or traditional radiotherapy.

5. **Roman Amphitheater**

   Building a distributed monitoring system to identify structural damage and reduce seismic risk of the Arena di Verona—one of the largest open-air opera theaters in the world built in the first century—involved creating a data acquisition system to communicate with existing measurement sensors in the theatrer that continuously monitors signals from sensor devices, tracks changes in a crack and the onset of structural vibrations related to environmental changes.

6. **Nucor Steel Plant**

   Employing LabVIEW programmable automation controllers has led to a tenfold increase in efficiency and drastically reduced the facility automation costs for steel recycling that reduces the amount of energy consumed while improving safety for workers.

7. **La Selva Rainforest**

   Employing a wide range of wireless environmental measurements using a single device that provides robotic control, remote configuration, and data sharing over the Web, researchers in Costa Rica are testing a hypothetical explanation of the exchange of CO_2 and other materials between the forest floor and the atmosphere—the so-called "gap theory."

8. **ARV Drone**

   A unique LabVIEW application is used to control an unmanned aerial vehicle, acquire the images taken by its cameras, and process the images with NI vision hardware in areas only viewable and accessible by air for fast detection of casualties in difficult-to-access areas, such as earthquake and collapsed-building sites.

9. **Tumor Treatment**

   Using NI hardware and software, a new medical device was created to treat breast tumors in a less invasive and nearly painless procedure, dramatically reducing the emotional and physical discomfort of patients undergoing tumor treatment.

10. **Autonomous Vehicle**

   A semiautonomous vehicle was developed using LabVIEW software and hardware that allows a blind driver to successfully navigate, control speed, and avoid collision through a secure driving course.

THE *LABVIEW STUDENT EDITION* SOFTWARE

The *LabVIEW Student Edition* software package is a powerful and flexible instrumentation, analysis, and control software platform for PCs running Microsoft Windows or Apple Macintosh OS X when installed using the MacOS Boot Camp application or a virtual machine environment. The student edition is designed to give students early exposure to the many uses of graphical programming. LabVIEW not only helps reinforce basic scientific, mathematical, and engineering principles, but it encourages students to explore advanced topics as well. Students can run LabVIEW programs designed to teach a specific topic, or they can use their skills to develop their own applications. LabVIEW provides a real-world, hands-on experience that complements the entire learning process.

WHAT'S NEW WITH THE LABVIEW STUDENT EDITION?

The demand for LabVIEW in colleges and universities has led to the development of *LabVIEW Student Edition* based on the industry version of LabVIEW. This is a new and significant software revision that delivers all of the graphical programming capabilities of the full edition. With the student edition, students can design graphical programming solutions for their classroom problems and laboratory experiments on their personal computers. The *LabVIEW Student Edition* features include the following:

- Express VIs that bring interactive, configuration-based application design for acquiring, analyzing, and presenting data.

- Interactive measurement assistants to make creating data acquisition and instrument control applications easier than ever.

- Full LabVIEW advanced analysis capability.

- Full compatibility with all National Instruments data acquisition and instrument control hardware.

- Support for all data types used in the LabVIEW Full Development System.

New LabVIEW software features introduced in this new edition of *Learning with LabVIEW* include:

- Updated editing and manipulations of wires including Owned Labels and Align/Distribute Wires, and you can now wire an error cluster directly to Boolean functions.

- Enhancements in creating subVIs including Create SubVI from Selection that automatically builds the connector pane and front panel of the subVI; and the icon and connector pane that is always visible for easier subVI wiring and icon editing.

- New editing features including new ways to view Control and Function palettes, the ability to change parameters of multiple objects at one time, and you can now right-click a cluster constant and select View Cluster As Icon from the shortcut menu to reduce the size of cluster constants on the block diagram.

- Improved programming constructs including automatically concatenating arrays leaving loops and conditionally processing loop outputs.

- A redesigned Getting Started window to allow ready access to the resources you need to use LabVIEW emphasizing common tasks, such as creating projects and opening existing files, and presents less common tasks, such as downloading additional drivers and products, through submenus.

This latest edition of *Learning with LabVIEW* also features:

- A myDAQ Building Blocks section in each chapter to provide the opportunity to apply the concepts introduced in the chapter using the myDAQ hardware.

- New relaxed readings that illustrate how students, engineers, and scientists are using LabVIEW to solve real-world problems.

- Information on how to become a certified LabVIEW user for career advancement and employment opportunities.

ORGANIZATION OF LEARNING WITH LABVIEW

This textbook serves as a LabVIEW resource for students. The pace of instruction is intended for both undergraduate and graduate students. The book is comprised of 11 chapters and should be read sequentially when first learning LabVIEW. For more experienced students, the book can be used as a reference book by using the index to find the desired topics. The 11 chapters are as follows:

CHAPTER 1: LabVIEW Basics—This chapter introduces the LabVIEW environment and helps orient students when they open a virtual instrument. Concepts such as windows, toolbars, menus, and palettes are discussed.

CHAPTER 2: Virtual Instruments—The components of a virtual instrument are introduced in this chapter: front panel, block diagram, and icon/connector pair. This chapter also introduces the concept of controls (inputs) and indicators (outputs) and how to wire objects together in the block diagram. Express VIs are introduced in the chapter.

CHAPTER 3: Editing and Debugging Virtual Instruments—Resizing, coloring, and labeling objects are just some of the editing techniques introduced in this chapter. Students can find errors using execution highlighting, probes, single-stepping, and breakpoints, just to name a few of the available debugging tools.

CHAPTER 4: SubVIs—This chapter emphasizes the importance of re-using code and illustrates how to create a VI icon/connector. It also shows parallels between LabVIEW and text-based programming languages.

CHAPTER 5: Structures—This chapter presents loops, case structures, and flat sequence structures that govern the execution flow in a VI. The Formula Node is introduced as a way to implement complex mathematical equations.

CHAPTER 6: Arrays and Clusters—This chapter shows how data can be grouped, either with elements of the same type (arrays) or elements of a different type (clusters). This chapter also illustrates how to create and manipulate arrays and clusters.

CHAPTER 7: Charts and Graphs—This chapter shows how to display and customize the appearance of single and multiple charts and graphs.

CHAPTER 8: Data Acquisition—The basic characteristics of analog and digital signals are discussed in this chapter, as well as the factors students need to consider when acquiring and generating these signals. This chapter introduces students to the Measurement and Automation Explorer (MAX) and the DAQ Assistant.

CHAPTER 9: Strings and File I/O—This chapter shows how to create and manipulate strings on the front panel and block diagram. This chapter also explains how to write data to and read data from files.

CHAPTER 10: NI LabVIEW MathScript RT Module—This chapter introduces the interactive MathScript environment, which combines a mathematics-oriented text-based language with the intuitive graphical dataflow programming of LabVIEW. Both the interactive MathScript environment for command line computation and the MathScript Node for integrating textual scripts within the LabVIEW block diagram are discussed.

CHAPTER 11: Analysis—LabVIEW can be used in a variety of ways to support analysis of signals and systems. Several important analysis topics are discussed in this chapter, including how to use LabVIEW for signal generation, signal processing, linear algebra, curve fitting, formula display on the front panel, differential equations, finding roots (zero finder), and integration and differentiation.

APPENDIX A: Instrument Control—The components of an instrument control system using a GPIB or serial interface are presented in this appendix. Students are introduced to the notion of instrument drivers and of using the Measurement and Automation Explorer (MAX) to detect and install instrument drivers. The Instrument I/O Assistant is introduced.

APPENDIX B: LabVIEW Developer Certification—Discusses the certification process to validate your expertise, beginning with the Certified LabVIEW Associate Developer (CLAD), continuing with the Certified LabVIEW Developer (CLD), and culminating with the Certified LabVIEW Architect (CLA). It includes a CLAD introductory-level certification practice test with complete answers, along with information on additional resources to help you prepare for the examination.

The important pedagogical elements in each chapter include the following:

1. A brief table of contents and a short preview of what to expect in the chapter.

2. A list of chapter goals to help focus the chapter discussions.

3. Margin icons that focus attention on a helpful hint or on a cautionary note.

Helpful hint   Cautionary note

4. An end-of-chapter summary and list of key terms.

KEY TERMS

5. Sections entitled **Building Blocks** near the end of each chapter present the continuous development and modification of a virtual instrument for calculating and generating a pulse-width modulated signal. The student is expected to construct the VIs based on the instructions given in the sections. The same VI is used as the starting point and then improved in each

subsequent chapter as a means for the student to practice with the newly introduced chapter concepts.

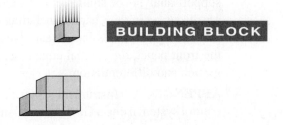

6. At the end of each chapter, we include a **myDAQ Building Blocks** section to provide the opportunity to apply the concepts introduced in the chapter using the myDAQ hardware. Each myDAQ Building Blocks exercise will help you build a project that will involve controlling LEDs and reading temperature from a thermistor to create a temperature monitoring system. As with the Building Blocks, the same VI is used as the starting point and then improved in each subsequent chapter as a means for the student to apply the chapter concepts in a my DAQ hardware environment.

7. Many worked examples are included in each chapter including several new examples introduced in this edition. In most cases, students construct the VIs discussed in the examples by following a series of instructions given in the text. In the early chapters, the instructions for building the VIs are quite specific, but in the later chapters, students are expected to construct the VIs without precise step-by-step instructions. Of course, in all chapters, working versions of the VIs are provided for all examples in the Learning directory included as part of the *LabVIEW Student Edition*. Here is a sample of the worked examples:

    - Temperature system demonstration.
    - Solving a set of linear differential equations.
    - Building your first virtual instrument.
    - Computing area, diameter, and radius of a circle.
    - Computing and graphing the time value of money.
    - Studying chaos using the logistic difference equation.
    - Acquiring data.
    - Writing ASCII data to a file.

8. A section entitled *Relaxed Reading* that describes how LabVIEW is being utilized to solve interesting real-world problems. The material is intended to give students a break from the technical aspects of learning LabVIEW and to stimulate thinking about how LabVIEW can be used in various other situations.

9. End-of-chapter exercises, problems, and design problems reinforce the main topics of the chapter and provide practice with LabVIEW.

ORIGINAL SOURCE MATERIALS

Learning with LabVIEW was developed with the aid of important references provided by National Instruments. The main references are the various Lab-VIEW help manuals found at the website www.ni.com/manuals. They provide information on LabVIEW programming concepts, step-by-step instructions for using LabVIEW, and reference information about LabVIEW VIs, functions, palettes, menus, and tools. You can access this same material in LabVIEW by selecting **Help≫LabVIEW Help** (see Chapter 1 of this book for more details on accessing the LabVIEW help). By design, there is a strong correlation between some of the material contained in the various LabVIEW help manuals and the material presented in this book. Our goal here has been to refine the information content and make it more accessible to students learning LabVIEW on their own.

OPERATING SYSTEMS AND ADDITIONAL SOFTWARE

It is assumed that the reader has a working knowledge of either the Windows or the Mac OS X operating system. If your computer experience is limited, you may first want to spend some time familiarizing yourself with your computer in order to understand the operation of your Mac or PC. You should know how to access pull-down menus, open and save files, download software from the Internet, and use a mouse. You will find previous computer programming experience helpful—but not necessary.

A set of virtual instruments has been developed by the author for this book. You will need to obtain the Learning directory from the companion website to this book at Prentice Hall:

http://www.pearsonhighered.com/bishop

For more information, you may also want to visit the NI Student Edition website at

http://www.ni.com/labviewse

All of the VI examples in this book were tested by the author an HP Elite-Book running Windows 7. Obviously, it is not possible to verify each VI on all the available Windows and Macintosh platforms that are compatible with Lab-VIEW so if you encounter platform-specific difficulties, please let us know.

If you would like information on upgrading to the LabVIEW Professional Version, please write to

National Instruments
att.: Academic Sales
11500 North Mopac Expressway
Austin, TX 78759

or visit the National Instruments website: **http://www.ni.com**

LIMITED WARRANTY

The software and the documentation are provided "as is," without warranty of any kind, and no other warranties, either expressed or implied, are made with respect to the software. National Instruments does not warrant, guarantee, or make any representations regarding the use, or the results of the use, of the software or the documentation in terms of correctness, accuracy, reliability, or otherwise and does not warrant that the operation of the software will be uninterrupted or error-free. This software is not designed with components and testing for a level of reliability suitable for use in the diagnosis and treatment of humans or as critical components in any life-support systems whose failure to perform can reasonably be expected to cause significant injury to a human. National Instruments expressly disclaims any warranties not stated herein. Neither National Instruments nor Pearson Education shall be liable for any direct or indirect damages. The entire liability of National Instruments and its dealers, distributors, agents, or employees are set forth above. To the maximum extent permitted by applicable law, in no event shall National Instruments or its suppliers be liable for any damages, including any special, direct, indirect, incidental, exemplary, or consequential damages, expenses, lost profits, lost savings, business interruption, lost business information, or any other damages arising out of the use, or inability to use, the software or the documentation even if National Instruments has been advised of the possibility of such damages.

ACKNOWLEDGMENTS

Thanks to all the folks at National Instruments for their assistance and input during the development of *Learning with LabVIEW*. A very special thanks to Gretchen Edelmon and Jayme Walton of NI for providing day-to-day support during the final months of the project. Thanks also go to the following reviewers: Austin B. Asgill, Southern Polytechnic State University; Jeff Doughty, Northeastern University; Buford Furman, San Jose State University; R. Glynn Holt, Boston University; Thomas Koon, Binghamton University; Milivoje Kostic, Northern Illinois University; Jay Porter, Texas A&M University; and Yi Wu, Penn State University. Finally, I wish to express my appreciation to Lynda Bishop for assisting me with the manuscript preparation, for providing valuable comments on the text, and for handling my personal day-to-day activities associated with the entire production.

KEEP IN TOUCH!

The author and the staff at Pearson and at National Instruments would like to establish an open line of communication with the users of the *LabVIEW Student Edition*. We encourage students to e-mail the author with comments and suggestions for this and future editions.

Keep in touch!

ROBERT H. BISHOP
robert.bishop@marquette.edu

GRETCHEN EDELMON
Senior Manager, Academic Courseware Systems
gretchen.edelmon@ni.com

LabVIEW™
Student Edition

LabVIEW Basics

Welcome to the *LabVIEW Student Edition*! **LabVIEW** is a powerful and complex programming environment. Once you have mastered the various concepts introduced in this book you will have the ability to develop applications in a graphical programming language and to develop virtual instruments for design, control, and test engineering. This introductory chapter provides a basic overview of LabVIEW and its components.

GOALS

1. Installation of the *LabVIEW Student Edition*.

2. Familiarization with the basic components of LabVIEW.

3. Introduction to front panels and block diagrams, shortcut and pull-down menus, palettes, VI libraries, and online help.

1.1   SYSTEM CONFIGURATION REQUIREMENTS

The *LabVIEW Student Edition* textbook bundle includes the LabVIEW Student Edition software for Windows 8.1/8/7/Vista/XP and Mac OS X, when installed using the MacOS Boot Camp application or a virtual machine environment. This textbook without software is available under the name *Learning with LabVIEW*.

Windows 8.1/8/7/Vista (32 bit and 64 bit), Windows XP SP3 (32 bit)

Processor:	Pentium III/Celeron 866 MHz or equivalent minimum
	Pentium 4/M or equivalent recommended
RAM:	256 MB minimum; 1 GB recommended
Screen Resolution:	1024 × 768 pixels
Operating System:	Windows 7/Vista/XP
Disk Space:	3.67 GB (includes default drivers from the NI Device Drivers DVD)

Macintosh OS X 10.3 or later

Windows Application:	Mac OS Boot Camp, virtual machine application, or Windows emulator
Processor:	Intel-based processor
RAM:	256 MB minimum; 1 GB recommended
Screen Resolution:	1024 × 768 pixels
Operating System:	Mac OS X (10.3 or later)
Disk Space:	1.2 GB on the Windows partition (for the complete installation excluding drivers)

1.2   INSTALLING THE *LABVIEW STUDENT EDITION*

Disable any automatic virus detection programs before you install. Some virus detection programs interfere with the installation program.

Windows

1. Log on as an administrator or as a user with administrator privileges.
2. Download and install LabVIEW using the detailed instructions found on this webpage: from http://www.ni.com/white-paper/13413/en/. If you plan

to use NI data acquisition hardware such as NI myDAQ, be sure to download and install the hardware drivers as described in step 3 of the installation instruction page. Use the serial number printed on the inside cover to activate the software.

3. Detailed instructions for activation are included in the installation instructions as well as on this page: http://www.ni.com/white-paper/14146/en/.

4. After installation, check your hard disk for viruses and enable any virus detection programs you disabled.

5. To use the *LabVIEW Help*, the Measurement & Automation Explorer (MAX) interactive help system, and the NI Example Finder, you must have Microsoft Internet Explorer 5.0 or later.

LabVIEW relies on a licensing activation process to obtain an activation code. The Activation Wizard is a part of the NI License Manager that takes you through the steps of enabling software to run on a machine. If you have questions regarding activation, visit the NI website at http://www.ni.com/labviewse/.

Macintosh

1. To install the downloaded LabVIEW Student Edition software on your Mac Computer, you will need a Mac Computer that supports the Mac OS Boot Camp application or a virtual machine environment, and a licensed copy of the Microsoft Windows operating system.

2. Boot the Mac OS computer into Windows or launch the Windows virtual machine.

3. Follow the instructions for Windows installation in the section above.

The LabVIEW installation folder might contain the following items:

- *cintools—Tools for calling C object code from LabVIEW*
- *AppLibs—(LabVIEW Professional Development System). Contains stub for stand-alone applications.*

Refer to LabVIEW Help for a complete list of the folders installed in LabVIEW directory structure.

1.3   THE LABVIEW ENVIRONMENT

LabVIEW is short for **Lab**oratory **V**irtual **I**nstrument **E**ngineering **W**orkbench. It is a powerful and flexible graphical development environment created by the folks at National Instruments—a company that creates hardware and software products that leverage computer technology to help engineers and scientists take measurements, control processes, and analyze and store data. National Instruments was founded over thirty-five years ago in Austin, Texas by James Truchard (known as Dr. T), Jeffrey Kodosky, and William Nowlin. At the time, all three were working on sonar applications for the U.S. Navy at the Applied Research Laboratories at The University of Texas at Austin. Searching for a way to connect test equipment to DEC PDP-11 computers, Dr. T decided to develop an interface bus. He recruited Jeff and Bill to join him in his endeavor, and together they successfully developed LabVIEW and the notion of a "virtual instrument." In the process they managed to infuse their new company—National Instruments—with an entrepreneurial spirit that still pervades the company today.

Engineers and scientists in research, development, production, test, and service industries as diverse as automotive, semiconductor, aerospace, electronics, chemical, telecommunications, and pharmaceutical have used and continue to use LabVIEW to support their work. LabVIEW is a major player in the area of testing and measurements, industrial automation, and data analysis. For example, SpaceX uses LabVIEW in a variety of mission applications, including the Falcon 9 rocket that carried the Dragon spacecraft into space. The Dragon is a free-flying reusable spacecraft that was the first commercially built vehicle to dock with the International Space Station and demonstrate a cargo resupply mission. This book is intended to help you learn to use LabVIEW as a programming tool and to serve as an introduction to the power of graphical programming and the myriad applications to which it can be applied.

LabVIEW programs are called **Virtual Instruments**, or VIs for short. LabVIEW is different from text-based programming languages (such as Fortran and C) in that LabVIEW uses a graphical programming language, known as the G programming language, to create programs relying on graphic symbols to describe programming actions. LabVIEW uses a terminology familiar to scientists and engineers, and the graphical icons used to construct programs in G are easily identified by visual inspection. You can learn LabVIEW even if you have little programming experience, but you will find knowledge of programming fundamentals helpful. If you have never programmed before (or maybe you have programming experience but have forgotten a few things) you may want to review the basic concepts of programming before diving into the G programming language.

LabVIEW provides an extensive library of virtual instruments and functions to help you in your programming. The MathScript RT Module environment is discussed in detail in Chapter 10. The MathScript RT Module provides a text-based command line environment complementing the LabVIEW graphical programming environment. Users have the capability to make quick calculations or computations at a LabVIEW command line prompt. The MathScript RT Module also enables users to integrate their scripts with LabVIEW block diagrams, easily mixing graphical programming and powerful user interfaces with text-based scripts. LabVIEW contains application-specific libraries for data acquisition (discussed in Chapter 8), file input/output (discussed in Chapter 9), and data analysis (discussed in Chapter 11). It includes conventional program debugging tools with which you can set breakpoints, single-step through the program, and animate the execution so you can observe the flow of data. Editing and debugging VIs is the topic of Chapter 3. LabVIEW has a set of VIs for data presentation on various types of charts and graphs. Chapter 7 discusses the process of presenting data on charts and graphs.

The LabVIEW system consists of LabVIEW application executable files and many associated files and folders. LabVIEW uses files and directories to store information necessary to create your VIs. Some of the more important files and directories are:

1. The LabVIEW executable. Use this to launch LabVIEW.
2. The vi.lib directory. This directory contains libraries of VIs such as data acquisition, instrument control, and analysis VIs; it must be in the same directory as the LabVIEW executable. Do not change the name of the vi.lib directory, because LabVIEW looks for this directory when it launches. If you change the name, you cannot use many of the controls and library functions.
3. The examples directory. This directory contains many sample VIs that demonstrate the functionality of LabVIEW.
4. The user.lib directory. This directory is where you can save VIs you have created, and they will appear in the LabVIEW **Functions** palette.
5. The instr.lib directory. This directory is where your instrument driver libraries are placed if you want them to appear in the **Functions** palette.
6. The Learning directory. This file contains a library of VIs that you will use with the *Learning with LabVIEW* book.

The files in the Learning directory must be downloaded from the site http://www.pearsonhighered.com/bishop. You can access the Pearson Higher Education website through the Internet using any standard Web browser.

1.4   THE GETTING STARTED SCREEN

When you launch LabVIEW by double-clicking on its icon, the Getting Started screen appears as in Figure 1.1. The Getting Started window provides ready access to LabVIEW resources to enhance the ease of use. Creating projects and opening existing files—some of the more common tasks—can be easily

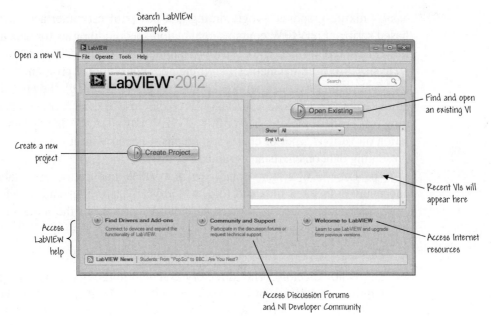

FIGURE 1.1
The Getting Started screen.

accomplished through the use of the handy buttons front and center in the window. Reaching online resources is made much easier by having access to submenus on the Getting Started window that link to NI resources to download additional drivers and products and connect to discussion forums and developer communities. The Getting Started screen contains a navigation dialog box that includes introductory material and common commands. The dialog box includes a menu that allows users to quickly create new projects, select among the most recently opened LabVIEW files, find examples, and search the LabVIEW Help. The upper section of the Getting Started screen contains *file* functions, such as creating a new VI or for opening an existing VI. Across the lower section are *resource* functions. Information and resources are reachable from the Getting Started screen to help you learn about LabVIEW using online manuals and tutorials, with convenient access to Internet links providing news, technical content, example programs, and training information from the National Instruments website. These categories are dynamically populated with a list of articles relating to LabVIEW.

On the Getting Started screen you can:

- Click on **File≫New** to open the **New** dialog box to create a blank VI or one based on a VI template.
- Click on **Open Existing** to open an existing VI by browsing to find the desired file, or select a VI from the list of recently used VIs.

On the lower section of the screen you can:

- **Find Drivers and Add-ons:** Link to resources to search for NI Device Drivers, connect to the NI Instrument Device Driver to install drivers for connected instruments, and search for LabVIEW add-ons to expand functionality.

- **Community and Support:** Access NI discussion forums and developer communities.

- **Welcome to LabVIEW:** Access resources to help with LabVIEW and learn more about new features as they become available.

 Throughout this book, use the left mouse button (if you have one) unless we specifically tell you to use the right one.

Searching the LabVIEW Examples

In this exercise you will search through the list of example VIs and demonstrations that are included with the *LabVIEW Student Edition*. Open the LabVIEW application and get to the Getting Started screen. The search begins at the LabVIEW Getting Started screen by selecting **Find Examples**, as shown in Figure 1.2. The NI Example Finder screen displays the numerous examples available with LabVIEW, as illustrated in Figure 1.3. The examples can be

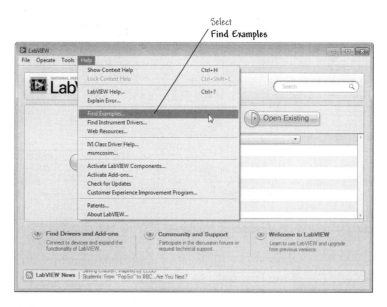

FIGURE 1.2
LabVIEW examples.

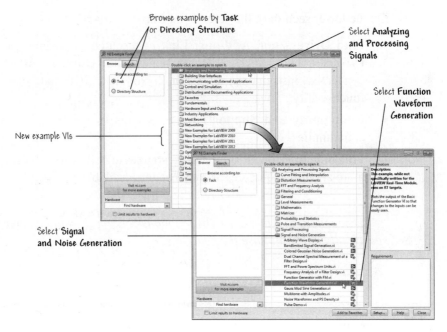

FIGURE 1.3
The NI Example Finder.

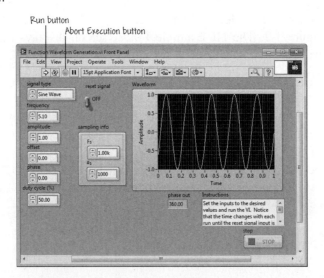

FIGURE 1.4
The Function Waveform Generation front panel.

browsed by Task or by Directory Structure. In Figure 1.3, the examples are sorted by Task. As indicated in Figure 1.3, on the **Browse** tab of the **NI Example Finder** you can find **New Examples for LabVIEW** associated with each new update to LabVIEW. These VIs are examples demonstrating new features of the current version of LabVIEW.

To reach the desired example—in this case, we are searching for the Function Waveform Generation—select **Analyzing and Processing Signals**, as shown in Figure 1.3. Selecting **Signal and Noise Generation** and **Function Waveform Generation** opens up the associated virtual instrument (VI) (more on VIs in Chapter 2). Just for fun, you can start the VI running and see what happens. Start the VI by clicking on the **Run** button, as shown in Figure 1.4. Stop the VI by clicking on the **Stop** button. Try it!   ◆

1.5   PANEL AND DIAGRAM WINDOWS

An untitled front panel window appears when you select **File≫New VI** from the Getting Started screen. The front panel window is the interface to your VI code and is one of the two LabVIEW windows that comprise a virtual instrument. The other window—the block diagram window—contains program code that exists in a graphical form (such as icons, wires, etc.).

Front panels and block diagrams consist of graphical objects that are the G programming elements. Front panels contain various types of controls and indicators (that is, inputs and outputs, respectively). Block diagrams contain terminals corresponding to front panel controls and indicators, as well as constants, functions, subVIs, structures, and wires that carry data from one object to another. Structures are program control elements (such as For Loops and While Loops).

Figure 1.5 shows a front panel and its associated block diagram. You can find the virtual instrument First VI.vi shown in Figure 1.5 in the Chapter 1 folder within the directory Learning. That VI can be located by choosing **Open Existing** on the Getting Started screen and navigating to the Chapter1 folder in the Learning directory and then selecting First VI.vi. Once you have the VI front panel open, find the **Run** button on the panel toolbar and click on it. Your VI is now running. You can turn the knob and vary the different inputs and watch the output changes reflected in the graph. Give it a try! If you have difficulty getting things to work, then just press ahead with the material in the next sections and come back to this VI when you are ready.

1.5.1   Front Panel Toolbar

A toolbar of command buttons and status indicators that you use for controlling VIs is located on both the front panel and block diagram windows. The front panel toolbar and the block diagram toolbar are different, although they do each contain some of the same buttons and indicators. The toolbar that appears at the top of the front panel window is shown in Figure 1.6.

While the VI is executing, the **Abort Execution** button appears. Although clicking on the abort button terminates the execution of the VI, as a general rule you should avoid terminating the program execution this way and either let the

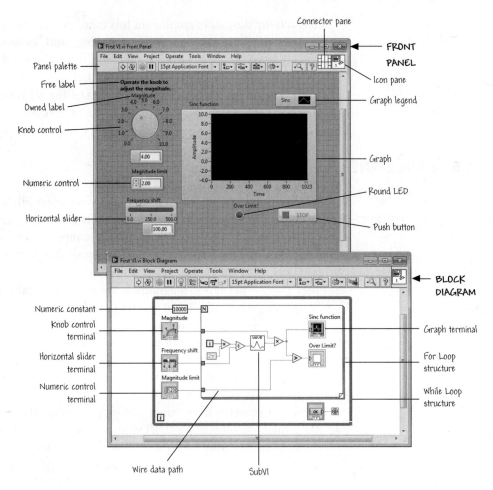

FIGURE 1.5
A front panel and the associated block diagram.

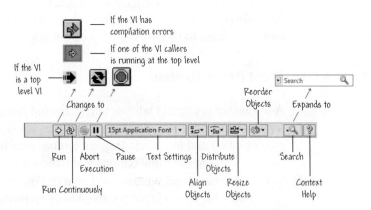

FIGURE 1.6
The front panel toolbar.

VI execute to completion or incorporate a programmatic execution control (that is, an on-off switch or button) to terminate the VI from the front panel.

The **Broken Run** button replaces the **Run** button when the VI cannot compile and run due to coding errors. If you encounter a problem running your VI, just click on the **Broken Run** button, and a window will automatically appear on the desktop that lists all the detected program errors. And then, if you double-click on one of the specific errors in the list, you will be taken automatically to the location in the block diagram (that is, to the place in the code) where the error exists. This is a great debugging feature! More discussion on the issue of debugging VIs can be found in Chapter 3.

Clicking on the **Run Continuously** button leads to a continuous execution of the VI. Clicking on this button again disables the continuous execution—the VI stops when it completes normally. The behavior of the VI and the state of the toolbar during continuous run is the same as during a single run started with the **Run** button.

The **Pause/Continue** button pauses VI execution. To continue program execution after pausing, press the button again, and the VI resumes execution.

The **Text Settings** pull-down menu, shown in Figure 1.7, sets font options— font type, size, style, and color.

The **Align Objects** pull-down menu sets the preferred alignment of the various objects on either the front panel or the block diagram. After selecting the desired objects for alignment, you can set the preferred alignment for two or more objects. For example, you can align objects by their left edges or by their top edges. The various alignment options are illustrated in the **Align Objects** pull-down menu shown in Figure 1.8. Aligning objects is very useful in organizing the VI front panel (and the block diagram, for that matter). On the surface, it may appear that aligning the front panel objects, while making things "neat and pretty," does not contribute to the goal of a functioning VI. As you gain experience with constructing VIs, you will find that they are easier to debug and

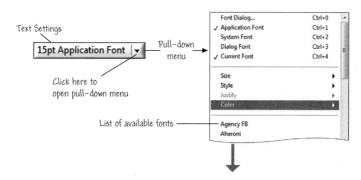

FIGURE 1.7
The **Text Settings** pull-down menu.

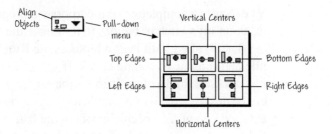

FIGURE 1.8
The **Align Objects** pull-down menu.

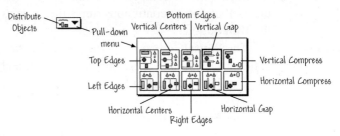

FIGURE 1.9
The **Distribute Objects** pull-down menu.

verify if the interface (front panel) and the code (block diagram) are organized to allow for easy visual inspection.

The **Distribute Objects** options pull-down menu, shown in Figure 1.9, sets the preferred distribution for two or more objects. For example, you can evenly space selected objects, or you can remove all the space between the objects.

The **Resize Objects** objects pull-down menu, shown in Figure 1.10, is used to resize multiple front panels to the same size. For example, this feature allows you to select multiple objects and resize them all to the same height as the object within the selected group with the maximum height.

The **Reorder** pull-down menu, shown in Figure 1.11, is used to move objects relative to each other. This is important because placing an object (such as a label) over or partially covering a control or indicator slows down screen updates and can make the control or indicator flicker. You can also use the **Reorder** pull-down menu to group objects so they can be moved as a single unit, or lock objects so they cannot be moved. Objects that are grouped maintain their relative arrangement and size when they are moved and resized and locked objects maintain their position on the front panel.

The **Search** button (in Windows) provides an application-wide search capability returning results from the Controls and Functions palettes, the help system,

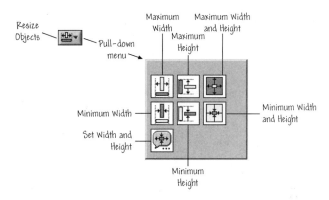

FIGURE 1.10
The **Resize Objects** pull-down menu.

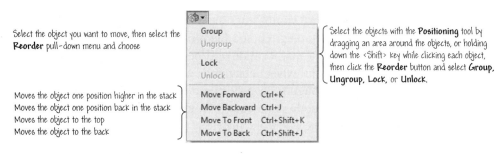

FIGURE 1.11
The **Recorder** pull-down menu.

and ni.com. The search text box appears in the upper-right corner of the Getting Started window as well as on the front panel and the block diagram.

Clicking on the **Context Help** button displays the **Context Help** window. When activated, this feature shows information about LabVIEW objects, such as functions, VIs, constants, structures, and palettes, as you move the cursor over the object.

1.5.2 Block Diagram Toolbar

The block diagram toolbar contains many of the same buttons as the front panel toolbar. Five additional program debugging features are available on the block diagram toolbar and are enabled via the buttons, as shown in Figure 1.12.

Clicking on the **Highlight Execution** button enables execution highlighting. As the program executes, you will be able to see the data flow through the code on the block diagram. This is extremely helpful for debugging and

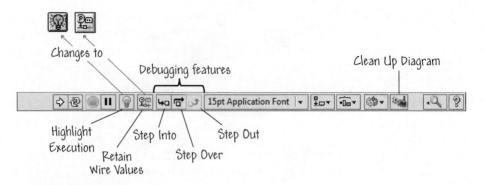

FIGURE 1.12
The block diagram toolbar.

verifying proper execution. In the execution highlighting mode, the button changes to a brightly lit light bulb.

Clicking on the **Retain Wire Values** button allows you to create probes after the VI runs. In other words, after the VI executes successfully, you can place a probe on the block diagram and the most recent value of the data that flowed through the wire at that location will be displayed in the probe. You may find this capability useful when debugging complex block diagrams.

LabVIEW debugging capabilities allow you to single-step through the VI node to node. A **node** is an execution element, such as a For Loop or subVI. You will learn more about the different types of nodes as you proceed through the book, but for the time being you can think of a node as a section of the computer code that you want to observe executing. Each node blinks to show it is ready to execute.

The **Step Into** button allows you to step into a node. Once you have stepped into the node, you can single-step through the node.

The **Step Over** button steps over a node. You are in effect executing the node without single-stepping through the node.

The **Step Out** button allows you to step out of a node. By stepping out of a node, you can complete the single-stepping through the node and go to the next node.

The **Warning** indicator only appears when there is a potential problem with your block diagram. The appearance of the warning does not prevent you from executing the VI. You can enable the Warning indicator on the Options...≫ Debugging menu in the **Tools** pull-down menu.

The **Clean Up Diagram** button is used to automatically clean up the block diagram by rerouting all existing wires and reordering objects. For example, you can configure LabVIEW to automatically move certain objects to desired locations on the block diagram, place a given number of pixels between block diagram objects, and compact the block diagram layout. Chapter 3 discusses further the editing of the block diagram.

1.6 SHORTCUT MENUS

LabVIEW has two types of menus—**pull-down** menus and **shortcut** menus. We will focus on shortcut menus in this section and on pull-down menus in the next section. Our discussions here in Chapter 1 are top-level; we reserve the detailed discussions on each menu item for later chapters as they are used.

To access a shortcut menu, position the cursor on the desired object on the front panel or block diagram and click the right mouse button on a PC-compatible or hold down the <command> key and then click the mouse button on the Mac. In most cases, a shortcut menu will appear since most LabVIEW objects have shortcut menus with options and commands. You will also find that you can **right-click** on the empty front panel and block diagram space, giving you access to the **Controls** and **Functions** palettes and other important palettes. The options presented to you on shortcut menus depend on the selected object—right-clicking on a numeric control will open a different shortcut menu than right-clicking on a For Loop. When you construct a program, you will use shortcut menus extensively!

Many shortcut and pull-down menus contain submenus, as shown in Figures 1.13 and 1.14.

On a PC-compatible, right-click on the object to open a shortcut menu. On a Mac, press <command> and simultaneously click on the object.

Menu items that expand into submenus are called **hierarchical** menus and are denoted by a right arrowhead on the menu. Hierarchical menus will present you with different types of options and commands. One typical option is the so-called mutually exclusive option. This means that if you select the option (by

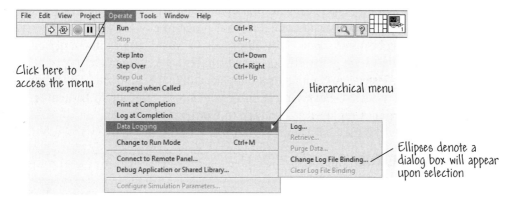

FIGURE 1.13
An example of a pull-down menu expanding into a submenu.

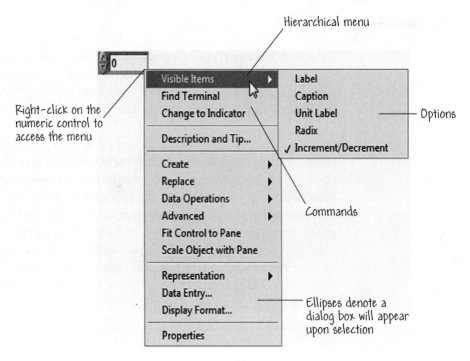

FIGURE 1.14
An example of a shortcut menu.

clicking on it), a check mark will appear next, indicating the option is selected; otherwise the option is not selected.

Right-clicking on different areas of an object may lead to different shortcut menus. If you right-click and do not see the anticipated menu selection, then right-click somewhere else on the object.

Another type of menu item opens dialog boxes containing options for you to use to modify and configure your program elements. Menu items leading to dialog boxes are denoted by ellipses (...). Menu items without right arrowheads or ellipses are generally commands that execute immediately upon selection. In some instances, commands are replaced in the menu by their inverse commands when selected. For example, after you choose **Change to Indicator**, the menu selection is replaced by **Change to Control**.

1.7 PULL-DOWN MENUS

The menu bar at the top of the LabVIEW screen, shown in Figure 1.15, contains the important pull-down menus. In this section we will introduce the pull-down menus: **File**, **Edit**, **View**, **Project**, **Operate**, **Tools**, **Window**, and **Help**.

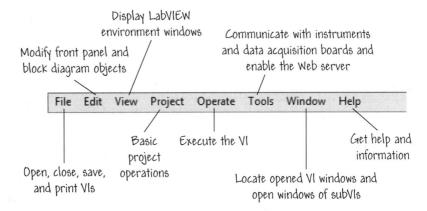

FIGURE 1.15
The menu bar.

    Some menu items are unavailable while a VI is in Run mode.

1.7.1   File Menu

The **File** pull-down menu, shown in Figure 1.16, contains commands associated with file manipulations. For example, you can create new VIs or open existing ones from the **File** menu. You use options in the **File** menu primarily to open, close, save, and print VIs. As you observe each pull-down menu, notice that selected commands and options have shortcuts listed beside them. These shortcuts are keystroke sequences that can be used to choose the desired option without pulling down the menu. For example, you can open a new VI by pressing <Ctrl-O> on the keyboard, or you can access the **File** pull-down menu and select **Open**.

1.7.2   Edit Menu

The **Edit** menu, shown in Figure 1.17, is used to modify front panel and block diagram objects of a VI. The **Undo** and **Redo** options are very useful when you are editing because they allow you to undo an action after it is performed, and once you undo an action you can redo it. By default, the maximum number of undo steps per VI is ninety-nine—you can increase or decrease this number if desired.

1.7.3   View Menu

The **View** menu, shown in Figure 1.18, contains items that display LabVIEW environment information. For example, you can easily access the Getting Started

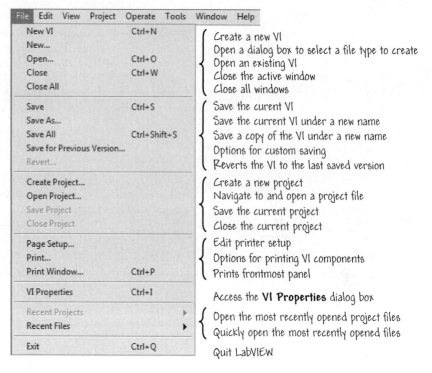

FIGURE 1.16
Pull-down menus—**File**.

screen and display the **Tools**, **Controls**, and **Functions** palettes. If you want to show the palettes on the desktop, you can use the **View** menu to select either palette (or both) (more on palettes in the next section).

*Some menu items are available only on specific operating systems and with specific LabVIEW development systems. For example, the **Navigation** window is available only in the LabVIEW Full and Professional Development Systems.*

Selecting **Error List, VI Hierarchy,** and **LabVIEW Class Hierarchy** opens display windows that list the errors in the current VI and let the user view the subVIs and other nodes that make up the VIs in memory, view the hierarchy of LabVIEW classes in memory, and search the LabVIEW class hierarchy, respectively. **Browse Relationships** contains items for viewing aspects of the current VI and its hierarchy. These functions are helpful when developing and debugging complex VIs. Selecting the **Probe Watch Window** opens a dialog box to manage probes. Probes are used to check intermediate values on a wire as a VI runs and are very helpful if you have a complicated block diagram with a series of operations that might return incorrect data. Chapter 3 gives more information on the use of probes.

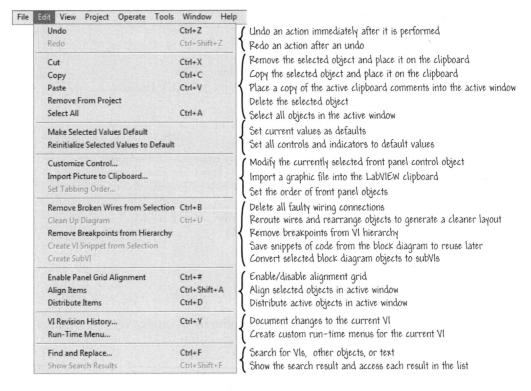

FIGURE 1.17
Pull-down menus—**Edit**.

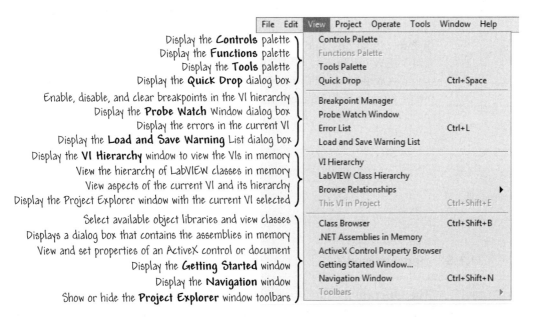

FIGURE 1.18
Pull-down menus—**View**.

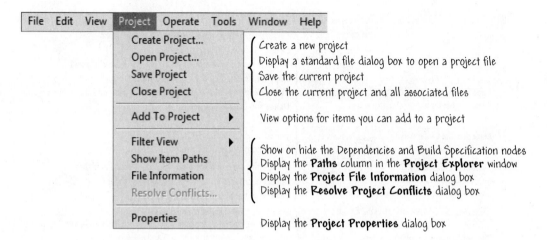

FIGURE 1.19
Pull-down menus—**Project**.

Class Browser, .NET Assemblies in Memory, ActiveX Control Property Browser, and Toolbars are helpful for advanced users of LabVIEW, and not discussed in detail in this book.

1.7.4   Project Menu

The **Project** menu, shown in Figure 1.19, contains items used for basic LabVIEW project operations, such as opening, closing, and saving projects; building specifications; and editing deployments. Projects are used to group together LabVIEW (and non-LabVIEW) files and to deploy or download files to targets, such as real-time (RT) or FPGA targets. The use of projects is most likely to be of interest to advanced students and users of LabVIEW. This book does not present topics associated with projects in any detail.

   A target is any device that can run a VI.

1.7.5   Operate Menu

The **Operate** menu, shown in Figure 1.20, can be used to run or stop your VI execution, change the default values of your VI, and switch between the run mode and edit mode.

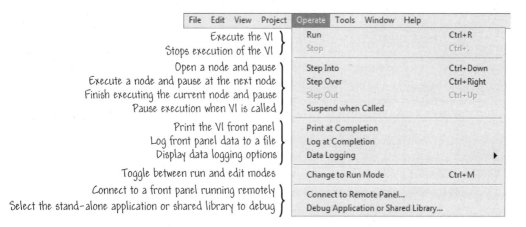

FIGURE 1.20
Pull-down menus—**Operate**.

1.7.6   Tools Menu

The **Tools** menu, shown in Figure 1.21, is used to communicate with instruments and data acquisition boards, compare VIs, build applications, enable the Web server, and access other options of LabVIEW.

The appearance and behavior of LabVIEW applications can be customized in the Tools≫Options dialog box, which is organized to enhance usability. The **Options** dialog box is shown in Figure 1.22. Select **New and Changed** in the menu on the left-hand side of the box to discover the latest enhancements. Some of the options you may want to modify include Colors, Fonts, and Debugging on the **Environment** page. On that page, for example, you can set the number of steps for the **Maximum undo steps per VI** option (the default number is 99). One key option for the **Block Diagram** is **Use transparent name labels**, which, if selected, allows you to create free labels without a border or background color (more on this in Chapter 3).

An important link is to the main National Instruments website, where you can obtain general information about the company and its products. If you use LabVIEW to control external instruments, you will be interested in the **Visit Instrument Driver Network** link which connects you to thousands of LabVIEW-ready instrument drivers. This can be found in the Tools≫ Instrumentation hierarchical menu. Refer to Appendix A for more information on instrument drivers.

1.7.7   Window Menu

The **Window** menu, shown in Figure 1.23, is used for a variety of activities. You can toggle between the front panel and block diagram, and you can "tile" both windows so you can see them at the same time (one above the other or

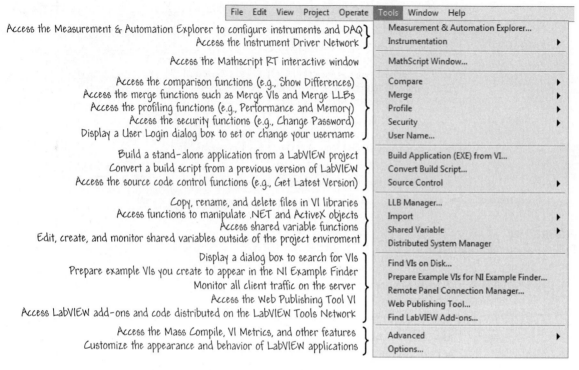

Access the Measurement & Automation Explorer to configure instruments and DAQ
Access the Instrument Driver Network

Access the Mathscript RT interactive window

Access the comparison functions (e.g., Show Differences)
Access the merge functions such as Merge VIs and Merge LLBs
Access the profiling functions (e.g., Performance and Memory)
Access the security functions (e.g., Change Password)
Display a User Login dialog box to set or change your username

Build a stand-alone application from a LabVIEW project
Convert a build script from a previous version of LabVIEW
Access the source code control functions (e.g., Get Latest Version)

Copy, rename, and delete files in VI libraries
Access functions to manipulate .NET and ActiveX objects
Access shared variable functions
Edit, create, and monitor shared variables outside of the project environment

Display a dialog box to search for VIs
Prepare example VIs you create to appear in the NI Example Finder
Monitor all client traffic on the server
Access the Web Publishing Tool VI
Access LabVIEW add-ons and code distributed on the LabVIEW Tools Network

Access the Mass Compile, VI Metrics, and other features
Customize the appearance and behavior of LabVIEW applications

FIGURE 1.21
Pull-down menus—**Tools**.

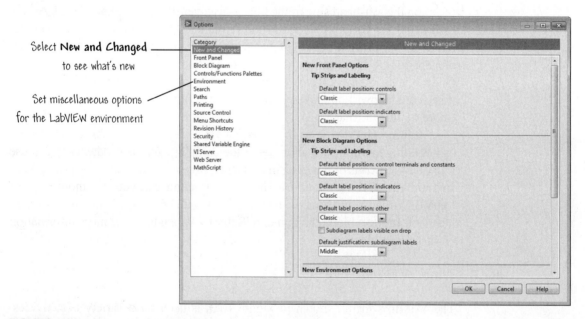

Select **New and Changed** to see what's new

Set miscellaneous options for the LabVIEW environment

FIGURE 1.22
The Options dialog box is accessed under Tools≫Options.

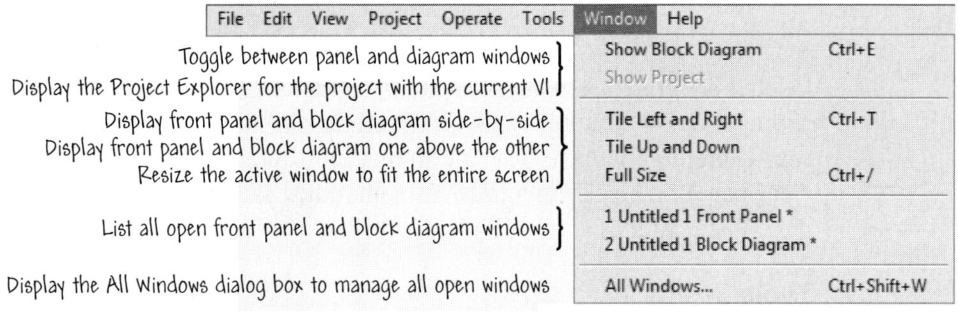

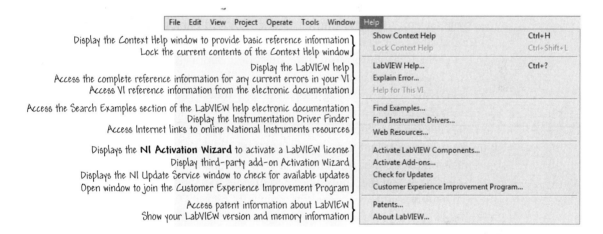

FIGURE 1.23
Pull-down menus—**Windows**.

FIGURE 1.24
Pull-down menus—**Help**.

side-by-side). All the open VIs are listed in the menu (at the bottom), and you can switch between the open VIs.

1.7.8   Help Menu

The **Help** menu, shown in Figure 1.24, provides access to the extensive Lab-VIEW online help facilities. You can view information about front panel or block diagram objects, activate the online reference utilities, and view information about your LabVIEW version number and computer memory. The direct pathways to the Internet are provided by **Web Resources** which connect you directly to the main source of information concerning LabVIEW.

1.8   PALETTES

Palettes are graphical panels that contain various tools and objects used to create and operate VIs. You can move the palettes anywhere on the desktop that you want—preferably off to one side so that they do not block objects on either the front panel or block diagram. It is sometimes said that the palettes float. The three main palettes are the **Tools**, **Controls**, and **Functions** palettes.

1.8.1   Tools Palette

A **tool** is a special operating mode of the mouse cursor. You use tools to perform specific editing functions, similar to how you would use them in a standard paint program. You can create, modify, and debug VIs using the tools located in the floating **Tools** palette, shown in Figure 1.25. If the **Tools** palette is not visible, select **Tools Palette** from the **View** pull-down menu to display the palette. After you select a tool from this palette, the mouse cursor changes to the appropriate shape.

When the **Automatic Tool Selection** button (located at the top center of the **Tools** palette, as shown in Figure 1.25) is enabled, LabVIEW automatically selects the corresponding tool from the **Tools** palette as you move the cursor over objects on either the front panel or the block diagram. If necessary, you can disable the automatic tool selection feature by clicking on the **Automatic Tool Selection** button to toggle the state. The green light on the button will turn off to indicate that the automatic tool selection is off. You also can manually select a tool on the **Tools** palette to disable the automatic tool selection feature. This allows you to select a tool manually by clicking the tool you want on the **Tools** palette.

One way to access the online help is to place any tool found in the **Tools** palette over the object of interest in the block diagram window. If online help exists for that object, it will appear in a separate Help window. This process

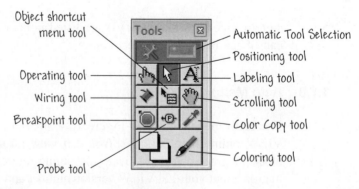

FIGURE 1.25
The **Tools** palette.

requires that you first select **Show Context Help** from the **Help** pull-down menu.

*On a **Windows** platform, a shortcut to accessing the **Tools** palette is to press the <Shift> and the right mouse button on the front panel or the block diagram. On a **Macintosh** platform, you can access the **Tools** palette by pressing <command–shift> and the mouse button on the front panel or block diagram.*

1.8.2   Controls Palette

The **Controls** palette, shown in Figure 1.26, consists of top-level icons representing subpalettes that contain a full range of available objects that you can use to create front panels. Controls simulate input devices and provide a pipeline

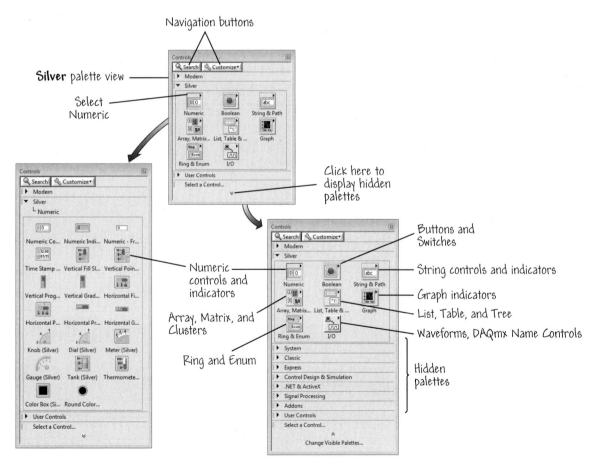

FIGURE 1.26
The **Controls** palette.

to move data to the block diagram. Examples of controls include knobs, push buttons, and dials. The controls are located on subpalettes based on the types of controls and the desired appearance. The palette in Figure 1.26 shows the **Modern, Silver,** and **User Controls**. Clicking on the double arrows at the bottom of the palette displays all the hidden palettes, including **System, Classic, Express**, and others. You can also click on **Customize** at the top of the palette as shown in Figure 1.27 and select **Change Visible Palettes** to open a dialog box to select the desired categories. In Figure 1.27, **Modern, Silver,** and **User Controls** are selected. Front panel objects generally have a high-color appearance, thus most likely the **Silver** view will provide the most appealing appearance.

*The palette view can be altered in the **Tools≫Advanced≫Edit Palette Set** dialog box.*

You can access the subpalettes by clicking on the desired top-level icon. As illustrated in Figure 1.26, the subpalette **Numeric**, when opened, will reveal the various numeric controls (by control we mean "input") and indicators (by indicator we mean "output") that you will utilize on your front panel as a way to move data into and out of the program code (on the block diagram). The topic of numeric controls and indicators is covered in Chapter 2. Each item on the

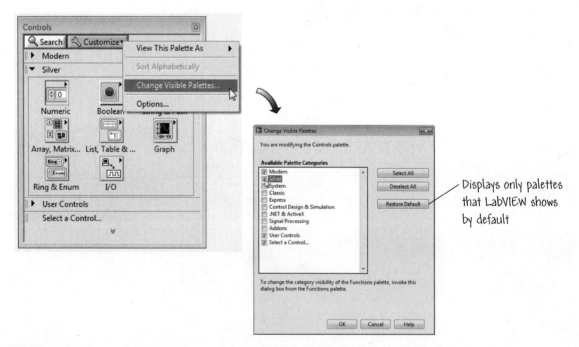

FIGURE 1.27
Changing the visible palettes on the **Controls** palette.

Controls palette is discussed in more detail in the chapter in which it is first utilized.

If the **Controls** palette is not visible, you can open the palette by selecting **Controls Palette** from the **View** pull-down menu. You can also access the **Controls** palette by right-clicking on an open area in the front panel window.

 *The **Controls** palette is available only when the front panel window is active.*

1.8.3 Functions Palette

The **Functions** palette, shown in Figure 1.28, works in the same general way as the **Controls** palette. It consists of top-level icons representing subpalettes, which contain a full range of available objects that you can use in creating the

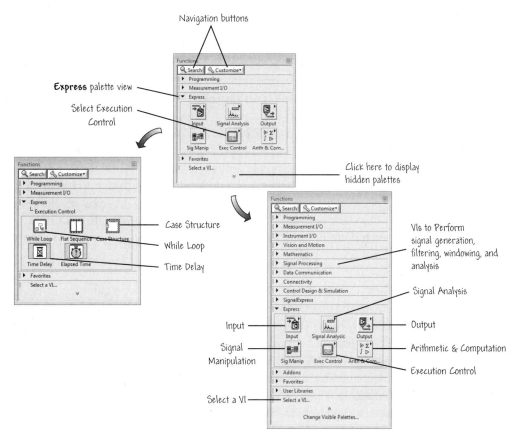

FIGURE 1.28
The **Functions** palette.

block diagram. The **Functions** palette contains functions (the vital elements of VIs), as well as many other LabVIEW VIs. The **Express** palette presents key express VIs, such as signal simulation. You access the subpalettes by clicking on the desired top-level icon. Many of the program elements are accessed through the **Functions** palette. For example, the subpalette **Execution Control** (see Figure 1.28) contains Case Structures, While Loops, and Time Delays—all of which are common elements of VIs. As with the **Controls** palette, clicking on the double arrows at the bottom of the palette displays all the hidden palettes, including **Programming, Mathematics, Measurement I/O,** and **Instrument I/O**. You can also click on **Customize** at the top of the palette and select **Change Visible Palettes** to open a dialog box to select the desired categories.

If the **Functions** palette is not visible, you can open the palette by selecting **Functions Palette** from the **View** pull-down menu. You can also access the **Functions** palette by right-clicking on an open area in the block diagram window.

*The **Functions** palette is available only when the block diagram window is active.*

1.8.4   Searching the Palettes and Quick Drop

If you are not sure on which palette a desired object is found, you can click the **Search** button on the palette toolbar (see Figures 1.26 and 1.28) to perform text-based searches for any VI, control, or function. Clicking the **Search** button on the **Controls** or **Functions** palette toolbar will display the **Search Palettes** dialog box, as shown in Figure 1.29. Clicking the **Search** button on the **Controls** palette will display the Controls tab by default. Similarly, clicking the **Search** button on the **Functions** palette displays the Functions tab. You can toggle between tabs as desired by selecting the desired tab with the cursor.

When the **Search Palette** dialog box appears, you can type in the text box the name of the object you want to search. The search will start as you begin to type, and any matches will be displayed. In Figure 1.29, we are searching for a function to generate a random number. We type "random" in the text box, and the function Random Number (0-1) appears. If two or more search results have the same name and are located on different subpalettes, the name of the palette that contains the object is shown in brackets to the right of the object name. For example, in Figure 1.29, the function Random Number (0-1) is found on two palettes.

Once the search results are available, you can click on the desired object and drag it to the front panel window or block diagram window (more on front

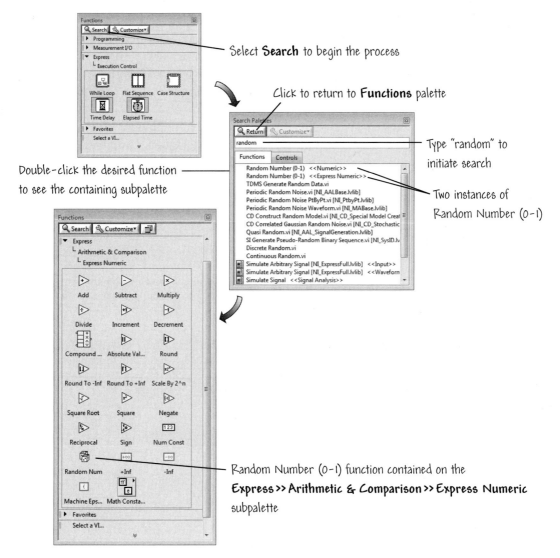

Select **Search** to begin the process

Click to return to **Functions** palette

Type "random" to initiate search

Two instances of Random Number (0-1)

Double-click the desired function to see the containing subpalette

Random Number (0-1) function contained on the **Express >> Arithmetic & Comparison >> Express Numeric** subpalette

FIGURE 1.29
Searching the **Controls** and **Functions** palettes.

panels and block diagrams in Chapter 2). You can also click on the object in the list to select it, and if you have **Context Help** (see Section 1.10), selected information about the object appears in the **Context Help Window**. Finally, double-clicking the desired object in the list will highlight its location on the palette where it exists. In Figure 1.29 we see that the Random Number (0-1) function can be found on the **Express≫Arithmetic & Comparison≫Express Numeric** palette.

To exit the **Search Palette** dialog box and return to the palette, click the **Return** button. The next time you click the **Search** button, the text box contains the last string you entered in the text box, which is useful if you double-click a search result and you want to return to the search results.

Another way to search for specific functions or controls is to use the **Quick Drop** dialog box to specify a palette object by name, as shown in Figure 1.30. Once the desired object is located, you can place it on the block diagram or front panel by double-clicking on the object. You can access the **Quick Drop** dialog box under the **View** pull-down menu (see Figure 1.18) or you can use the shortcut <Ctrl-Space>. If you access the **Quick Drop** dialog box from the

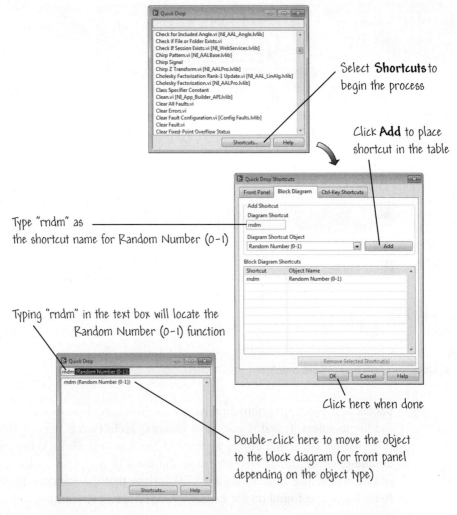

FIGURE 1.30
Configuring shortcuts for the **Quick Drop** dialog box (shortcut to access **Quick Drop:** <Ctrl-Space> in Windows and <Command-Shift-Space> in Mac OS).

front panel window, you will be able to search for objects for the front panel (such as the Waveform Chart). Similarly, if you access the **Quick Drop** dialog box from the block diagram window, you will be able to search for objects for the block diagram (such as the Random Number (0-1) function).

A helpful feature of the **Quick Drop** dialog box is the **Shortcuts** button at the bottom (see Figure 1.30), which lets you create your own shortcuts. On the **Quick Drop** dialog box click **Shortcuts** to access the **Quick Drop Shortcuts** dialog box. Type the shortcut name of your choice into the text box, select the front panel or block diagram object to which you want to assign the shortcut, and select **Add** to add the object to the table. For example, in Figure 1.30 the shortcut "rndm" is configured for the function Random Number (0-1). Each time you access the **Quick Drop** dialog box afterward, you can type in "rndm" into the text box to quickly access the Random Number (0-1) function.

1.9   OPENING, LOADING, AND SAVING VIS

When you click the **New** link on the LabVIEW Getting Started screen (see Figure 1.1), the **New** dialog box appears, as shown in Figure 1.31. Selecting the Blank VI from the **Create New** list opens a blank VI front panel and block diagram.

*Remember that from the VI pull-down menu, you can select File≫New to display the **New** dialog box (refer to Section 1.7.1). You can then open a blank by selecting **Blank VI**.*

The **New** dialog box can be employed to help you create LabVIEW applications by utilizing VI templates. If you prefer not to start with a blank VI to build your VI from scratch, you can start with a VI template. This may simplify your programming task. When you select a template in the **Create New** list, previews of the VI and a description of the template appear in the **Description** section. Figure 1.31 shows the Generate and Display VI template.

Making a selection from the **VI from Template** opens a front panel and block diagram with components you need to build different types of VIs. For example, the DAQ VI template opens a front panel and a block diagram with the components you need to measure or generate signals using the DAQ Assistant Express VI and NI-DAQmx. The Instrument I/O VI template opens a front panel and block diagram with the components you need to communicate with an external instrument attached to the computer through a port, such as a serial or GPIB-enabled device. The **User≫Browse...** button displays a dialog box so you can navigate to a VI or VI template.

You load a VI into memory by choosing the **Open** option from the **File** menu. When you choose that option, a dialog box similar to the one in

Choose Blank VI
to start from scratch

VI template
preview

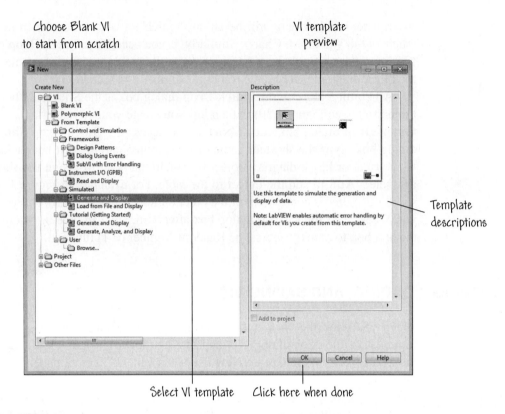

Template
descriptions

Select VI template   Click here when done

FIGURE 1.31
New dialog box.

Figure 1.32 appears. VI directories and VI libraries appear in the dialog box next to a representative symbol. VI libraries contain multiple VIs in a compressed format.

You can open a VI library or directory by clicking on its icon and then on **Open**, or by double-clicking on the icon. The dialog box opens VI libraries as if they were directories. Once the directory or library is opened, you can locate your VI and load it into memory by clicking on it and then on **OK**, or by double-clicking on the VI icon.

If LabVIEW cannot immediately locate a subVI called by the VI, it begins searching through all directories specified by the VI Search Path (Tools≫ Options≫Paths). A status dialog box will appear (sometimes the status box disappears so fast that you cannot see it was even there) as the VI loads. The *Searching* field in the status box lists directories or VIs as LabVIEW searches through them. The *Loading* field lists the subVIs of your VI as they are loaded into memory. If a subVI cannot be found, you can have LabVIEW ignore the subVI by clicking on **Ignore SubVI** in the status box, or you can click on **Browse** to search for the missing subVI.

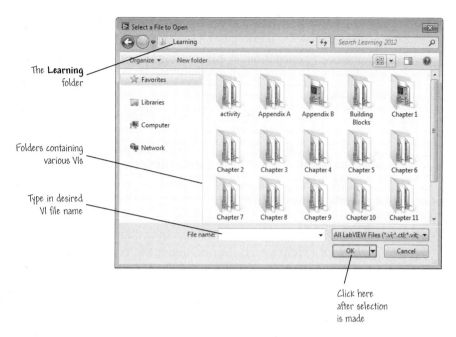

FIGURE 1.32
Locating the desired VI.

You can save your VI to a regular directory or VI library by selecting **Save**, **Save As**, or **Save All** from the **File** menu. You also can transfer VIs from one platform to another (for example, from LabVIEW for Macintosh to LabVIEW for Windows). LabVIEW automatically translates and recompiles the VIs on the new platform. Because VIs are files, you can use any file transfer method or utility to move your VIs between platforms.

1.10   LABVIEW HELP OPTIONS

The two common help options that you will use as you learn about LabVIEW programming are the **Context Help** and the **LabVIEW Help**. Both help options can be accessed in the **Help** pull-down menu.

1.10.1   Context Help Window

To display the help window, choose **Show Context Help** from the **Help** pull-down menu. If you have already placed objects on the front panel or block diagram, you can find out more about those objects by placing the cursor on the front panel or block diagram. This process causes the **Context Help Window** to appear showing the icon associated with the selected object and displaying the wires attached to each terminal. As you will discover, some icon terminals must be wired and others are optional. To help you locate terminals that require wiring, in the help window required terminals are labeled in bold, recommended connections in plain text, and optional connections are gray. The example in Figure 1.33 displays a help window in the so-called **Simple Context Help** mode.

On the lower left-hand side of the help window is a button to switch between the simple and detailed context help modes. The simple context emphasizes the important terminal connections—de-emphasized terminals are shown by wire stubs. The detailed help displays all terminals, as illustrated in Figure 1.34.

On the lower left-hand side of the help window is a lock icon that locks the current contents of the help window, so that moving the tool over another function or icon does not change the display. To unlock the window, click again on the lock icon at the bottom of the help window.

The **More Help** icon is the question mark located in the lower left-hand portion of the context help window. This provides a link to the description of the object in the online reference documentation, which features detailed descriptions of most block diagram objects.

You can readily access the data type associated with each terminal of a VI or function using the **Context Help** window. As illustrated in Figure 1.35, when we hover over the output terminal on the Generate Waveform.vi in the block diagram with **Context Help** selected, the terminal data type is given as Waveform Point (double [64-bit real($\sim$15 digit precision)]) in the help window.

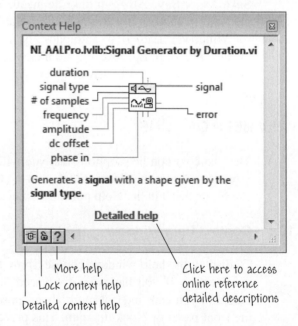

FIGURE 1.33
A simple Context Help window.

Detailed context help
displays all terminals

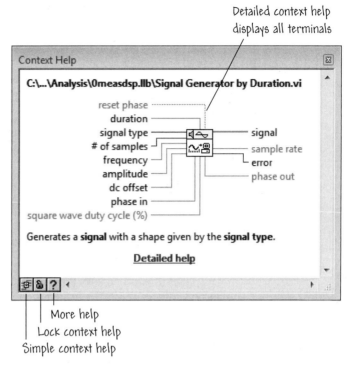

More help
Lock context help
Simple context help

FIGURE 1.34
A detailed Context Help window.

Generate waveform.vi **Context Help** selected

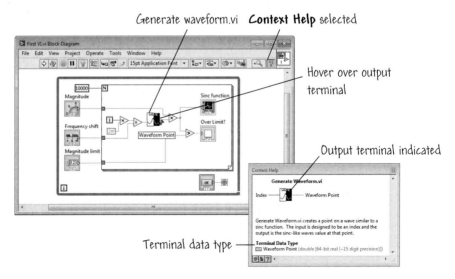

Hover over output
terminal

Output terminal indicated

Terminal data type

FIGURE 1.35
Viewing data types of parameters in the **Context Help** window.

The connected wire data type can be determined similarly by hovering over the wire.

1.10.2 LabVIEW Help

The LabVIEW online reference contains detailed descriptions of most block diagram objects. This information is accessible by clicking on the Detailed Help icon in the Context Help window and choosing **Contents** or **Index** from the **Help** menu.

> **BUILDING BLOCK**

1.11 BUILDING BLOCKS: PULSE WIDTH MODULATION

At the end of each chapter in this book, you will find a Building Blocks section. The purpose of this section is to give you the opportunity to apply the main principles and techniques learned in the chapter. You will apply the knowledge gained in the chapter to improve a VI of your design that has continuously evolved from chapter to chapter. This notion of a continuous design project employs the interesting application of **pulse width modulation**.

Pulse width modulation (PWM) is a method used to digitally encode an analog signal level through modulation of the duty cycle. PWM is used in a variety of situations, from controlling the intensity of a light bulb with a dimmer switch to transmitting data across a fiber optic line. Through the use of this technique, analog circuits can be controlled digitally, thus reducing system costs and power consumption.

Figure 1.36(a) depicts a signal with a 50% duty cycle. The duty cycle of a signal is defined to be the ratio of pulse width to total cycle time. As illustrated in Figure 1.36(a), the output is "on" for 50% of each cycle. The duty cycle is increased to 80% in Figure 1.36(b). The length of the cycle remains the same, but the longer pulse length, or "on" time results in an increased duty cycle. The signal in Figure 1.36(c) is always "on," so the duty cycle is 100%. In a dimmer switch, the length of the "on" time determines how intense the light coming from the light bulb will be. A 50% duty cycle on a dimmer switch causes the light to be only at half brightness, while 100% causes the light to be at maximum.

In this first building block, you will open and run a VI that embodies the characteristics of a VI that you will develop in the subsequent chapters. The VI you should open is called **PWM with Square Wave VI.vi** and is included in the **Building Blocks** folder of the **Learning** directory. The VI is shown in Figure 1.37. Run the VI and observe the change resulting from varying the period and the duty cycle. Keep the period constant and adjust the duty cycle. Observe that the pulse length, or amount of time that the signal is asserted, changes, but

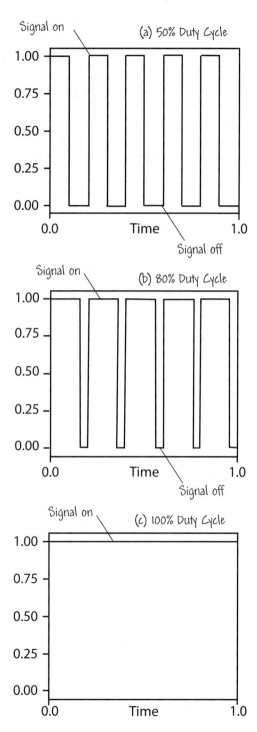

FIGURE 1.36
Signal duty cycle: (a) 50%, (b) 80%, and (c) 100%.

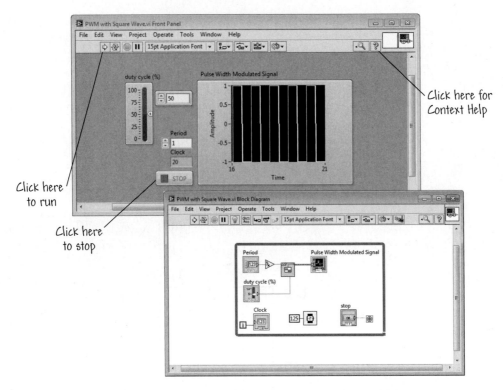

FIGURE 1.37
The PWM with Square Wave front panel and block diagram.

the signal still toggles at the same rate. Now keep the duty cycle constant and adjust the period. You should see that the ratio of on time to off time remains the same, but the overall signal frequency changes. Open the **Context Help** window and hold your cursor over the VI icon, located in the upper right hand corner on the front panel, to see a description of the VI and a diagram showing all of its inputs and outputs.

1.12 RELAXED READING: REVOLUTIONARY CANCER TREATMENT MINIMIZES DAMAGE TO HEALTHY TISSUE

In this reading we consider the problem of maximizing the destruction of cancerous tumor cells while minimizing healthy tissue damage in patients. LabVIEW helps advance a unique and more accurate form of cancer treatment where accelerated charged particles, such as protons or carbon ions, can precisely target deep-seated cancer cells. The application of accelerated proton and carbon beams to a wide spectrum of deep tumors opens new perspectives for patients with respect to surgery or traditional radiotherapy.

Today, tumor treatment successes are due mainly to the efficacy of surgery and radiotherapy. The more familiar forms of non-invasive cancer treatment of chemotherapy and radiation can have a crippling effect on the human body with severe damage to both healthy and tumor cells alike. The use of accelerated charged particles, such as protons and carbon ions, is a form of radiation therapy, known as hadron therapy, that is a step toward developing more targeted and effective cancer treatment that spares healthy tissues from being damaged. This is critical when cancer develops near vital organs in the body. The protons and ions are accelerated by means of a synchrotron, illustrated in Figure 1.38(a).

Although hadron therapy has been around since the early 1950s, recent advances in the technology have accelerated research on the technique at the Italian National Center for Oncological Hadron therapy, a clinical facility located in Pavia, Italy that was created and financed by the Ministry of Health. Less energy is deposited into the healthy tissue surrounding the target tissue by aiming the energetic ionizing particles accurately at the targeted tumor. Depending on the particular formation of each tumor, oncologists must frequently adjust the physical characteristics of the particle beams to optimize the efficacy of the procedure requiring precise control systems. The patient treatment room is depicted in Figure 1.38(b).

Directing the beam at the tumor requires systems to prepare the beam and then measure and control the beam intensity and position while distributing it

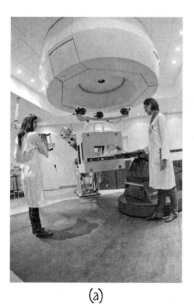

(a)

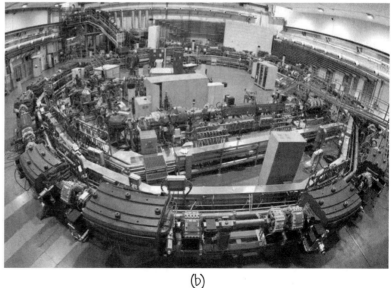

(b)

FIGURE 1.38
(a) The patient treatment room. (b) Overview of the synchrotron.

evenly across the tumor. Indeed, administering hadron therapy is complex and requires nearly 300 networked devices to control the operation of the machine and the enclosing room. The system to create and control the particle beam requires user interfaces connected to real-time and FPGA-based devices for control. LabVIEW system design software helped simplify this problem by integrating and abstracting the complexity of these multiple heterogeneous computing devices in a single development environment. For secure access to the treatment rooms during the emission of nuclear radiation, a safety interlock system was developed using the LabVIEW FPGA Module and PXI hardware. Since timing and synchronization are critical for safely creating and controlling the particle beams, a unique Ethernet-based messaging protocol was developed using the LabVIEW Real-Time Module and PXI. The controlled system measures beam intensity every microsecond and beam position every 100 μs with 100 to 200 micrometer accuracy. The beam controller system delivers the accurate measurements, real-time control, and data visualization needed by the scientists operating the beam.

Statistics from the Italian Association of Radiotherapists and Oncologists estimate more than 3% of Italian radiotherapy patients, that is, more than 3,000 new patients per year will be treated with hadron therapy and this number will steadily increase.

For further information, visit http://sine.ni.com/cs/app/doc/p/id/cs-10007 and http://home.web.cern.ch/about/updates/2012/11/clinical-trials-using-carbon-ions-begin-cnao.

1.13 myDAQ BUILDING BLOCKS

At the end of each chapter we will include a **myDAQ Building Blocks** section to provide the opportunity to apply the concepts introduced in the chapter using the myDAQ hardware. Each myDAQ Building Blocks exercise will help you build a project that will involve controlling LEDs and reading temperature from a thermistor to create a temperature monitoring system. Temperature monitoring systems are part of more complex systems. Examples you encounter every day include refrigerators, ovens, HVAC systems, computers and even cars. Before continuing with this exercise, you must install the NI myDAQ hardware and the myDAQ and NI-ELVISmx software

For tips on setting up, running, and verifying your configuration of your NI myDAQ hardware and software, visit http://www.ni.com/mydaq/support/.

1.13.1   Building a Basic LED Circuit

Light emitting diodes, known as LEDs, are often used to indicate if something is turned on, what power level a machine is running at, and for low-cost lighting solutions in remote locations. It is important to know that a LED is a diode, which only allows current to flow in one direction. This means that we have to connect the positive input to the positive lead of the lead and the negative input to the negative lead. Also, a LED can handle only a certain amount of current, so it is necessary to connect a resistor in series to regulate the current flow. A basic 5-mm LED is a low-cost solution for emitting light for various applications, including the myDAQ Building Blocks presented in this textbook.

As illustrated in Figure 1.39, the LED circuit is comprised of an LED wired in series with a 470 Ω resistor. Assuming the LED has negligible resistance, the total resistance of the circuit is 470 Ω. According to Ohm's Law, the current drive with a 5 V DC power supply from the digital line would result in 5 V ÷ 470 Ω = 10.64 mA. The LED can handle a current up to 30 mA, but is recommended to be used with 10 to 20 mA. At 10.64 mA, the expected current is in the desired range. The LED requires a positive input on the longer lead side and a negative input on the shorter side. The power is supplied to the circuit using Digital Line 0 (DIO0) of the myDAQ and the circuit is grounded to the digital ground of the myDAQ.

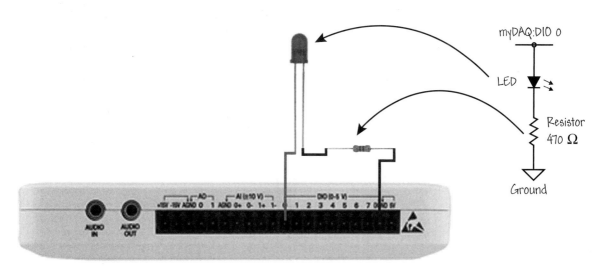

FIGURE 1.39
The LED circuit for the myDAQ Building Blocks.

1.13.2   NI ELVISmx Instrument Launcher

NI myDAQ can be used with NI ELVISmx software front panels that turn NI myDAQ into eight different measurement and signal generation tools.[1] The NI ELVISmx Instrument Launcher, illustrated in Figure 1.40, opens automatically when a NI myDAQ is connected to a system that has NI ELVISmx installed. You will be using the Digital Multimeter (DMM) and Digital Writer (DigOut) to test the LED circuit.

You can find the Instrument Launcher in the Windows Start menu at **Start≫ All Programs≫National Instruments≫NI ELVISmx for NI ELVIS & NI myDAQ≫NI ELVISmx Instrument Launcher.**

1.13.3   Testing with the ELVISmx Instrument Panel

Launch the digital multimeter by selecting the DMM icon as depicted in Figure 1.40. The goal is to use the digital multimeter to confirm the resistance of the resistor. As illustrated in Figure 1.41, on the DMM under Measurement Settings, select "Ω", then specify the range to be 2 kΩ. This is an acceptable range since we expect the measured value to be near 470 Ω. Once you have configured the instrument, select **Run** to begin taking measurements and select **Stop** when finished. To measure the resistance, use the DMM probes provided with the myDAQ and connect as shown in Figure 1.41 on the DMM front panel in the Banana Jack Connectors boxed area. In this one case, the measured resistance was 461 Ω, which is near our expected value of 470 Ω.

Next, launch the Digital Writer instrument found on the NI ELVISmx Instrument Launcher as DigOut as illustrated in Figure 1.40. Select the **Run** button and use **Toggle switch 0** to control the state of DIO0 and the LED as illustrated

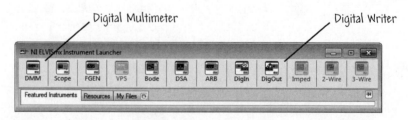

FIGURE 1.40
The NI ELVISmx Instrument Launcher.

---

1. You can get more information about all of the instruments at http://www.ni.com/white-paper/11420/en.

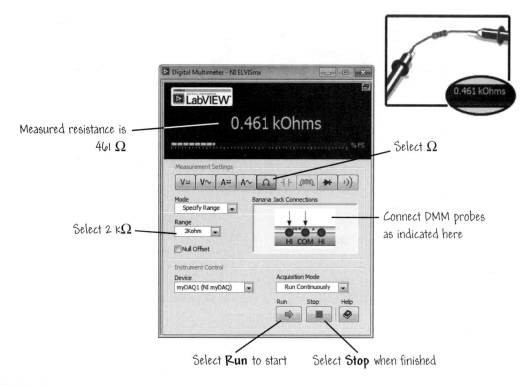

FIGURE 1.41
The digital multimeter in NI ELVISmx showing the measured value of the resistor in the LED circuit.

in Figure 1.42. Selecting **Toggle switch 0** will turn the LED on. Try it! Press the **Stop** button when you are finished.

In this first myDAQ Building Block, you created an LED circuit with the myDAQ hardware and a few simple components (LED and resistor) and used NI ELVISmx Instrument Launcher to access the Digital Multimeter and the Digital Writer. The DMM was used to measure the actual resistance of the 470 Ω resistor in the circuit and the Digital Writer was used to toggle the LED on and off. In the remaining chapters, we will extend this building block example to highlight the various LabVIEW topics within the chapters using the myDAQ.

1.14   SUMMARY

LabVIEW is a powerful and flexible development tool designed specifically for the needs of scientists and engineers. It uses the graphical programming language to create programs called virtual instruments (VIs) in a flowchart-like form called a block diagram. The user interacts with the program through the front panel. LabVIEW has many built-in functions to facilitate the programming process. The next chapters will teach you how to make the most of LabVIEW's many features.

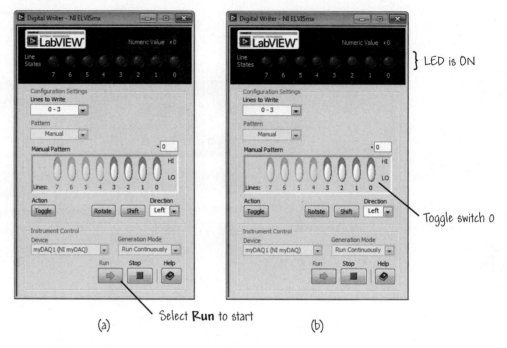

(a)   Select **Run** to start   (b)

FIGURE 1.42
The digital writer in NI ELVISmx to control the state of the LED using the toggle switch 0; (a) Off; (b) On.

KEY TERMS

Block diagram: Pictorial representation of a program or algorithm. The block diagram, which consists of executable icons called nodes and wires that carry data between the nodes, is the source code for the VI.

Context Help window: Special window that displays the names and locations of the terminals for a function or subVI, the description of controls and indicators, the values of universal constants, and descriptions and data types of control attributes. The window also accesses the **LabVIEW Help**.

Controls palette: Palette containing front panel controls and indicators.

Front panel: The interactive interface of a VI. Modeled from the front panel of physical instruments, it is composed of switches, slides, meters, graphs, charts, gauges, LEDs, and other controls and indicators.

Functions palette: Palette containing block diagram structures, constants, and VIs.

Hierarchical menus: Menu items that expand into submenus.

LabVIEW: **Lab**oratory **V**irtual **I**nstrument **E**ngineering **W**orkbench. It is a powerful and flexible instrumentation and analysis software development application.

Nodes: Execution elements of a block diagram consisting of functions, structures, and subVIs.

Palette: Menu of pictures that represent possible options.

Pull-down menu: Menu accessed from a menu bar. Pull-down menu options are usually general in nature.

Right-click: To call up a special menu by clicking an object with the right mouse button (on **Windows** platforms) or with the command key and the mouse button (on **Macintosh** platforms).

Shortcut menu: Menu accessed by right-clicking, usually on an object. Menu options pertain to that object specifically.

Tool: A special operating mode of the mouse cursor.

Tools palette: Palette containing tools you can use to edit and debug front panel and block diagram objects.

Toolbar: Bar containing command buttons to run and debug VIs.

Virtual instrument (VI): Program in LabVIEW; so-called because it models the appearance and function of a physical instrument.

EXERCISES

E1.1 On the Getting Started screen, select **Open**. In the LabVIEW directory, navigate to the examples folder. Within the examples folder, navigate to the apps subfolder. Select the file demos.llb and upon opening, select Vibration Analysis.vi. The front panel should look like the one shown in Figure E1.1.

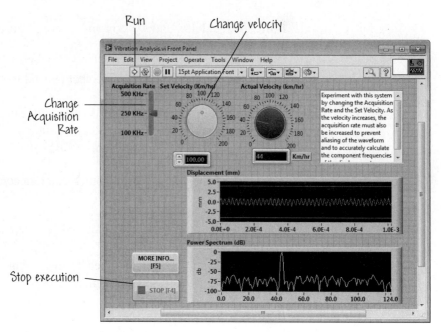

FIGURE E1.1
The Vibration Analysis.vi front panel.

(a) Run the VI by clicking on the **Run** button.

(b) Vary the Acquisition Rate on the vertical pointer slide control.

(c) Vary the desired velocity on the Set Velocity [km/hr] dial and verify that the actual velocity, as indicated on the Actual Velocity [km/hr] gauge, matches the desired velocity.

E1.2 Referring to Vibration Analysis.vi from E1.1, we can inspect the block diagram and watch it execute using **Highlight Execution**. Under the **Window** pull-down menu, select **Show Block Diagram**. The panel should switch to the block diagram shown in Figure E1.2.

(a) Click on the **Highlight Execution** button.

(b) Run the VI by clicking on the **Run** button.

(c) Watch as the data flows through the code.

E1.3 In this exercise, we want to open and run an existing VI. In LabVIEW, go to the Getting Started screen and select **Help≫Find Examples**. Click the **Search** tab and type "filter." Select **filter** to display the example VIs that include filter in the title. Find the VI titled Express Filter.vi and open it. The VI front panel is shown in Figure E1.3.

(a) Run the VI by clicking on the **Run** button.

(b) Vary the Simulated frequency and watch the values change.

Select Execution Highlighting

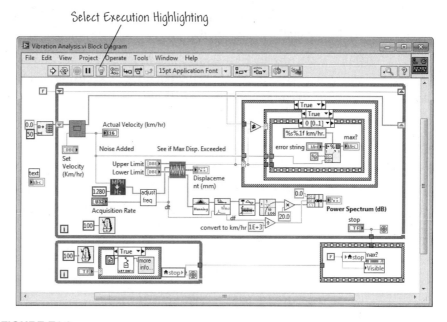

FIGURE E1.2
The Vibration Analysis.vi block diagram.

Three inputs to vary

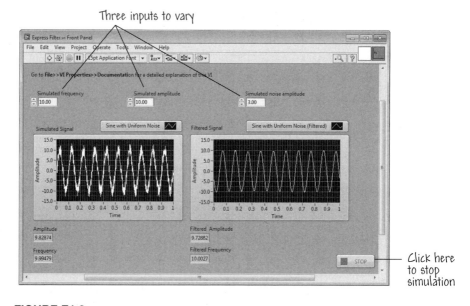

FIGURE E1.3
The Express Filter.vi front panel.

(c) Vary Simulated amplitude and Simulated noise amplitude and verify that the value on the indicator matches the graph.

E1.4 Referring to Express Filter.vi from E1.3, we can inspect the block diagram and watch it execute using **Highlight Execution**. Under the **Window** pull-down menu, select **Show Block Diagram**. The panel should switch to the block diagram shown in Figure E1.4.

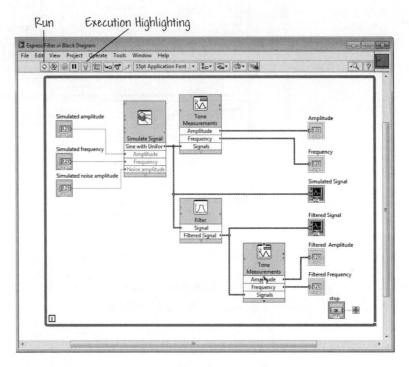

FIGURE E1.4
The Express Filter.vi block diagram.

(a) Click on the **Highlight Execution** button.

(b) Run the VI by clicking on the **Run** button.

(c) Watch as the data flows through the code and the Express VIs.

(d) Stop the VI and return to the block diagram.

(e) Double click on the Simulate Signal Express VI. Change the signal type from Sine to Square. Click **OK** and return to the front panel. Run the VI again. Notice it is now plotting a square wave instead of a sine wave.

E1.5 Open up a new blank VI. Navigate to the **Help** pull-down menu and select **Find Examples** and view the examples available in other categories. Look at some of the examples in Browse≫Analyzing and Processing Signals. The NI Example Finder is shown in Figure E1.5. Run several other VIs.

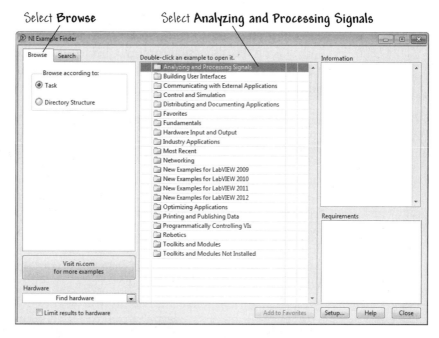

FIGURE E1.5
Select **Analyzing and Processing Signals**.

E1.6   On the LabVIEW Getting Started screen, select **File≫New**. On this screen, we have the option to open a blank VI or a VI from a template. With a template, you will not have to start building your application from scratch. Browse through the available templates and then open From Template≫Tutorial (Getting Started)≫ Generate, Analyze, and Display. Look at the front panel and block diagram of this VI and then run the VI. What signal is displayed in the Waveform Graph? Change the signal type to a sawtooth and run the VI.

E1.7   In a web browser, go to http://www.ni.com/training/ and select **Browse Certifications** under the topic **NI Certification: Validate Your Expertise**. Read about the benefits of becoming a certified LabVIEW Associate Developer (CLAD). Where are the closest testing centers for you?

The testing center list can be found at www.pearsonvue.com/ni. Click on **Locate a Test Center** *on the menu on the right hand side page and fill in your location to find the nearest test center.*

E1.8 Suppose you need to create a random number from 1 to 100. The random number VI in LabVIEW creates numbers in the range of [0,1], so how would you convert the random number in the [0,1] range so that each number in the [0,100] range is equally likely? Whenever you are not sure how to do something a good place to start is the Discussion Forums on the NI website where LabVIEW users post problems and other users post solutions. Go to www.ni.com/support and click on the discussion forums. Click on the LabVIEW message board and search for "Random Modification". Did that help you solve your answer? Was the answer what you would have expected?

*It is a good idea to remember how to generate random numbers as you will be using this technique in other problems in this book. The random number generator can be found on the **Functions≫Programming≫Numeric** palette.*

E1.9 Reconsider the problem of creating a random number from 1 to 100 discussed in E1.8. Go to www.ni.com/support and click on Knowledge Base under Browse by Resource. Then select LabVIEW Development Systems and search for Random Numbers. Can you find how to Generate Random Numbers outside of the 0-1 range. Was this help different than when you used the discussion forums? When would you use each method?

PROBLEMS

P1.1 In this problem, you will open and run an existing VI. On the Getting Started screen, click **Help≫Find Examples**. In the **Example Finder**, open the Waveform Min Max example.vi, located in **Fundamentals≫Waveforms**.

(a) Run the VI by clicking on the **Run** button.

(b) Vary the amplitude and frequency by typing in new values and pressing **Enter**, or by clicking the arrows on the left side of the controls. Observe how the outputs change for the new input values.

(c) Under the **Window** pull-down menu, select **Show Block Diagram**. Open the **Context Help** window and hold the cursor over the objects on the block diagram to see terminal connections and a brief description of the functionality of various block diagram objects.

P1.2 In this problem we want to open an existing VI from the Learning directory. You can open the VI by either selecting **Open** from the Getting Started screen, or if you are already in LabVIEW, you can use the **File** pull-down menu (see Figure 1.16) and select **Open**. In both cases, you must navigate through your local file structure to find the desired VI. Find, open, and run Memory.vi located in the Learning directory. Navigate to the Instructional VIs folder and open games.llb.

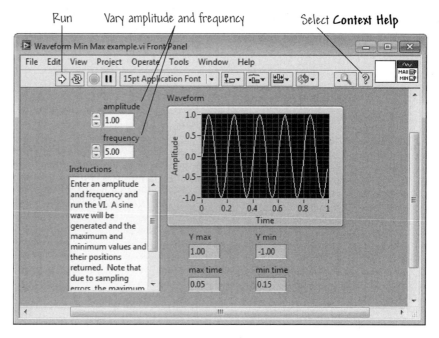

Run Vary amplitude and frequency Select **Context Help**

FIGURE P1.1
The front panel for the Waveform Min Max example VI.

P1.3 You can construct games using LabVIEW. In this problem, you will download the LabVIEW game of chess from LabVIEW Zone. Go to the LabVIEW games by visiting the website http://www.ni.com/devzone/lvzone/games.htm. At this location, you can select the Two-Player Chess game to download. Once you have the chess game open in LabVIEW, investigate the block diagram to see the code. Using LabVIEW **Context Help** (found in the **Help** menu) locate the While Loop by scanning over the block diagram with the **Positioning** tool. What function does the While Loop perform?

P1.4 Find the following objects. Write down the palette on which each can be found.

(a) Round LED

(b) Add function

(c) Or function

(d) String control

P1.5 Open a new VI. On the front panel place a Thermometer, a Horizontal Toggle button, and a Waveform Chart. Right click on each of the icons to determine the number of associated hierarchical menus:

(a) Thermometer on the front panel

(b) Thermometer on the block diagram

(c) Boolean Push Button on the block diagram

(d) Waveform Chart on the front panel

*You can locate the Thermometer on the **Silver≫Numeric** subpalette, the Push Button on the **Silver≫Boolean** subpalette, and the Waveform Chart on the **Silver≫Graph** subpalette.*

P1.6 On which pull-down menus can you find the following:

(a) Mathscript Window

(b) Explain Error

(c) Find Examples

(d) Tools Palette

(e) Breakpoint Manager

(f) Options

(g) Tile Left and Right

P1.7 Open a new VI. Right click on the front panel to see the **Controls** palette. Click on the thumbtack in the upper left corner of the palette. To change the way your palette looks, click on the **Customize** (see Figure 1.25) and select **View This Palette As**. Experiment with the various options. Choose the option that you like the best. Repeat this for the **Function** palette on the block diagram as well.

DESIGN PROBLEMS

D1.1 On the Getting Started screen, select **Help≫Find Examples** to open the NI Example Finder. Locate Express Comparison.vi found in the subfolder Mathematics within the Analyzing and Processing Signals folder. The front panel should resemble the front panel in Figure D1.1.

(a) Run the VI by selecting **Run** on the **Operate** menu, or by clicking the **Run** button.

(b) Move the slides on the controls on the front panel and observe how the corresponding values on the chart change.

(c) Verify that the signal Greater is high (that is, it takes a value of 1) when Value A is greater than Value B.

(d) Open the block diagram and double-click the Express Comparison VI, which is currently configured to perform the Greater Than function. Change the settings in the configuration window so that the Express VI will instead perform the Less Than function. Note that the name on the chart of the result of the Express Comparison VI automatically changes from Greater to Less.

(e) Run the VI again and verify that the signal Less is high when Value A is less than Value B.

Save the VI as Less Than Comparison.vi in the Users Stuff folder in the Learning directory.

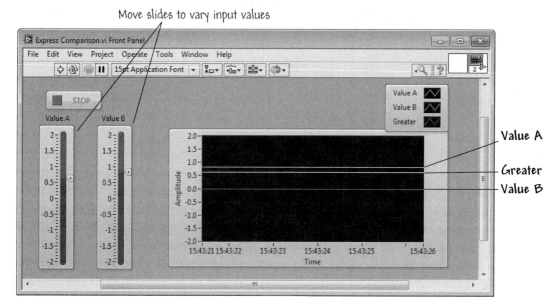

FIGURE D1.1
The front panel for the Express Comparison VI.

D1.2 In this design problem, you will take the first step to construct your own VI. On the Getting Started screen, select **File≫New** and open a **Blank VI**. Go to the block diagram and place a Compound Arithmetic function, found in the **Programming≫Numeric** subpalette. By default, this function returns the sum of two inputs.

(a) Using LabVIEW **Context Help**, find out how to configure the function to compute other operations besides addition.

(b) Using what you learned in Step (a), configure the function to multiply two inputs and display the result on a Numeric Indicator.

(c) Using LabVIEW **Context Help**, find out how to change the Compound Arithmetic function to compute operations on more than two inputs.

(d) Using what you learned in Step (c), configure the function to multiply three inputs together.

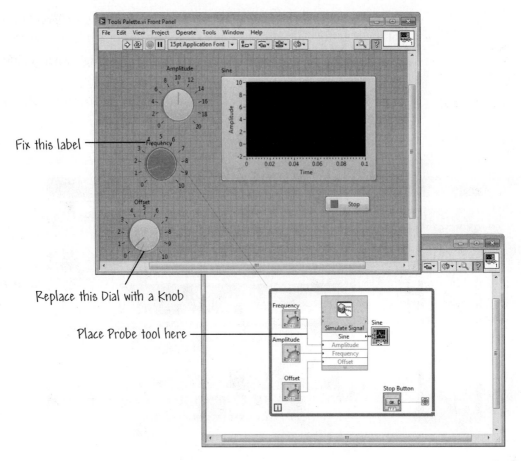

Fix this label

Replace this Dial with a Knob

Place Probe tool here

FIGURE D1.3
Using the **Tools** palette to edit a VI.

D1.3   Consider the VI shown in Figure D1.3. Using only the **Tool** palette options, perform the following actions:

(a) Use the **Positioning** tool to align all three controls and move the Frequency label off of the Knob (the knob in the middle).

(b) Using the **Object Shortcut Menu** tool to replace the Dial with a Knob.

(c) To make the Frequency knob the exact same color as the others, use the **Get Color** tool and click on the Amplitude knob. Notice the colors changed in the **Set Color** toolbox. Click on the **Set Color** tool now and change the color of the knobs.

(d) Switch to the block diagram. Notice that the labels are mislabeled. Change the labels by using the **Edit Text** tool.

(e) Change the value of the Frequency to be 18 using the **Operate Value** tool.

(f) Place a probe on the **Frequency** wire using the **Probe Data** tool.

(g) Run the VI and watch the probe.

(h) Use the **Operate Value** tool to change the value of the knobs and investigate the effect on the sine wave.

You can locate the VI in Figure D1.3 in the **Learning** *directory in* **Chapter 1** *of the folder* **Exercises&Problems**. *The VI is named* **Tools Palette.vi**.

D1.4 Consider the VI shown in Figure D1.4. You can locate the Pythagorean Theorem VI in the **Learning** directory in **Chapter 1** of the folder **Exercises&Problems**. The VI is named **Pythagorean-Messy.vi**. Open the VI and switch to the block diagram. Can you follow the code and readily explain what function this VI

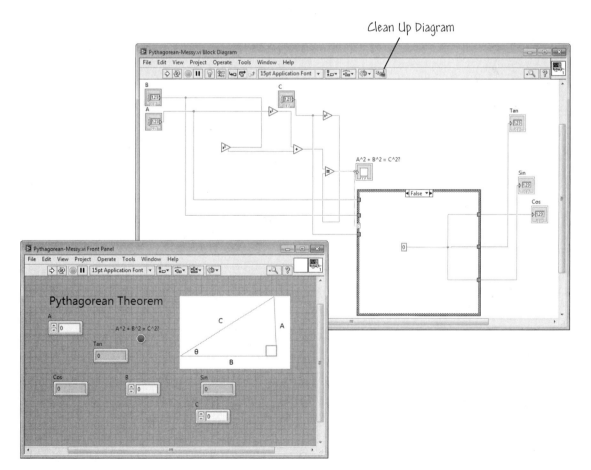

FIGURE D1.4
Cleaning up a messy block diagram.

performs? Click on the Block Diagram Clean Up icon and re-consider the code again. Describe the visual differences in the block diagram. It should be easier to explain the code once it is cleaned up.

Switch to the front panel and arrange the controls and indicators to a logical arrangement with the controls on the left side and the indicators on the right side. Using the **Edit Text** tool, enter inputs in the controls A, B & C and click the **Run** button on the VI. Change the values of the controls and experiment with the VI. Click the **Run Continuously** button and experiment with changing the inputs as the VI runs. (Hint: Try A=3, B=4, and C=5).

CHAPTER 2

Virtual Instruments

Virtual instruments (VI) are the building blocks of LabVIEW programming. We will see in this chapter that VIs have three main components: the front panel, the block diagram, and the icon and connector pane pair. We will revisit the front panel and block diagram concepts first introduced in Chapter 1. An introduction to wiring the elements together on the block diagram is presented, although many of the debugging issues associated with wires are left to Chapter 3. The important notion of data flow programming is also discussed in this chapter. Finally—you will have the opportunity to build your first VI!

GOALS

1.   Gain experience by running more worked examples.
2.   Understand the three basic components of a virtual instrument.
3.   Begin the study of programming in LabVIEW.
4.   Understand the notion of data flow programming.
5.   Build your first virtual instrument.

2.1 WHAT ARE VIRTUAL INSTRUMENTS?

LabVIEW programs are called virtual instruments (VIs) because they often have the look and feel of physical systems or instruments. Figure 2.1 shows an example of a front panel. A VI and its components are analogous to main programs and subroutines from text programming languages like C and Fortran. VIs have both an interactive user interface—known as the front panel—and the source code—represented in graphical form on the block diagram. LabVIEW provides mechanisms that allow data to pass easily between the front panel and the block diagram.

The block diagram is a pictorial representation of the program code. The block diagram associated with the front panel in Figure 2.1 is shown in Figure 2.2. The block diagram consists of executable icons (called nodes) connected (or **wired**) together. We will discuss wiring later in this chapter. The important concept to remember is that in LabVIEW the block diagram is the source code.

The art of successful programming in LabVIEW is an exercise in **modular programming**. After dividing a given task into a series of simpler subtasks (these subtasks are called subVIs), you then construct a virtual instrument to accomplish each subtask. Chapter 4 focuses on building subVIs. The resulting subtasks (remember, these are called subVIs) are then assembled on a top-level block diagram to form the complete program. Modularity means that you can execute each subVI independently, thus making debugging and verification easier. Furthermore, if your subVIs are general purpose programs, you can use them in other programs.

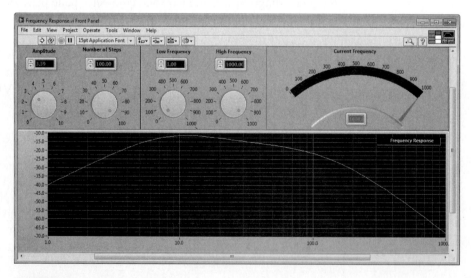

FIGURE 2.1
A virtual instrument front panel.

This is a formula node

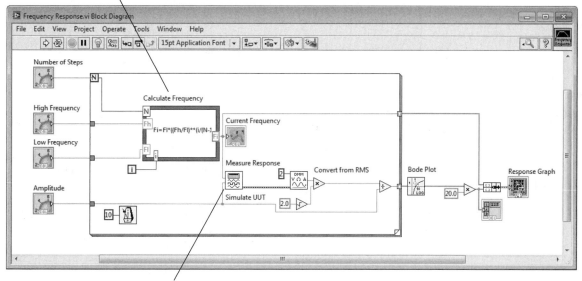

This is a subVI called Demo Tek FG 5010.vi

FIGURE 2.2
The virtual instrument block diagram associated with the front panel in Figure 2.1.

VIs (and subVIs) have three main parts: the front panel, the block diagram, and the icon/connector pane. The front panel is the interactive user interface of a VI—a window through which the user interacts with the code. When you run a VI, you must have the front panel open so you can pass inputs to the executing program and receive outputs (such as data for graphical display). The front panel is indispensable for viewing the program outputs. It is possible, as we will discuss in Chapter 9, to write data out to a file for subsequent analysis, but generally you will use the front panel to view the program outputs. The front panel contains knobs, push buttons, graphs, and many other controls (the term *controls* is interchangeable with *inputs*) and indicators (the term *indicators* is interchangeable with *outputs*).

The block diagram is the source code for the VI made up of graphical icons, wires, and such, rather than traditional "lines of code." The block diagram is actually the executable code. The **icons** of a block diagram represent lower-level VIs, built-in functions, and program control structures. These icons are wired together to allow the data flow. As you will learn later in this chapter, the execution of a VI is governed by the data flow and not by a linear execution of lines of code. This concept is known as **data flow programming**.

The **icons** and **connector panes** specify the pathways for data to flow into and out of VIs. The icon is the graphical representation of the VI in the block diagram and the connector defines the inputs and outputs. All VIs have an icon

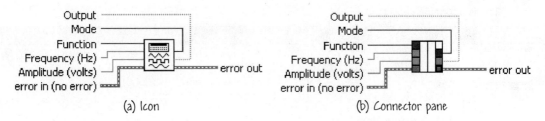

(a) Icon   (b) Connector pane

FIGURE 2.3
The icon and connector of the Demo Tek FG 5010.vi subVI shown in Figure 2.2.

and a connector pane. As previously mentioned, VIs are hierarchical and modular. You can use them as top-level (or calling) programs or as subprograms (or subVIs) within other programs. The icon and connector are shown in Figure 2.3 for the subVI Demo Tek FG 5010.vi. This subVI can be found in the center of the Frequency Response.vi diagram in Figure 2.2. It simulates a Textronix FG 5010 function generator.

2.2   SEVERAL WORKED EXAMPLES

Before you construct your own VI, we will open several existing LabVIEW programs and run them to see how LabVIEW works. The first VI example—Moonlanding.vi—can be found in the suite of examples provided as part of LabVIEW. The second VI example—ODE Example.vi—illustrates how LabVIEW can be used to simulate linear systems. In this example, the motion of a mass-spring-cart system is simulated, and you can observe the effects of changing any of the system parameters on the resulting motion of the cart.

**Moonlanding
Demo**

In this example, you will open and run the virtual instrument called Moonlanding.vi. At the Getting Started window, select **Help≫Find Examples** and when the NI Example Finder appears, select the **Browse according to: Directory Structure** button on the upper left-hand side. Click on the express folder and upon opening find Moonlanding.vi and double click on the VI. The front panel window appears and should resemble the one shown in Figure 2.4. The front panel contains a tank, a slide control, dials, a chart, and several numeric indicators.

Run the VI by clicking on the **Run** button. The front panel toolbar changes as the VI switches from edit mode to run mode. For example, once the VI begins executing, the **Stop** button will change appearance on the front panel toolbar (it changes from a shaded symbol to a red stop sign). Also, the **Run** button changes appearance to indicate that the VI is running.

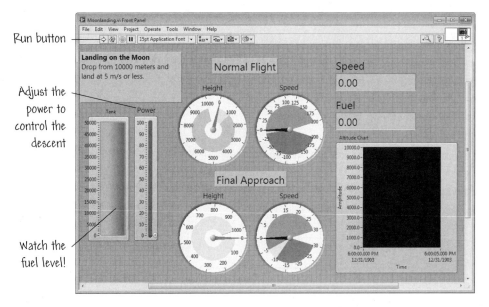

Run button

Adjust the
power to
control the
descent

Watch the
fuel level!

FIGURE 2.4
Moonlanding front panel.

In this VI, you get to practice landing on the Moon. Initially the altitude is 10,000 meters. For a successful landing you need to have a velocity of 5 m/s or less when the altitude is zero (that is, when you have landed). You control the engine power by adjusting the Power setting in the slide control. While the VI is running use the **Operating** tool to change the power setting in the vertical slide control. The amount of fuel remaining is displayed in the fuel tank indicator, as well as in the numeric indicator. The speed of the lander is displayed in the indicator dial and in a numeric indicator. This illustrates two methods for displaying data to the user.

From 10,000 m to 1000 m you are in the Normal Flight regime. Once you reach 1000 m you enter the Final Approach. The dial indicators for the Final Approach begin operation once 1000 m is reached. The flight path is updated in the Altitude Chart.

Switch to the block diagram by choosing **Show Block Diagram** from the **Window** pull-down menu. The block diagram shown in Figure 2.5 is the underlying code for the VI. At this point in the learning process, you may not understand all of the block diagram elements depicted in Figure 2.5—but you will eventually! Note that this VI employs Express VIs as key elements of the code.

When you are finished experimenting with the Moonlanding.vi, close the VI and subVI by selecting **Close** from the **File** pull-down menu on each open front panel. Remember—do not save any changes!

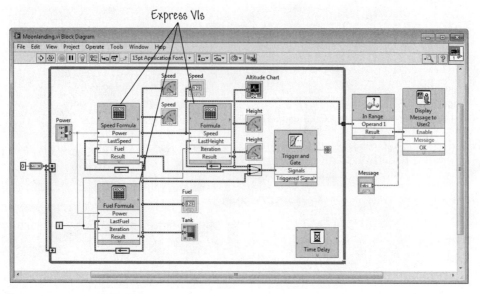

FIGURE 2.5
Moonlanding demonstration block diagram—the code.

> *Selecting **Close** from the **File** pull-down menu of a block diagram closes the block diagram window only. Selecting **Close** on a front panel window closes both the front panel and the block diagram.* ◆

◆ A Linear System of Equations Example

In this example we use LabVIEW to solve a set of linear, constant coefficient, ordinary differential equations. Many physical systems can be modeled mathematically as a set of linear, constant coefficient differential equations of the form

$$\dot{\mathbf{x}} = \mathbf{A}\mathbf{x}$$

with initial conditions $\mathbf{x}(0) = \mathbf{x}_o$, where the matrix $\mathbf{A}$ is a constant matrix.

Suppose we want to model the motion of a mass-spring-damper system, as shown in Figure 2.6. Let m represent the mass of the cart, k represent the spring constant, and b the damping coefficient. The position of the cart is denoted by y, and the velocity is the time derivative of the position, that is, the velocity is $\dot{y}$. Equating the sum of forces to the mass times acceleration (using Newton's Second Law) we obtain the equation of motion:

$$m\ddot{y}(t) + b\dot{y}(t) + ky(t) = 0,$$

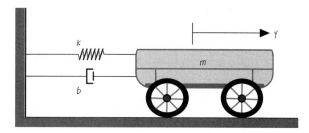

FIGURE 2.6
A simple mass-spring-damper system.

with the initial conditions $y(0) = y_o$ and $\dot{y}(0) = \dot{y}_o$. The motion of the cart is described by the solution of the second-order linear differential equation above. For this simple system we can obtain the solution analytically. For more complex systems it is usually necessary to obtain the solution numerically using the computer. In this exercise we seek to obtain the solution numerically using LabVIEW.

It is sometimes convenient for obtaining the numerical solution to rewrite the second-order linear differential equation as two first-order differential equations. We first define the state vector of the system as

$$\mathbf{x}(t) = \begin{pmatrix} x_1(t) \\ x_2(t) \end{pmatrix} = \begin{pmatrix} y(t) \\ \dot{y}(t) \end{pmatrix},$$

where the components of the state vector are given by

$$x_1(t) = y(t) \quad \text{and} \quad x_2(t) = \dot{y}(t).$$

Using the definitions of the state vector and the equation of motion, we obtain

$$\dot{x}_1(t) = x_2(t)$$
$$\dot{x}_2(t) = -\frac{k}{m}x_1(t) - \frac{b}{m}x_2(t).$$

Writing in matrix notation yields

$$\dot{\mathbf{x}}(t) = \mathbf{A}\mathbf{x}(t),$$

where

$$\mathbf{A} = \begin{bmatrix} 0 & 1 \\ -k/m & -b/m \end{bmatrix}.$$

In this example, let

$$\frac{k}{m} = 2 \quad \text{and} \quad \frac{b}{m} = 4.$$

Choose the initial conditions as

$$x_1(0) = 10 \quad \text{and} \quad x_2(0) = 0.$$

Now we are ready to compute the solution of the system of ordinary differential equations numerically with LabVIEW.

- Select **Open Existing** from the Getting Started screen or use the **File** menu to open the VI.

- Open the ODE Example.vi located in the folder Chapter 2 in the directory Learning.

The front panel depicted in Figure 2.7 will appear. Verify that the initial conditions $\mathbf{x}_o$ are set correctly to $x_1(0) = 10$ and $x_2(0) = 0$. Run the VI by clicking on the **Run** button. About how long does it take for the cart position to come to rest? Change the initial position to $x_1(0) = 20$. Now how long does it take for the cart to come to rest? What is the maximum value of the cart velocity?

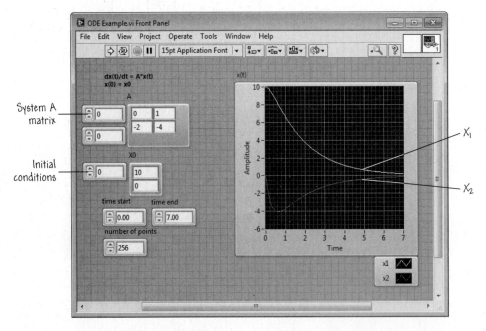

FIGURE 2.7
Linear system of equations front panel.

LabVIEW does not accept values in numeric displays until you press the **Enter** *button on the toolbar or click the mouse in an open area of the window.*

Open and examine the block diagram by choosing **Show Block Diagram** from the **Window** menu. The code is shown in Figure 2.8. If you need to solve a set of linear ordinary differential equations, you can start with this VI and modify it as necessary. The idea of starting with a VI that solves a related problem is a good approach in the early stages of learning LabVIEW. Access the **Help** pull-down menu and select **Show Context Help**. Move the cursor over various objects on the block diagram and read what the online help has to say.

Most VIs are hierarchical and modular. After creating a VI, you can (with a little work configuring the icon and connector pane) use the VI as a subVI (similar to a subroutine) in the block diagram of another VI. By creating subVIs, you can construct modular block diagrams that make your VIs easier to debug. The ODE Example.vi uses a subVI named ODE Linear System Numeric.vi. Open the subVI by double-clicking on the appropriate subVI icon (see Figure 2.8). The icon for the subVI is labeled. The front panel shown in Figure 2.9 should appear.

The icon and connector pane provide the graphical representation and parameter definitions required to use a VI as a subVI in the block diagrams of other VIs. The icon and connector pane are located in the upper right corner of the VI front panel (see Figure 2.9). The icon is a graphical representation of the VI

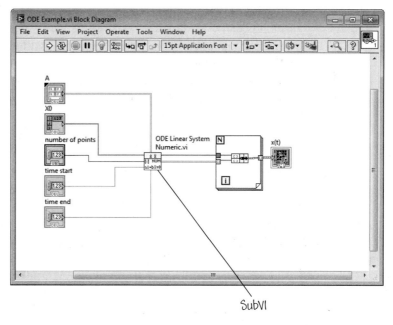

FIGURE 2.8
Linear system of equations block diagram.

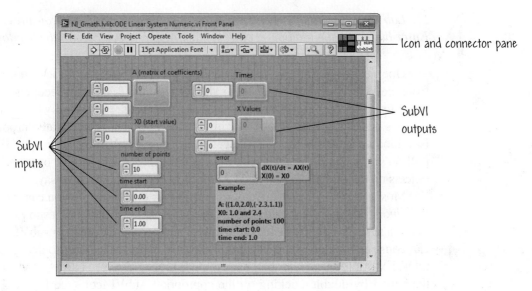

FIGURE 2.9
ODE Linear System Numeric subVI.

when used as a component in a LabVIEW program, that is, when used in the block diagram of other VIs. An icon can be a pictorial representation or a textual description of the VI, or a combination of both. The icon for the ODE Linear System Numeric subVI includes both text and graphics.

Every VI (and subVI) has a connector pane. The connector pane is a set of terminals that correspond to its controls and indicators. In Chapter 4, you will learn how to associate front panel controls and indicators with connector pane terminals. The connector pane terminals determine where you must wire the inputs and outputs on the icon. The icon and connector pane of the ODE Linear System Numeric subVI are shown in Figure 2.10.

Every VI has a default icon and a default connector pane displayed in the upper right corner of the front panel, as shown in Figure 2.11. You will learn how to edit the VI icon in Chapter 4. You may want to personalize VI icons so that they transmit information by visual inspection about the contents of the underlying VI.

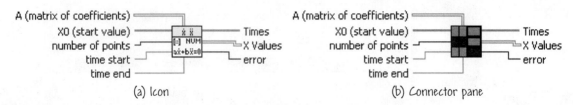

FIGURE 2.10
The icon and connector pane of the ODE Linear System Numeric subVI.

FIGURE 2.11
The default icon.

When you are finished experimenting, close the VI by selecting **Close** from the **File** menu.   ◆

2.3   THE FRONT PANEL

The front panel of a VI is a combination of controls and indicators. Controls simulate the types of input devices you might find on a conventional instrument, such as knobs and switches, and provide a mechanism to move input from the front panel to the underlying block diagram. On the other hand, indicators provide a mechanism to display data originating in the block diagram back on the front panel. Indicators include various kinds of graphs and charts (more on this topic in Chapter 7), as well as numeric, Boolean, and string indicators. Thus, when we use the term *controls* we mean "inputs," and when we say *indicators* we mean "outputs."

You place controls and indicators on the front panel by selecting and "dropping" them from the **Controls** palette. Once you select a control (or indicator) from the palette and release the mouse button, the cursor will change to a "hand" icon, which you then use to carry the object to the desired location on the front panel and "drop" it by clicking on the mouse button again. Once an object is on the front panel, you can easily adjust its size, shape, and position (see Chapter 3). If the **Controls** palette is not visible, right-click on an open area of the front panel window to display the **Controls** palette.

2.3.1   Numeric Controls and Indicators

You can access the numeric controls and indicators from the **Numeric** sub-palette. A complete list of numeric controls and indicators can be found under the **Modern** subpalette of the **Controls** palette, as shown in Figure 2.12. The more common controls and indicators can be found under the **Silver** subpalette of the **Controls** palette. This is a preferred appearance used in this book whenever available. There are quite a large number of available numeric controls and indicators. The two most commonly used numeric objects are the numeric control and the numeric indicator. When you construct your first VI later in this

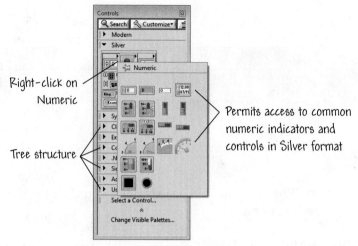

Right-click on Numeric

Tree structure

Permits access to common numeric indicators and controls in Silver format

(a) Commonly used numeric controls

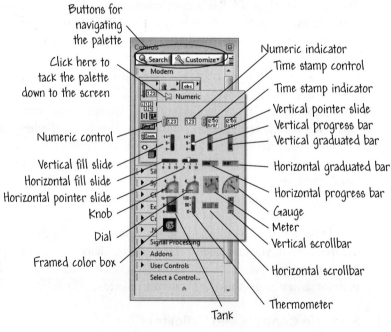

Buttons for navigating the palette

Click here to tack the palette down to the screen

Numeric control

Vertical fill slide
Horizontal fill slide
Horizontal pointer slide
Knob
Dial
Framed color box

Numeric indicator
Time stamp control
Time stamp indicator
Vertical pointer slide
Vertical progress bar
Vertical graduated bar
Horizontal graduated bar
Horizontal progress bar
Gauge
Meter
Vertical scrollbar
Horizontal scrollbar
Thermometer

Tank

(b) All numeric controls

FIGURE 2.12
Numeric controls and indicators.

chapter, you will get the chance to practice dropping numeric controls and indicators on the front panel. Once a numeric control is on the front panel, you click on the increment buttons (that is, the up and down arrows on the left-hand side of the control) with the **Operating** tool to enter or change the displayed numerical values. Alternatively, you can double-click on the current value of the numeric control with the **Auto** tool, which will highlight the value, and you can then enter a different value.

 *You can tack down the **Numeric** palette (and most other palettes) to the screen so they are visible at all times by clicking on the thumbtack on the top left corner of the palette.*

You can use the tree structure (see Figure 2.12) on the **Controls** and **Functions** palettes to navigate and search for controls, VIs, and functions. When you right-click the cursor over a subpalette icon, the palette expands to display the subpalette you selected and the **Controls** (or **Functions**) palette remains in view. An example of a subpalette icon on the **Controls** palette is the **Numeric** subpalette. The **Controls** and **Functions** palettes contain two navigation buttons (as illustrated in Figure 2.13):

- **Search**—Changes the palette to search mode. In search mode, you can perform text-based searches to locate controls, VIs, or functions in the palettes.

- **Customize**—Allows you to configure the appearance of the palettes.

Suppose you need to search for a function that provides a random number between 0 and 1. On the **Functions** palette, select the **Search** button and in the dialog box that appears, enter the word "random," as illustrated in Figure 2.14. In the results list of the search, select **Random Number (0-1)≫Express Numeric** The **Express Numeric** palette will appear and the Random Number (0–1) function is located, as shown in Figure 2.14. See Section 1.8.4 for more detailed information on searching the palettes.

2.3.2 Boolean Controls and Indicators

You can access the Boolean controls and indicators from the **Boolean** subpalette located under the **Silver** subpalette of the **Controls** palette, as shown in Figure 2.15. As with the numeric controls and indicators, there are quite a large number of available Boolean controls and indicators. Boolean controls and

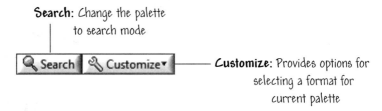

FIGURE 2.13
Navigating the **Functions** and **Controls** palettes.

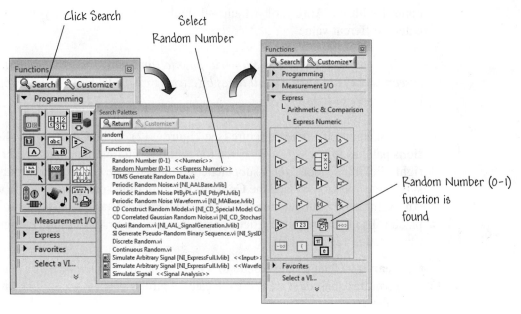

FIGURE 2.14
Searching the **Functions** palette for a Random Number VI.

indicators simulate switches, buttons, and LEDs and are used for entering and displaying Boolean (True-False) values. For example, you might use a Boolean LED in a temperature monitoring system as a warning signal that a measured temperature has exceeded some predetermined safety limit. The measured temperature is too high, so an LED indicator on the front panel turns from green to red! Before continuing, take a few moments to familiarize yourself with the types of available numeric and Boolean controls and indicators shown in Figures 2.12 and 2.15, so that when you begin to construct your own VIs, you will have a feel for what is available on the palettes.

Mechanical actions associated with Booleans represent their behavior when pressed or released. The two types of mechanical action are switching and latching. Switching returns the Boolean to its default state when directed by the user. Latching returns the Boolean to its default state when directed by the user or when LabVIEW has read its value.

The Boolean switching mechanical actions are of three kinds:

- **Switch when pressed**—Changes the control value each time you click it with the **Operating** tool.

- **Switch when released**—Changes the control value only after you release the mouse button during a mouse click within the graphical boundary of the control.

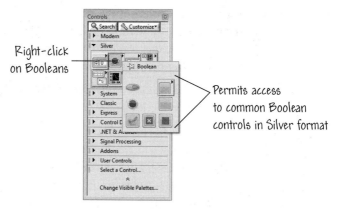

(a) Commonly used Boolean controls

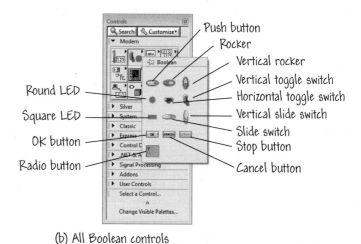

(b) All Boolean controls

FIGURE 2.15
Boolean controls and indicators.

- **Switch until released**—Changes the control value when you click it and retains the new value until you release the mouse button. At this time, the control reverts to its default value. (This behavior is like that of a door buzzer.)

The frequency with which the VI reads the control does not affect the switching mechanical action behavior.

The Boolean latching mechanical actions also are of three kinds:

- **Latch when pressed**—Changes the control value when you click it and retains the new value until the VI reads it once. The control then reverts to its default value, even if you keep pressing the mouse button. (This behavior is similar to that of a circuit breaker.)

- **Latch when released**—Changes the control value only after you release the mouse button within the graphical boundary of the control. When the VI reads it once, the control reverts to its default value. (This behavior is like that of dialog box buttons and system buttons.)

- **Latch until released**—Changes the control value when you click it and retains the value until the VI reads it once or you release the mouse button, whichever occurs last.

*You cannot select the **Switch until released** or any of the three latching mechanical actions for radio buttons control. (See Figure 2.15 for the location of the radio button on the **Boolean** subpalette.)*

To select the mechanical action, right-click on the Boolean object on the front panel to access the pull-down menu and select **Mechanical Action**, as shown in Figure 2.16. You can choose the desired mechanical action from the list of six possibilities.

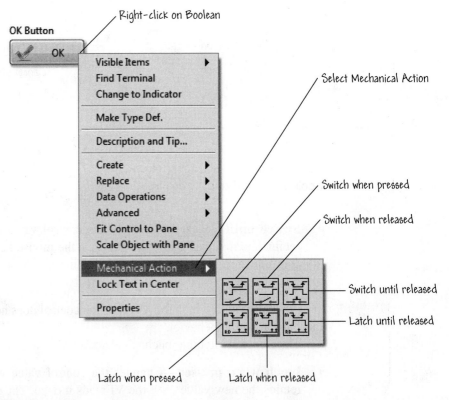

FIGURE 2.16
Selecting the mechanical action of Booleans.

*To learn more about mechanical action for Boolean controls, experiment with the Mechanical Action of Booleans VI. In the **NI Example Finder**, select the **Search** tab, type "boolean" as the keyword in the text window, and click Search.*

2.3.3   Configuring Controls and Indicators

Right-clicking on a numeric control displays the shortcut menu, as shown in Figure 2.17. You can change the defaults for controls and indicators using options from the shortcut menus. For example, under the submenu **Representation** you will find that you can choose from 15 representations of the numeric control or indicator, including 32-bit single precision, 64-bit double precision, signed integer 8-bit, and more. The representation indicates how much memory is used to represent a number. This choice is important if you are displaying large amounts of data and you want to conserve computer memory. Another useful item on the shortcut menu is the capability to switch from control and indicator using **Change to Indicator** and vice versa. We will discuss each item in the shortcut menu on an as-needed basis.

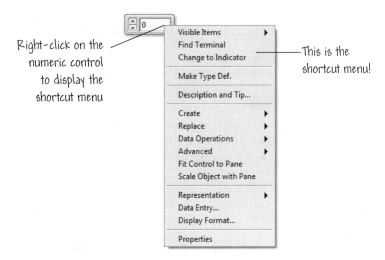

FIGURE 2.17
Configuring a numeric control using the shortcut menu.

2.4   THE BLOCK DIAGRAM

The graphical objects comprising the block diagram together make up what is usually called the source code. The block diagram (visually resembling a computer program flowchart) corresponds to the lines of text found in text-based

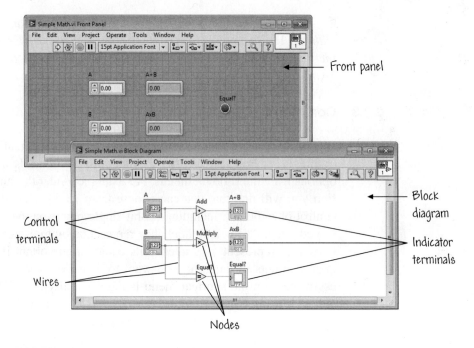

FIGURE 2.18
A typical VI illustrating nodes, terminals, and wires.

programming languages. In fact, the block diagram is the actual executable code that is compiled while you are programming, allowing instant feeedback when you make mistakes, such as wiring incompatible data types. The block diagram is built by wiring together objects that perform specific functions.

The components of a block diagram (a VI is depicted in Figure 2.18) belong to one of three classes of objects:

- **Nodes**: Program execution elements.

- **Terminals**: Ports through which data passes between the block diagram and the front panel and between nodes of the block diagram.

- **Wires**: Data paths between terminals.

2.4.1   VIs and Express VIs

An important element of LabVIEW is the Express VI. These VIs are provided to allow for quick construction of VIs designed to accomplish common tasks, such as acquisition, analysis, file reading, and writing. Express VIs are nodes (see Section 2.4.2 for more information on nodes) that require minimal wiring because they are configured with dialog boxes. The positive implications of minimizing the required wiring will become more evident as we proceed through the forthcoming chapters.

LabVIEW views a VI placed on the block diagram to be a subVI (more on subVIs in Chapter 4). A subVI is itself a VI that can be used as an element on a block diagram. In general, a block diagram can have VIs and Express VIs as elements. One difference in their use is that when you double-click a subVI, its front panel and block diagram appear. When you double-click an Express VI a dialog box appears in which you can configure the VI to meet your needs. The idea is that the Express VI allows you to quickly configure a VI by interacting with a dialog box rather than reconfiguring the code in a subVI block diagram. When an Express VI is available for the task you need, it will be better, in general, to consider the use of the Express VI.

VIs and Express VIs are distinguishable on the block diagram through the use of colored icons. By default, icons for Express VIs appear on the block diagram as expandable nodes (see Section 2.4.2) with icons surrounded by a blue field, while icons for VIs have white backgrounds. The other common element on the block diagram is the function (or primitive), and you can easily identify functions since their icons have pale yellow backgrounds, meaning you cannot open the code behind the function.

2.4.2   Nodes

Nodes are analogous to statements, functions, and subroutines in text-based programming languages. There are three node types—**functions**, **subVI nodes**, and **structures**. Functions are the built-in nodes for performing elementary operations such as adding numbers, file I/O, or string formatting. Functions are the fundamental operating element of a block diagram. The Add and Multiply functions in Figure 2.18 represent one type of node. SubVI nodes are VIs that you design and later call from the diagram of another VI. You can also create subVIs from Express VIs. Structures—such as For Loops and While Loops—control the program flow.

Express VIs and subVIs can be displayed as either icons or as expandable nodes. By default, most subVIs appear as icons that are not expandable. On the other hand, most Express VIs appear as expandable icons. Figure 2.19 depicts the various possibilities for displaying subVIs and Express VIs.

Expandable nodes appear as icons surrounded by a colored field. SubVIs appear with a yellow field, and Express VIs appear with a blue field. By default, subVIs appear as icons on the block diagram, and Express VIs appear as expandable nodes. To display a subVI or Express VI as an expandable node, right-click the subVI or Express VI and select **View As Icon** from the shortcut menu.

You can resize the expandable node to make wiring even easier. When you place the **Positioning** tool over an expandable node, resizing handles will appear at the top and bottom of the node. Placing the cursor over a resizing handle will transform the cursor to the resizing cursor, which can be used to drag the lower border of the node down to display input and output terminals. In Figure 2.19,

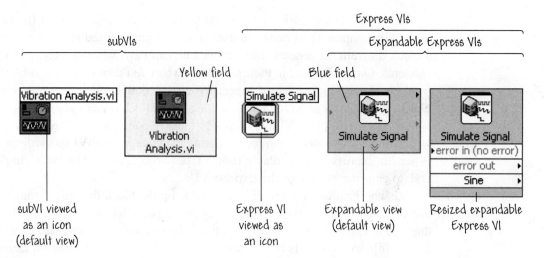

FIGURE 2.19
Expandable Icons versus Icons.

the Simulate Signal Express VI is resized to display the inputs "error in (no error)" and outputs "error out" and "Sine." For Express VIs with several inputs and outputs, the inputs and order in which they appear can be selected by clicking on the input and selecting the input or output you want in that position. Resizing the expandable node takes a larger amount of space on the block diagram which can then clutter a complex program, but it does aid in the readability of your code. Therefore, use icons if you want to conserve space on the block diagram and use expandable nodes to make wiring easier and to aid in documenting block diagrams.

2.4.3   Terminals

Terminals are analogous to parameters and constants in text-based programming languages. There are different types of terminals—control and indicator terminals, node terminals, constants, and specialized terminals that you will find on various structures. In plain words, a terminal is any point to which you can attach a wire to pass data.

For example, in the case of control and indicator terminals, numeric data entered into a numeric control passes to the block diagram via the control terminals when the VI executes. When the VI finishes executing, the numeric output data passes from the block diagram to the front panel through the indicator terminals. Data flows in only one direction—from a "source" terminal to one or more "destination" terminals. In particular, controls are source terminals and indicators are destination terminals. The data flow direction is from the control terminal to the indicator terminal, and not vice versa. Clearly, controls and indicators are not interchangeable.

Control and indicator terminals belong to front panel controls and indicators and are automatically created or deleted when you create or delete the corresponding front panel control or indicator. The block diagram of the VI in Figure 2.18 shows terminals belonging to five front panel controls and indicators. The Add and Multiply functions shown in the figure also have node terminals through which the input and output data flow into and out of the functions.

You can configure front panel controls or indicators to appear as icon terminals or data type terminals on the block diagram. The terminals shown in Figure 2.19 are all icon terminals. To display a terminal as a data type on the block diagram, right-click the terminal and select **View As Icon** from the shortcut menu to remove the checkmark.

Control terminals have thick *borders and indicator terminals have* thin *borders. It is important to distinguish between thick and thin borders since they are not functionally equivalent. Additionally, small arrows point out of (to the right) controls and into (from the left) indicators to depict data flow.*

Data types indicate what objects, inputs, and outputs you can wire together. You cannot wire together objects with incompatible data types. For example, a switch has a green border (Boolean data type) so you can wire a switch to any input with a green label on an Express VI. Similarly, a knob has an orange border (Numerical data type) so you can wire a knob to any input with an orange label. However, you cannot wire an orange knob (Numeric) to an input with a green label (Boolean).

Wires are the same color as the terminal.

Associated with the Express VI is the **dynamic data type**. The dynamic data type stores information generated or acquired by an Express VI and appears on the block diagram as a dark blue terminal. You can wire the dynamic data type to any indicator or input that accepts numeric, waveform, or Boolean data. The objective is to wire the dynamic data type to an indicator (graph, chart, or numeric) that can best present the data.

Most subVIs and functions do not accept the dynamic data type directly. To use a subVI or function to process the dynamic data type data, you must convert the dynamic data type using the Convert from Dynamic Data Express VI to change the dynamic data type to numeric, Boolean, waveform, or array data types. As illustrated in Figure 2.20, the Convert to Dynamic Data Express VI and

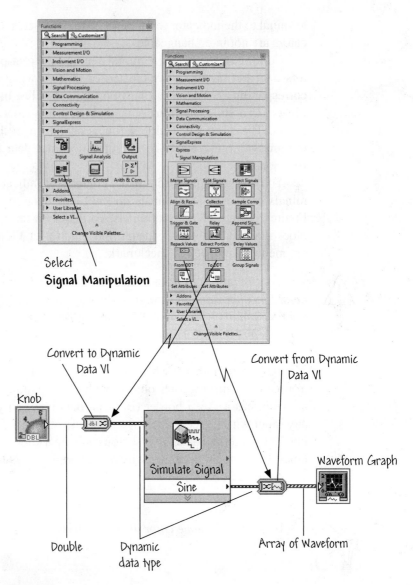

FIGURE 2.20
The Dynamic data type.

the Convert from Dynamic Data Express VI are found on the **Signal Manipula-tion** palette. When you place the Convert from Dynamic Data Express VI on the block diagram, a dialog box appears and displays options that let you specify how you want to format the data that the Convert from Dynamic Data Express VI returns. In Figure 2.20 the Convert from Dynamic Data Express VI is config-ured to transfer double numeric data from a knob to the dynamic data type used by the Simulate Signal Express VI.

When you wire a dynamic data type to an array indicator, LabVIEW automatically places the Convert from Dynamic Data Express VI on the block diagram. Double-click the Convert from Dynamic Data Express VI to open the dialog box to control how the data appears in the array.

Transforming data for use in Express VIs occurs in a similar fashion. You use the Convert to Dynamic Data Express VI to convert numeric, Boolean, waveform, and array data types to the dynamic data type for use with Express VIs. When you place the Convert to Dynamic Data Express VI on the block diagram, a dialog box appears that allows you to select the kind of data to convert to the dynamic data type.

2.4.4 Wiring

Wires are data paths between terminals and are analogous to variables in conventional languages. How then can we represent different data types on the block diagram? Since the block diagram consists of graphical objects, it seems appropriate to utilize different wire patterns (shape, line style, color, etc.) to represent different data types. In fact, each wire possesses a unique pattern depending on the type of data (numeric, Boolean, string, etc.) that flows through the wire. Each data type appears in a different color for emphasis. To determine the data types on a given wire, match up the colors and styles with the wire types as shown in Table 2.1.

The hot spot of the **Wiring** tool is the tip of the unwound wiring segment as seen in Figure 2.21. To wire from one terminal to another, click the hot spot of the **Wiring** tool on the first terminal (you can start wiring at either terminal), move the tool to the second terminal, and click on the second terminal. There is

TABLE 2.1   Common wire types

	Scalar	**1D array**	**2D array**	**Color**
Numeric	——————	▬▬▬▬	▭▭▭▭	Orange (floating point) & Blue (integer)
Boolean	··············	▬▬▬▬	▬▬▬▬	Green
String	∿∿∿∿∿	▨▨▨▨	▨▨▨▨	Pink

FIGURE 2.21
The hot spot on the **Wiring** tool.

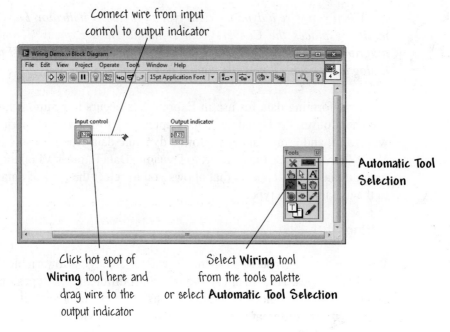

FIGURE 2.22
Wiring terminals.

no need to hold down the mouse button while moving the **Wiring** tool from one terminal to the other. The wiring process is illustrated in Figure 2.22. When you wire two terminals, notice that moving the **Wiring** tool over one of the terminals causes that terminal to blink. This is an indication that clicking the mouse button will make the wire connection.

The VI shown in Figure 2.22 is easy to construct. It consists of one numeric control and one numeric indicator wired together. Open a new VI and try to build the VI! A working version can be found in the **Chapter 2** folder in the **Learning** directory—it is called **Wiring Demo.vi**. The function of the VI is to set a value for **Input control** on the front panel and to display the same input at the numeric indicator **Output indicator**.

To delete a wire as you are wiring:

- **Windows**—Click the right mouse button or click on the origination terminal.

- **Macintosh**—Hold down <option> and click, or click on the origination terminal.

When wiring two terminals together, you may want to bend the wire to avoid running the wire under other objects. This is accomplished during the wiring process by clicking the mouse button to tack the wire down at the desired location of the bend, and moving the mouse in a perpendicular direction to continue the wiring to the terminal. Another way to change the direction of a wire while wiring is to press the space bar while moving the **Wiring** tool.

Automatic wiring is also available. Instead of tacking down the wire, simply connect the two terminals and LabVIEW will choose the best path. If you have an existing wire that you would like to fix, right-click on the wire and choose **Clean Up Wire** on the pull-down menu.

Windows: All wiring is performed using the left *mouse button.*

Tip strips make it easier to identify function and node terminals for wiring. When you move the **Wiring** tool over a terminal, a tip strip appears, as illustrated in Figure 2.23. Tip strips are small text banners that display the name of each terminal. When you place the **Wiring** tool over a node, each input and output will show as a wire stub—a dot at the end of the wire stub indicates an input. Tip strips should help you wire the terminals correctly.

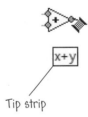

Tip strip

FIGURE 2.23
Tip strip.

It is possible to have objects wired automatically. A feature of LabVIEW is the capability to automatically wire objects when you first drop them on the diagram. After you select a node from the **Functions** palette, move that node close to another node to which you want to wire the first node. Terminals containing similar datatypes with similar names will automatically connect. You can disable the automatic wiring feature by pressing the space bar. You can adjust the auto wiring settings from the Tools≫Options≫Block Diagram window.

Since it is important to correctly wire the terminals on functions, LabVIEW provides an easy way to show the icon connector pane to make the wiring task easier. This is accomplished by right-clicking on the function and choosing Visible Items≫Terminals from the shortcut menu, as illustrated in Figure 2.24. To return to the icon, right-click on the function and deselect Visible Items≫ Terminals.

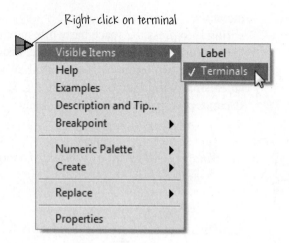

FIGURE 2.24
Showing terminals.

2.5   BUILDING YOUR FIRST VI

In this section you will create your first virtual instrument to perform the following functions:

- Add two input numbers and display the result.
- Multiply the same two input numbers and display the result.
- Compare the two input numbers and turn on an LED if the numbers are equal.

Begin by considering the front panel shown in Figure 2.25. It has two numeric control inputs for the numbers A and B, two numeric indicator outputs to display the results $A + B$ and $A \times B$, respectively, and a round LED that will turn on when the input numbers A and B are identical.

As with the development of most sophisticated computer programs, constructing VIs is an art, and you will develop your own style as you gain experience with programming in LabVIEW. With that in mind, you should consider the following steps as only one possible path to building a working VI that carries out the desired calculations and displays the results.

1. Open a new front panel by choosing **New VI** from the **File** menu.
2. **Create the numeric controls and indicators**. The two front panel controls are used to enter the numbers, and the two indicators are used to display the results of the addition and multiplication of the input numbers.
   (a) Select **Numeric Controls** from the **Controls≫Silver** palette. If the **Controls** palette is not visible, right-click in an open area of the front panel to gain access to the palette.

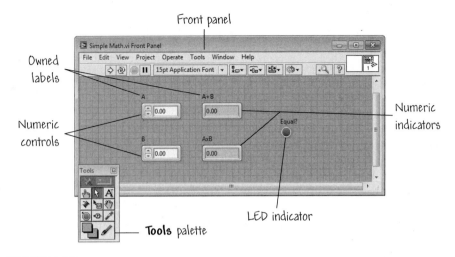

FIGURE 2.25
The front panel for your first VI.

        (b) Drop the control on the front panel, as illustrated in Figure 2.26. Drag the control to the desired location and then click the mouse button to complete the drop.

        (c) Type the letter **A** inside the label box (which appears above the control) and press the **Enter** button on the front panel toolbar. If you do not

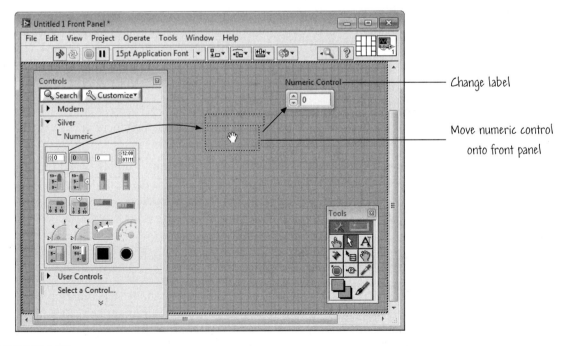

FIGURE 2.26
Placing the controls and indicators on the front panel.

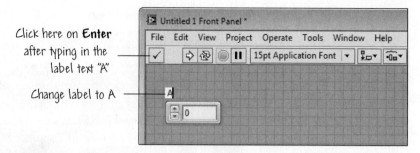

FIGURE 2.27
Labeling the numeric control and indicators on the front panel.

type the control label before starting other programming actions (such as dropping the other control on the front panel), the label box will remain labeled with the default label Numeric Control. If the control or indicator does not have a label, you can right-click on the control and select **Label** from the **Visible Items** menu. The label box appears, and you can then edit the text using the **Labeling** tool (see Figure 2.27).

(d) Repeat the above process to create the second numeric control and the two numeric indicators. You can arrange the controls and indicators in any manner that you choose—although a neat and orderly arrangement is preferable. Add the labels to each control and indicator using Figure 2.25 as a guide.

3. **Create the Boolean LED**. This indicator will turn on if the two input numbers are identical, or remain off if they do not match.

(a) Select **Round LED** from the **Silver≫Boolean** subpalette of the **Controls** palette. Place the indicator on the front panel, drag it to the desired location, and then click the mouse button to complete the process.

(b) Type Equal? inside the label box and click anywhere outside the label when finished, or click on the **Enter** button.

Each time you create a new control or indicator, LabVIEW automatically creates the corresponding terminal in the block diagram. When viewed as icons, the terminals are graphical representations of the controls or indicators.

The Block Diagram

1. Switch your center of activity to the block diagram by selecting **Show Block Diagram** from the **Window** pull-down menu. The completed block diagram is shown in Figure 2.28. It may be helpful to display the front panel and block diagram simultaneously using either the **Tile Left and Right** or the **Tile Up and Down** options found in the **Window** pull-down menu. For this

example, the up and down option works better in the sense that all the block diagram and front panel objects can be displayed on the screen without having to use the scrollbars.

2. Now we want to place the addition and multiplication functions on the block diagram. Select the Add function from the **Programming≫Numeric** sub-palette of the Functions palette. If the Functions palette is not visible, right-click on an open area of the block diagram to gain access to the palette. Drop the Add function on the block diagram in approximately the same position as shown in Figure 2.28. The label for the Add function can be displayed using the shortcut menu and selecting Visible Items≫Label. This is illustrated in Figure 2.29. Following the same procedure, place the Multiply function on the block diagram and display the label.

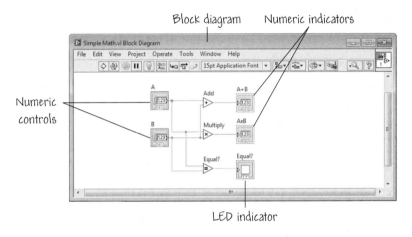

FIGURE 2.28
The block diagram window for your first VI.

3. Select the Equal? function from the **Programming≫Comparison** sub-palette of the **Functions** palette and place it on the block diagram, as shown in Figure 2.30. The Equal? function compares two numbers and returns True if they are equal or False if they are not. To get more information on this function, you can activate the online help by choosing **Show Context Help** from the **Help** menu. Then placing the cursor over the Equal? function (or any of the other functions on the block diagram) leads to the display of the online help information.

4. Using the **Wiring** tool, wire the terminals as shown in Figure 2.28. As seen in Figure 2.31, to wire from one terminal to another, click the **Wiring** tool on the first terminal, move the tool to the second terminal, and click on the second terminal. Remember that it does not matter on which terminal you initiate the wiring. To aid in wiring, right-click on the three functions and

Function label

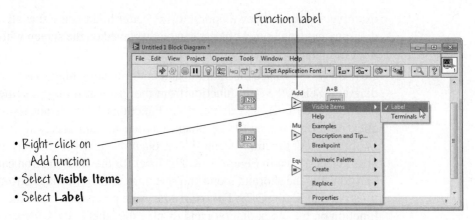

• Right-click on
  Add function
• Select **Visible Items**
• Select **Label**

FIGURE 2.29
Right-click to select the Visible Items≫Label option.

Control terminals
have thick borders   Indicator terminals have thin borders

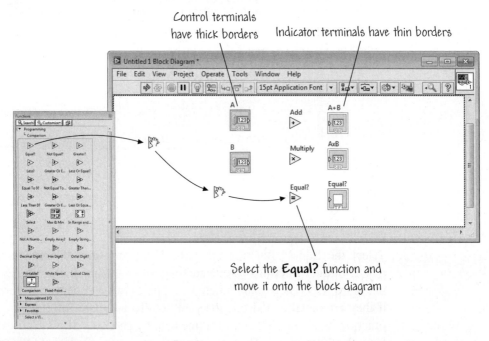

Select the **Equal?** function and
move it onto the block diagram

FIGURE 2.30
Adding the Equal? function to the VI.

choose **Visible Items≫Terminals.** Having the terminals shown explicitly
helps to wire more quickly and accurately. Once the wiring is finished for
a given function, it is best to return to the icon by right-clicking on the
function and choosing **Visible Items**, and deselecting **Terminals**.

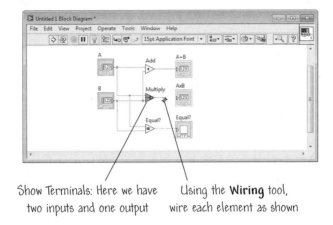

Show Terminals: Here we have   Using the **Wiring** tool,
two inputs and one output   wire each element as shown

FIGURE 2.31
Wiring from one terminal to another.

5. Switch back to the front panel window by clicking anywhere on it or by choosing **Show Front Panel** from the **Window** menu.

6. Save the VI as Simple Math.vi. Select **Save** from the **File** menu and make sure to save the VI in the Users Stuff folder within the Learning directory.

In case you cannot get your VI to run properly, a working version of the VI (called Simple Math.vi*) is located in the* Chapter 2 *folder within the* Learning *directory.*

7. **Enter input data.** Enter numbers in the numeric controls utilizing the **Auto** tool by double-clicking in the numeric control box and typing in a number. The default values for A and B are 0 and 0, respectively. You can run the VI using these default values as a first try! When you use the default values, the LED should light up since $A = B$.

8. Run the VI by clicking on the **Run** button.

9. Experiment with different input numbers—make A and B identical and verify that the LED does indeed turn on.

10. When you are finished experimenting, close the VI by selecting **Close** from the **File** menu.

2.6 DATA FLOW PROGRAMMING

The principle that governs VI execution is known as data flow. Unlike most sequential programming languages, the executable elements of a VI execute only when they have received all required input data—in other words, data

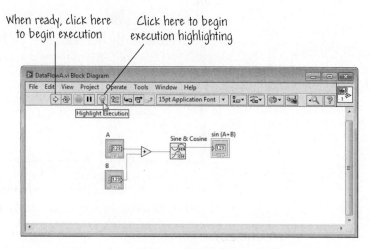

FIGURE 2.32
Block diagram that adds two numbers and then computes the sine of the result.

flows out of the executable element only after the code is finished executing. The notion of data flow contrasts with the control flow method of executing a conventional program, in which instructions execute sequentially in the order specified by the programmer. Another way to say the same thing is that the flow of traditional sequential code is instruction driven, while the data flow of a VI is data driven.

Consider the VI block diagram, shown in Figure 2.32, that adds two numbers and then computes the sine of the result. In this case, the block diagram executes from left to right, not because the objects are placed in that order, but because one of the inputs of the Sine & Cosine function is not valid until the Add function has added the numbers together and passed the data to the Sine & Cosine function. Remember that a node (in this case, the Sine function) executes only when data is available at all of its input terminals, and it supplies data to its output terminals only when it finishes execution. Open DataFlowA.vi located in the Chapter 2 folder of the Learning directory, click on execution highlighting, and then run the VI.

Consider the example in Figure 2.33. Which code segment would execute first—the one on the left or the one on the right? You cannot determine the answer just by looking at the segments. The one on the left does not necessarily execute first. In a situation where one code segment must execute before another, and there is no type of dependency between the functions, you must use a Sequence structure to force the order of execution. To observe the data flow on the code in Figure 2.33, open DataFlowB.vi located in the folder Chapter 2 in the Learning directory. Before running the VI, click on the **Highlight Execution** button and then watch the flow of the execution.

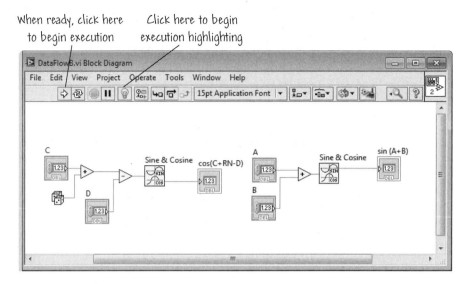

FIGURE 2.33
Which code executes first?

2.7   BUILDING A VI USING EXPRESS VIS

In this section, you will create your first VI using Express VIs. The objective is to construct a VI that generates a sawtooth signal at an amplitude that we prescribe on the front panel and displays the sawtooth signal graphically on the front panel. LabVIEW provides a VI template containing information that will help you in building this VI.

At the LabVIEW Getting Started screen (see Figure 1.1), select **VI from Template** under **File≫New** to display the **New** dialog box. Select **Generate and Display** from the **VI≫ From Template≫ Simulated** folder in the **Create New** list. This template VI generates and displays a signal.

1. Click the **OK** button to open the template. You also can double-click the name of the template VI in the **Create New** list to open the template.

2. Examine the front panel of the VI, as shown in Figure 2.34. The front panel appears with a waveform graph and a stop button. The title bar of the front panel indicates that this window is the front panel for the Generate and Display [GenerateDisplay1.vi] VI.

If the front panel is not visible, you can display the front panel by selecting Window≫Show Front Panel. *Press the* <Ctrl-E> *keys to switch from the front panel to the block diagram or from the block diagram to the front panel.*

3. Examine the block diagram of the VI shown in Figure 2.34. The block diagram appears with the Simulate Signal Express VI, a waveform graph icon, a stop button icon, and a While Loop (more on loops in Chapter 5). The title bar of the block diagram indicates that this window is the block diagram for the Generate and Display [GenerateDisplay1.vi] VI.

If the block diagram is not visible, you can display the block diagram by selecting Window ≫ Show Block Diagram.

4. On the front panel toolbar, click the **Run** button and verify that a sine wave appears on the graph.

5. Stop the VI by clicking the Stop button located at the bottom right-hand side of the front panel.

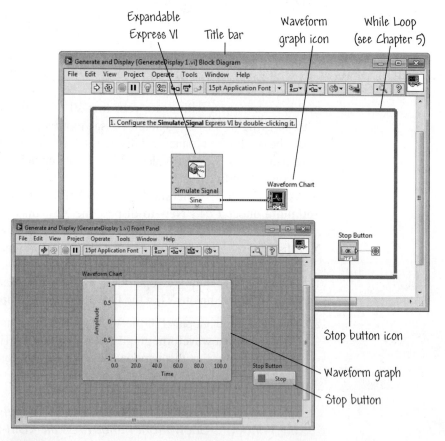

FIGURE 2.34
Building a VI using the Simulate Signal Express VI.

*Although the **Abort Execution** button may seem, at first glance, to operate like a stop button, the **Abort Execution** button does not always properly close the VI. It is recommended to always stop your VIs using the **Stop** button on the front panel. Use the **Abort Execution** button only when errors prevent you from terminating the application using the **Stop** button.*

We can now add other elements to the VI template to begin the process of constructing the VI. First, we will add a control to the front panel to use to vary the sawtooth signal amplitude. In our VI, the control will supply the sawtooth amplitude data to the block diagram.

6. On the front panel, right-click in an open area to display the **Controls** palette.

7. Move the cursor over the icons on the **Controls** palette to locate the **Numeric** subpalette under the **Silver** palette. Notice that when you move the cursor over icons on the **Controls** palette, the subpalette appears with the name in the gray space above all the icons on the palette. When you idle the cursor over any icon on any palette, the full name of the control or indicator appears in the blue space above the icons.

8. Click the **Numeric** icon to access the **Numeric** palette.

9. Since many physical instruments have knobs to vary the operational parameters of the instrument, we will select the **Knob** control on the **Numeric** palette and place it on the front panel to the left of the waveform graph, as shown in Figure 2.35. This knob will provide control over the amplitude of the sawtooth signal once it is properly wired.

10. Select File≫Save As and save this VI as Acquiring a Signal.vi in the Users Stuff folder in the Learning directory. Notice that the VI name appears in the closed brackets in the title bar (see Figure 2.35).

*Save the VIs you edit or create using this book, in the **Users Stuff** folder in the **Learning** directory.*

Examine the block diagram of the VI as it currently is configured. Notice that the block diagram has a blue icon labeled Simulate Signal. This icon represents the Simulate Signal Express VI, which simulates a sine wave by default. To meet our objectives, we must reconfigure the Express VI to simulate a sawtooth signal.

11. Display the block diagram by selecting Window≫Show Block Diagram. The Simulate Signal Express VI depicted in Figure 2.36 simulates a signal

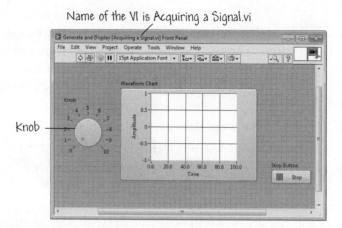

FIGURE 2.35
Adding a **Knob** to the Generate and Display template.

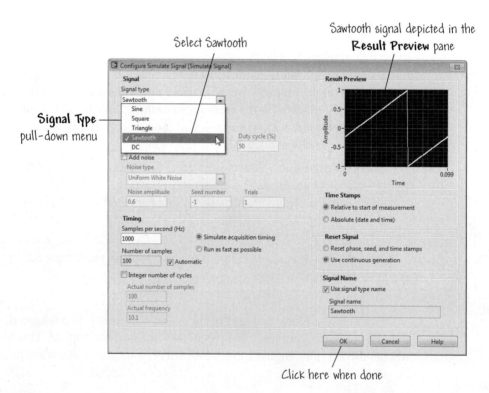

FIGURE 2.36
The **Configure Simulate Signal** dialog box.

based on the configuration that you specify. We must interact with the Express VI dialog box to output a sawtooth signal, since the default is a sine wave signal.

12. Right-click the Simulate Signal Express VI and select **Properties** from the shortcut menu to display the **Configure Simulate Signal** dialog box, as shown in Figure 2.36. You can also double-click the Express VI to access the dialog box.

13. Select **Sawtooth** from the **Signal Type** pull-down menu. Notice that the waveform on the graph in the **Result Preview** section changes to a sawtooth wave.

14. Click the **OK** button to apply the current configuration and close the **Configure Simulate Signal** dialog box.

15. Now we want to expand the Simulate Signal Express VI to show the inputs and outputs to make the wiring easier. Move the cursor over the down arrow at the bottom of the Simulate Signal Express VI as illustrated in Figure 2.37.

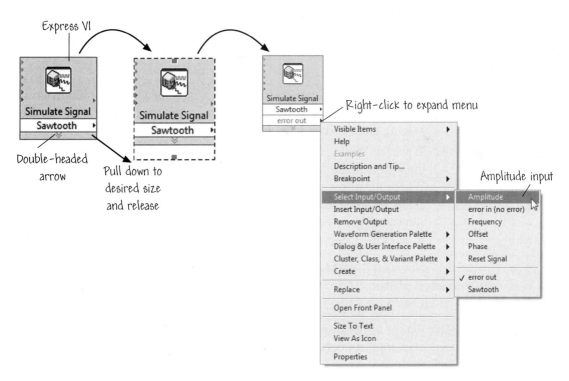

FIGURE 2.37
Input and Output of the Simulate Signal Express VI.

16. When a double-headed arrow appears, click and drag the border of the Express VI down to show one additional input. By default, **error out** appears. Click on **error out** and select **Amplitude** from the pull-down menu. Because the **Amplitude** input appears on the block diagram, you can configure the amplitude of the sawtooth wave on the block diagram.

When inputs appear on the block diagram and in the configuration dialog box, you can configure the inputs in either location.

Now we can complete the VI by finishing up the wiring. To use the **Knob** control to change the amplitude of the signal, the **Knob** must be wired to the **Amplitude** input on the Simulate Signal Express VI.

17. Move the cursor over the **Knob** terminal until the **Positioning** tool appears (see Figure 1.25 for review of tools). The **Positioning** tool, represented by an arrow, is used to select, position, and resize objects.

18. Click the **Knob** terminal to select it, then drag the terminal to the left of the Simulate Signal Express VI. Make sure the **Knob** terminal is inside the While Loop. Deselect the **Knob** terminal by clicking a blank space on the block diagram.

The cursor does not switch to another tool while an object is selected.

19. Move the cursor over the arrow on the right-hand side of the **Knob** terminal. This will result in the cursor becoming the **Wiring** tool. Now we can use the **Wiring** tool to wire the **Knob** to the Express VI.

20. When the **Wiring** tool appears, click the arrow and then click the **Amplitude** input of the Simulate Signal Express VI to wire the two objects together. When the wire appears and connects the two objects, then data can flow along the wire from the **Knob** to the Express VI. The final block diagram is shown in Figure 2.38.

21. Select File≫Save to save this VI.

Now that the VI is ready for execution, we can see if we have successfully achieved the goal of generating a signal and displaying it graphically on the front panel.

22. Return to the front panel by selecting Window≫Show Front Panel and click the **Run** button.

23. Move the cursor over the knob control. Notice how the cursor becomes the **Operating** tool. The **Operating** tool can be used to change the value of the sawtooth amplitude.

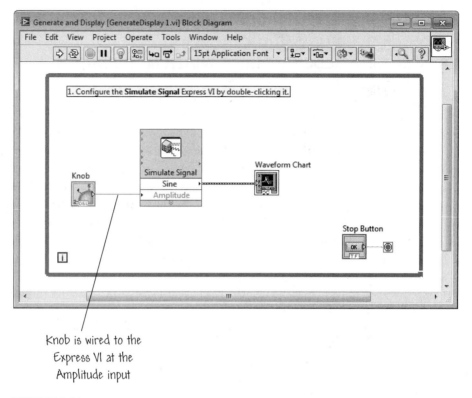

Knob is wired to the
Express VI at the
Amplitude input

FIGURE 2.38
The block diagram of the completed VI to generate and display a signal.

24. Using the **Operating** tool, turn the knob to adjust the amplitude of the saw-
    tooth wave. Notice how the amplitude of the sawtooth wave changes as you
    turn the knob. Also notice that the *y*-axis on the graph autoscales to account
    for the change in amplitude. To indicate that the VI is running, the **Run** but-
    ton changes to a darkened arrow. You cannot edit the front panel or block
    diagram while the VI runs.

25. Click the Stop button to stop the VI when you are finished experimenting.

   Your VI development is now finished. The utility of using the provided VI
templates is evident in this example. We were able to start from a point much
closer to the desired final VI using the Generate and Display VI than starting
with a blank VI. Also, the utility of the Express VIs is demonstrated with the
Simulate Signal VI. This VI was easily configured to simulate a sawtooth signal
and provided a quick and easy solution to our problem. Clearly, it would have
taken more effort to construct the VI from scratch.

BUILDING BLOCK

2.8   BUILDING BLOCKS: PULSE WIDTH MODULATION

As we progress through the building blocks in each chapter, we will create a pulse width modulation VI to output a periodic pulse whose length and magnitude are prescribed by controls on the front panel. Recall from your experimentation with the VI in the Chapter 1 Building Block section that an adjustment to either the period or the duty cycle impacts the pulse length or on time. Increasing the period will increase the total on time as well as the total off time. Increasing the duty cycle will increase the on time and decrease the off time such that the total cycle time remains constant. In this Building Block, we will construct a VI to detect the **Falling Edge** of the signal, which will determine whether the output should toggle to low or remain high on the next cycle. This VI will be used in subsequent Building Block VIs. In Chapter 4 you will learn how to modify this VI to be used within another VI as a subVI.

Open a new VI and save it as **Falling Edge.vi**. Define the output of the VI as a Boolean indicator called **Falling Edge**. This indicator will be True when a "falling edge" is to occur upon the next clock cycle. The inputs to this VI are **Clock**, **Period**, and **Duty Cycle**. The input **Clock** is an integer value denoting the current value of the clock. The time units of the clock are not relevant at this time. The input **Period** is an integer value greater than or equal to 100 specifying the desired cycle duration. The input **Duty Cycle** is a value between 0 and 1 describing the ratio of on time to total cycle time.

The resolution of the clock limits the precision with which we can specify **Period** and **Duty Cycle**. In this VI, we will limit the **Period** to a minimum value of 100 so that duty cycle can be expressed to the nearest hundredth. For example, if we set **Duty Cycle** = 0.25 and set **Period** = 100, the clock will increment 25 times before the signal toggles. If we had instead specified **Duty Cycle** as 0.333, then the clock would need to increment 33.3 times before the output would need to change. Since this is not an integer value, this is not a feasible set of inputs. A similar problem would be encountered for any period of less than 100. Assume for now that the user will always provide valid inputs for **Duty Cycle** and **Period**.

Figure 2.39 shows a completed version of **Falling Edge.vi**. The output of the VI should be True whenever the percentage of the period specified by **Duty Cycle** has elapsed. In order to determine whether this event has occurred, subtract the pulse length (determined by the multiplication of the duty cycle and the period) from the current value of the clock and check to see if this value is evenly divisible by the period length. If this division produces an integer, then **Falling**

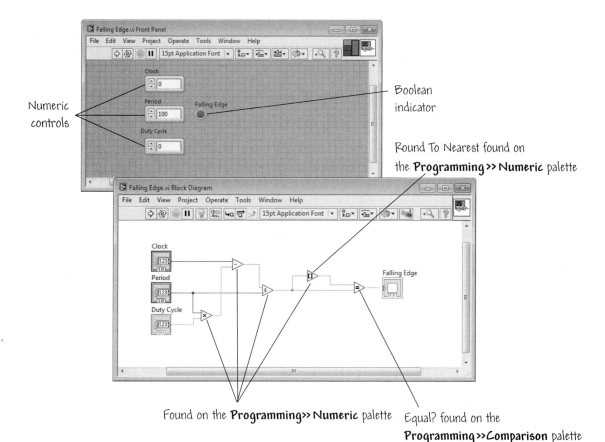

Numeric controls

Boolean indicator

Round To Nearest found on the **Programming >> Numeric** palette

Found on the **Programming >> Numeric** palette

Equal? found on the **Programming >> Comparison** palette

FIGURE 2.39
The Falling Edge VI.

Edge is True and the output should go low on the next clock cycle. To check if the result of the division is an even multiple of the period, we use the Round To Nearest function and the Comparison function as shown in Figure 2.39.

Experiment with different inputs to the VI and verify that it produces the expected outputs. Try running the VI in **Highlight Execution** with Duty Cycle = 0.5 and Period = 100. Set the Clock to 0. What happens? The Falling Edge should not be detected. Now try the input Clock = 50. Leaving **Highlight Execution** on, what happens this time? You should see that the Falling Edge is detected.

When you are done experimenting with your new VI, save it as Falling Edge.vi in the Users Stuff folder in the Learning directory. You will use this VI as a Building Block in later chapters—so make sure to save your work!

A working version of **Falling Edge.vi** *can be found in the* **Building Blocks** *folder of the* **Learning** *directory.*

Next we need to design a way to detect the **Rising Edge**. This task is fully defined in design problem D2.1.

2.9   RELAXED READING: BUILDING A SEMIAUTONOMOUS VEHICLE DRIVEN BY THE VISUALLY IMPAIRED

In this reading we consider the problem of developing a semiautonomous vehicle that allows a blind driver to successfully navigate, control speed, and avoid collision through a secure driving course. Using NI CompactRIO and LabVIEW software, faculty and students at Virginia Tech took on the challenge to build and program the world's first functional prototype of a blind driver vehicle.

In an effort to promote the often underestimated capabilities of the visually impaired and to inspire innovation in the development of blind access technologies, the National Federation of the Blind proposed a challenge to design a system capable of providing the blind with the ability to drive. The Robotics and Mechanisms Laboratory at Virginia Tech took on the challenge and defined the initial goals for the world's first working prototype of a blind driver vehicle. A blind driver would be expected to safely perform the three fundamental driving tasks: (1) navigate through a curved driving course defined by a single lane of traffic cones, (2) regulate speed within a predefined limit, and (3) exhibit sufficient emergency-stop capability to avoid colliding with an obstacle.

The current blind driver system consists of sensors and novel nonvisual driver interfaces attached as a modular system to a modified dune buggy, as depicted in Figure 2.40(a). A single-plane laser rangefinder is used for environmental perception to scan the driving environment for cones and other obstacles and feed the information to an onboard CompactRIO real-time controller and its real-time field-programmable gate array processing targets. A laptop running LabVIEW software provides USB hosting capability to the CompactRIO controller and also allows a sighted passenger to passively monitor the operation of all hardware and software and easily modify any heuristic-based programming for quick calibration during field testing. Additional sensors gather important information regarding the state of the vehicle, such as speed from a Hall effect sensor and steering angle from a string potentiometer. The data is acquired from the sensors and processed directly using the high-speed FPGA on the CompactRIO real-time controller.

(a)

(b)

FIGURE 2.40
(a) Blind drivers are provided a combination of auditory and vibrotactile cues to effectively and efficiently drive the vehicle and navigate obstacles. (b) Blind 15 year-old pictured in the drivers seat drives the world's first blind driver vehicle developed by Virginia Tech.

After collecting an image of the driving environment using the various sensors, the information is processed and transmitted to the driver through nonvisual cues. A blind driver is pictured in the world's first blind driver vehicle developed by Virginia Tech in Figure 2.40(b). The goal when developing a nonvisual driver interface is to effectively and efficiently provide information to a driver to maximize situational awareness and allow the driver to make quick and precise driving decisions. The array of nonvisual driver interfaces on the first iteration of the vehicle is a combination of informational and instructional cues for safety and redundancy.

For speed regulation, the driver can operate at a comfortable speed until reaching a maximum speed limit at which point a vibrotactile vest on the seat belt informs the driver what degree of braking is necessary to return to a safe operating speed. If the vehicle detects an unavoidable collision with an obstacle, the vest cues the driver to stop the vehicle immediately. For steering guidance, a potential field algorithm provides the path generation. After calculating a path, the system instructs the driver where to steer to stay in the lane and avoid obstacles. The driver is told how many *clicks* to turn the steering wheel via a pair of headphones and LabVIEW text-to-speech software. A mechanism attached to the steering column clicks every five degrees to provide precise audible feedback. The intuitive graphical programming interface of LabVIEW made it easy for a team of undergraduate mechanical engineering students to quickly and efficiently create custom embedded software without the need for expertise in specific text-based programming language.

Additionally, a prototype for a tactile map was developed that is conceptually similar to a high-resolution grid of regenerative braille. The map places an image of the surrounding environment literally in the hands of the driver. Similar to the tiny holes on an air hockey table, a physical map is generated by passing compressed air through small pixels to depict the surrounding obstacles detected by the laser range finder. This device allows the driver to *see* the surroundings and navigate safely through them. The audio and vibrotactile nonvisual driver interfaces are still necessary for redundancy, but using the driver's high-bandwidth sense of tactation through this tactile map technology makes data pathways available for other driving uses, such as listening and interacting with a GPS through voice-recognition software for higher-level path planning.

The Virginia Tech program has provided more than 30 blind and visually impaired people of all ages from around the country with the opportunity to drive a vehicle. Whether it was their first time behind the wheel or a long-awaited reunion with an automobile, their reactions were overwhelmingly positive and filled with hope. Resulting national and international media coverage is raising tremendous awareness to the capabilities of the blind as well as generating interest in collaboration for the research and development of novel blind access technologies in various applications.

The numerous potential spin-off technologies were a major emphasis throughout the design process. As these devices are proven to be sufficient in allowing a blind person to drive a vehicle, one can imagine the benefits for drivers who have bad vision, are talking or sending text messages on a cell phone, or are drowsy and otherwise distracted on the road. Early warning devices and collision mitigation systems could be created for all driving environments, especially in bad weather or low-visibility conditions.

Other than automotive applications, there is also potential for advances with haptic human interface devices, especially for blind pedestrians. The nonvisual interfaces could easily be deployed in aircraft cockpits in which the current technology state relies heavily on the pilot's visual capabilities. Sending the high-bandwidth information from the highly saturated visual environment across the other senses will greatly increase a pilot's situational awareness, which is a critical aspect when operating any vehicle.

We may not see blind drivers on the road for many years—however, the potential spin-off technologies are suitable for use in countless applications. For further information, visit http://sine.ni.com/cs/app/doc/p/id/cs-12294 and http://www.unirel.vt.edu/bdc.html.

2.10   myDAQ BUILDING BLOCKS

In this myDAQ Building Block, you will build and test the thermistor circuit that will become part of your temperature monitoring system. The thermistor—a contraction of *thermal* and *resistor*—is a two-terminal semiconductor device whose resistance varies with temperature. Most thermistors (including the one used here) are of the negative temperature coefficient (NTC) type, meaning their resistance varies inversely with temperature. Constructing a voltage divider from a thermistor and a fixed-value resistor offers an easy-to-build yet effective interface circuit. The voltage readings, along with the known value of the fixed resistor, can be used to calculate the resistance of the thermistor as it changes.

2.10.1   Building a Basic Thermistor Circuit

To build the circuit, you will need the following:

1. Thermistor, 10 kΩ
2. Ceramic disk capacitor, 0.1 μF
3. Resistor, 10 kΩ
4. Breadboard
5. Jumper wires

Before building the circuit, use the ELVIS Instrument DMM to measure the actual resistance of the 10 kΩ resistor and record the value for later use.

2.10.2   The Thermistor Circuit Test VI

You will use Thermistor Circuit Test.vi to test the thermistor circuit. The VI is shown in Figure 2.42. Open the block diagram to explore how the thermistor resistance is calculated.

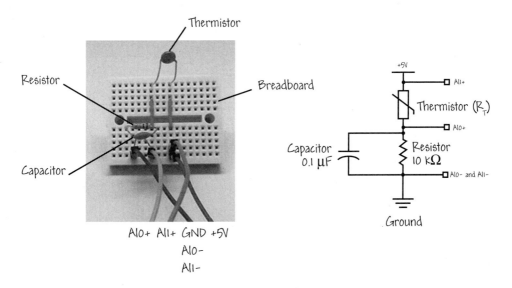

FIGURE 2.41
The thermistor circuit.

A working version of Thermistor Circuit Test.vi can be found in the myDAQ Building Blocks folder in the Learning directory.

2.10.3   Testing the Thermistor Circuit

On the Thermistor Circuit Test VI, enter the measured value of the fixed resistor on the front panel. Run the VI. It will display the measured resistance of the thermistor; expect to see a value close to 10 kΩ at room temperature. Try heating

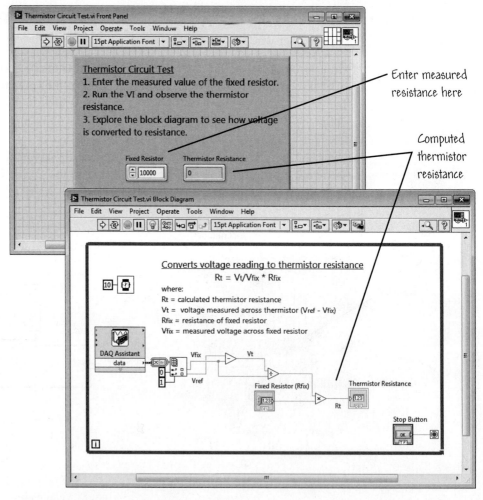

FIGURE 2.42
The Thermistor Circuit Test VI.

the thermistor by gently pinching the thermistor body with your fingertips. You may also use a straw to blow warm air on the thermistor. You should observe the resistance going down. How low can you make the resistance? Use the **Stop** button to stop the VI when you are done.

In this second myDAQ Building Block, you created a thermistor circuit with the myDAQ hardware and a few simple components (thermistor, capacitor, and resistor). The DMM (see Chapter 1) was used to measure the actual resistance of the 10 kΩ resistor that was used in the circuit. The Thermistor Circuit Test VI was used to compute the thermistor resistance and required the measured resistor resistance (near 10 kΩ) as an input. In the remaining chapters, we will extend the LED circuits and the thermistor circuits to highlight the various LabVIEW topics within the chapters using the myDAQ.

2.11   SUMMARY

Virtual instruments (VIs) are the building blocks of LabVIEW. The graphical programming language is known as the G programming language. VIs have three main components: the front panel, the block diagram, and the icon and connector pane pair. VIs follow a data flow programming convention in which each executable node of the program executes only after all the necessary inputs have been received. Correspondingly, output from each executable node is produced only when the node has finished executing.

KEY TERMS

Boolean controls and indicators: Front panel objects used to manipulate and display or input and output Boolean (True or False) data.

Terminals: Part of the VI or function node through which data passes to and from the node. The terminals provide connection to the inputs and outputs of the VI or function node.

Connector pane: Region in the upper right corner of the front panel that displays the VI terminal pattern. It underlies the icon pane.

Data flow programming: Programming system consisting of executable nodes in which nodes execute only when they have received all required input data and produce output automatically when they have executed.

G programming language: Graphical programming language used in LabVIEW.

Icon: Graphical representation of a node on a block diagram.

Icon pane: Region in the upper right corner of the front panel and block diagram that displays the VI icon.

Input terminals: Terminals that emit data. Sometimes called source terminals.

Modular programming: Programming that uses interchangeable computer routines.

Numeric controls and indicators: Front panel objects used to manipulate and display numeric data.

Output terminals: Terminals that absorb data. Sometimes called destination terminals.

String controls and indicators: Front panel objects used to manipulate and display or input and output text.

Terminals: Objects or regions on a node through which data passes.

Tip strips: Small yellow text banners that identify the terminal name and make it easier to identify function and node terminals for wiring.

Wire: Data path between nodes.

Wiring tool: Tool used to define data paths between source and sink terminals.

EXERCISES

E2.1 In this first exercise you get to play a drawing game. Open Drawing.vi located in Chapter 2 in the Exercises & Problems folder of the Learning directory. The front panel should look like the one shown in Figure E2.1. To play the drawing game, run the VI and experiment by varying the controls. After clicking on the **Run** button to start the VI execution, click on the Begin drawing push button on the front panel. Once the drawing starts, the Stop drawing label appears in the push button. The drawing is halted by clicking on the Stop drawing button. Clicking on the Stop button halts execution of the VI.

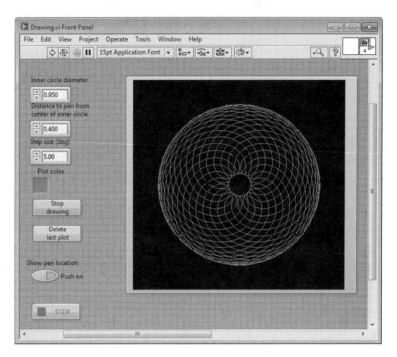

FIGURE E2.1
The Drawing.vi front panel.

E2.2 Construct a VI that uses a round push-button control to turn on a round light indicator whenever the push button is depressed. The front panel is very simple and should look something like the one shown in Figure E2.2.

E2.3 Open Temperature System Demo.vi, located in the tempsys.llb found in the subfolder apps within the examples folder in the LabVIEW directory. The front panel and block diagram are shown in Figure E2.3. Using the **Context Help** (Ctrl-H), determine the inputs and outputs of the Temperature Status subVI. Sketch the subVI icon and connector pane showing the inputs and outputs.

E2.4 Open the VI you created in Section 2.5 called Simple Math.vi.

(a) Locate all of the controls in this VI, specifying their label and type (numeric, Boolean, or string).

(b) Locate all of the indicators in this VI, specifying their label and type (numeric, Boolean, or string).

FIGURE E2.2
A front panel using a round push-button control and a round light indicator.

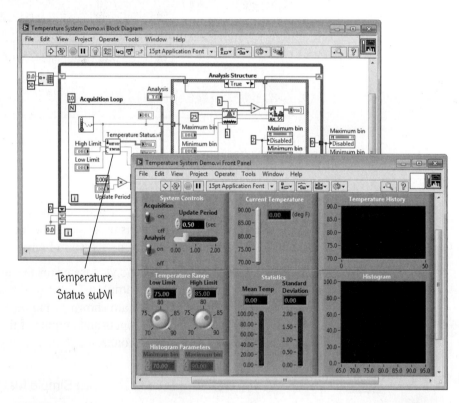

Temperature
Status subVI

FIGURE E2.3
The Temperature System Demo VI.

(c) Locate and identify all of the nodes in this VI.

(d) Locate and identify all of the terminals in this VI.

(e) Replace the two numeric controls labeled A and B with dials by right-clicking on each of the two dials, and on the shortcut menu under **Visible Items** select **Digital Display**.

(f) Run the program continuously and experiment changing values in both the dials and the digital displays.

E2.5 Create a VI which compares the inputs on two numeric controls and turns on an LED if these values are equal. The front panel should resemble Figure E2.5.

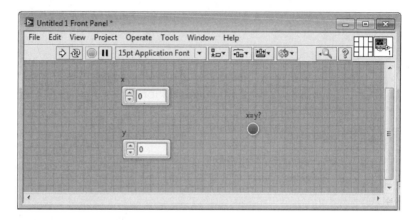

FIGURE E2.5
The Number Comparison VI front panel.

E2.6 Design a VI which determines whether a number input on a floating point numeric control is an integer. Place an LED on the front panel that lights whenever the input is an integer.

E2.7 On which palettes are the following controls located? What is the purpose of these controls?

(a) OK Button

(b) Square Push Button

(c) String

(d) Intensity Graph

(e) RealMatrix

*Use the **Search** button on the **Controls** palette as described in Section 2.3.1 to locate the controls.*

E2.8   On which palettes are the following functions located? What is the purpose of these functions?

(a) Mathscript Node

(b) Derivative x(t)

(c) Sine Wave

(d) Select

(e) Wait (ms)

*Use the **Search** button on the **Functions** palette as described in Section 2.3.1 to locate the finctions.*

E2.9   Consider the block diagram in Figure E2.9. Using the concept of data flow, answer the following questions:

(i) Which function executes first?

(a) Add

(b) Multiply

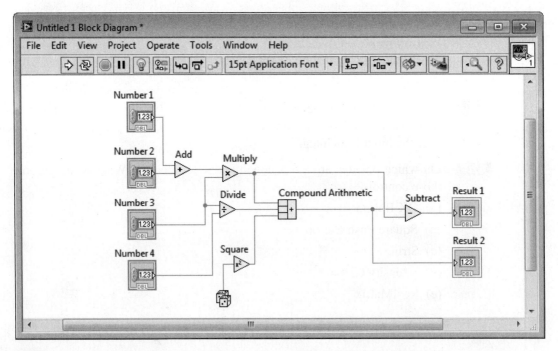

FIGURE E2.9
Investigations into data flow.

(c) Subtract

(d) Unknown

(ii) Which function executes first?

(a) Add

(b) Divide

(c) Random Number

(d) Unknown

(iii) Which function executes last?

(a) X squared

(b) Multiply

(c) Compound Arithmetic

(d) Unknown

E2.10 Open Data Flow.vi located in Chapter 2 in the Exercises & Problems folder of the Learning directory. Consider the block diagram in Figure E2.10. Using the concept of data flow, list the functions in the order that they execute. Some of the functions may execute at the same time; if so, list them in a single group.

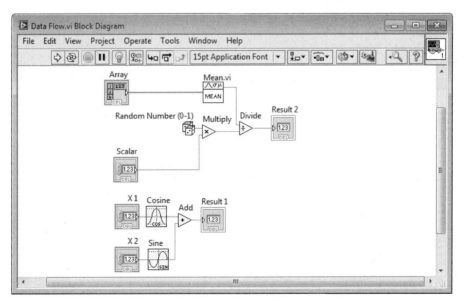

FIGURE E2.10
Investigations into data flow.

PROBLEMS

P2.1 Open the VI created in Section 2.7, **Acquiring a Signal.vi**. Modify this VI so that it meets the following specifications:

- Two sine waves are generated using the Simulate Signal Express VI and displayed on corresponding graphs on the front panel.

- The amplitude of each sine wave can be adjusted to any value using numeric controls on the front panel.

Save the VI as **Compare Signals.vi** in the **Users Stuff** folder in the **Learning** directory. The front panel should resemble the front panel seen in Figure P2.1.

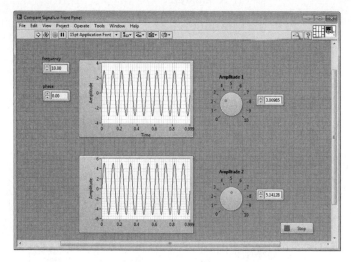

FIGURE P2.1
The front panel for the Compare Signals VI.

P2.2 Construct a VI that performs the following tasks:

- Takes two floating-point numbers as inputs on the front panel: X and Y.

- Subtracts Y from X and displays the result on the front panel.

- Divides X by Y and displays the result on the front panel.

- If the input $Y = 0$, a front panel LED lights up to indicate division by zero.

Name the VI **Subtract and Divide.vi** and save it in the **Users Stuff** folder in the **Learning** directory.

P2.3 Construct a VI that uses a vertical pointer slide control for input and a meter indicator for output display. A front panel and block diagram that can be used as a guide are shown in Figure P2.3. Referring to the block diagram, you see a pair of dice, which is the icon for a Random Number function. You will find the

Random Number function on the **Programming≫Numeric** palette. Using the **Labeling** tool, change the maximum value of the range of the Vertical Pointer Slide control and the Meter indicator. Set the maximum value to 100 by selecting the default value and typing in "100" in the text box on both the Vertical Pointer Slide and Meter. When running the VI, any input you provide via the vertical pointer slide will be reflected on the meter indicator. The Random Number function adds "noise" to the input so that the meter output will not be exactly the same as the input. Run the VI in **Run Continuously** mode and vary the slide input.

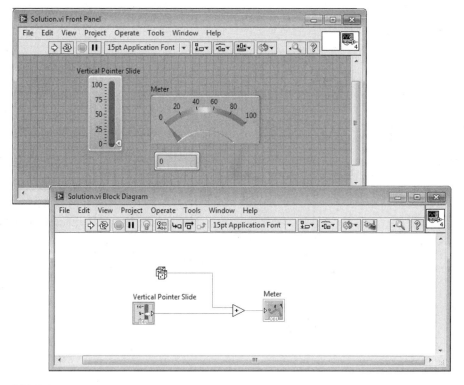

FIGURE P2.3
Using a vertical pointer slide control and a meter indicator.

P2.4 Create a VI that has a numeric control to input a number x and uses the Add and Multiply functions to calculate $3x^2 + 2x + 5.0$. Display the output using a numeric indicator.

P2.5 Use a VI template and modify Express VIs to create a program that generates a Triangle Wave with a frequency of 125 Hz and added noise.

P2.6 Construct a VI that accepts an input in feet and converts this value to its equivalent in both meters and miles, displaying each on an indicator on the front panel. Recall that 1 mile = 5,280 feet and 1 meter = 3.281 feet.

Using the VI, confirm that 10,000 feet is equivalent to 1.89 miles and 3047.85 meters. Save the VI as Convert.vi in the Users Stuff folder in the Learning directory.

P2.7 Design a VI that computes both the sum and the difference of two numeric inputs, labeled x and y. It should then light one of three LEDs, depending on whether or not the sum is greater than, equal to, or less than the difference between the two inputs. The front panel should resemble the front panel seen in Figure P2.7.

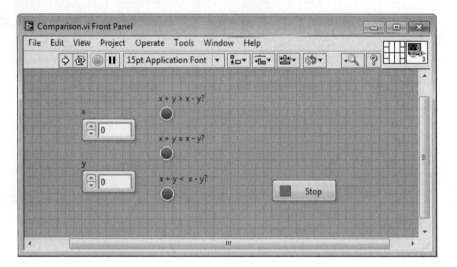

FIGURE P2.7
The front panel for the Comparison VI.

Using the VI, test the following three conditions:

1. $x = 4$ and $y = 1$

2. $x = 3$ and $y = 0$

3. $x = 2$ and $y = -3$

Save the VI as Comparison.vi in the Users Stuff folder in the Learning directory.

P2.8 Create a VI that will calculate your age in a future year. The VI should have three numeric controls labeled Current Year, Current Age, and Future Year and one numeric indicator labeled Age in Future Year.

Use the VI to solve the following: If the year is 2014 and you are 24 years old, how old will you be in 2052? Save the VI as Age in Future.vi in the Users Stuff folder in the Learning directory.

P2.9 Create a VI that will convert seconds into a display consisting of hours, minutes, and seconds. The VI should have one numeric control labeled seconds and three numeric indicators labeled hours, minutes, and seconds.

Use the VI to solve the following: How many hours, minutes, and seconds are there in 86400 seconds? Save the VI as Converting Time.vi in the Users Stuff folder in the Learning directory.

DESIGN PROBLEMS

D2.1 In the Building Blocks section at the end of this chapter, you created a VI called Falling Edge.vi to be used in determining where falling edges should occur on the output of the pulse width modulation VI. Now create a VI called Rising Edge.vi that returns a value of True on its Boolean output, Rising Edge, whenever a full output cycle has completed and a rising edge is to occur on the next clock cycle. Unlike the Falling Edge VI, this VI will only have two inputs: Period and Clock. The input Duty Cycle is not necessary in this VI since a change in the duty cycle of a signal changes only the location of the falling edge and not the location of the rising edge.

Save the VI as Rising Edge.vi in the Building Blocks folder in the Learning directory. The front panel should resemble the front panel seen in Figure D2.1.

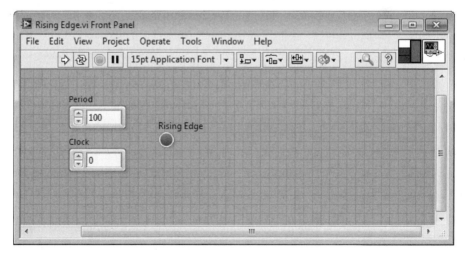

FIGURE D2.1
The front panel for the Rising Edge VI.

D2.2 In this design problem you will create a VI that simulates a vending machine. The vending machine sells three items:

1. Candy bars for $0.80 each,

2. Potato chips for $0.60 a bag, and

3. Chewing gum for $0.40.

The vending machine accepts only five dollar bills, one dollar bills, quarters, dimes, and nickels.

Inputs on the front panel should include a numeric control for the user to enter the amount of money inserted into the vending machine and three more integer numeric controls that designate how many of each item the user wishes to purchase from the vending machine. Your VI should check to see if the amount of money input is greater than or equal to the total cost of the selected purchase. If there is not enough money, display a message notifying the customer that more money is needed to meet the total, using the Display Message to User Express VI. Then light an LED indicator on the front panel and display the amount needed on a numeric indicator. If enough money is inserted into the vending machine based on the user selection, output the change the user will receive, showing the quantity of dollar bills, quarters, dimes, and nickels to be dispensed by the vending machine.

D2.3 Construct a VI in which the user guesses an integer between 0 and 5 using a numeric control. Generate a random integer in this range, display it with a numeric indicator, and compare it to the input. If the numbers are equal, light an LED.

Save the VI as **Guessing Game.vi** in the **Users Stuff** folder in the **Learning** directory. The front panel should resemble the front panel seen in Figure D2.3.

D2.4 Suppose that you are holding US currency (denoted by USD) and are planning a trip to visit either Brazil or Bolivia. To assist in computing the currency conversions, create a VI that uses the currency exchange rate for the Brazilian Real (denoted by BRL) and the Bolivian Boliviano (denoted by BOB). The VI should convert the currencies to/from USD to BRL/BOB. It should have a Boolean switch to select the desired country and another Boolean switch to specify whether you are converting to or from US dollars. You will always exchange from USD to BRL/BOB or from BRL/BOB to USD. Assume that the current exchange rate is 1 USD = 1.96 BRL and 1 USD = 7.02 BOB.

Use the front panel in Figure D2.4 as a guide. Save the VI as **Currency Exchange.vi** in the **Users Stuff** folder in the **Learning** directory.

*Use the Select function on the **Programming≫Comparison** palette.*

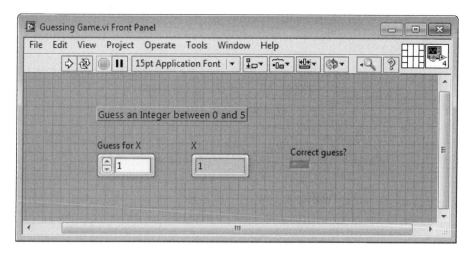

FIGURE D2.3
The front panel for the Guessing Game VI.

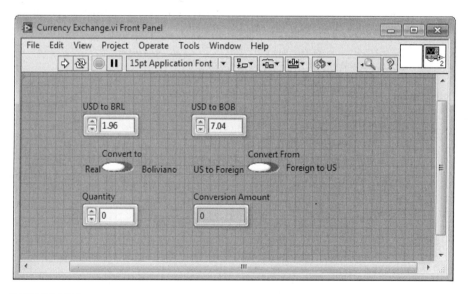

FIGURE D2.4
Currency converter front panel.

D2.5   Develop a VI to find the roots of the quadratic equation

$$ax^2 + bx + c = 0$$

where the constants a, b, and c are real numbers and $a > 0$. The two roots of the equation can be either real or complex numbers. The VI should have three numeric controls and two numeric indicators. Save the VI as Quadratic Formula.vi in the Users Stuff folder in the Learning directory.

Hint: After you place the numeric indicators on the front panel, right click on each indicator and in the shortcut menu select **Representation≫Complex Single**. *This will allow you to display the roots as complex numbers (as well as real numbers).*

D2.6 Develop a VI that computes the surface area and volume of a sphere. The only input is the sphere diameter, denoted by d. The VI should have one numeric control and two numeric indicators. The volume of a sphere is

$$V = \frac{4}{3}\pi r^3$$

where $r = d/2$ is the radius of the sphere. The surface area is given by

$$S = 4\pi r^2$$

Save the VI as Sphere.vi in the Users Stuff folder in the Learning directory.

Hint: To compute r^3 in the formula for the volume, it might be useful to use the Power Of X function.

D2.7 Develop a VI that computes

$$f = 5x^2 + 9y - z$$

where the inputs x, y, and z are real numbers. Provide two solution methods. For the first solution use only arithmetic functions (such as Square, Add, Multiply, and Subtract). For the second solution use the Formula Express VI. Which solution is easier to program? Which solution provides a quicker way to change the formula? Which solution executes faster?

You can find the Formula Express VI on the **Express≫Arithmetic & Comparison** *palette.*

CHAPTER 3

Editing and Debugging Virtual Instruments

Like text-based computer programs, virtual instruments are dynamic. VIs change as their applications evolve (usually increasing in complexity). For instance, a VI that initially only performs addition may at a later time be updated to add a multiplication capability. You need debugging and editing tools to verify and test VI coding changes. Since programming in LabVIEW is graphical in nature, editing and debugging are also graphical, with options available in pull-down and shortcut menus and various palettes. We learn here how to programmatically read and write the properties of objects using the Property Node. The important topics of cleaning up the block diagram, routing wires, and reusing snippets of code are presented. Debugging subjects covered include execution highlighting (you can watch the code run), single-stepping through code, and inserting probes to view data as the VI executes.

GOALS

1. Learn to access and practice with VI editing tools.
2. Learn to access and practice with VI debugging tools.

3.1   EDITING TECHNIQUES

3.1.1   Creating Controls and Indicators on the Block Diagram

As discussed in previous chapters, when building a VI you can create controls and indicators on the front panel and know that their terminals will automatically appear on the block diagram. Switching to the block diagram, you can begin wiring the terminals to functions (such as addition or multiplication functions), subVIs, or other objects. In this section we present an alternative method to create and wire controls and indicators in *one* step on the block diagram. For the following discussions, you should open a new VI and follow along by repeating the steps as presented.

1.   Open a new VI and switch to the block diagram.

2.   Place the Square Root function located in the **Programming≫Numeric** subpalette of the **Functions** palette on the block diagram, as shown in Figure 3.1.

3.   Now we want to add (and wire) a control terminal to the Square Root function. As shown in Figure 3.2, you right-click on the left side of the Square Root function and select **Control** from the **Create** menu. The result is that a control terminal is created and automatically wired to the Square Root function. Switch to the front panel and notice that a numeric control has appeared!

*To change a control to an indicator (or vice versa), right-click on the terminal (on the block diagram) or on the object (on the front panel) and select **Change to Indicator** (or **Change to Control**).*

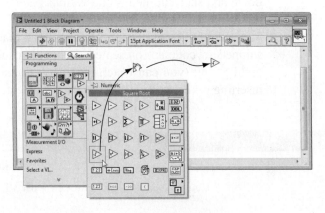

FIGURE 3.1
Add the Square Root function to the block diagram.

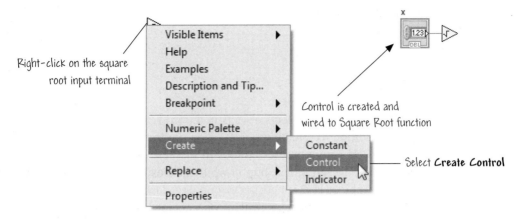

FIGURE 3.2
Front panel control created for the Square Root function.

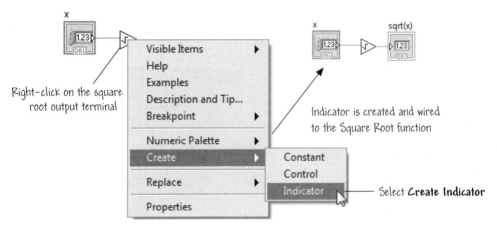

FIGURE 3.3
Front panel indicator created for the Square Root function.

4.   In fact, you can wire indicators and constants by right-clicking on the node terminal and choosing the desired selection. For example, right-clicking on the right side of the Square Root function and selecting **Indicator** from the **Create** menu creates and automatically wires a front panel indicator, as illustrated in Figure 3.3. In many situations, you may wish to create and automatically wire a constant to a terminal, and you accomplish this by right-clicking and choosing **Constant** from the **Create** menu.

*To change a control or indicator (for example, from a knob to a dial), right-click on the object on the front panel and select **Replace**. The **Controls** palette will appear, and you can navigate to the desired new object.*

5. The resulting control and indicator for the Square Root function are shown in Figure 3.4.

3.1.2 Selecting Objects

The **Positioning** tool selects objects in the front panel and block diagram windows. In addition to selecting objects, you use the **Positioning** tool to move and resize objects (more on these topics in the next several sections). To select an object, click the left mouse button while the **Positioning** tool is over the object. When the object is selected, a surrounding dashed outline appears, as shown in Figure 3.5. To select more than one object, shift-click (that is, hold down <Shift> and simultaneously click) on each additional object you want to select. You also can select multiple objects by clicking in a nearby open area and dragging the cursor until all the desired objects lie within the selection rectangle that appears (see Figure 3.6).

Sometimes after selecting several objects, you may want to deselect just one of the objects (while leaving the others selected). This is accomplished by shift-clicking on the object you want to deselect—the other objects will remain selected, as desired.

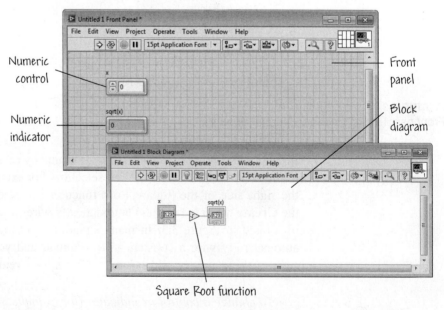

FIGURE 3.4
Front panel and block diagram for the Square Root function.

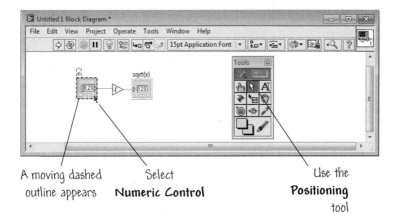

A moving dashed
outline appears

Select
Numeric Control

Use the
Positioning
tool

FIGURE 3.5
Selecting an object using the **Positioning** tool.

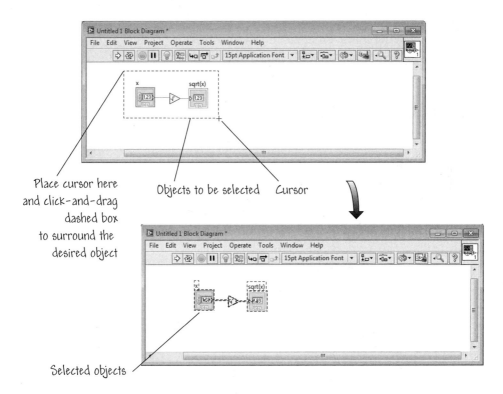

Place cursor here
and click-and-drag
dashed box
to surround the
desired object

Objects to be selected

Cursor

Selected objects

FIGURE 3.6
Selecting a group of objects.

3.1.3 Moving Objects

You can move an object by clicking on it with the **Positioning** tool and dragging it to a desired location. The objects of Figure 3.6 are selected and moved as shown in Figure 3.7. Selected objects can also be moved using the up/down and right/left arrow keys. Pressing the arrow key once moves the object one pixel; holding down the arrow key repeats the action. Holding down the <Shift> key while using the arrow keys moves the object several pixels at a time. In this manner, you can move and locate your objects very precisely.

The direction of movement of an object can be restricted to being either horizontal or vertical by holding down the <Shift> key when you move the object. The direction you initially move decides whether the object is limited to horizontal or vertical motion.

If you change your mind about moving an object while you are in the midst of dragging it to another location, continue to drag until the cursor is outside all open windows and the dashed line surrounding the selected object disappears—then release the mouse button. This will cancel the move operation and the object will not move. Alternatively, if the object is dragged and dropped to an undesirable location, you can select **Undo Move** from the **Edit** menu to undo the move operation.

3.1.4 Deleting and Duplicating Objects

You can delete objects by selecting the object(s) and choosing **Remove From Project** from the **Edit** menu or pressing <Backspace> or <Delete> (Windows) or <delete> (Macintosh). Most objects can be deleted; however you cannot delete certain components of a control or indicator, such as the label or digital display. You must hide these components by right-clicking and deselecting

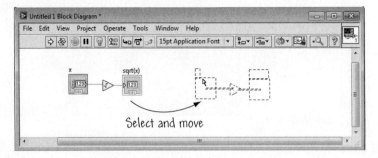

FIGURE 3.7
Selecting and moving a group of objects.

Visible Items≫ Label or Visible Items≫Digital Display from the shortcut menu.

Most objects can be duplicated, and there are three ways to duplicate an object—by copying and pasting, by cloning, and by dragging and dropping. In all three cases, a complete new copy of the object is created, including, for example, the terminal belonging to a front panel control and the control itself.

You can copy text and pictures from other applications and paste them into LabVIEW.

To clone an object, click the **Positioning** tool over the object while pressing <Ctrl> for Windows (or <option> on the Mac) and drag the object to its new location. After you drag the selection to a new location and release the mouse button, a copy of the object appears in the new location, and the original object remains in the original location. When you clone or copy objects, the copies are labeled by the same name as the original with an incrementing number at the end of each label (copy*2*, copy*3*, and so forth).

You also can duplicate objects using Edit≫Copy and then Edit≫Paste from the **Edit** menu. First, select the desired object using the **Positioning** tool and choose Edit≫Copy. Then click at the location where you want the duplicate object to appear and choose Edit≫Paste from the **Edit** menu. You can use this process to copy and paste objects within a VI or even copy and paste objects between VIs.

The third way to duplicate objects is to use the drag-and-drop capability. In this way, you can copy objects, pictures, or text between VIs and from other applications. To drag and drop an object, select the object, picture, or text file with the **Positioning** tool and drag it to the front panel or block diagram of the target VI. You can also drag VIs from the file system (in Windows and on the Mac) to the active block diagram to create subVIs (we will discuss subVIs in Chapter 4).

3.1.5   Resizing Objects

You can easily resize most objects. **Resizing handles** appear when you move the **Positioning** tool over a resizable object, as illustrated in Figure 3.8. On rectangular objects the resizing handles appear at the corners of the object; resizing circles appear on circular objects. Passing the **Positioning** tool over a resizing handle transforms the tool into a resizing cursor. To enlarge or reduce the size of an object, place the **Positioning** tool over the resizing handle and click and drag the resizing cursor until the object is the desired size. When you release the mouse button, the object reappears at its new size. The resizing process is illustrated in Figure 3.8.

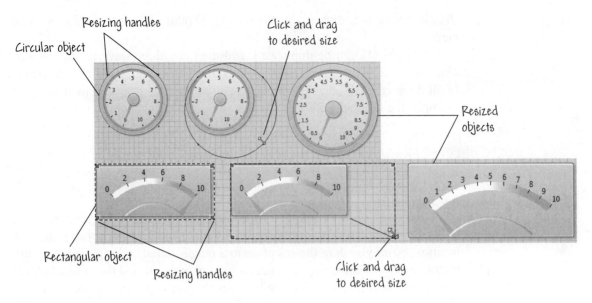

FIGURE 3.8
Resizing rectangular and circular objects.

To cancel a resizing operation, continue dragging the frame corner outside the active window until the dotted frame disappears—then release the mouse button and the object will maintain its original size. Alternately, if the object has already been resized and you want to undo the resizing, you can use the **Undo Resize** command found in the **Edit** pull-down menu.

Some objects only change size horizontally or vertically, or keep the same proportions when you resize them (e.g., a knob). In these cases, the resizing cursor appears the same, but the dotted resize outline moves in only one direction. To restrict the resizing of any object in the vertical or horizontal direction (or to maintain the current proportions of the object) hold down the <shift> key as you click and drag the object.

You may want to resize multiple objects to have similar dimensions, either horizontally, vertically, or both. To do this, use the **Resize Objects** menu on the LabVIEW toolbar as will be described in Section 3.1.10.

3.1.6 Labeling Objects

Labels are blocks of text that annotate components of front panels and block diagrams. There are two kinds of labels—free labels and owned labels. Owned labels belong to and move with a particular object and describe that object only. You can hide these labels but you cannot copy or delete them independently of their owners. Free labels are not attached to any object, and you can create, move, or delete them independently. Use free labels to annotate your front panels and block diagrams.

Free labels are one way to provide accessible documentation for your VIs.

You use the **Labeling** tool to create free labels or to edit either type of label. To create a free label, double-click anywhere in an open area and type the desired text in the bordered box that appears. An example of creating a free label is shown in Figure 3.9.

When finished entering the text, click on the **Enter Text** button on the toolbar, which appears on the toolbar to remind you to end your text entry. You can also end your text entry by pressing the <Enter> key on the numeric keypad (if you have a numeric keypad). If you do not type any text in the label, the label disappears as soon as you click somewhere else.

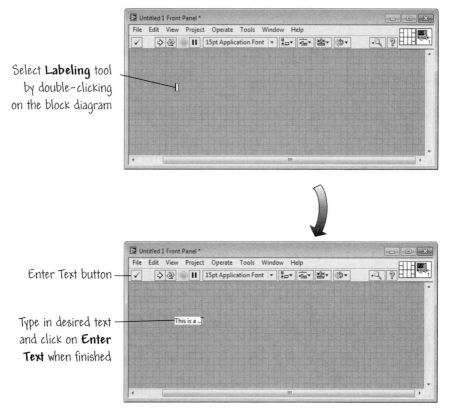

Select **Labeling** tool by double-clicking on the block diagram

Enter Text button

Type in desired text and click on **Enter Text** when finished

FIGURE 3.9
Creating a free label.

When you add a control or an indicator to the front panel, an owned label automatically appears. The label box is ready for you to enter the desired text. If you do not enter text immediately, the default label remains. To create an owned label for an existing object, right-click on the object and select Visible Items≫Label from the shortcut menu (see Figure 1.14). You can then enter your text in the bordered box that appears. If you do not enter the text immediately, the label disappears.

You can use an owned label to label a wire and identify its use. To create an owned label on a wire, right-click the wire and select **Visible Items≫Label** from the shortcut menu, as illustrated in Figure 3.10. Using the **Labeling** tool, you can add the label of your choice. You can move an owned label anywhere along the wire; however, you cannot lock an owned label to a wire so that the label remains in the same position on the wire.

Labeling wires is useful for wires coming from shift registers and for long wires that span the entire block diagram.

You can copy the text of a label by double-clicking on the text with the **Labeling** tool or dragging the **Labeling** tool across the text to highlight the

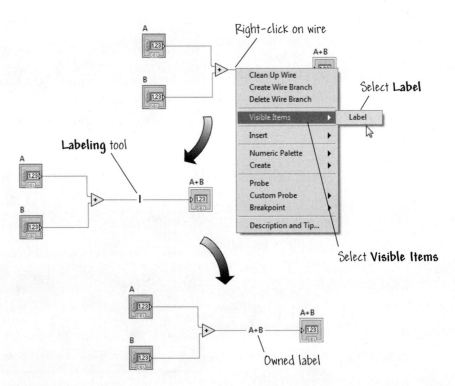

FIGURE 3.10
Creating an owned label on a wire.

desired text. When the desired text is selected, choose Edit≫Copy to copy the text onto the clipboard. You can then highlight the text of a second label and use Edit≫Paste to replace the highlighted text in the second label with the text from the clipboard. To create a new label with the text from the clipboard, double-click on the screen where you want the new label positioned and then select Edit≫Paste.

The resizing technique described in the previous section also works for labels. You can resize labels as you do other objects by using the resizing cursor. Labels normally autosize; that is, the label box automatically resizes to contain the text you enter. If for some reason you do not want the labels to automatically resize to the entered text, right-click on the label and select **Size to Text** to toggle autosizing off.

The text in a label remains on one line unless you enter a carriage return to resize the label box. By default, the <Enter> (or <return>) key is set to add a new line. This can be changed in Tools≫Options≫Environment so that the text input is terminated with the <Enter> (or <return>) key.

3.1.7   Changing Font, Style, and Size of Text

Using the **Text Settings** in the toolbar, you can change the font, style, size, and alignment of any text displayed in a label or on the display of controls and indicators. Certain controls and indicators display text in multiple locations— for example, on graphs (a type of indicator), the graph axes scale markers are made up of many numbers, one for each axes tick. You have the flexibility to modify each text display independently.

Text settings were discussed in Chapter 1 and are shown in Figure 1.7. Notice the word **Application** showing in the **Text Settings** pull-down menu, which also contains the **System**, **Font Dialog**, and **Current** options. The last option in the list—the **Current Font**—refers to the last font style selected. The predefined fonts are used for specific portions of the various interfaces:

- The **Application** font is the default font. It is used for the **Controls** and **Functions** palettes.

- The **System** font is the font used for menus.

- The **Font Dialog** is the font used for text in dialog boxes.

These fonts are predefined so that they map "best" when porting your VIs to other platforms.

Text Settings has size, style, justify, and color options. Selections made from any of these submenus (that is, size, style, etc.) apply to all selected objects. For example, if you select a new font while you have a knob selected, the labels, scales, and digital displays all change to the new font. Figure 3.11 illustrates the

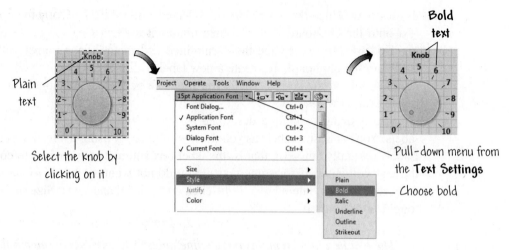

FIGURE 3.11
Changing the font style on a knob control.

situation of changing the style of all the text associated with a knob from plain text to bold text.

If you select any objects or text and make a selection from the **Text Settings** pull-down menu, the changes apply to everything selected. The process of selecting just the text of an owned label of a knob is illustrated in Figure 3.12. Once the desired text is selected (in this case, click on the label once to select it), you can make any changes you wish by selecting the proper pull-down submenu from **Text Settings**. If no text is selected, the font changes apply to the default font, so that labels created from that point on will reflect the new default font, while not affecting the font of existing labels.

When working with objects that have multiple pieces of text (e.g., slides and knobs), remember that text settings selections affect the objects or text currently selected. For example, if you select the entire knob while selecting bold text, the

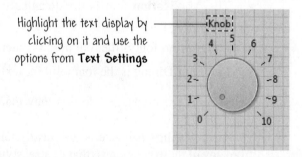

FIGURE 3.12
Selecting text for modification of style, font, size, and color.

scale, digital display, and label all change to a bold font, as shown in Figure 3.11. As shown in Figure 3.13(a), when you select the knob label, followed by selecting bold text from the **Style** submenu of the **Text Settings** pull-down menu, only the knob label changes to bold. Similarly, when you select text from a scale marker while choosing bold text, all the markers change to bold, as shown in Figure 3.13(b).

If you select **Font Dialog** in the **Text Settings** while a front panel is active, the dialog box shown in Figure 3.14 appears. If a block diagram is active instead, the **Diagram default** option at the bottom of the dialog box is checked. With

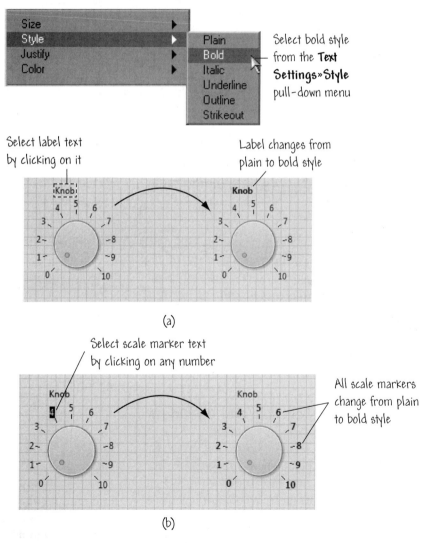

FIGURE 3.13
Changing text attributes on a knob.

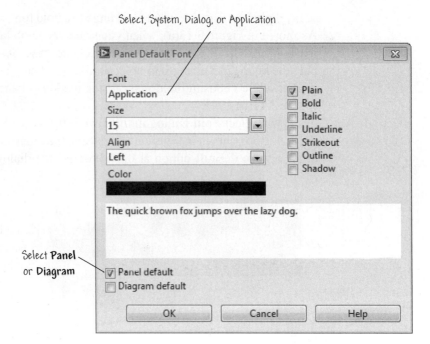

FIGURE 3.14
Using the **Font Dialog** box to change font, size, alignment, color, and style.

either the **Panel default** or **Diagram default** checkboxes selected, the other selections made in this dialog box will be used with new labels on the front panel or block diagram. In other words, if you click the **Panel default** and/or **Diagram default** checkboxes, the selected font becomes the current font for the front panel, the block diagram, or both. The current font is used on new labels. The checkboxes allow you to set different fonts for the front panel and block diagram. For example, you could have a small font on the block diagram and a large one on the front panel.

3.1.8   Selecting and Deleting Wires

A single horizontal or vertical piece of wire is known as a **wire segment**. The point where three or four wire segments join is called a **junction**. A **wire branch** contains all the wire segments from one junction to another, from a terminal to the next junction, or from one terminal to another if there are no junctions in between. You select a wire segment by clicking on it with the **Positioning** tool. Clicking twice selects a branch, and clicking three times selects the entire wire. See Figure 3.15 for an example of selecting a branch, segment, or an entire wire.

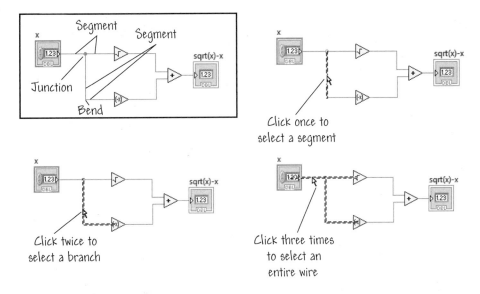

FIGURE 3.15
Selecting a segment, branch, or an entire wire.

3.1.9   Wire Stretching and Broken Wires

Wired objects can be moved individually or in groups by dragging the selected objects to a new location with the **Positioning** tool. The wires connecting the objects will stretch automatically. If you want to move objects from one diagram to another, the connecting wires will not move with the selected objects, unless you select the wires as well. After selecting the desired objects and the connecting wires, you can cut and paste the objects as a unit to another VI.

An example of stretching a wire is shown in Figure 3.16. In the illustration, the object (in this case, a numeric indicator) is selected with the **Positioning** tool and stretched to the desired new location.

Wire stretching occasionally leaves behind loose ends. Loose ends are wire branches that do not connect to a terminal. Your VI will not execute until you remove all loose ends. An example of a loose end is shown in Figure 3.17. Loose ends can be easily removed using the Edit≫Remove Broken Wires command.

When you make a wiring mistake, a broken wire (indicated by a dashed line with a red X) appears. Figure 3.18 shows a dashed line that represents a broken wire. It is inevitable that broken wires will occur in the course of programming in LabVIEW. One common mistake is to attempt to connect two control terminals together or to connect a control terminal to an indicator terminal when the data types do not match (for example, connecting a Boolean to a numeric). If you have a broken wire, you can remove it by selecting it with the **Positioning** tool and eliminate it by pressing the <Delete> key. If you want to remove all

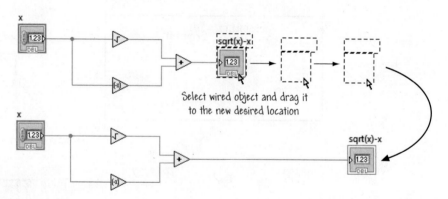

FIGURE 3.16
Moving wired objects.

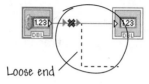

FIGURE 3.17
Loose ends.

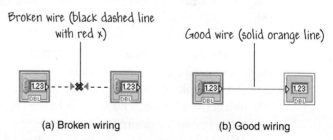

FIGURE 3.18
Locating broken wires.

broken wires on a block diagram at one time, choose **Remove Broken Wires** from the **Edit** menu.

There are many different conditions leading to the occurrence of broken wires. Some examples are:

- **Wire type, dimension, unit, or element conflicts**—A wire type conflict occurs when you wire two objects of different data types together, such as a numeric and a Boolean, as shown in Figure 3.19(a).

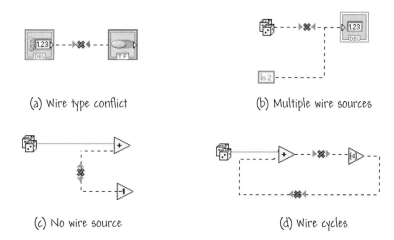

(a) Wire type conflict   (b) Multiple wire sources

(c) No wire source   (d) Wire cycles

FIGURE 3.19
Typical wiring errors leading to broken wire indications.

- **Multiple wire sources**—You can wire a control to multiple output destinations (or indicators), but you cannot wire multiple data sources to a single destination. In the example shown in Figure 3.19(b), we have attempted to wire two data sources (that is, the random number and the constant ln 2) to one indicator. This produces a broken wire and must be fixed by disconnecting the random number (represented by the dice) or disconnecting the ln 2 constant. Another example of a common multiple sources error occurs during front panel construction when you inadvertently place a numeric control on the front panel when you meant to place a numeric indicator. If you do this and then try to wire an output value to the terminal of the front panel control, you will get a multiple sources error. To fix this error, just right-click on the terminal and select **Change To Indicator**.

- **No wire source**—An example of a wire with no source is shown in Figure 3.19(c). This problem is addressed by providing a control. Another example of a situation leading to a no-source error is if you attempt to wire two front panel indicators together when one should have been a control. To fix this error, just right-click on a terminal and select **Change to Control**.

- **Wire cycles**—In most cases, wires must not form closed loops of icons or structures, as shown in Figure 3.19(d). These closed loops are known as cycles. LabVIEW will not execute cycles because each node waits on the other to supply it data before it executes. Feedback nodes (discussed in Chapter 5) provide the proper mechanism to feed back data in a repetitive calculation or loop.

Sometimes you have a faulty wiring connection that is not visible because the broken wire segment is very small or is hidden behind an object. If the

Broken Run button appears in the toolbar, but you cannot see any problems in the block diagram, select Edit≫Remove Broken Wires—this will remove all broken wires in case there are hidden, broken wire segments. If the **Run** button returns, you have corrected the problem. If the wiring errors have not all been corrected after selecting Edit≫Remove Broken Wires, click on the **Broken Run** button to see a list of errors. Click on one of the errors listed in the **Error List** dialog box, and you will automatically be taken to the location of the erroneous wire in the block diagram. You can then inspect the wire, detect the error, and fix the wiring problem.

LabVIEW automatically finds a wire route around existing objects on the block diagram when you are wiring components. The natural inclination of the automatic routing is to decrease the number of bends in the wire. When possible, automatically routed wires from control terminals exit the right side of the terminal, and automatically routed wires to indicator terminals enter the left side of the terminal.

If you find that you have a messy wiring situation on your block diagram you can easily fix it. You can right-click any wire and select **Clean Up Wire** from the shortcut menu to automatically route an existing wire. This may help in debugging your VI since it will be easier to see the wire routes on the block diagram. You can also clean up wires on the entire block diagram by selecting the **Clean Up Diagram** button on the toolbar (see Figure 1.12).

If for some reason you want to temporarily disable automatic wire routing and route a wire manually, first use the **Wiring** tool to click a terminal and release the mouse. Then press the <A> key to temporarily disable automatic wire routing for the current wire. Click another terminal to complete the wiring. Once the wiring is complete, the automatic wire routing is resumed. You also can temporarily disable automatic routing after you click to start wiring by holding down the mouse button while you wire to another terminal and then releasing the mouse button. After you release the mouse button, automatic wire routing resumes.

3.1.10   Aligning, Distributing, and Resizing Objects

To align a group of objects, first select the desired objects. Then choose the axis along which you want to align them from the **Align Objects** pull-down menu in the toolbar. You can align objects along the vertical axis using left, center, or right edge. You also can align objects along a horizontal axis using top, center, or bottom edge. Open a new VI and place three objects (such as, three numeric controls) on the front panel and experiment with different aligning options. Figure 3.20 illustrates the process of aligning three objects by their left edges.

In a similar fashion, you can distribute a group of objects by selecting the objects and then choosing the axis along which you want to distribute the selected

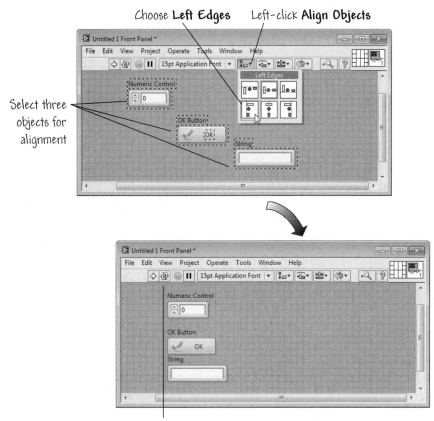

FIGURE 3.20
Aligning objects.

objects from the **Distribute Objects** pull-down menu in the toolbar (see Figure 1.9). In addition to distributing selected objects by making their edges or centers equidistant, four menu items at the right side of the ring let you distribute the gaps between the objects, horizontally or vertically. For example, the three objects shown in Figure 3.21 are distributed at different distances from each other. If we want them to be equally spaced by their top edges, we can use the **Distribute Objects** pull-down menu, as shown in Figure 3.21, to rearrange the elements at equal spacing.

You can use the **Resize Objects** pull-down menu to resize multiple front panel objects to the same size. For example, in Figure 3.22, three objects of different widths are selected. Then using the **Resize Objects** pull-down menu, the **Maximum Width** icon is selected resulting in the three objects taking the width of the String, which originally had the largest width.

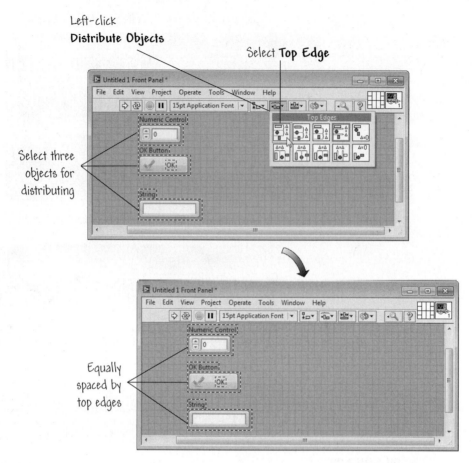

FIGURE 3.21
Distributing objects.

3.1.11   Coloring Objects

You can customize the color of many LabVIEW objects. However, the coloring of objects that convey information via their color is unalterable. For example, block diagram terminals of front panel objects and wires use color codes for the type and representation of data they carry, so you cannot change their color.

To change the color of an object (or the background of a window), right-click on the object of interest with the **Coloring** tool from the **Tools** palette, as seen in Figure 3.23. Choose the desired color from the selection palette that appears. If you keep the mouse button pressed as you move through the color selection palette, the object or background you are coloring redraws with the color the cursor currently is touching. This gives you a "real-time" preview of the object in the new color. If you release the mouse button on a color, the selected object retains the chosen color. To cancel the coloring operation, move

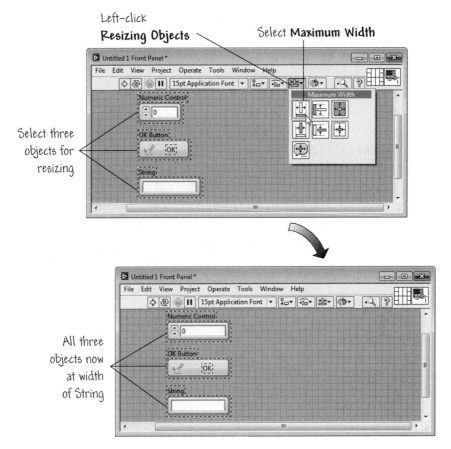

FIGURE 3.22
Resizing objects.

the cursor out of the color selection palette before releasing the mouse button, or use the Edit≫Undo Color Change command after the undesired color change has been made.

If you select the box with a T in it, the object is rendered transparent. One use of the T (transparent) option is to create numeric controls without the standard three-dimensional border. Transparency affects the appearance but not the function of the object.

Some objects have both a foreground and a background color that you can set separately. The foreground color of a knob, for example, is the main dial area, and the background color is the base color of the raised edge. On the **Coloring** tool, you select the foreground or background, as illustrated in Figure 3.23. Clicking the upper left box on the **Coloring** tool and then clicking on an object on the front panel will color the object's foreground. Similarly, clicking the lower right box on the **Coloring** tool and then clicking on an object on the front panel will color the object's background.

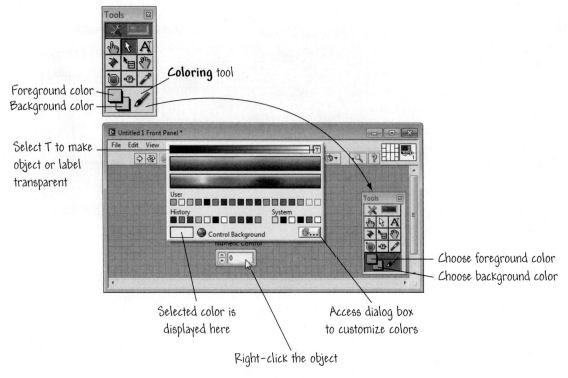

FIGURE 3.23
Customizing the color of objects.

Selecting the button on the lower right-hand side in the palette (see Figure 3.24) accesses a dialog box with which you can customize the colors. Each of the three color components, red, green, and blue, describes eight bits of a 24-bit color (in the lower right-hand side of the **Color** dialog box). Therefore, each component has a range of 0 to 255. The last color you select from the palette becomes the current color. Clicking on an object with the **Color** tool sets that object to the current color.

Using the **Color Copy** tool, you can copy the color of one object and transfer it to a second object without using the **Color** palette. To accomplish this, click with the **Color Copy** tool on the object whose color you want to transfer to another object. Then, select the **Color** tool and click on another object to change its color.

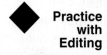

Practice with Editing

In this exercise you will edit and modify an existing VI to look like the panel shown in Figure 3.25. After editing the VI, you will wire the objects in the block diagram and run the program.

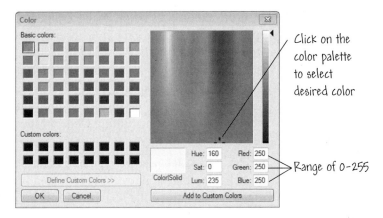

Click on the
color palette
to select
desired color

Range of 0-255

FIGURE 3.24
The **Color** dialog box.

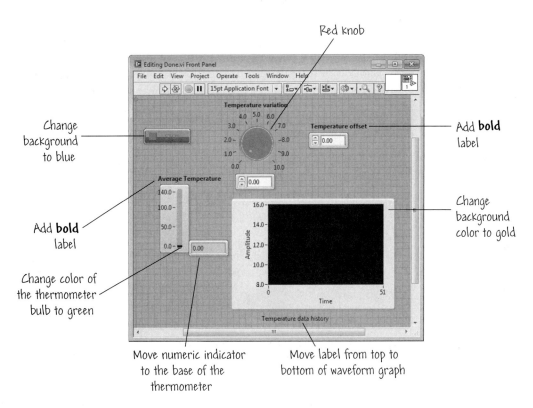

Red knob

Change
background
to blue

Add **bold**
label

Add **bold**
label

Change
background
color to gold

Change color of
the thermometer
bulb to green

Move numeric indicator
to the base of the
thermometer

Move label from top to
bottom of waveform graph

FIGURE 3.25
The front panel for the Editing VI.

1. Open the Editing VI by choosing **Open** from the **File** menu (or using the **Open Existing** button on the Getting Started screen) and searching in the Chapter 3 folder in the Learning directory. The front panel of the Editing VI contains a number of objects depicted in Figure 3.26. The objective of this exercise is to make the front panel of the Editing VI look like the one shown in Figure 3.25. If you have a color monitor, you can see the final VI in color by opening the Editing Done VI located in the Chapter 3 folder in the Learning directory.

2. Add an owned label to the numeric control using the **Positioning** tool by right-clicking on the numeric control and selecting Visible Items≫Label from the menu. Type the text Temperature offset inside the bordered box and click the mouse outside the label or click the **Enter Text** button on the left-hand side of the toolbar.

3. Reposition the waveform graph and numeric control. Click on the waveform graph and drag it to the lower right of the front panel (see Figure 3.25 for the approximate location). Then click on the numeric control and drag it to the upper right of the front panel (see Figure 3.25 for approximate location).

   Notice that as you move the numeric control, the owned label moves with the control. If the control is currently selected, click on a blank space on the front panel to deselect the control and then click on the label and drag it to another location. Notice that in this case the control does not follow the move. You can position an owned label anywhere relative to the control. If

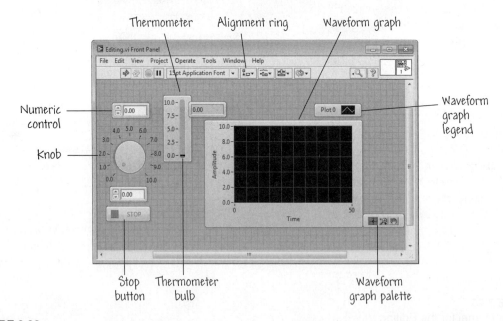

FIGURE 3.26
The Editing VI with various objects: knob, waveform chart, thermometer, and numeric control.

you move the owned label to an undesirable location, you can move it back by selecting Edit≫Undo Move to place the label back above the numeric control.

4. Reposition the stop button, the knob, and the thermometer to the approximate locations shown in Figure 3.25.

5. Add labels to the knob and to the thermometer. To accomplish this task, right-click on the object and choose Visible Items≫Label. When the label box appears, type in the desired text. In this case, we want to label the knob Temperature variation and the thermometer Average Temperature.

6. Move the numeric indicator associated with the thermometer to a location near the bottom of the thermometer. Then reposition the thermometer label so that it is better centered above the thermometer (see Figure 3.25).

7. In this next step, we will reformat the waveform graph.

   (a) To remove the waveform graph legend, right-click on the waveform graph and select **Properties**. On the **Appearance** menu, deselect the check mark next to **Show Plot Legend**.

   (b) To remove the waveform graph palette, deselect **Show Graph Palette** from the **Appearance** menu.

   (c) Remove the waveform x-axis scale by selecting the **Scales** menu. On the **Scales** menu, be sure to select the x-axis from the drop-down box. Deselect the **Show Scale** option.

   (d) Add a label to the wavefrom graph—label the object Temperature data history. Select the owned label and move it to the bottom of the waveform graph, as illustrated in Figure 3.25. Note that this can also be done on the **Properties** screen.

8. Align the stop button and the thermometer.

   (a) Select both the stop button and the thermometer using the **Positioning** tool. Pick a point somewhere to the upper left of the stop button and drag the dashed box down until it encloses both the stop button and the thermometer. Upon release of the mouse button, both objects will be surrounded by moving dashed lines.

   (b) Click on **Align Objects** and choose **Left Edges**. The stop button and the thermometer will then align to the left edges. If the objects appear not to move at all, this indicates that they were essentially aligned to the left edges already.

9. Align the stop button, the knob, and the numeric control horizontally by choosing the **Vertical Centers** axis from the **Align Objects** pull-down menu in the toolbar. Remember to select all three objects beforehand.

10. Space the stop button, the knob, and the numeric control evenly by choosing the **Horizontal Centers** axis from the **Distribute Objects** pull-down menu in the toolbar. Again, remember to select all three objects beforehand.

11. Change the color of the stop button.

    (a) Using the **Coloring** tool, right-click on the stop button to display the color palette.

    (b) Choose a color from the palette. The object will assume the last color you selected. In the Editing Done VI you will see that a dark blue color was selected—you can choose a color that you like.

12. Change the color of the knob. Using the **Coloring** tool, right-click on the knob to display the color palette and then choose the desired color from the color palette. In the Editing Done VI you will find that a red/pink color was selected—you can choose a color that suits you.

13. Change the color of the waveform graph. In the Editing Done VI you will see that a gold color was selected.

14. Change the color of the thermometer bulb. Using the **Coloring** tool, right-click on the thermometer bulb (at the bottom of the thermometer) to display the color palette and then choose the desired color. In the Editing Done VI you will find that a green color was selected.

15. Change the font style of the owned labels. Use the cursor to select each of the labels and then select Style≫Bold from **Text Settings**. See if you can do this for all three of the owned labels at once: thermometer, knob, and numeric control.

At this point, the front panel of the Editing VI should look very similar to the front panel shown in Figure 3.25.

The block diagram for the Editing VI is shown in Figure 3.27 after the editing of the front panel is finished—but before wiring the block diagram.

We can now wire together the various objects to obtain a working VI. Notice that several additional objects are on the block diagram: a uniform white noise subVI, a mean subVI that computes the mean (or average) of a signal, a While Loop, and an Add function. In later chapters, you will learn to use While Loops and learn how to use preexisting subVIs packaged with LabVIEW. For instance, the Uniform White Noise.vi is located on the **Functions** palette under the **Signal Processing≫Signal Generation** subpalette.

Go ahead and wire the Editing VI block diagram so that it looks like the block diagram shown in Figure 3.28. If you run into difficulties, you can open and examine the block diagram for the Editing Done VI (which can be found in Chapter 3 of the Learning directory).

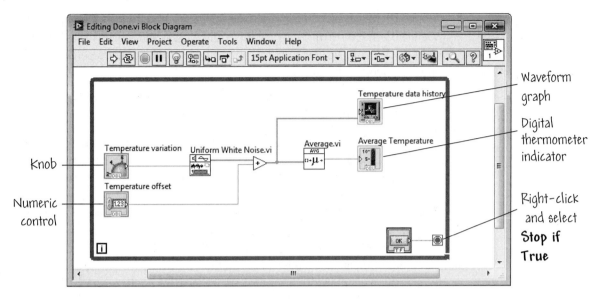

FIGURE 3.27
The block diagram for the Editing VI—before wiring.

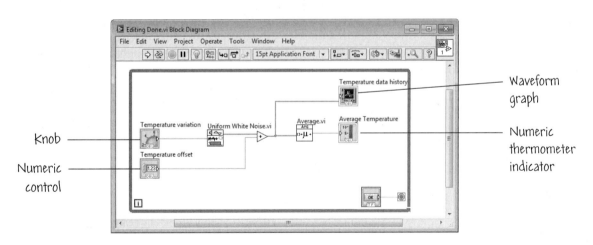

FIGURE 3.28
The block diagram for the Editing Done VI—after wiring.

A few wiring tips:

1. To wire the objects together with the **Wiring** tool, click and release on the source terminal and drag the **Wiring** tool to the destination terminal. When the destination terminal is blinking, click and release the left mouse button.

2. To identify terminals on the Add function, right-click on the icon and select Visible Items≫Terminal to see the connector. When the wiring is finished, right-click again on the function and choose Visible Items≫Terminal to show the icon once again.

3. To bend the wires as you connect two objects, click with the left mouse button on the bend location with the **Wiring** tool.

After you have finished wiring the objects together, switch to the front panel by selecting **Show Front Panel** from the **Window** menu. Use the **Operating** tool to change the value of the front panel controls. Run the VI by clicking on the **Run** button on the toolbar.

The Average Temperature indicator should be approximately the value that you select as the Temperature offset. The amount of variation in the temperature history, as shown on the Temperature data history waveform graph, should be about the same as the setting on the Temperature variation knob.

When you are finished editing and experimenting with your VI, save it by selecting **Save As** from the **File** menu. Remember to save all your work in the Users Stuff folder in the Learning directory. Close the VI by selecting **Close** from the **File** menu. ◆

3.1.12   Cleaning Up the Block Diagram

It is very important that the block diagram be logically arranged so that the code is readable, making it easier to understand and easier to debug (see Section 3.2 for more discussion on debugging a VI). It is possible to clean up the entire block diagram by rerouting all wires and reordering all objects in a single action. It is also possible to clean up only selected selections of the block diagram.

You can reroute all existing wires and reorder all objects on the block diagram in a single action by the following steps:

1. First you can configure the cleanup options. Select Tools≫Options to display the **Options** dialog box and select **Block Diagram** from the **Category** list. On the **Block Diagram** page, you can set the configuration to automatically move controls to the left side of the block diagram and indicators to the right side, place a given number of pixels between block diagram objects and wires, and compact the block diagram layout.

2. It is also an option to exclude from the cleanup process the reorganizing of contents of selected structures (see Chapter 5 for more information on structures such as the Case Structure). To accomplish this, right-click on the structure and select **Exclude from Diagram Cleanup**. The structure may move as part of the cleanup process, but its contents will not be reorganized.

3. Once the desired cleanup options are chosen and all structures to be excluded from the process are appropriately accommodated, you can clean up the

block diagram in a single action by selecting **Clean Up Diagram** from the **Edit** pull-down menu. You also can click the **Clean Up Diagram** button on the block diagram toolbar or press the <Ctrl-U> keys in Windows or the <Command-U> keys in Mac OS.

The process is illustrated in Figure 3.29. The block diagram in Figure 3.29(b) is more logically configured for readability than the same block diagram shown in Figure 3.29(a). Notice that the objects inside the Case Structure are not reorganized, even though the Case Structure has moved.

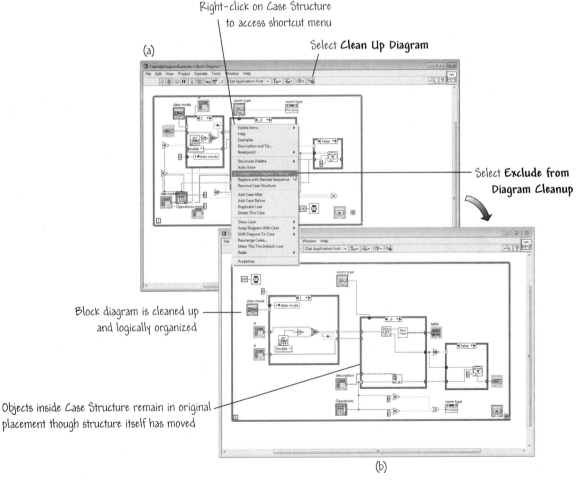

FIGURE 3.29
Using the **Clean Up Diagram** feature.

Rerouting only selected wires and reordering only selected objects on the block diagram is accomplished as follows:

1. First, if you want to exclude reorganizing the contents of certain structures, right-click the desired structure and select **Exclude from Diagram Cleanup** as discussed above.

2. Click and drag a selection rectangle around the items you want to clean up. If you want to add additional objects to the cleanup process, press the <Shift> key and click and drag a selection rectangle around the additional objects to create multiple selections. You also can press the <Shift> key while you click objects you want to deselect.

 The block diagram is not cleaned up if you do not select any structures and none of the objects you select overlap each other or if the selection contains only wires without corresponding inputs and outputs.

3. Select **Clean Up Selection** from the **Edit** pull-down menu.

The process is illustrated in Figure 3.30. Notice that the objects inside the three Case Structures are not reorganized. In this case, we excluded all three Case Structures from the cleanup. Once you have selected the desired objects to clean up, the **Clean Up Diagram** button on the block diagram toolbar will automatically change modes to **Clean Up Selection**, and you can use this to clean up the selection as well as **Clean Up Selection** from the **Edit** pull-down menu.

3.1.13 Routing Wires

As you wire objects on the block diagram, LabVIEW will automatically find a route for the wire around existing objects. The automatic wire routing also seeks to find a path such that the number of bends in the wire is minimized, wires from control terminals exit the right side of the terminal, and wires to indicator terminals enter the left side of the terminal. Even so, the wires on the block diagram may still require some cleaning up. As discussed in Section 3.1.12, you can reroute existing wires and rearrange existing objects on the block diagram automatically using Edit≫Clean Up Diagram or Edit≫Clean Up Selection. However, you can reroute only specific wires using the **Clean Up Wire** feature.

To reroute an existing wire automatically, right-click the wire and select **Clean Up Wire** from the shortcut menu, as illustrated in Figure 3.31. In this case, only the selected wire is rerouted.

You can align and distribute wires on the block diagram using the **Align Objects** and **Distribute Objects** pull-down menus. Select the wires to be aligned, as illustrated in Figure 3.31, then select **Align Objects** in the pull-down menu. As seen in Figure 3.32, we choose the **Horizontal Centers** which leaves

Select the desired section of
the block diagram to clean up

Select **Clean Up Selection**

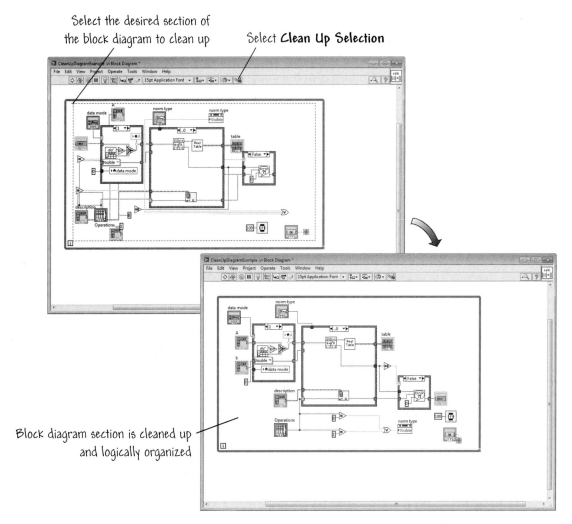

Block diagram section is cleaned up
and logically organized

FIGURE 3.30
Using the **Clean Up Selection** feature.

the block diagram wiring much cleaner. In complex block diagrams this may improve greatly the readability of the diagram.

*If the automatic wire routing and various clean-up processes do not lead (in your opinion) to a satisfactory block diagram arrangement, such that the code is readable and easy to debug, you can always use the **Positioning** tool and manually rearrange the wires and objects.*

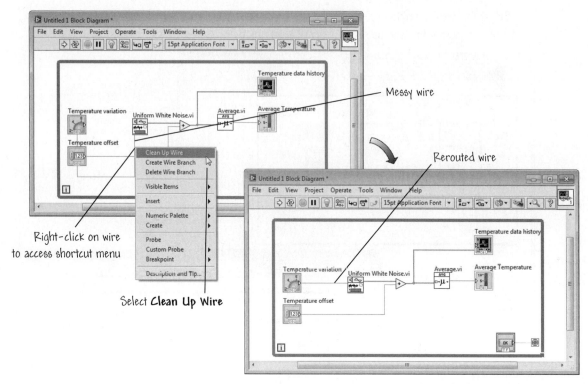

FIGURE 3.31
Using the **Clean Up Wire** feature.

3.1.14   Reusing Snippets of Code

You can save snippets of code from the block diagram to reuse later or to share with other LabVIEW users. Snippets of code can be opened in the version of LabVIEW in which they were created and newer versions. LabVIEW embeds the code into a .png image file. The image file actually contains the code so that when you drop the image file onto a block diagram the code will emerge.

To save a snippet of code as a .png image file, first select the code section of interest, as illustrated in Figure 3.33. Then select Edit≫Create VI Snippet from Selection to open a dialog box to navigate to the desired folder and enter the file name of your choice. In Figure 3.33, the file name entered is Temperature and the destination folder is VI Snippets. After you save the VI snippet as a .png file, you can drag the file from the directory where you saved it and drop the file onto a block diagram as illustrated in Figure 3.34. The code emerges from the image file and is fully functional.

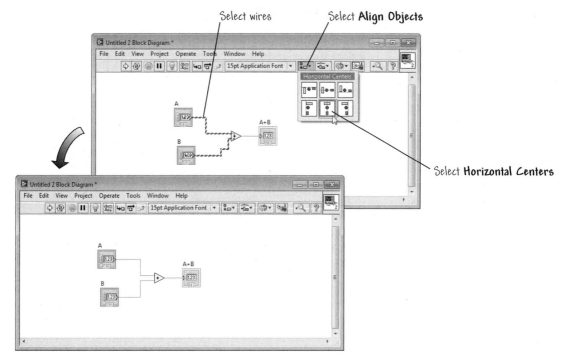

FIGURE 3.32
Using the **Align Objects** and **Distribute Objects** to assist in cleaning up the wiring diagram.

Here are three suggestions for creating VI snippets:

- Stick to simple VIs that only include LabVIEW native SubVIs.

- Remove data from arrays, graphs, or charts when possible to minimize the file size of your code snippet.

- Be wary of programs that might remove or corrupt the VI encoded in the image, such as photo editing tools.

When using VI snippets from the web, they can be dragged to the block diagram from Internet Explorer. Other web browsers may require the saving of the image that can then be dragged and dropped on your block diagram.

3.2 DEBUGGING TECHNIQUES

In this section we discuss LabVIEW's basic debugging elements, which provide an effective programming debugging environment. Most features commonly associated with good interactive debugging environments are provided—and in keeping with the spirit of graphical programming, the debugging features

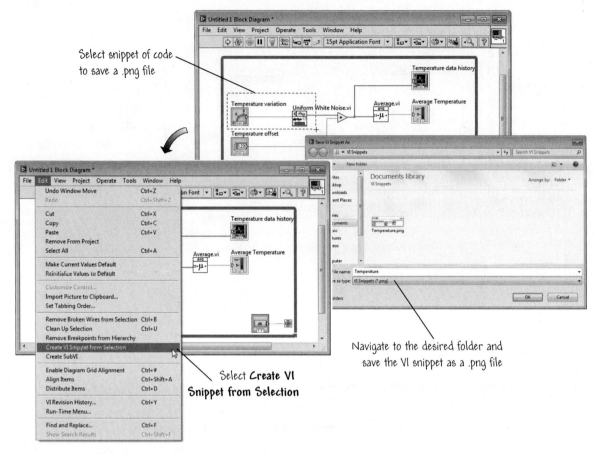

FIGURE 3.33
Creating snippets of code as a .png file.

are accessible graphically. Execution highlighting, single-stepping, breakpoints, and probes help debug your VIs easily by tracing the flow of data through the VI. You can actually watch your program code as it executes!

3.2.1 Finding Errors

When your VI cannot compile or run due to a programming error, a **Broken Run** button appears on the toolbar. Programming errors typically appear during VI development and editing stages and remain until you properly wire all the objects in the block diagram. You can list all your program errors by clicking on the **Broken Run** button. An information box called **Error List** appears listing all the errors. This box is shown in Figure 3.35 for the Editing Done VI with a broken wire.

Warnings make you aware of potential problems when you run a VI, but they do not inhibit program execution. If you want to be notified of any warnings,

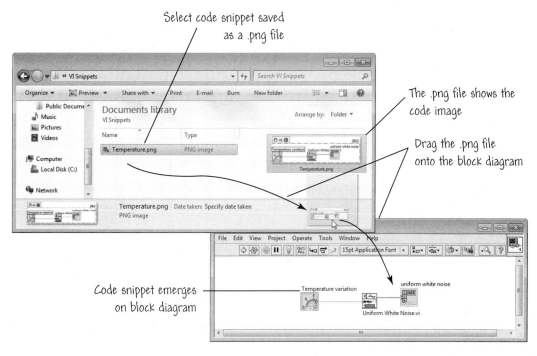

FIGURE 3.34
Dropping the code snippet saved as a .png file onto the block diagram.

click the **Show Warnings** checkbox in the **Error List** dialog box. A warning button then appears on the toolbar whenever a warning condition occurs.

If your program has any errors that prevent proper execution, you can search for the source of a specific error by selecting the error in the **Error List** (by clicking on it) and then clicking on **Show Error** (lower right-hand corner of the **Error List** dialog box). This process will highlight the object on the block diagram that reported the error, as illustrated in Figure 3.35. Double-clicking on an error in the error list will also highlight the object reporting the error.

Some of the most common reasons for a VI being broken during editing are:

1. A function terminal requiring an input is unwired. For example, an error will be reported if you do not wire all inputs to arithmetic functions.

2. The block diagram contains a broken wire because of a mismatch of data types or a loose, unconnected end.

3. A subVI is broken.

3.2.2 Highlight Execution

You can animate the VI block diagram execution by clicking on the **Highlight Execution** button located in the block diagram toolbar, shown in Figure 3.36.

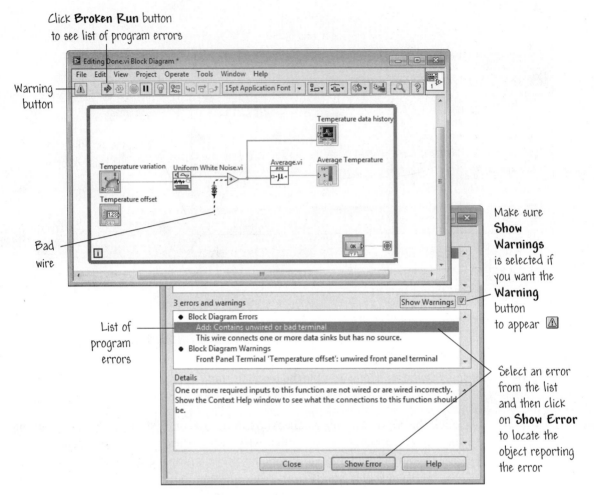

Click **Broken Run** button to see list of program errors

Warning button

Bad wire

Make sure **Show Warnings** is selected if you want the **Warning** button to appear

List of program errors

Select an error from the list and then click on **Show Error** to locate the object reporting the error

FIGURE 3.35
Locating program errors.

For debugging purposes, it is helpful to see an animation of the VI execution in the block diagram, as illustrated in Figure 3.37. When you click on the **Highlight Execution** button, it changes to a bright light to indicate that the data flow will be animated for your visual observation when the program executes. Click on the **Highlight Execution** button at any time to return to normal running mode.

Execution highlighting is commonly used with single-step mode (more on single stepping in the next section) to trace the data flow in a block diagram in an effort to gain an understanding of how data flows through the block diagram. Keep in mind that when you utilize the highlight execution debugging feature, it greatly reduces the performance of your VI—the execution time increases significantly. The data flow animation shows the movement of data from one node to another using "bubbles" to indicate data motion along the wires. This

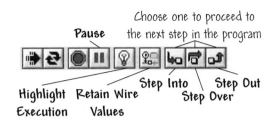

FIGURE 3.36
The **Highlight Execution** and step buttons located on the toolbar.

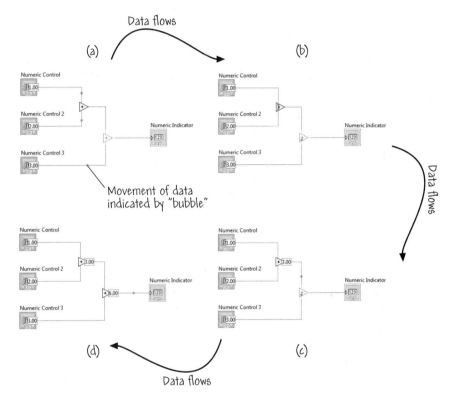

FIGURE 3.37
Using the highlight execution mode to watch the data flow through a VI.

process is illustrated in Figure 3.37. In Figure 3.37(a) we see the data as it flows out of the three controls labeled **Numeric Control**, **Numeric Control 2**, and **Numeric Control 3**. The movement of the data is indicated by the "bubbles." The data from the controls **Numeric Control** and **Numeric Control 2** enters the first Add function and the data from the **Numeric Control 3** control waits at the second Add function, as shown in Figure 3.37(b). The data then flows out the first Add function and heads toward the second, where data from **Numeric Control 3** awaits, as shown in Figure 3.37(c). The final summation then occurs in Figure 3.37(d), and the result flows to the indicator labeled

Numeric Indicator. Additionally, in single-step mode, the next node to be executed blinks until you click on the next step button.

3.2.3 Single-Stepping Through a VI and Its SubVIs

For debugging purposes, you may want to execute a block diagram node by node. This is known as single-stepping. To run a VI in single-step mode, press any of the debugging step buttons on the toolbar to proceed to the next step. The step buttons are shown on the toolbar in Figure 3.36. The step button you press determines where the next step executes. You click on either the **Step Into** or **Step Over** button to execute the current node and proceed to the next node. If the node is a structure (such as a While Loop) or a subVI, you can select the **Step Over** button to execute the node, but not single-step through the node. For example, if the node is a subVI and you click on the **Step Over** button, you execute the subVI and proceed to the next node, but cannot see how the subVI node executed internally. To single-step through the subVI you would select the **Step Into** button.

Click on the **Step Out** button to finish execution of the block diagram nodes or finish up the single-step debugging session. When you press any of the step buttons, the **Pause** button is pressed as well. You can return to normal execution at any time by releasing the **Pause** button.

If you place your cursor over any of the step buttons, a tip strip will appear with a description of what the next step will be if you press that button.

You might want to use highlight execution as you single-step through a VI, so that you can follow data as it flows through the nodes. In single-step mode and highlight execution mode, when a subVI executes, the subVI appears on the main VI diagram with either a green or red arrow in its icon, as illustrated in Figure 3.38. The diagram window of the subVI is displayed on top of the main VI diagram. You then single-step through the subVI or let it complete executing.

By default, debugging has been disabled on LabVIEW analysis functions. For this reason, you cannot step into the Uniform White Noise VI in Figure 3.38. However, Average VI is not a LabVIEW analysis function, so debugging is enabled.

You can save a VI without single-stepping or highlight execution capabilities. This compiling method typically reduces memory requirements and increases performance by 1–2%. To do this, right-click in the icon pane (upper right corner of the front panel window) and select **VI Properties**. As in Figure 3.39, from the **Execution** menu, deselect the **Allow debugging** option to hide the **Highlight Execution** and **Single Step** buttons.

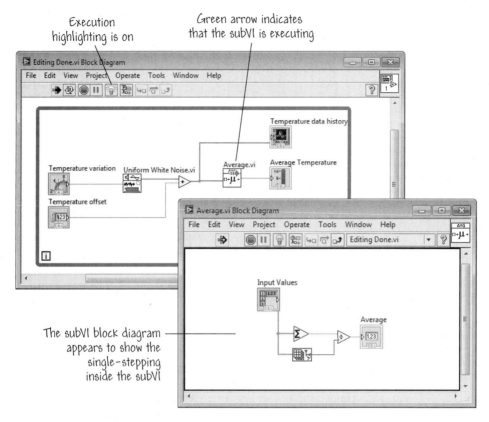

FIGURE 3.38
Single stepping into a subVI with execution highlighting selected.

3.2.4 Breakpoints and Probes

You may want to halt execution (set **breakpoints**) at certain locations of your VI (for example, subVIs, nodes, or tool, wires). Using the **Breakpoint** tool click on any item in the block diagram where you want to set or clear a breakpoint. Breakpoints are depicted as red frames for nodes and red dots for wires.

You use the **Probe** tool to view data as it flows through a block diagram wire. To place a **probe** in the block diagram, click on any wire in the block diagram where you want to place the probe. Probes are depicted as numbered yellow boxes. You can place probes all around the block diagram to assist in the debugging process. The **Probe Watch Window** is a unified tool to view and manage all probes. There are three ways to access the **Probe Watch Window**. You can either select View≫Probe Watch Window, right-click the wire that contains the data you want to check and select **Probe** from the shortcut menu or use the **Probe** tool. If you use the **Probe** tool to place a probe on the block diagram, the **Probe Watch Window** will automatically appear on the desktop.

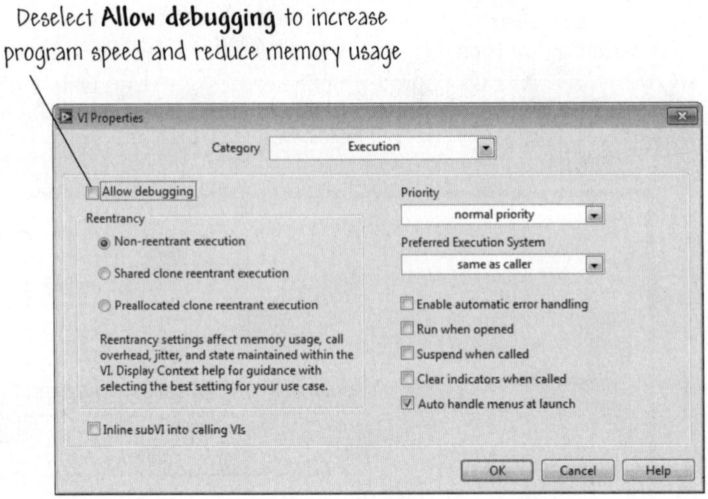

Deselect **Allow debugging** to increase
program speed and reduce memory usage

FIGURE 3.39
Turning off the debugging options using the **VI Setup** menu.

The **Probe Watch Window** shown in Figure 3.40 lists each VI and its corresponding probe(s) in the Probe(s) column. The Value column displays the last known value to pass through the probed wire. The Last Update column displays the timestamp of the last time data passed through the probe. The Probe Display on the right side of the **Probe Watch Window** displays the data that last passed through the selected probe.

In the **Probe Watch Window** shown in Figure 3.40 there are two VIs (Editing Done.vi and Debug.vi) each containing two probes. The probes in the Editing Done VI report the data in Temperature offset (probe #1) and Average (probe #2). The probes in the Debug VI report the data in Temperature bias (probe #3) and Thermometer (probe #4). The last data to pass through the Temperature offset wire and indicated by probe #1 is 3.67 as seen in the Probe Display on the right side of the **Probe Watch Window**.

Conditional probes allow you to set conditional breakpoints for common cases of each data type. For example, creating a conditional probe on a numeric allows you to set a pause condition using Equal to, Greater than, or Less than. This gives you flexibility in your debugging.

A feature of LabVIEW is the ability to retain wire values. This feature can be enabled and disabled with the **Retain Wire Values** button on the toolbar of the block diagram, as shown in Figure 3.36. Once enabled, if you create a probe on the block diagram *after* executing your VI, the probe displays the data that flowed through the wire during the last VI execution. The VI must run sucessfully at least once before you can collect data from any wires using probes. Retaining wire values is a handy feature when you need to debug a VI with a complex block diagram.

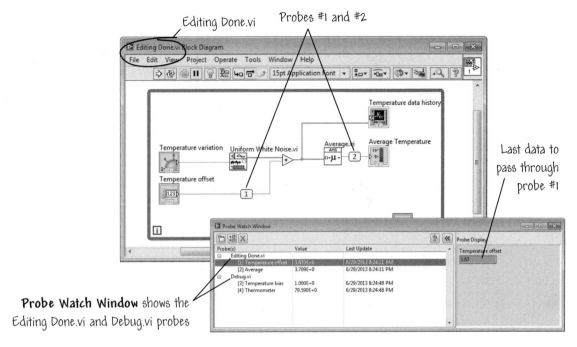

Editing Done.vi

Probes #1 and #2

Last data to pass through probe #1

 **Probe Watch Window** *shows the Editing Done.vi and Debug.vi probes*

FIGURE 3.40
The **Probe Watch Window**.

Practicing with Debugging

A nonexecutable VI—named Debug.vi—has been developed for you to debug and to fix. You will get to practice using the single-step and execution highlighting modes to step through the VI and inserting breakpoints and probes to regulate the program execution and to view the data values.

1. Open the Debug VI in Chapter 3 of the Learning directory by choosing **Open** from the **File** menu. Notice the **Broken Run** button in the toolbar indicating the VI is not executable.

2. Switch to the block diagram by choosing **Show Block Diagram** from the **Window** pull-down menu. You should see the block diagram shown in Figure 3.41.

   (a) The Random Number (0-1) function (represented by the two die) can be found in the **Functions≫Programming≫Numeric** subpalette and returns a random number between zero and one.

   (b) The Add function (which can also be found in the **Functions≫ Programming≫Numeric** subpalette) adds the random number to a bias number represented by the variable Temperature bias, which accepts input by the user on the front panel.

   (c) The subVI Simulated Temperature (deg F).vi found in the activity folder in the Learning directory simulates acquiring temperature data.

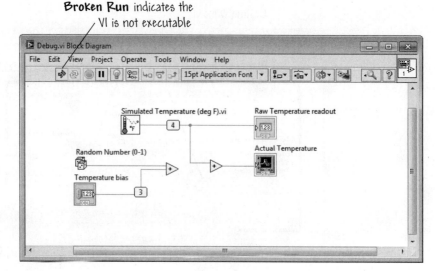

FIGURE 3.41
Block diagram showing the debugging exercise.

As you single step through the code, you will see the program execution inside the subVI.

(d) The second Add function adds the simulated temperature data point to the random number plus the bias number to create a variable named **Actual Temperature**.

3. Investigate the source of the programming error by clicking on the **Broken Run** button to obtain a list of the programming errors. You should find one error—**Add: contains unwired or bad terminal**.

4. Highlight the error in the **Error List** and then click on **Show Error** to locate the source of the error within the block diagram. You should find that the second Add function (the one on the right side) is shown to be the source of the programming error.

5. Fix the error by properly wiring the two Add functions together. Once this step is successfully completed, the **Broken Run** button should reappear as the **Run** button.

6. Select **Highlight Execution** and then run the VI by pressing the **Run Continuously** button. You can watch the data flow through the code. Returning to the front panel, you will see the waveform chart update as the new temperature is calculated and plotted. Notice that the simulation runs very slowly in highlight execution mode. Click the **Highlight Execution** button off to see everything move much faster.

7. Terminate the program execution by clicking on the **Abort Execution** button.

8. Enable single-stepping by clicking on one of the step buttons. You can enable **Highlight Execution** if you want to see the data values as they are computed at each node.

9. Use the step buttons to single-step through the program as it executes. Remember that you can let the cursor idle over the step buttons and a tip strip will appear with a description of what the next step will be if you press that button. Press the **Pause** button at any time to return to normal execution mode.

10. Enable the probe by right-clicking on the wire connecting the two Add functions and selecting **Probe**. Continue to single-step through the VI and watch how the probe displays the data carried by the wire. Alternatively, you could have used the **Probe** tool to place the probe on the wire. Try placing a probe on the block diagram (say between the Simulated Temperature subVI and the Add function) using the **Probe** tool.

11. Place a few more probes around the VI and repeat the single-stepping process to see how the probes display the data.

12. When you are finished experimenting with the probes, close all open probe windows.

13. Set a breakpoint by selecting the **Breakpoint** tool from the **Tools** palette and clicking on the wire between the two Add functions. You will notice that a red ball appears on the wire indicating that a breakpoint has been set at that location.

14. Run the VI by clicking on the **Run Continuously** button. The VI will pause each time at the breakpoint. To continue VI execution, click on the **Pause** button. It helps to use highlight execution when experimenting with breakpoints; otherwise the program executes too fast to easily observe the data flow.

15. Terminate the program execution by pressing the **Abort Execution** button. Remove the breakpoint by clicking on the breakpoint (that is, on the red ball) with the **Breakpoint** tool.

16. Save the working VI by selecting **Save** from the **File** menu. Remember to save the working VI in the folder Users Stuff within the Learning directory. Close the VI and all open windows by selecting **Close** from the **File** menu. ◆

3.2.5 Navigation Window

When working with complicated VIs, the block diagrams and front panels may be too unwieldy to easily navigate. LabVIEW provides the feature of the **Navigation** window as a tool to display an overview of the active front panel in edit mode or the active block diagram. To activate the **Navigation** window, select View≫Navigation Window. Once the **Navigation** window is open, you click

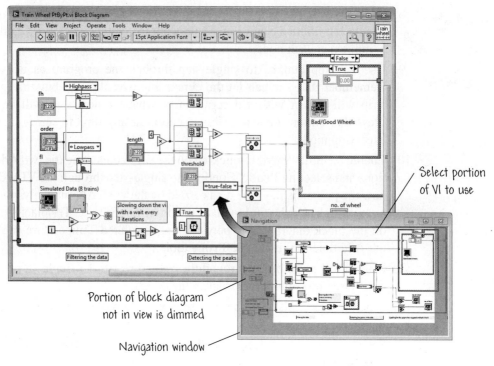

FIGURE 3.42
Using the Navigation window.

an area of the image in the **Navigation** window to display that area in the front panel or block diagram window.

Consider the Train Wheel PtByPt VI depicted in Figure 3.42 with the **Navigation** window in view. The portion of the block diagram in the center of the **Navigation** window has been selected and the associated block diagram code appears in the block diagram window. You can also click and drag the image in the **Navigation** window to scroll through the front panel or block diagram. As illustrated in Figure 3.42, portions of the front panel or block diagram that are not visible appear dimmed in the **Navigation** window.

3.3 PROPERTY NODES

Property Nodes allow you to *set* the properties and to *get* the properties of objects. In some applications, you might want to programmatically modify the appearance of front panel objects in response to certain inputs. For example, a front panel object can be made to vanish from the front panel while the VI is running if a Boolean input is True and made to reappear when the Boolean input is False. As another example, you might want an LED to start blinking if a user enters an invalid password. In Chapter 7 we will show how to use Property Nodes to change the color of a trace on a chart when data points are above a certain value. Property Nodes can also be used to programmatically resize front panel objects, hide parts of the front panel, move controls and indicators on the front panel, and for many other uses.

You create a property from a front panel object by right-clicking the object, selecting Create≫Property Node, and selecting a property from the shortcut menu, as illustrated in Figure 3.43. A Property Node is created on the block diagram that is implicitly linked to the front panel object. If the object has a label, the Property Node has the same label. In Figure 3.43, the Property Node is labeled Input, since the front panel control has the label Input. You can change the label after you create the node. When you create a Property Node, it initially has one terminal representing a property you can modify for the corresponding front panel object. Using this terminal on the Property Node, you can either set (write) the property or get (read) the current state of that property.

Some properties, such as the Label property, are read only and some properties, such as the Value (Signaling), are write only.

In Figure 3.43, when the Property Node was created for the digital numeric control labeled Input using the **Visible** property, a small arrow appeared on the right side of the Property Node terminal, indicating that the property value was being read. You can change the action to write by right-clicking the Property Node and selecting **Change To Write** from the shortcut menu.

In Figure 3.43, the simple VI has one control and one indicator. The Property Node for the control has been set to **Visible**, so that wiring a False Boolean value to the **Visible** property terminal causes the numeric control to vanish from the front panel when the Property Node receives the data. Wiring a True Boolean value causes the control to reappear. This is an example of writing a property using the Property Node.

*The VI shown in Figure 3.43 can be found in **Chapter 3** of the **Learning** directory. Open and run the VI in **Run Continuously** mode. Click the push button labeled **Visible** to make the control **Input** vanish and then reappear upon a second click. Also, you can change the value of the **Position:Left** numeric control to move the control **Input** horizontally on the front panel.*

You can create multiple Property Nodes for the same front panel object. To add terminals to the node, right-click the white area of the node and select **Add Element** from the shortcut menu or use the **Positioning** tool to resize the node. Then you can associate each Property Node terminal with a different property from its shortcut menu. Property Nodes execute in order from top to bottom. If

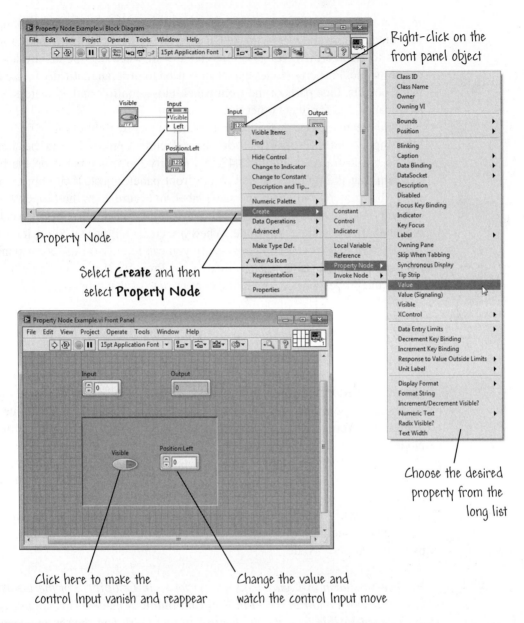

Right-click on the front panel object

Property Node

Select **Create** and then select **Property Node**

Choose the desired property from the long list

Click here to make the control Input vanish and reappear

Change the value and watch the control Input move

FIGURE 3.43
Using the Property Node.

an error occurs on a terminal, the node stops at that terminal, returns an error, and does not execute any further terminals. In Figure 3.43, the two properties selected are **Visible** and **Position: Left**.

Using a large number of Property Nodes in subVIs can negatively impact program execution speed. Property Nodes can be memory intensive and if invoked in a subVI will load its front panel into memory leading to unnecessary memory usage. Use local variables (covered in Chapter 5) for reading and writing object values when possible.

3.4   A FEW SHORTCUTS

Frequently used menu options have equivalent command key shortcuts. For example, to save a VI you can choose **Save** from the **File** menu, or press the control key equivalent <Ctrl-S> (Windows) or <command-S> (Macintosh). Some of the main key equivalents are shown in Table 3.1.

Shortcut access to the **Tools** palette on the front panel and block diagram is given by:

- **Windows**—Press <shift> and the right mouse button
- **Macintosh**—Press <command-shift> and the mouse button

TABLE 3.1   Frequently Used Command Key Shortcuts

Windows	Macintosh	Function
<Ctrl-S>	<command-S>	Save a VI
<Ctrl-R>	<command-R>	Run a VI
<Ctrl-E>	<command-E>	Toggle between the front panel and the block diagram
<Ctrl-H>	<command-H>	Toggle the **Context Help** window on and off
<Ctrl-B>	<command-B>	Remove all bad wires
<Ctrl-W>	<command-W>	Close the active window
<Ctrl-F>	<command-F>	Find objects and VIs

BUILDING BLOCK

3.5   BUILDING BLOCKS: PULSE WIDTH MODULATION

The Rising Edge and Falling Edge VIs that you constructed in Chapter 2 will be improved with the editing techniques learned in this chapter. Navigate to the Users Stuff in the Learning directory and open both VIs.

Working versions of Falling Edge.vi and Rising Edge.vi can be found in the Building Blocks folder of the Learning directory. These working VIs are provided in case you did not save them in the Users Stuff folder after working in Chapter 2.

It is important that you create documentation of VIs you make so that information is available which may not be readily apparent to users interacting with your VI for the first time. You may wish to use free text labels to clarify features of either the front panel or the block diagram. To write a VI description for the Falling Edge VI, go to the **File** pull-down menu, select **VI Properties**, and enter the following description under the category **Documentation** in the **VI Description** dialog box:

This VI is used in Pulse Width Modulation to determine when a high output signal should switch to low, based on the current input values of Duty Cycle, Period, and Clock.

Controls such as knobs and slides are useful in situations where there is a preferred or ideal input range for any given variable, since each control limits the user to its specified input range. Replace the numeric control for Duty Cycle with a vertical pointer slide in the Falling Edge VI. Open the **Properties** window for the slide pointer control to edit the data range and scale range. We want to set the limits so that the user can only enter values between 0 and 1 on the control. Click the **Data Entry** tab and set the increment value to 0.01. Then select **Coerce to nearest** so that any input value specified past the hundredths place will be rounded to the nearest hundredth. On the **Appearance** tab, make the Digital Display visible. This gives the user the flexibility of adjusting the duty cycle by sliding the pointer or by entering a new value in the digital display.

Recall that in Chapter 2 Building Blocks we assumed the user will always provide valid inputs for Duty Cycle and Period. Now we can guarantee that the user does not input invalid values for Period and Clock by enforcing the limits on the inputs. To adjust the data range for Period and Clock so that no negative value can be input to Clock and the minimum value that can be input to Period is 100, right-click on the respective input to open the **Properties** window and click the **Data Entry** tab. In the case of the Clock, deselect **Use Default Limits** and set the minimum value to zero. In the case of the Period, follow the same steps, except set the minimum value to 100.

After modifying the front panel, you may find that you need to rearrange some objects for ease of use of the inputs. Use the **Align Objects** pull-down menu to organize objects on the front panel and the **Clean-Up Diagram** button to organize objects on the block diagram. A smartly organized VI makes debugging more straightforward and enhances the ease with which others can understand and utilize the program.

Your VI should now resemble the one shown in Figure 3.44. Run the VI with the new Duty Cycle control. Note that as you slide the control, the val-

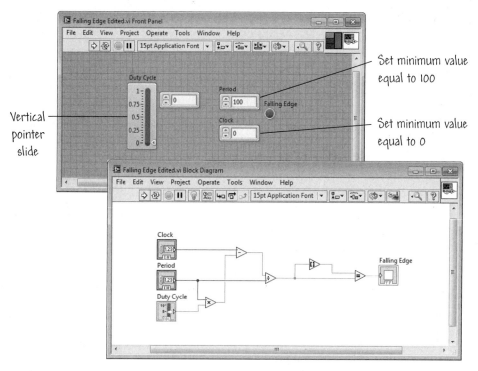

FIGURE 3.44
Editing the Falling Edge VI.

ues increment by hundredths, and if you enter an invalid input on the numeric display, it is rounded to a valid value.

When you are done experimenting with your new VI, save it as Falling Edge Edited.vi in the Users Stuff folder in the Learning directory. You will use this VI as a building block in later chapters—so make sure to save your work!

We now edit the Rising Edge VI to set the minimum range of the Period to 100. Once Rising Edge.vi is open, right-click the control labeled Period to access the **Properties** window. Click the **Data Entry** tab and set the minimum value of Period to 100. Repeat the editing process for the control labeled Clock and check that **Use Default Range** is selected. When you are done editing the Rising Edge VI, save it as Rising Edge Edited.vi in the Users Stuff folder in the Learning directory.

Working versions of Falling Edge Edited.vi and Rising Edge Edited.vi can be found in the Building Blocks folder of the Learning directory.

3.6 RELAXED READING: USING GRAPHICAL SYSTEM DESIGN FOR TUMOR TREATMENTS

In this reading, we learn about a new medical device, the Visica 2, used to treat breast tumors in a less invasive and nearly painless procedure, dramatically reducing the emotional and physical discomfort of patients undergoing tumor treatment. Using National Instruments hardware and software, the Visica 2 was developed in a flexible and reliable manner under extreme time-to-market pressures.

Imagine a potentially revolutionary product that would let doctors treat benign tumors by freezing and killing them in an almost painless outpatient procedure. This would be a dramatic change from an in-patient surgical solution or a "wait-and-see" approach. Sanarus, a small medical device start-up company, created a well-executed design and development plan to produce a new device of this sort that could have an important impact on breast cancer treatment. The resulting product is the Visica Treatment System (V2) shown in Figure 3.45. Amazingly, a working prototype of the V2 system was produced within four months to meet the product-release schedule.

The V2 is an instrument for use in a doctor's office or clinic. It involves local anesthesia and a real-time, ultrasound-guided approach that is virtually painless. During the 10- to 20-minute treatment the targeted tissue is frozen and destroyed through an incision so small that it does not require stitches.

Writing firmware and designing a custom circuit board for the device would have been time consuming, and any error at the firmware or software level could have created delays threatening the entire project. Also, because V2 is a medical device, errors that could compromise system performance were unacceptable. If the device had failed any part of the exhaustive testing, the entire project would have failed and the V2 would not have made it to market. Based on these requirements, it was clear that an extremely reliable development option for V2 was required.

CompactRIO was selected because of its mix of programmability and integrated I/O development. Designing a prototype using CompactRIO showed that the V2 could be developed reliably in a short period of time. While a custom solution would have taken months to develop, the NI solution took only weeks.

With custom firmware, too, "late game" changes would require new and difficult revisions, whereas with the CompactRIO platform the code could be revised if needed with minimal effort. The flexibility afforded by the National Instruments solution accommodated new feature requests without causing delays in the development schedule. The user interface was developed with the LabVIEW for Windows graphical programming environment for a PanelPC, as

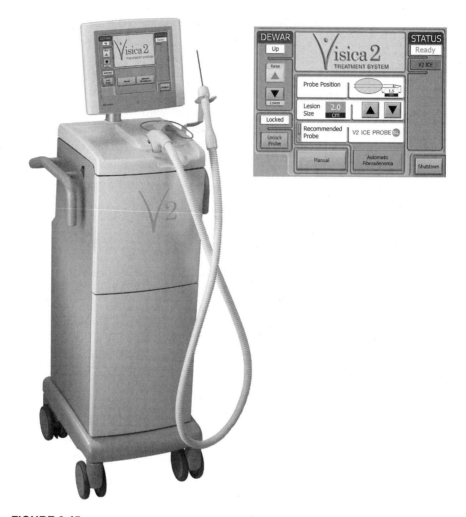

FIGURE 3.45
The CompactRIO platform from National Instruments enabled Sanarus to quickly develop a working prototype. (Photo courtesy of Sanarus Medical.)

shown in Figure 3.45. Communications between the GUI and the CompactRIO real-time controller were managed using LabVIEW shared variables. LabVIEW Real-Time was used to implement a state machine on the CompactRIO real-time controller, and on the PID, LabVIEW Real-Time regulated loops controlled the temperature of the tip of the probe. This was done by prescribing control inputs to the liquid nitrogen pump for cooling as well as a simple resistive heating element. The LabVIEW FPGA was used to manage the interfacing to the I/O signals necessary to control these devices. With LabVIEW, the controller was designed and coded in-house, then prototyped and deployed. The design and development cycle was short but it was very successful.

Long-term studies have shown the technique to be highly effective in destroying common tumors. The V2 is now available at selected centers throughout the United States.

The V2's product design, prototype, and eventual deployment timelines were met because the National Instruments solution enabled efficient development of an embedded control system with a user-friendly GUI while maintaining the highest quality, and, ultimately, ensuring the safety of patients. For more information on the Visica 2 system, please visit http://www.sanarus.com/visica.html. Additional information can be found at the NI website http://sine.ni.com/cs/app/doc/p/id/cs-11023.

3.7   myDAQ BUILDING BLOCKS

In this myDAQ Building Block, you will use the editing and debugging techniques to complete the first two VIs that will become part of your final temperature monitoring system. The first VI you will modify is the LED Control unedited.vi illustrated in Figure 3.46. Open the VI found in the Chapter 3 folder of the myDAQ Building Blocks folder in the Learning directory and display the block diagram by selecting **Window≫Show Block Diagram**. On the block diagram, note that the **Run** arrow is broken. Click on the **Broken Run** arrow to see the list of errors.

Follow the steps below to edit the LED Control unedited.vi:

1. To the front panel, add a **Push Button** control and **LED** from the **Silver≫Boolean** subpalette of the **Controls** palette.

2. Using the **Labeling** tool, name the **Push Button** control LED Control and name the **LED** indicator LED Status.

3. Switch to the block diagram to wire the output from LED Control to the Build Array function and wire the Build Array function to the DAQ Assistant input labeled data.

4. Wire the output of LED Control to LED Status. This displays the status of the LED on the front panel.

5. On the front panel, toggle the LED Control and run the VI. Note that the VI does not run continuously, so to change the state of the LED Control and run the VI again.

6. Note how the LED diode in the LED circuit (see Chapter 1 for more details on the LED circuit) turns on and off depending on the setting of the LED Control.

7. Save the edited VI in the Chapter 3 folder in the myDAQ Building Blocks folder within the Learning directory and name it LED Control.vi. Close the VI.

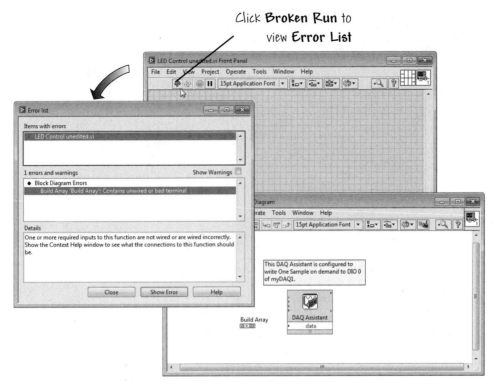

FIGURE 3.46
The LED Control unedited.vi.

Following a similar path, edit the **Read Temp unedited.vi** illustrated in Figure 3.47. The **Read Temp unedited.vi** can be found in the **Chapter 3** folder of the **myDAQ Building Blocks** folder in the **Learning** directory. Open the VI and display the block diagram by selecting **Window≫Show Block Diagram**.

This VI is intended to work with the thermistor circuit described in the Chapter 2 myDAQ Building Blocks. Select **Context Help** on the block diagram, as illustrated in Figure 3.47, and use the terminal names and descriptions to connect the appropriate terminals of the elements on the block diagram. Add indicators to display the resistance of the thermistor and the temperature. Run the VI several times while changing the conditions of the thermistor. Do you get the results you expected? If you are using a different thermistor model, Temp Conversion.vi may not return correct temperature values. Consult the data sheet or follow the calibration instructions in the myDAQ Building Block in Chapter 11. Save and close your VI when you are finished.

*Working versions of **LED Control.vi** and **Read Temp.vi** can be found in the **Chapter 3** folder of the **myDAQ Building Blocks≫Solutions** folder in the **Learning** directory.*

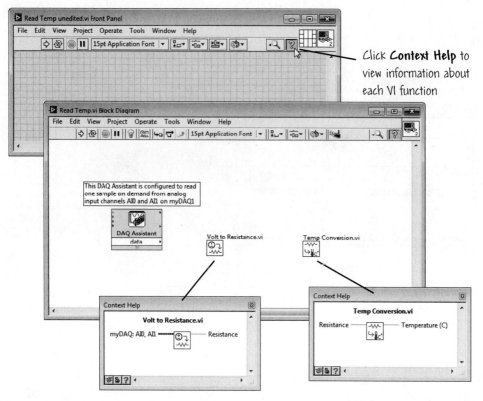

FIGURE 3.47
The Read Temp unedited.vi.

In this myDAQ Building Blocks exercise, you edited two VIs. The first VI allows programmatic control of the LED circuit attached to the myDAQ and created in Chapter 1. The LED Control.vi employs a push button to change the state of the LED and displays the status of the LED on the front panel. The second VI Read Temp.vi displays the resistance of the thermistor in the thermistor circuit created in Chapter 2. In the remaining chapters, we will continue to build upon these VIs to highlight the various LabVIEW topics within the chapters using the myDAQ.

3.8 SUMMARY

The main subjects of this chapter were the editing and debugging of VIs. Just as you would edit a C program or debug a Fortran subroutine, you must know how to edit and debug a VI. We discussed how to create, select, delete, move, and arrange objects on the front panel and block diagram. We developed the important topics of cleaning up the block diagram, routing wires, and reusing snippets of code. We learned how to programmatically read and write the properties of objects using the Property Node. Discussion of program debugging included execution highlighting, how to step into, through, and out of the code, and how to use probes to view data on the **Probe Watch Window** as it flows through the code.

KEY TERMS

Breakpoint: A pause in execution used for debugging. You set a breakpoint by clicking a VI, node, or wire with the **Breakpoint** tool.

Breakpoint tool: Tool used to set a breakpoint on a node or wire.

Broken VI: A VI that cannot compile and run.

Coloring tool: Tool used to set foreground and background colors.

Execution highlighting: Debugging feature that animates the VI execution to illustrate the data flow within the VI.

Label: Text object used to name or describe other objects or regions on the front panel or block diagram.

Labeling tool: Tool used to create labels and enter text.

Navigation window: Displays an overview of the active front panel in edit mode or the active block diagram.

Operating tool: Tool used to enter data into controls and operate them—resembles a pointing finger.

Positioning tool: Tool used to move, select, and resize objects.

Probe: Debugging feature for checking intermediate values in a VI during execution.

Probe tool: Tool used to create probes on wires.

Probe Watch Window: A unified tool to view and manage all probes.

Property Node: Feature that allows you to programmatically set (write) and get (read) the properties of objects.

> **Resizing handles**: Angled handles on the corners of objects that indicate resizing points.
>
> **VI Snippets**: Sections of code from the block diagram saved in a .png image file to reuse later or to share with other users.

EXERCISES

E3.1 Construct a VI to accept five numeric inputs, add them up and display the result on a gauge, and light up a round light if the sum of the input numbers is less than 8.0. The light should light up in green, and the gauge dial should be yellow. The VI in Figure E3.1 can be used as a guide.

 *Hint: Use the Compound Arithmetic function found on the **Programming≫ Numeric** palette to add the five inputs together.*

E3.2 Open Change to Indicator.vi found in the Chapter 3 subfolder in the Exercises&Problems folder within the Learning directory. Click on the **Broken Run** button. Select the error, and click on the **Show Error** button. This wire connects more than one data source. The wire between the output of the Add function and the control result is highlighted and has an "x" over it. Both the output and the control are "sources of data." You can only connect a data source to a data display. Right-click on the control result and select **Change to Indicator**. The error disappears and you can run the VI. Input $x = 2$ and $y = 3$ and run the VI. Verify that the output result is 5.

E3.3 Open the virtual instrument Multiple Controls-1 Terminal.vi. You can find this VI in the Chapter 3 subfolder in the Exercises&Problems folder within the Learning directory. Click on the **Broken Run** button, select the error **This Wire Connects to More than One Data Source**, and click on the **Show Error** button. The wires from the two controls are highlighted. If you look closely, you will see that both wires are going to the same terminal. That means two controls are defining the value of one input value. Delete the wire from the y control. Wire the y control to the other input terminal on the Add function. Both errors disappear and you can run the VI. Input $x = 1$ and $y = 4$ and run the VI. Verify that the output $x + y = 5$.

E3.4 Open the virtual instrument Crazy Wires.vi. You can find this VI in the Chapter 3 subfolder in the Exercises&Problems folder within the Learning directory. Go to the block diagram and see the wires running everywhere. There

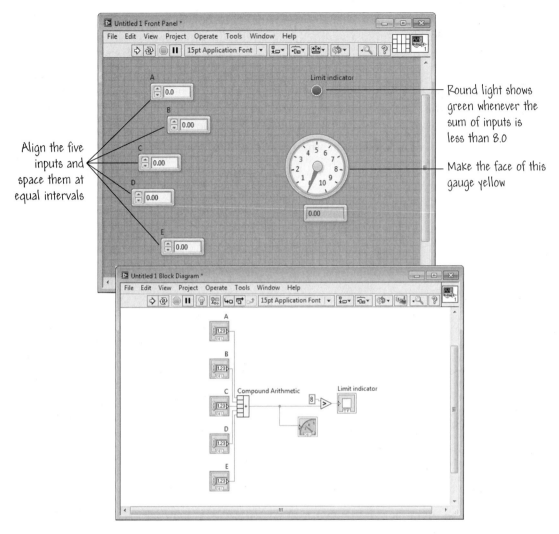

FIGURE E3.1
A VI to add five numeric inputs and light up a round LED if the sum is less than 8.0.

is an easy way to fix this! Right-click on the wire running from Knob to the top terminal of the Comparison function. Select **Clean Up Wire** and watch the wire get straightened instantly. You can repeat the rerouting process for each wire in the block diagram by right-clicking on each wire individually and selecting **Clean Up Wire**. Another way to tidy up the block diagram is to select the **Clean Up Diagram** button in the block diagram toolbar (see Section 3.1.9). Besides rerouting the wires, this button will reorder the objects on the block diagram. Use the **Clean Up Diagram** button and clean up the remaining wires on the block diagram.

E3.5 Open the Oil Change VI in the **Chapter 3** subfolder in the **Exercises& Problems** folder within the **Learning** directory and shown in Figure E3.5. The VI assists in determining whether or not a car requires an oil change based on the mileage and elapsed time since the last oil change. The VI assumes that a car needs an oil change after either 6 months or 5,000 miles, whichever comes first. Can you see the problem with the VI by inspection of the block diagram? To confirm and fix the problem, perform the following steps:

(a) Click the broken arrow to see what errors exist in the VI.

(b) Fix the error and run the VI for various values of months and miles.

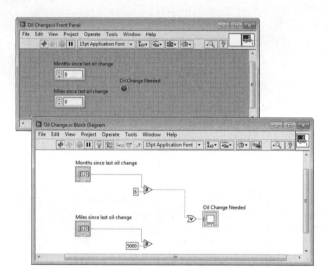

FIGURE E3.5
The Oil Change VI is broken.

E3.6 Open Broken VI found in the **Chapter 3** subfolder in the **Exercises&Problems** folder within the **Learning** directory. Click the broken run arrow to see a list of errors in the program. Determine why this VI fails to compile and run. Modify the VI as necessary so that the errors are repaired.

PROBLEMS

P3.1 Construct a VI that generates two random numbers (between 0 and 1) and displays both random numbers on meters. Label the meters **Random number 1** and **Random number 2**, respectively. Make the face of one meter blue and the face of the other meter red. When the value of the random number on the red meter is greater than the random number on the meter with the blue face, have a square LED show green; otherwise have the LED show black. Run the VI several times and observe the results. On the block diagram select **Highlight Execution** and watch the data flow through the code.

P3.2 In this problem you will construct a stop light display. Create a dial control that goes from 0 to 2, with three LED displays: one green, one yellow, and one red. Have the VI turn the LED green when the dial is on 0, yellow when the dial is on 1, and red when the dial is on 2.

P3.3 Open the block diagram of the Waveform Min Max example VI located in the library Operations.llb. in the Waveform subfolder within the examples folder. See Figure P3.3 for reference.

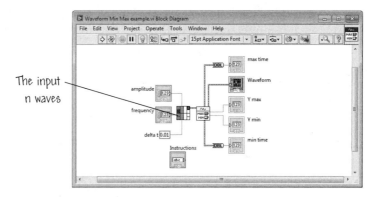

FIGURE P3.3
The block diagram of the Waveform Min Max example.vi showing the Simple Sine Waveform subVI terminals.

(a) On the block diagram, delete the numeric control labeled frequency by clicking on it and then selecting **Remove From Project** on the **Edit** pull-down menu. Clean up the broken wire left behind by right-clicking it and selecting **Delete Wire Branch** from the shortcut menu or by selecting **Remove Broken Wires** from the **Edit** menu.

(b) The control can be replaced two different ways. On the block diagram, right-click the node labeled n waves on the Simple Sine Waveform subVI and select Create≫Control from the shortcut menu. Alternatively, in the **Controls** palette on the front panel, go to the **Numeric** subpalette. Select a numeric control and place it on the front panel. Then wire the control to the n waves input to the Simple Sine Waveform VI on the block diagram. Using one of these methods, replace the numeric control and label it frequency.

(c) Move to the front panel and right-click the frequency control. Select Replace≫Silver≫Numeric≫Knob to change the control to a knob. Try moving different components around on the front panel to make space for it, using the **Align Objects** tools if necessary.

(d) Run the VI and experiment with different input values.

If you want to save your changes, be sure not to replace the original VI with your edited one. Save the VI you create with a new name in the Users Stuff *folder in the* Learning *directory.*

P3.4 In this problem, we want to open and run an existing VI. In LabVIEW, go to Help≫Find Examples and click the **Search** tab and type "probes." Select "probes" to display the example VIs that include "probes" in the title. Find the VI entitled Using Supplied Probes.vi and open it.

(a) Change to the block diagram. In this problem, you will create custom probes by right-clicking the data wires and selecting **Custom Probe**. For each probe (numbered 1 through 9) select the appropriate probe from the shortcut menu according to the instructions at the bottom of the block diagram. These custom probes also provide conditional breakpoint functionality, and you will experiment with the conditional settings to pause execution.

(b) What probes are available on the block diagram?

(c) On the Conditional Double Array probe, what happens if you set the condition to Pause if the number of elements equal to 250?

(d) To continue after a break, press the **Pause** button on the menu bar.

(e) Continue examining the **Probe Watch Window**. What information is supplied to you in the Dynamic Data Type probe?

(f) What happens if you implement a conditional breakpoint of zero on the Error Probe?

(g) With the VI running, stop the VI from the front panel. What is the state of the Boolean probe?

P3.5 Open the Temperature Converter VI in the Chapter 3 subfolder in the Exercises&Problems folder within the Learning directory. The block diagram is shown in Figure P3.5. This VI takes an input in degrees Celsius and converts the input to Kelvins and degrees Fahrenheit, displaying the converted temperatures on numeric indicators. Explain how this VI could have been documented differently to make its purpose more readily evident to other programmers modifying the code and also more user friendly. Then fix the VI to reflect these changes.

Save the VI as Temperature Converter.vi in the Users Stuff folder in the Learning directory.

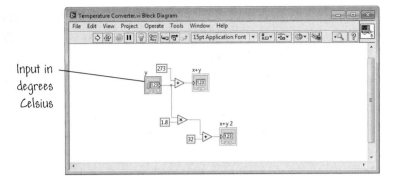

FIGURE P3.5
The block diagram for the Temperature Converter VI.

P3.6 Develop a VI that converts an input value in degrees to radians with four digits of precision. Refer to Figure P3.6 for help on changing the output display to four digits.

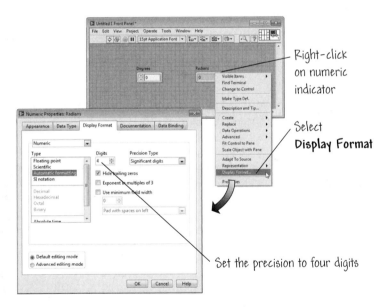

FIGURE P3.6
Converting degrees to radians with four digits of output precision.

P3.7 In this problem you will debug a VI that contains a logic error. Consider the VI shown in Figure P3.7. It is named **Logic Error.vi** and can be found in the **Chapter 3** subfolder in the **Exercises&Problems** folder within the **Learning** directory. The VI's objective is to test whether the input number (labeled **X** on the front panel) is even. If it is not even, a warning message will pop up; other-

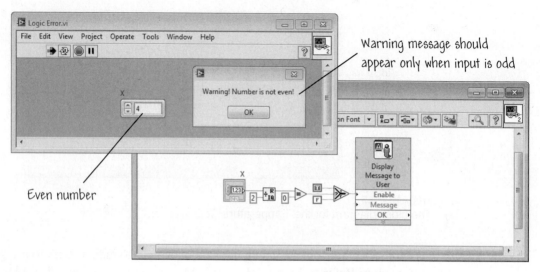

FIGURE P3.7
Fixing a logic error.

wise the VI executes without displaying a warning message. However, as the VI is currently coded, a warning message when the input X is even indicates a logic error in the code. Why is this VI not working properly? Open the VI and test it out to confirm that the logic is flawed. Then debug the code and fix the logic error.

Save the debugged VI as Logic Error.vi in the Users Stuff folder in the Learning directory.

DESIGN PROBLEMS

D3.1 In this problem you will use the editing techniques from this chapter to improve the appearance of both the front panel and block diagram of Averaging.vi. This will aid in making the purpose and functionality of the Average VI more readily apparent to users who are unfamiliar with the VI.

Open the Averaging VI in the Chapter 3 subfolder in the Exercises& Problems folder within the Learning directory. The block diagram is shown in Figure D3.1.

Perform the following editing steps:

(a) Resize the knobs on the front panel so that they are all the same size. You may wish to use some of the tools on the **Resize Objects** menu.

(b) Evenly align and distribute the objects on both the front panel and the block diagram. It is conventional to place inputs (controls) on the left side and outputs (indicators) on the right side. As you rearrange objects on the block diagram, you will most likely need to clean up the wires.

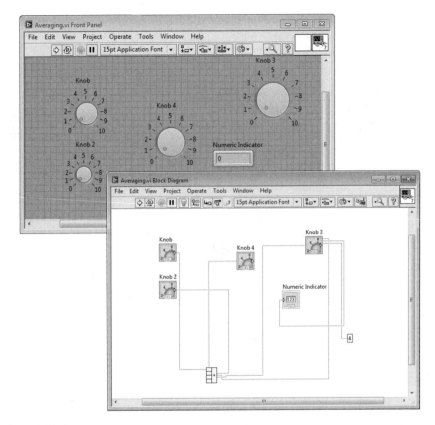

FIGURE D3.1
The unedited Average VI.

(c) Hide the labels on the knob controls and change the label on the indicator to read **average**.

(d) Place a free label on the front panel which briefly explains the purpose of the VI.

*You can also use the **Clean Up Diagram** button on the block diagram toolbar to organize the objects and wires without individually aligning the objects and cleaning up the wires.*

D3.2 Create a front panel that has 8 LED indicators and a vertical slider control that is an 8 bit unsigned integer. Display a digital indicator for the slider, and make sure that the LEDs are evenly spaced and aligned at the bottom. The problem is to turn the 8 LEDs into a binary (base 2) representation for the number in the slider. For example, if the slider is set to the number 10 (which in base 2 is $00001010 = 1 * (2^3) + 1 * (2^1)$), the LEDs 1 and 3 should be on. To test your solution, check the number 131. LEDs 0, 1 and 7 should be on since 131 is 10000011 in base 2.

D3.3   Construct a VI that generates a random number between -10 and 10. The VI should display the random number on an indicator on the front panel and then light an LED corresponding to the range in which the number falls. Use one LED for negative values of the random number, one LED for values in the range of 0 to 4, and one LED for values greater or equal to 5. Change the properties of the LEDs so that they are bright yellow when turned on and dark yellow when off. Use descriptive labels for the indicators and use the tools on the **Align Objects** and **Distribute Objects** menus to space the objects evenly on the block diagram and front panel.

D3.4   In this design problem you will edit six gauges on the front panel of a VI and then wire the block diagram to take an average of the inputs from five input gauges and display the average on the remaining indicator gauge.

Consider the VI shown in Figure D3.4. The VI is named **Gauge Properties.vi** and can be found in the **Chapter 3** subfolder in the **Exercises&**

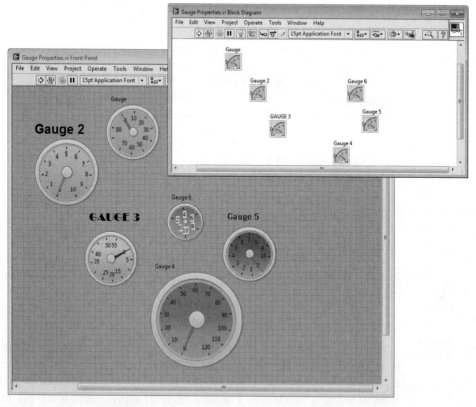

FIGURE D3.4
Practice with editing and wiring.

Problems folder within the Learning directory. Open the VI and edit the front panel so that all the gauges look the same (size, color, font, and needles). Align the gauges in a logical order. Then switch to the block panel and wire up the gauges to take an average of the inputs from gauges 1 through 5 and display it on the indicator gauge 6.

Save the edited VI as Gauge Properties.vi in the Users Stuff folder in the Learning directory.

D3.5 Develop a VI that has two inputs, X and Y, where X is always negative and Y always positive. The VI should compute the ratio X/Y and output the quotient and the remainder.

Consider the VI shown in Figure D3.5. It is named Logic Debug.vi and can be found in the Chapter 3 subfolder in the Exercises&Problems folder within the Learning directory. Open the VI and make two runs. For the first run, have $|X| > Y$ with $X < 0$ and $Y > 0$. Record the resulting quotient and remainder. For the second run, have $|X| < Y$ with $X < 0$ and $Y > 0$. Record the resulting quotient and remainder. Are they what you expected? Why or why not? Switch to the block diagram and edit the VI appropriately to obtain the desired results when $|X| > Y$ with $X < 0$ and $Y > 0$ and when $|X| < Y$ with $X < 0$ and $Y > 0$.

Save the edited VI as Logic Debug.vi in the Users Stuff folder in the Learning directory.

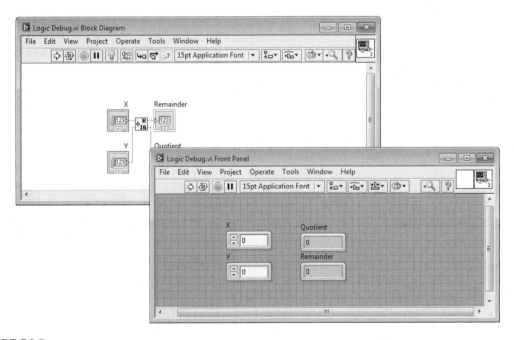

FIGURE D3.5
Debugging a logic error and designing a VI to compute the quotient and remainder of X/Y when $X < 0$ and $Y > 0$.

CHAPTER 4

SubVIs

In this chapter we learn to build subVIs. SubVIs are VIs used by other VIs. The hierarchical design of LabVIEW applications depends on the use of subVIs. We will discuss creating subVIs from VIs and creating subVIs from selections. Automatic error handling, manual error handling, and error clusters are presented as three mechanisms to tackle the important tasks of VI and subVI error checking and error handling. The **Icon Editor** is used to personalize subVI icons. The **VI Hierarchy** window will be introduced as a helpful tool for managing your programs.

GOALS

1. Learn to build and use subVIs.
2. Understand the hierarchical nature of VIs.
3. Practice with the **Icon Editor** and with assigning terminals.
4. Understand the importance of error checking and handling.

182

4.1   WHAT IS A SUBVI?

It is important to understand and appreciate the hierarchical nature of virtual instruments. **SubVIs** are critical components of a hierarchical and modular VI that is easy to debug and maintain. A subVI is a stand-alone VI that is called by other VIs—that is, a subVI is used in the block diagram of a **top-level VI**. SubVIs are analogous to subroutines in text-based programming languages like C or Fortran, and the subVI node is analogous to a subroutine call statement. The **pseudocode** and block diagram shown in Figure 4.1 demonstrate the analogy between subVIs and subroutines. There is no limit to the number of subVIs you can use in a calling VI. Using subVIs is an efficient programming technique in that it allows you to reuse the same code in different situations. The hierarchical nature of programming follows from the fact that you can call a subVI from within a subVI.

You can create subVIs from VIs, or create them from selections (selecting existing components of a VI and placing them in a subVI). When creating a new subVI from an existing VI, you begin by defining the inputs and outputs of the subVI, and then you "wire" the subVI connector properly. This allows

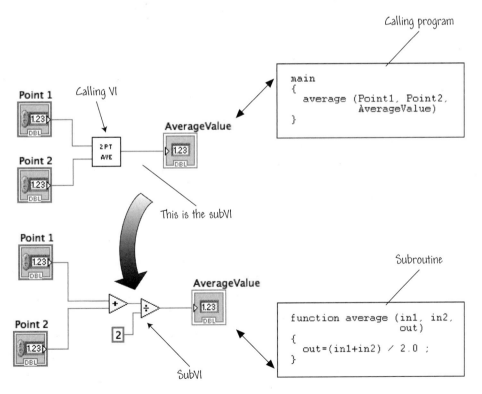

FIGURE 4.1
Analogy between subVIs and subroutines.

calling VIs to send data to the subVI and to receive data from the subVI. On the other hand, if an existing complex block diagram has a large number of icons, you may choose to group related functions and icons into a lower-level VI (that is, into a subVI) to maintain the overall simplicity of the block diagram. This is what is meant by using a modular approach in your program development.

4.2 REVIEW OF THE BASICS

Before moving on to the subject of subVIs, we present a brief review of some of the basics presented in the first three chapters as part of an exercise in constructing a VI. Your VI will ultimately be used as a subVI later in the chapter.

Three parameters associated with a circle of given radius are circumference, diameter, and area. For a radius of length r we have the following relationships:

$$C = 2\pi r$$
$$A = \pi r^2$$
$$D = 2r$$

where C denotes circumference, A denotes area, D denotes diameter, and π is a constant. In this chapter we will build a subVI that computes and outputs C, A, and D, given the circle radius r as input. To begin the design process, we will construct a VI that performs the desired calculations. Then, later in the chapter we will modify the VI so that it can be used as a subVI in other applications.

The task is to construct a VI to utilize the formulas given above to compute C, A, and D, given r. The following list serves as a step-by-step guide. Feel free to deviate from the provided list according to your own style. As you proceed through this book you will find that the step-by-step lists will slowly disappear, and you will have to construct your VIs on your own (with help, of course!).

1. Open a new VI. Save the untitled VI as **Circle Measurement.vi**. In this case you are saving the VI before any programming has actually occurred. This is a matter of personal choice, but saving a VI frequently during development may save you a great deal of rework if your system freezes up or crashes.

2. Add a numeric control to the front panel and label it **r**. This will be where the radius is input.

3. Place a numeric indicator on the front panel and label it **C**. This will be where the circumference is displayed.

4. Make two duplicates (or clones) of the numeric indicator. Using the **Positioning** tool, select the numeric control while pressing the <Ctrl> (Windows) or <option> (Macintosh) key and drag the numeric control to a new location. When you release the mouse button, a copy labeled **C 2** appears in the new location. Repeat this process so that three numeric indicators are visible on the front panel.

5. Label the new numeric indicators **A** and **D**. These indicators will display area and diameter.

6. Place the numeric indicators in a column on the right side of the front panel. Using the **Align Objects** tool, align the three numeric indicators by their left edges.

7. Switch to the block diagram. Arrange the control and indicator terminals so that the three indicator terminals appear aligned in a vertical column on the right side and the control appears on the left side.

8. Use functions and constants from the **Programming≫Numeric** palette to program the formulas for circumference, area, and diameter. You can use the VI shown in Figure 4.2 as a guide.

9. Once the block diagram has been wired, switch back to the front panel and input a value into the radius control.

10. Run the program with **Run Continuously** and select **Highlight Execution** to watch the data flow through the program. Change the value of the radius and see how the outputs change. When you are finished experimenting, you can stop the program execution.

11. Practice single-stepping through the program using the step buttons on the block diagram.

12. Once the VI execution is complete, switch back to the front panel and observe the default connector, as illustrated in Figure 4.2. When we are done, it will indicate one input terminal and three output terminals. We will be learning how to edit the icon and connector in the next section.

13. Save the VI by selecting **Save** from the **File** menu.

This exercise provided the opportunity to practice placing objects on the front panel, wiring them on the block diagram, using editing techniques to arrange the objects for easy visual inspection of the code, and using debugging tools to verify proper operation of the VI. The Circle Measurements VI can be

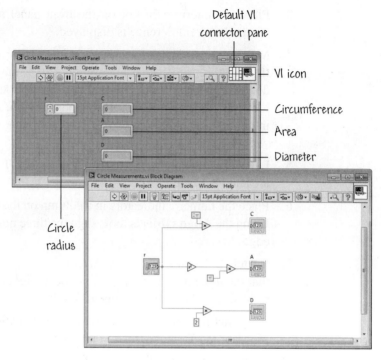

FIGURE 4.2
The Circle Measurements VI default VI connector.

used as a subVI (requiring some effort to wire the VI connector properly), and that is the subject of the following sections.

A working version of the Circle Measurements VI can be found in the **Chapter 4** *folder of the* **Learning** *directory. It is called* **Circle Measurements.vi.**

4.3 EDITING THE ICON AND CONNECTOR

A subVI is represented by an icon on the block diagram of a calling VI and must have a connector with terminals properly wired to pass data to and from the top-level VI. The connector pane is situated next to the VI icon in the upper right corner of the front panel window, as illustrated in Figure 4.3. The side-by-side placement of the connector pane and icon allows you to readily assign front panel controls and indicators as inputs and outputs without switching back and forth between the connector pane and icon (as in previous versions of Lab-VIEW). In this section we discuss icons and connectors.

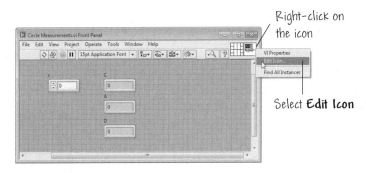

FIGURE 4.3
Activating the **Icon Editor**.

4.3.1   Icons

Every VI has a default icon displayed in the upper right corner of the front panel and block diagram windows. The default icon is a picture of the LabVIEW logo. You use the **Icon Editor** to customize the icon. To activate the editor, right-click on the default icon in the front panel and select **Edit Icon** as illustrated in Figure 4.3. You can also display the **Icon Editor** by double-clicking on the default icon.

*You can only open the **Icon Editor** when the VI is not in **Run** mode. If necessary, you can change the mode of the VI by selecting **Change to Edit Mode** from the **Operate** menu.*

The **Icon Editor** dialog box provides editing tools for creating icons. As shown in Figure 4.4, it provides icon templates, many useful glyphs, options for adding and formatting icon text, and support for editing with layers. The **Icon Editor** dialog box saves icons you create in both 256-color and black-and-white format. You also can use it to create and save icon templates or glyphs for later use as 256-color .png files.

The tools on the palette at the right of the dialog box are used to create the icon design in the pixel editing area. A true-size image of the icon appears just right of center in the dialog box and to the left of the pixel editing area (see Figure 4.4). The tools in the **Icon Editor** palette can be used to perform many functions, as described in Table 4.1. If you have used other paint programs, you will be familiar with these tools. You can cut, copy, and paste images from and to the icon. When you paste an image and a portion of the icon is selected, the image is resized to fit into the selection.

Add text to the icon

Glyphs library

Display **Layers** page

Pixel editing area

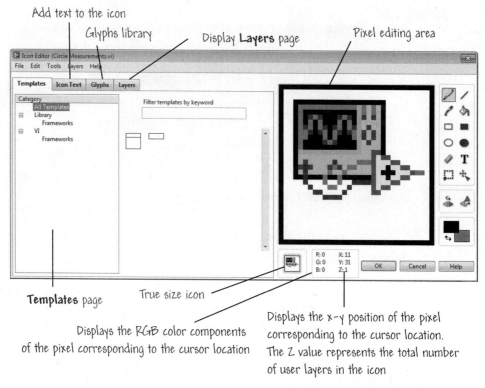

Templates page

True size icon

Displays the RGB color components
of the pixel corresponding to the cursor location

Displays the x–y position of the pixel
corresponding to the cursor location.
The Z value represents the total number
of user layers in the icon

FIGURE 4.4
The **Icon Editor**.

*The edited icon you create might appear slightly different from the corresponding icon on the block diagram, since the **Icon Editor** dialog box displays icons in 24-bit color but LabVIEW displays icons only in 8-bit color on the block diagram.*

The **Icon Editor** dialog box includes four main pages: **Templates, Icon Text, Glyphs,** and **Layers**. Selecting **Templates** displays icon templates you can use as background for the icon. This page displays all .png, .bmp, and .jpg files in the Icon Templates directory of the LabVIEW Data folder. Lists of icon templates by category are presented with names that correspond to the subfolders in the Icon Templates directory. You can also filter templates by keyword so that the **Icon Editor** dialog box displays all icon templates whose name contains the given keyword.

Selecting **Icon Text** displays a page that allows you to place text on the icon. You can specify four lines of text, the color of the text in each line, the font and font size, alignment (left, center, or right), and several other characteristics of the text.

TABLE 4.1   Icon Editor Tools

Draws individual pixels in the **Line Color** you specify		Draws a line in the **Line Color** you specify
Sets the **Line Color** or the **Fill Color**		Fills all connected pixels of the same color with **Line Color**
Draws a rectangular border in the **Line Color**		Draw a rectangle with a border in the **Line Color** and filled in the **Fill Color**
Draws an elliptical border in the **Line Color**		Draws an ellipse with a border in the **Line Color** and filled in the **Fill Color**
Draws individual pixels as transparent		Enters text at the location you specify
Selects an area of the icon to cut, copy, or move		Moves all pixels in the user layer you select
Horizontally flips the user layer you select		Rotates the user layer you select in a clockwise direction
Specifies the color to use for lines or borders and for fill areas		——————Line color ——————Fill color

Selecting **Glyphs** displays a library from which you can select glyphs to include in the icon. All .png, .bmp, and .jpg files in the Glyphs directory of the LabVIEW Data folder are displayed. You can synchronize the glyphs available on this page with the most recent glyphs in the **Icon Library** at ni.com by selecting Tools≫Synchronize with ni.com Icon Library.

*In the **Icon Editor** dialog box, select **Edit≫Import Glyph from File** to display a file dialog box where you can locate and import a graphic file to use as an icon or template.*

You can use layers in the design of the icon. Selecting the **Layers** page displays all layers of the icon. On that page you can display the preview, name, opacity, and visibility of the **Icon Text** layer and the **Icon Template** layer. You can change only the opacity and visibility of these two layers. On the **Layer** page you also find the preview, name, opacity, and visibility of all user layers. You can add layers, remove layers, and move up and down between layers.

Select Layers≫Show Layers Page *to display the* **Layers** *page if it is not showing.*

When you have finished editing, click on the **OK** button at the bottom right of the **Icon Editor** to save your edited icon and return to the front panel window, or click on the **Cancel** button to return to the front panel window without saving any changes.

The **Show Terminals** *option (accessed in* Edit≫Show Terminals *in the* **Icon Editor** *pull-down menu) can be used to display the terminal pattern of the connector overlaid on the icon.*

As an exercise, open Circle Measurements.vi and edit the icon. Remember that you developed the Circle Measurements.vi at the beginning of this chapter—it should be saved in Users Stuff. In case you did not save it before or cannot find it now, you can open the VI located in the **Chapter 4** folder in the **Learning** directory. An example of an edited icon is shown in Figure 4.5. Can you replicate the new icon? Check out the Circle Measurements Edited.vi to see the edited icon shown in Figure 4.5.

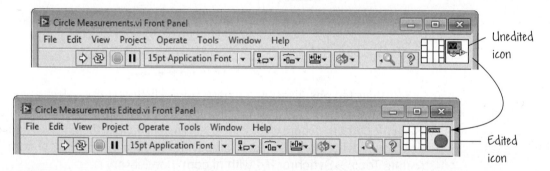

FIGURE 4.5
Editing the Circle Measurements VI icon.

4.3.2 Connectors

The connector is a set of terminals that correspond to the VI controls and indicators. This is where the inputs and outputs of the VI are established so that the VI can be used as a subVI. A connector receives data at its input terminals and passes the data to its output terminals when the VI is finished executing. Each terminal corresponds to a particular control or indicator on the front panel.

The connector terminals act like parameters in a subroutine parameter list of a function call.

If you use the front panel controls and indicators to pass data to and from subVIs, these controls or indicators need terminals on the connector pane. In this section, we will discuss how to define connections by choosing the number of terminals you want for the VI and assigning a front panel control or indicator to each of those terminals.

You can view and edit the connector pane from the front panel only.

The connector pane appears by default in the upper right hand corner of the front panel window just to left of the icon, as shown in Figure 4.6. By default, LabVIEW displays a terminal pattern shown in Figure 4.6. Control terminals are on the left side of the connector pane, and indicator terminals are on the right. If desired, you can select a different terminal pattern for your VI. In the early stage of creating a new VI or subVI, if you think you may make changes that would require additional inputs or outputs, you should consider keeping the default connector pane pattern shown in Figure 4.6, leaving the extra terminals unassigned. This will not adversely affect the function of the code but will provide flexibility so you can make changes without breaking existing wire connections.

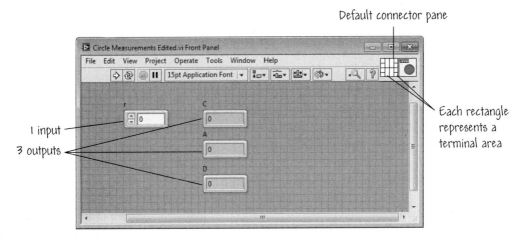

FIGURE 4.6
Defining a connector.

4.3.3   Selecting and Modifying Terminal Patterns

To select a different terminal pattern for your VI, right-click on the connector and choose **Patterns**. This process is illustrated in Figure 4.7. To change the pattern, click on the desired pattern on the palette. If you choose a new pattern,

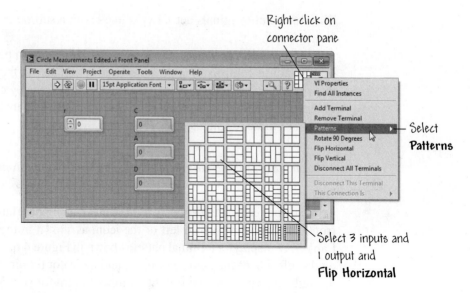

FIGURE 4.7
Changing your terminal pattern.

you will lose any assignment of controls and indicators to the terminals on the old connector pane.

The maximum number of terminals available for a subVI is 28.

If you want to change the spatial arrangement of the connector terminal patterns, choose one of the following commands from the connector pane menu: **Flip Horizontal**, **Flip Vertical**, or **Rotate 90 Degrees**. If you want to add a terminal to the connector pattern, place the cursor where the terminal is to be added, right-click on the connector pane window, and select **Add Terminal**. If you want to remove an existing terminal from the pattern, right-click on the terminal and select **Remove Terminal**.

Think ahead and plan your connector patterns well. For example, select a connector pane pattern with extra terminals if you think that you will add additional inputs or outputs at a later time. With these extra terminals, you do not have to change the connector pane for your VI if you find you want to add another input or output. This flexibility enables you to make subVI changes with minimal effect on your hierarchical structure. Another useful hint is that if you create a group of subVIs that are often used together, consider assigning the subVIs a consistent connector pane with common inputs. You then can easily remember each input location without using the **Context Help** window—this will save you time. If you create a subVI that produces an output that is used as the input to another subVI, try to align the input and output connections. This technique simplifies your wiring patterns and makes debugging and program maintenance easier.

> *Place inputs on the left and outputs on the right of the connector when linking controls and indicators—this prevents complex wiring patterns.*

4.3.4   Assigning Terminals to Controls and Indicators

Front panel controls and indicators are assigned to the connector Terminals using the **Wiring** tool. The following steps are used to associate the connector pane with the front panel controls and indicators.

1. Click on the connector terminal with the **Wiring** tool. The terminal turns black, as illustrated in Figure 4.8(a).

2. Click on the front panel control or indicator that you want to assign to the selected terminal. As shown in Figure 4.8(b), a dotted-line **marquee** frames the selected control.

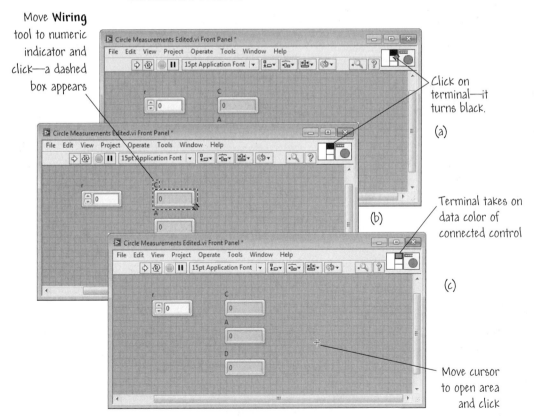

FIGURE 4.8
Assigning connector terminals to controls and indicators.

3. Position the cursor in an open area of the front panel and click. The marquee disappears, and the selected terminal takes on the data color of the connected object, indicating that the terminal is assigned. This process is illustrated in Figure 4.8(c).

The connector terminal turns white to indicate that a connection was not made. If this occurs, you need to repeat steps 1 through 3 until the connector terminal takes on the proper data color.

4. Repeat steps 1–3 for each control and indicator you want to connect.

As already discussed, you can choose a pattern with more terminals than you need since unassigned terminals do not affect the operation of the VI. It is also true that you can have more front panel controls or indicators than termi-

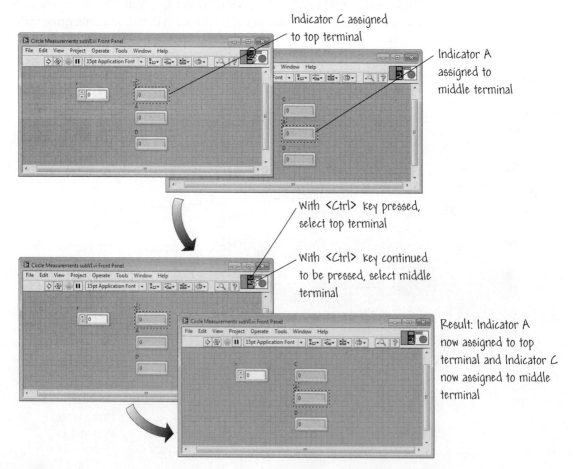

FIGURE 4.9
Switching terminals on the connector pane.

nals. Once the terminals have been connected to controls and indicators, you can disconnect them all at one time by selecting **Disconnect All Terminals** in the connector pane shortcut menu. Note that although the **Wiring** tool is used to assign terminals on the connector to front panel controls and indicators, no wires are drawn.

It is also possible to readily switch the location of connector pane terminals without disconnecting and rewiring the terminals. To switch two connector pane terminals, press and hold the <Ctrl> key and use the **Positioning** tool to select the two terminals you want to switch. In Figure 4.9, the numeric indicator C is assigned to the top terminal and the numeric indicator A is assigned to the middle terminal. We can readily switch the top connector pane terminal with the middle terminal to switch indicator A to the top terminal and indicator C to the middle terminal. To accomplish this, using the **Positioning** tool, the top connector pane terminal is selected (while holding the <Ctrl> key), then the middle connector terminal is selected (while continuing to hold the <Ctrl> key).

On Mac OS, press and hold the <Option> key.

4.4   THE HELP WINDOW

You enable the **Context Help** window by selecting Help≫Show Context Help. When you do this, you find that whenever you move an editing tool across a subVI node, the **Context Help** window displays the subVI icon with wires attached to each terminal. An example is shown in Figure 4.10 in which the cursor was moved over the subVI Temp & Vol.vi and the **Context Help** window appeared and showed that the subVI has two outputs: Temp and Volume.

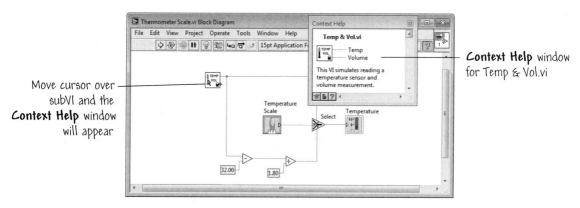

FIGURE 4.10
The **Context Help** window.

LabVIEW has a help feature that can keep you from forgetting to wire subVI connections—indications of required, recommended, and optional connections in the connector pane and the same indications in the **Context Help** window. For example, by classifying an input as **Required**, you can automatically detect whether you have wired the input correctly and prevent your VI from running if you have not wired correctly. To view or set connections as **Required, Recommended**, or **Optional**, right-click a terminal in the connector pane on the VI front panel and select **This Connection Is**. A check mark indicates its status, as shown in Figure 4.11. By default, inputs and outputs of VIs you create are set to **Recommended**—if a change is desired, you must change the default to either **Required** or **Optional**.

When you make a connection **Required**, then you cannot run the VI as a subVI unless that connection is wired correctly. In the **Context Help** window, **Required** connections appear in bold text. When you make a connection **Recommended**, then the VI can run even if the connection is not wired, but the error list window will list a warning. In the **Context Help** window, **Recommended** connections appear in plain text. A VI with **Optional** connections can run without the optional terminals being wired. In the **Context Help** window, **Optional** connections are grayed text for detailed diagram help and hidden for simple diagram help.

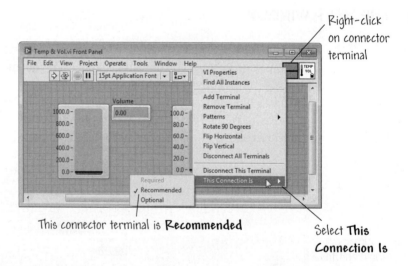

FIGURE 4.11
Showing the status of terminal connectors.

Building a SubVI

A VI designed to compute the radius, circumference, and area of a circle was presented in Section 4.2. In this exercise you will assign the connector terminals of that VI to the numeric controls and numeric indicator so that the VI can be used as a subVI by other programs. Begin by opening **Circle Measurements.vi** in the **Chapter 4** folder of the **Learning** directory. Recall that the difference between the two VIs Circle Measurements Edited and Circle Measurements is that the edited version has a custom icon rather than the default icon. A subVI with an edited icon (rather than the default icon) enables easier visual determination of the purpose of the subVI because you can inspect the icon as it exists in the code.

When you have completed assigning the connector terminals, the front panel connector pane should resemble the one shown in Figure 4.12. Right-click on the default connector pane and choose **Patterns**. Select a connector with 3 inputs and 1 output. Right-click on the connector pane again and choose **Flip Horizontal**. The resulting connector looks like the connector in Figure 4.12.

Using the **Wiring** tool, assign the input terminal to the numeric control r. Similarly, assign the three output terminals to the numeric indicators C, A, and D.

There are two levels of code documentation that you can pursue. It is important to document your code to make it accessible and understandable to other users. You can document a VI by choosing File≫VI Properties≫Documentation. A **VI description** box will appear as illustrated in Figure 4.13. You type a description of the VI in the dialog box, and then whenever **Show Context Help** is selected and the cursor is placed over the VI icon, your description will appear in the **Context Help** window.

You can also document the objects on the front panel by right-clicking on an object and choosing **Description and Tip** from the object shortcut menu. Type the object description in the dialog box that appears as in Figure 4.14. You cannot edit the object description in Run Mode—change to Edit Mode.

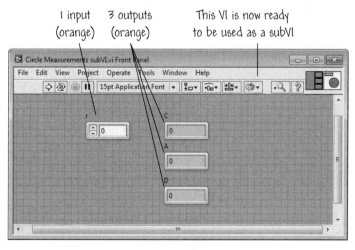

FIGURE 4.12
Assigning terminals for the Circle Measurements VI.

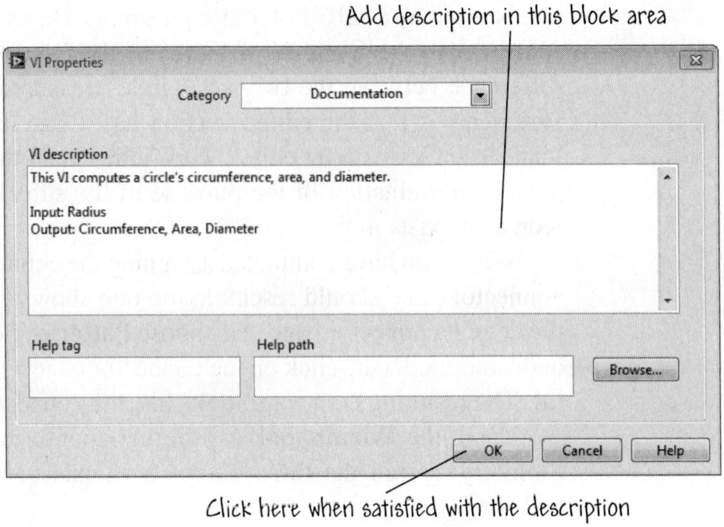

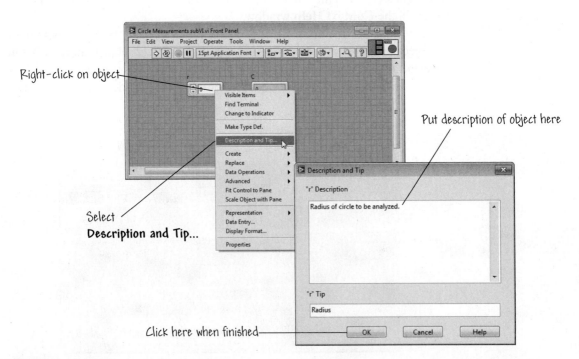

Add description in this block area

Click here when satisfied with the description

FIGURE 4.13
Typing the description for the subVI.

Right-click on object

Select
Description and Tip...

Put description of object here

Click here when finished

FIGURE 4.14
Documenting objects on a subVI.

Continuing with the exercise, you should add a description for each object on the front panel. Some suggested descriptions follow:

- *r*: Radius of circle to be analyzed.

- *C*: The circumference of a circle with radius equal to *r*.

- *A*: The area of a circle with radius equal to *r*.

- *D*: The diameter of a circle with radius equal to *r*.

When you have finished assigning the connector terminals to the controls and indicators, and you have documented the subVI and each of the controls and indicators, save the subVI as Circle Measurements subVI.vi. Make sure to save your work in the Users Stuff folder.

A working version of Circle Measurements subVI can be found in the **Chapter 4** *folder in the* **Learning** *directory. You may want to open the VI, read the documentation, and take a look at the icon and the connector just to verify that your construction of the same subVI was done correctly.* ◆

4.5   USING A VI AS A SUBVI

There are two basic ways to create and use a subVI: creating subVIs from VIs and creating subVIs from selections. In this section we concentrate on the first method, that is, using a VI as a subVI. Any VI that has an icon and a connector can be used as a subVI. In the block diagram, you can select VIs to use as subVIs from **Functions≫Select a VI** palette. Choosing this option produces a file dialog box from which you can select any available VI in the system, as shown in Figure 4.15.

LabVIEW comes with many ready-to-use VIs. In Chapter 1 you searched around the LabVIEW examples and found many of the example and demonstration VIs. In the following example you will use one of the preexisting VIs called Temp & Vol.vi as a subVI.

Using a VI as a SubVI

1. Open a new front panel.

2. Select a vertical slide switch control from the **Controls≫Modern≫ Boolean** palette and label it Temperature Scale. Place free labels on the vertical slide switch to indicate Fahrenheit and Celsius using the **Labeling** tool, as shown in Figure 4.16, and arrange the labels as shown in the figure.

3. Select a thermometer from the **Controls≫Silver≫Numeric** palette and place it on the front panel. Label the thermometer Temperature.

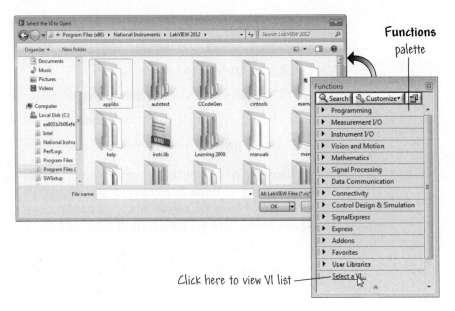

FIGURE 4.15
The **Select a VI** palette.

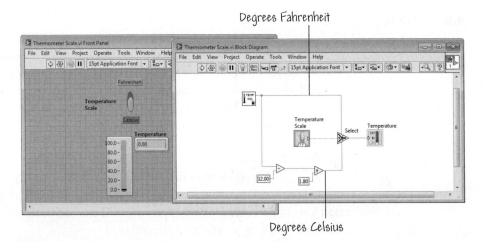

FIGURE 4.16
Calling the subVI Temp & Vol.vi.

4. Change the range of the thermometer to accommodate values ranging bet-ween 0.0 and 100.0. With the **Operating** tool, double-click on the high limit and change it to 100.0 if necessary.

5. Switch to the block diagram by selecting Window≫Show Block Diagram.

6. Right-click in a free area of the block diagram and choose **Functions≫ Select a VI** to access the dialog box. Select Temp & Vol.vi in the Chapter 4

folder of the **Learning** directory. Click **Open** in the dialog box to place **Temp & Vol.vi** on the block diagram.

7. Add the other objects to the block diagram using Figure 4.16 as a guide.

   • Place a Subtract function and a Divide function on the block diagram. These are located in the **Functions≫Programming≫Numeric** palette. On the Subtract function, add a numeric constant equal to 32. You can add the constant by right-clicking on an input of the Subtract function and selecting **Create Constant**. Then using the **Labeling** tool change the constant from the default 0.0 to 32.0. Similarly, add a numeric constant of 1.8 on the Divide function. These constants are used to convert from degrees Fahrenheit to Celsius according to the relationship

   $$°C = \frac{°F - 32}{1.8}.$$

   • Add the Select function (located on the **Programming≫Comparisons** subpalette of the **Functions** palette). The Select function returns the value wired to the True or False input, depending on the Boolean input value. Use **Show Context Help** for more information on how this function works.

8. Wire the diagram objects as shown in Figure 4.16.

9. Switch to the front panel and click the **Run Continuously** button in the toolbar. The thermometer shows the value in degrees Fahrenheit or degrees Celsius, depending on your selection.

10. Switch the scale back and forth to select either Fahrenheit or Celsius.

11. When you are finished experimenting with your VI, save it as **Thermometer Scale.vi** in the **Users Stuff** folder.   ◆

A working version of **Thermometer Scale.vi** *exists in the* **Chapter 4** *folder located in the* **Learning** *directory. When you open this VI, notice the use of color on the vertical switch (orange denotes Fahrenheit and blue represents Celsius).*

In the previous exercise, the stand-alone **Temp & Vol.vi** was used in the role of a subVI. Suppose that you have a subVI on your block diagram and you want to examine its contents—to view the code. You can easily open a subVI front panel window by double-clicking on the subVI icon. Once the front panel opens, you can switch to the subVI block diagram. At that point, any changes you make to the subVI code alter only the version in memory—until you save the subVI. Also note that, even before saving the subVI, the changes affect all calls to the subVI and not just the node you used to open the VI.

4.6   CREATING A SUBVI FROM A SELECTION

The second way to create a subVI is to select components of the main VI and group them into a subVI. You capture and group related parts of VIs by selecting the desired section of the VI with the **Positioning** tool and then choosing **Create SubVI** from the **Edit** pull-down menu. The selection is automatically converted into a subVI, and a default icon replaces the entire section of code. The controls and indicators for the new subVI are automatically created and wired to the existing wires. Using this method of creating subVIs allows you to modularize your block diagram, thereby creating a hierarchical structure.

Building a SubVI Using the Selection Technique

In this exercise you will modify the Thermometer Scale VI developed in Section 4.5 to create a subVI that converts Fahrenheit temperature to Celsius temperature. The subVI selection process is illustrated in Figure 4.17.

Open the Thermometer Scale VI by selecting **Open** from the **File** menu. The VI is located in the **Chapter 4** folder of the **Learning** directory. For reference, the front panel and block diagram are shown in Figure 4.16.

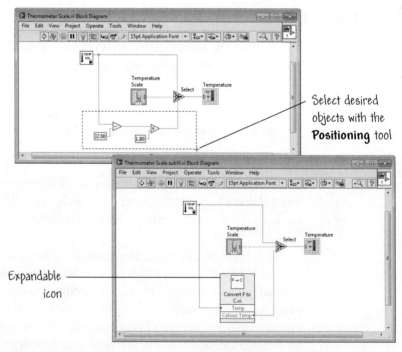

FIGURE 4.17
Selecting the code that converts from Fahrenheit to Celsius.

To create a subVI that converts Fahrenheit to Celsius, begin by switching to the block diagram window. The goal is to modify the existing block diagram to call a subVI created using the **Create SubVI** option.

Select the block diagram elements that comprise the conversion from Fahrenheit to Celsius, as shown in Figure 4.17. A moving dashed line will frame the chosen portion of the block diagram. Now select **Create SubVI** in the **Edit** menu. A default subVI icon will appear in place of the selected group of objects. You can use this selection method of creating subVIs to modularize your VI.

The next step is to modify the icon of the new subVI. Open the new subVI by double-clicking on the default icon called Untitled 1 (SubVI). Two front panel objects should be visible: one labeled Temp and one labeled Numeric Control. Using the **Labeling** tool, change the label from Numeric Control to Temp Celsius.

Use the **Icon Editor** to create an icon similar to the one shown in Figure 4.18. Invoke the editor by right-clicking in the **Icon Pane** in the front panel of the subVI and selecting **Edit Icon** from the menu. Erase the default icon by double-clicking on the **Select** tool and pressing <Delete>. Redraw the icon frame by double-clicking on the rectangle tool which draws a rectangle around the icon. The easiest way to create the text is using the **Text** tool and the font Small Fonts. The arrow can be created with the **Pencil** tool.

The connector is automatically wired when a subVI is created using the ***Create SubVI*** *option.*

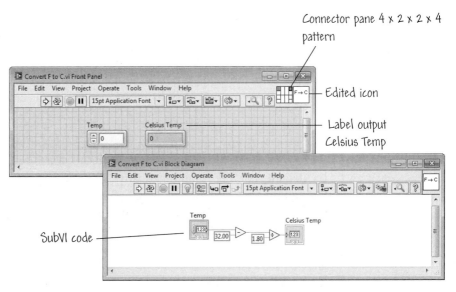

FIGURE 4.18
The new subVI to convert degrees Fahrenheit to degrees Celsius.

When creating a subVI from a block diagram selection in earlier versions of LabVIEW, the connector pane and front panel of the new subVI had to be cleaned up manually. Now, the connector pane is automatically created with a $4 \times 2 \times 2 \times 4$ connector pane pattern unless the subVI requires more terminals. The controls are automatically aligned on the left side and the indicators aligned on the right side of the front panel.

When you are finished editing the icon, close the **Icon Editor** by clicking on **OK**. The new icon appears in the upper right corner of the front panel, as shown in Figure 4.18. Right-click on the Convert F to C.vi and click on **View as Icon** to remove the checkmark. Then use the **Positioning** tool to expand the icon to show the input Temp and the output Celsius Temp, as shown in Figure 4.17. Save the subVI by choosing **Save** from the **File** menu. Name the VI Convert F to C and save it in the Users Stuff folder.

You cannot build a subVI from a section of code with more than 28 inputs and outputs because 28 is the maximum number of inputs and outputs allowed on a connector pane. ◆

4.7   ERROR CHECKING AND ERROR HANDLING

Error checking is a key element of a VI or subVI. If an error situation arises, it is important to know why and where the error occurred. Also it is often important to programmatically respond to errors. If an error occurs, you may want to avoid suspending the execution of the VI and instead to respond programmatically by taking another course of action within the code. You might, for example, pause the execution and wait for a user input. When you perform any kind of input or output (I/O) you should be ready to handle errors that might occur. Most I/O functions return errors and supply mechanisms to handle them appropriately.

There are three ways to tackle error checking and error handling: automatic error handling, manual error handling, and error clusters.

4.7.1   Automatic Error Handling

By default, LabVIEW automatically handles errors by suspending execution, highlighting the subVI or function where the error occurred, and displaying an error dialog box. Each error has a numeric code and a corresponding error message, supplying the information needed to investigate the source of the error. You can disable automatic error handling for the current VI, for all new VIs, and for subVIs. To disable it for the *current* VI, select File≫VI Properties. In the **VI Properties** dialog box, select **Execution** from the **Category** pull-down menu and deselect **Enable automatic error handling**. To disable automatic error handling for any *new* VIs you create, select Tools≫Options. In the

Options dialog box, select **Block Diagram** from the **Category** list and deselect **Enable automatic error handling in new VIs**. To disable automatic error handling for a subVI or function within a VI, wire its error out parameter to the error in parameter of another subVI or function or to an error out indicator.

4.7.2   Manual Error Handling

You can formulate error-handling decisions to manage errors on the block diagram using the error-handling VIs and functions on the **Dialog & User Interface** subpalette (found on the **Programming** palette) in conjunction with the error in and error out parameters of most VIs and functions and the debugging tools discussed in Chapter 3. For example, placing the Simple Error Handler.vi from the **Dialog & User Interface** palette on the block diagram provides a straightforward means to obtain a description of any error and optionally display a dialog box. You can then use the debugging tools, such as Single Stepping and Execution Highlighting, to delve deeper into the areas of the code with the error. Functions generally return errors with numeric error codes, and VIs return errors with an error cluster.

It is strongly recommended that you use some type of error handling in your VIs.

4.7.3   Error Clusters

Error-cluster controls and indicators are used to create error inputs and outputs in subVIs. The error in and error out clusters provide the following information:

- status is a Boolean value that reports TRUE if an error occurred.
- code is a 32-bit signed integer that identifies the error numerically. A non-zero error code coupled with a status of FALSE signals a warning rather than an error.
- source is a string that identifies where the error occurred.

To handle errors using logical operations, you can wire an error cluster directly to the Boolean functions such as And, Or, Exclusive Or, Implies, Not, Not And, Not Exclusive Or, and Not Or. If an error occurs, the error cluster passes a TRUE value to the Boolean function.

Consider the VI in Figure 4.19, which generates and plots one of the following signals: sine wave, sawtooth, square wave, and triangle wave. It also allows the user to vary the signal amplitude and frequency. On the block diagram in Figure 4.19, the error-cluster indicator on the front panel provides the above error information to the user and terminates execution of the VI. When an error occurs, right-click within the cluster border and select **Explain Error** from

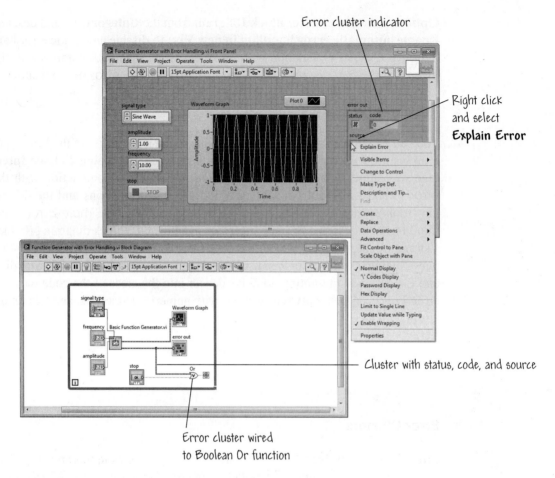

FIGURE 4.19
Using error clusters to assist with error checking and error handling.

the shortcut menu to open the **Explain Error** dialog box. The shortcut menu includes an **Explain Warning** option if the VI contains warnings but no errors.

The VI in Figure 4.19 can be found in **Chapter 4** *of the* **Learning** *directory. It is named* **Function Generation with Error Handling.vi**.

4.8   SAVING YOUR SUBVI

It is highly recommended that you save your subVIs to a file in a directory rather than in a library. While it is possible to save multiple VIs in a single file called a **VI library**, this is not desirable. Saving VIs as individual files is the most effective storage path because you can copy, rename, and delete files more easily than when using a VI library.

VI libraries have the same load, save, and open capabilities as other directories, but they are not hierarchical. That is, you cannot create a VI library inside of another VI library, nor can you create a new directory inside a VI library. After you create a VI library, it appears in the **File** dialog box as a file with an icon that is somewhat different from a VI icon.

4.9   THE VI HIERARCHY WINDOW

When you create an application, you generally start at the top-level VI and define the inputs and outputs for the application. Then you construct the subVIs that you will need to perform the necessary operations on the data as it flows through the block diagram. As discussed in previous sections, if you have a very complex block diagram, you should organize related functions and nodes into subVIs for desired block diagram simplicity. Taking a modular approach to program development creates code that is easier to understand, debug, and maintain.

The **VI Hierarchy** window displays a graphical representation of the hierarchical structure of all VIs in memory and shows the dependencies of top-level VIs and subVIs. There are several ways to access the **VI Hierarchy** window:

- You can select View≫VI Hierarchy to open the **Hierarchy Window** with the VI icon of the current **active window** surrounded by a thick red border.

- You can right-click on a subVI and select **Show VI Hierarchy** to open the **VI Hierarchy** window with the selected subVI surrounded by a thick red border.

- If the **VI Hierarchy** window is already open, you can bring it to the front by selecting it from the list of open windows under the **Window** menu.

The **VI Hierarchy** window for the Thermometer Scale subVI.vi is shown in Figure 4.20. The window displays the dependencies of VIs by providing information on VI callers and subVIs. As you move the **Operating** tool over objects in the window, the name of the VI is shown below the VI icon. This window also contains a toolbar, as shown in Figure 4.20, that you can use to configure several types of settings for displayed items.

You can switch the **VI Hierarchy** window display mode between horizontal and vertical display by pressing either the **Horizontal Layout** or **Vertical Layout** button in the hierarchy toolbar. In a horizontal display, subVIs are shown to the right of their calling VIs; in a vertical display, they are shown below their calling VIs. In either case, the subVIs are always connected with lines to their calling VIs. The window shown in Figure 4.20 is displayed vertically.

Arrow buttons and arrows beside nodes indicate what is displayed and what is hidden according to the following rules:

- A red arrow button pointing towards the node indicates some or all subVIs are hidden, and clicking the button will display the hidden subVIs.

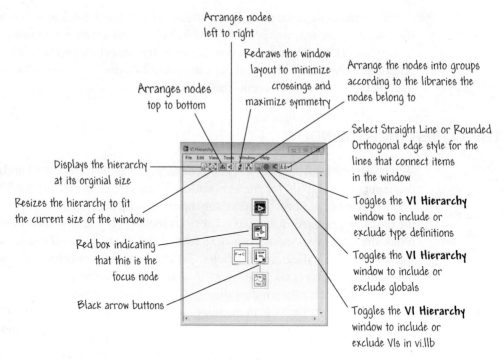

Arranges nodes
left to right

Redraws the window
layout to minimize
crossings and
maximize symmetry

Arrange the nodes into groups
according to the libraries the
nodes belong to

Arranges nodes
top to bottom

Select Straight Line or Rounded
Orthogonal edge style for the
lines that connect items
in the window

Displays the hierarchy
at its orginial size

Resizes the hierarchy to fit
the current size of the window

Toggles the **VI Hierarchy**
window to include or
exclude type definitions

Red box indicating
that this is the
focus node

Toggles the **VI Hierarchy**
window to include or
exclude globals

Black arrow buttons

Toggles the **VI Hierarchy**
window to include or
exclude VIs in vi.llb

FIGURE 4.20
The **VI Hierarchy** window for Thermometer Scale subVI.vi.

- A black arrow button pointing towards the subVIs of the node indicates all immediate subVIs are shown.

- A blue arrow pointing towards the callers of the node indicates the node has additional callers in this VI hierarchy but they are not shown at the present time. If you show all subVIs, the blue arrow will disappear. If a node has no subVIs, no red or black arrow buttons are shown.

Double-clicking on a VI or subVI opens the front panel of that node. You also can right-click on a VI or subVI node to access a menu with options, such as showing or hiding all subVIs, opening the VI or subVI front panel, editing the VI icon, and so on.

You can initiate a search for a given VI by simply typing the name of the node directly onto the **VI Hierarchy** window. When you begin to type in text from the keyboard a small search window will automatically appear displaying the text that has been typed and allowing you to continue adding text. The search for the desired VI commences immediately. The search window is illustrated in Figure 4.21. If the characters currently displayed in the search window do not match any node names in the search path, the system beeps, and no more characters can be typed. You can then use the <Backspace> or <Delete>

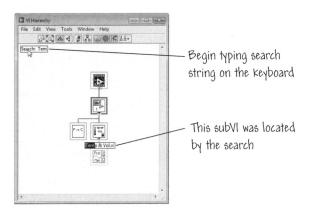

FIGURE 4.21
Searching the **VI Hierarchy** window for VIs.

key to delete one or more characters to resume typing. The search window disappears automatically if no keys are pressed for a certain amount of time, or you can press the <Esc> key to remove the search window immediately. When a match is found you can use the right or down arrow key, the <Enter> key on Windows, the <return> key on Macintosh to find the next node that matches the search string. To find the previous matching node, press the left or up arrow key, <Shift-Enter> on Windows, or <shift-return> on Macintosh.

BUILDING BLOCK

4.10   BUILDING BLOCKS: PULSE WIDTH MODULATION

The goal of this building block is to modify the VIs created in the previous Building Block sections to be utilized as subVIs. Open Rising Edge Edited.vi and Falling Edge Edited.vi. Both VIs should be saved in the Users Stuff folder in the Learning directory.

Working versions of both VIs can be found in the Building Blocks folder in the Learning directory.

It is very helpful to users of your subVIs if the icons visually convey the purpose of each VI. For this exercise, begin by creating an icon and a connector for the Rising Edge Edited VI and the Falling Edge Edited VI. Use Figure 4.22

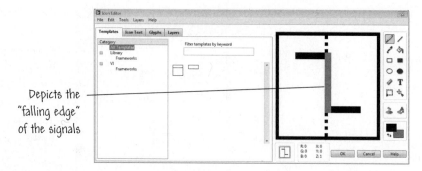

Depicts the
"falling edge"
of the signals

FIGURE 4.22
The edited icon for the Falling Edge Edited VI.

as a guide for a visual representation of the Falling Edge Edited VI. You can extend the idea to the Rising Edge Edited VI. When you are finished, save the VIs as **Rising Edge subVI.vi** and **Falling Edge subVI.vi** in the **Users Stuff** folder in the **Learning** directory.

Once the icon has been edited, it is time to assign the terminals to the inputs and outputs to establish the subVI. First, right-click on the icon and select **Show Connector**. Use Figure 4.23 as a guide to assign the three controls (one floating-point number and two integers) and the single output (a Boolean). Follow the procedure described in Section 4.3 to make the terminal assignments.

Once you have developed a subVI to perform a specific task, you will likely want to have it accessible from the **Functions** palette so that it is readily available. To accomplish this task for your new subVIs, you need to save them in the

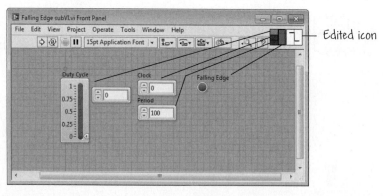

Edited icon

FIGURE 4.23
Connecting the Falling Edge Edited VI for use as a subVI.

folder User.lib. Go ahead and save the Falling Edge and Rising Edge subVIs in User.lib located in the LabVIEW directory.

After you have saved the subVIs in User.lib, you will need to exit LabVIEW and then open it again. Once this is done, open a new VI and switch to the block diagram. Navigate to **User Libraries** on the **Functions** palette and locate the icon for the Falling Edge subVI, as illustrated in Figure 4.24. The subVI is now readily accessible to you in your future programming endeavors.

*You can also access the subVIs through **Functions≫Select a VI** by navigating to Users Stuff where you previously saved the programs.*

When you are done experimenting with your new VI, save it as Falling Edge subVI.vi in the Users Stuff folder in the Learning directory. You will use this VI as a building block in later chapters—so make sure to save your work. Follow the same process and construct a subVI version of the Rising Edge Edited VI.

Working versions of the Falling Edge subVI and the Rising Edge subVI can be found in the Building Blocks folder of the Learning directory.

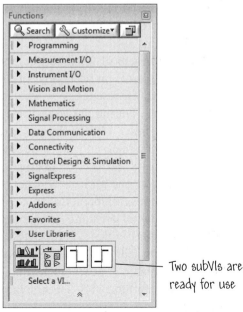

FIGURE 4.24
Finding subVIs in the User.lib folder.

4.11   RELAXED READING: MOBILE RESCUE OPERATIONS

In this reading we consider the problem of developing a system for fast detection of casualties in difficult–to-access areas, such as earthquake and collapsed-building sites. A unique LabVIEW application is used to control an unmanned aerial vehicle, acquire the images taken by its cameras, and process the images with NI vision hardware to detect casualties trapped in areas only viewable and accessible by air.

After a catastrophe such as an earthquake or a collapsed building, rescue teams are often working in or near inaccessible areas where people may be trapped and whose lives depend on rapid intervention. To meet the needs of the rescue teams, the mobile rescue system has two main subsystems, a mobile device — known as FIRST — to help the rescue teams locate casualties and determine their condition, and a system utilizing an unmanned aerial vehicle to fly over inaccessible areas to detect possible trapped casualties.

The FIRST mobile search device is carried by a rescuer to locate casualties or people that are trapped and monitor the dangerous conditions in the collapsed building or disaster site. The FIRST device, illustrated in Figure 4.25, is made of chemical sensors, heat cameras, visual cameras, highly sensitive microphones, and a backpack with a chemical processor and a CPU that processes the acquired signals and generates an alarm to the control center. LabVIEW is used to process the signals from the sensors to detect life signs in the inspected area, and generates alarms that inform the rescuer of these life signs and sends them to the control center.

A key element of the mobile rescue system is an unmanned aerial vehicle providing information accessible most readily by air. The aerial vehicle is an important tool for rescue teams when it comes to quickly locating casualties in disaster zones. During the last decade, the quadrotor has become the standard platform for micro aerial vehicles in research and robotic projects. A quadrotor is a small remote-controlled helicopter propelled by four rotors distributed in rotation pairs. Each rotor is driven by a battery-powered electric motor. A microcontroller and a range of sensors, including accelerometers, gyroscopes, and magnetometers, maintain the stability of the quadrotor. The aerial vehicle carries a series of onboard sensors for ground sensing, such as altimeters and video cameras.

Data recorded by the FIRST device and the aerial vehicle is downlinked to the control center. All communication between the equipment is carried out via WiFi. The communication between the FIRST device, the computer that controls the aerial vehicle, and the control center is at 5 GHz, while the communication between the computer and the aerial vehicle is at 2.4 GHz.

The LabVIEW interface and fast integration of the various acquisition components, as well as the ability to implement effective communication protocols,

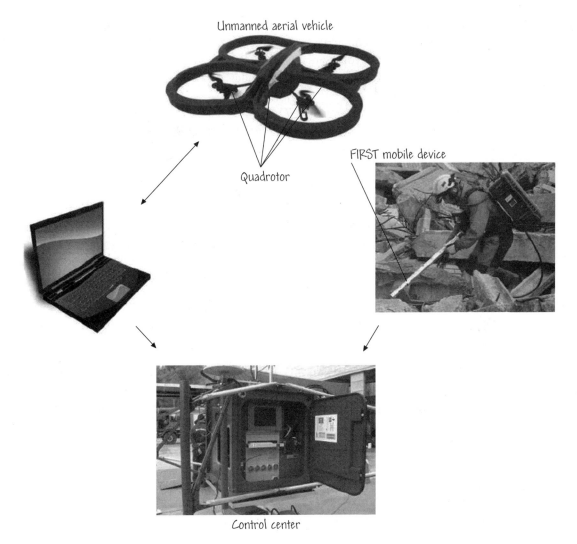

FIGURE 4.25
Communication between rescue devices.

enabled the innovative development of mobile recue systems at low cost. For further information, visit http://sine.ni.com/cs/app/doc/p/id/cs-15449.

4.12   myDAQ BUILDING BLOCKS

In this myDAQ Building Blocks exercise, you will modify the VIs you created in Chapter 3 to be used as subVIs. You will modify the LED Control.vi and the Read Temp.vi to be used as subVIs by wiring the inputs and outputs to the connector panes. Also, you will edit the icon to better illustrate the application of each subVI and add documentation appropriately. Begin by opening the **LED Control.vi** that you worked on in Chapter 3 and saved in the **Chapter 3** folder in the **myDAQ Building Blocks** folder within the **Learning** directory.

*Working versions of LED Control.vi and Read Temp.vi can be found in the **Chapter 3** folder of the **myDAQ Building Blocks**≫**Solutions** folder in the **Learning** directory.*

Open the two VIs and follow the steps below.

1. Wire the controls and indicators to the connector panes, as illustrated in Figure 4.26 and Figure 4.27 for **LED Control.vi** and **Read Temp.vi**, respectively. It is good practice to assign inputs on the left side of the connector pane and outputs on the right side.

2. Save the VIs as **LED Control subVI.vi** and **Read Temp subVI.vi**, respectively, in the **myDAQ Building Blocks**≫**Support VIs** folder in the **Learning** directory for easy access in later exercises.

3. Document each terminal with a one sentence description of the input or output. On the front panel, right-click on the control or indicator and select **Description and Tip** to edit the documentation. For the control labeled **LED Control** on the **LED Control subVI.vi**, use the text (see Figure 4.28(a))

    Boolean push button to change the state of the LED on the myDAQ LED circuit.

    For the indicator labeled **LED Status**, use the text

    Round LED to indicate the state of the LED on the front panel.

    For the indicator labeled **Resistance** on the **Temp Read subVI.vi**, use the text (see Figure 4.28(b))

    Numeric indicator to display the resistance of the thermistor as the ouput of the Volt to Resistance.vi.

    For the indicator labeled **Temperature (C)**, use the text

    Numeric indicator to display the temperature in deg C of the thermistor as the ouput of the Temp Conversion.vi.

4. For each of the two subVIs, with **Highlight Execution** selected on the front panel, hover the cursor over the controls and indicators and confirm that the description displayed in the **Context Help** window is as shown in Figure 4.28.

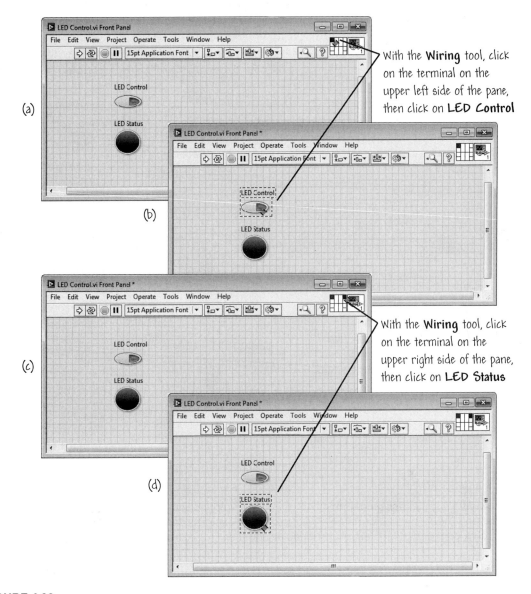

With the **Wiring** tool, click on the terminal on the upper left side of the pane, then click on **LED Control**

With the **Wiring** tool, click on the terminal on the upper right side of the pane, then click on **LED Status**

FIGURE 4.26
LED Control subVI.vi: (a) To assign a terminal to the input with the **Wiring** tool, click on the upper left terminal and then, (b) click on the LED Control; (c) to assign a terminal to the indicator with the **Wiring** tool, click on the upper right terminal connector and then, (d) click on the LED Status.

5. Document each subVI with a short description of its functionality as shown in Figure 4.28. To edit the description, select **File≫VI Properties** and choose **Documentation**. For the LED Control subVI.vi use the following documentation:

    This VI controls the LED connected to DIO0 of myDAQ1.

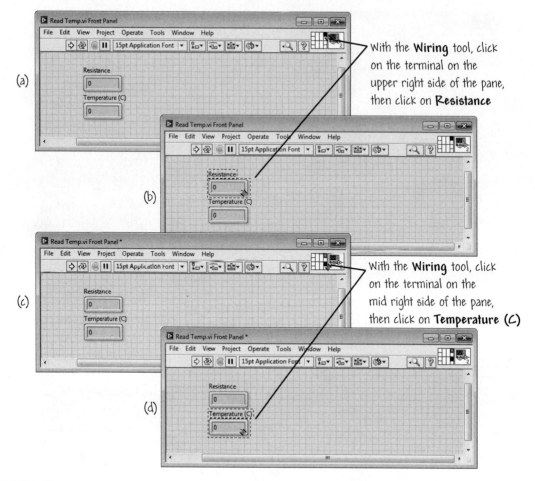

With the **Wiring** tool, click on the terminal on the upper right side of the pane, then click on **Resistance**

With the **Wiring** tool, click on the terminal on the mid right side of the pane, then click on **Temperature (C)**

FIGURE 4.27
Read Temp subVI.vi: (a) To assign a terminal to one of the outputs with the **Wiring** tool, click on the upper right terminal and then, (b) click on the Resistance; (c) to assign a terminal to the other output with the **Wiring** tool, click on the mid right terminal connector and then, (d) click on the Temperature (C).

For the Temp Read subVI.vi use the following documentation:

> *This VI reads from a thermistor in a voltage divider circuit connected to AI0 and AI1 of myDAQ1 and calculates resistance (Ohms) and temperature (deg C).*

6. Modify the subVI icon to graphically represent its functionality. Double-click on the icon to open the **Icon Editor**. Use Figure 4.28 as your guide.

*In the **Icon Editor**, select **Glyphs** to find many of the interesting elements of the icons shown in Figure 4.28.*

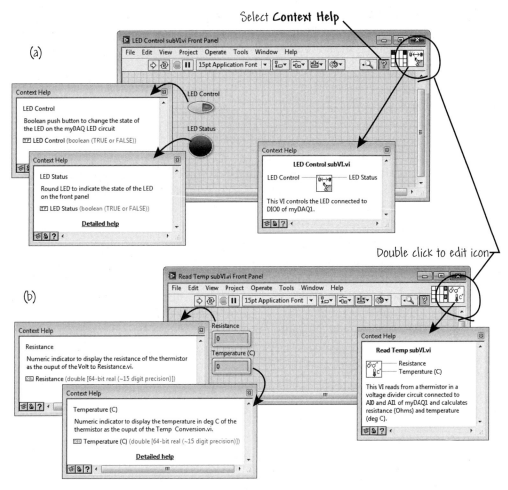

FIGURE 4.28
Using **Context Help** for the (a) LED Control subVI.vi, and (b) Read Temp subVI.vi.

The seven steps listed above provide only suggestions to wiring and documenting the subVIs. Figure 4.26 thru Figure 4.28 illustrate possible VI descriptions, wiring schemes, and icons. Your subVIs may be slightly different, but you should have the same inputs and outputs, even if they are connected to different terminals, and your icon may look different than proposed here.

Working versions called LED Control subVI.vi *and* Read Temp subVI.vi *can be found in the* Chapter 4 *folder of the* myDAQ Building Blocks≫ Solutions *folder in the* Learning *directory.*

In this myDAQ Building Blocks exercise, two important subVIs, the LED Control subVI.vi and the Read Temp subVI.vi, were created for use in subsequent exercises. You gained experience wiring inputs and outputs to the connector pane, adding documentation that is accessible through **Context Help**, and editing the subVI icon.

4.13   SUMMARY

Constructing subVIs was the main topic of this chapter. One of the keys to constructing successful LabVIEW programs is understanding how to build and use subVIs. SubVIs are the primary building blocks of modular programs, that are easy to debug, understand, and maintain. They are analogous to subroutines in text-based programming languages like C or Fortran. Two methods of creating subVIs were discussed—creating subVIs from VIs and creating subVIs from VI selections. A subVI is represented by an icon in the block diagram of a calling VI and must have a connector with terminals properly wired to pass data to and from the calling VI. Automatic error handling, manual error handling, and error clusters were presented as three mechanisms to tackle the important tasks of VI and subVI error checking and error handling. The **Icon Editor** was discussed as the way to personalize the subVI icon so that information about the function of the subVI is readily apparent from visual inspection of the icon. The editor has a tool set similar to that of most common paint programs. The **VI Hierarchy** window was introduced as a helpful tool for managing the hierarchical nature of your programs.

KEY TERMS

Active window: Window that is currently set to accept user input, usually the frontmost window. You make a window active by clicking on it or by selecting it from the **Window** menu.

Automatic error handling: Default error handling in which LabVIEW automatically suspends execution, highlights the subVI or function where the error occurred, and displays an error dialog box.

Control: Front panel object for entering data to a VI interactively or to a subVI programmatically.

Description box: Online documentation for VIs.

Dialog box: An interactive screen with prompts in which you describe additional information needed to complete a command.

Error checking: A process to tell you why and where errors occur.

Error cluster: A convention for managing error inputs and outputs that enable error status to passed between subVIs and displayed on the front panel.

Function: A built-in execution element, comparable to an operator or statement in a conventional programming language.

Icon Editor: Interface similar to that of a paint program for creating VI icons.

Indicator: Front panel object that displays output.

Manual error handling: Error handling in which decisions are made to manage errors on the block diagram using the error-handling VIs and functions on the **Dialog & User Interface** palette.

Marquee: Moving, dashed border surrounding selected objects.

Pseudocode: Simplified, language-independent representation of programming code.

SubVI: A VI used in the block diagram of another VI—comparable to a subroutine.

Top-level VI: The VI at the top of the VI hierarchy. This term distinguishes a VI from its subVIs.

VI Hierarchy window: Window that graphically displays the hierarchy of VIs and subVIs.

VI library: Special file that contains a collection of related VIs.

EXERCISES

E4.1 Open a new VI and switch to the block diagram window. Select Write to Text File.vi from the **Functions≫Programming≫File I/O** palette and place it on the block diagram. Notice that the **Run** button breaks when this subVI is placed on the block diagram. Click the **Broken Run** button to access the **Error List**. The one listed error states that the VI has a required input. As you know from the chapter, each input terminal can be categorized as required, recommended, or optional. You must wire values to required inputs before you can run a VI. Recommended and optional inputs do not prevent the VI from running and use the default value for that input.

Select the error from the **Error List** dialog box by clicking on it and then choose **Show Error**. Notice that the Write to Text File.vi icon is highlighted by a marquee around the icon. Open **Context Help** (Help≫Show Context Help or <Ctrl-H>) to read about the VI, paying special attention to the inputs listed as **Required** (shown in bold). You should see that text is a required input. Right-click on the center terminal on the right side of the icon and select **Create Control**. Notice that once the string control is attached to the icon, the error disappears and you can run the VI.

Navigate to the **Functions≫Mathematics≫Integ & Diff** palette and select the Numeric Integration VI and place it on the block diagram. Notice that the **Run** button does not break. Use **Context Help** again and notice that none of the inputs to this VI are **Required**. You will be able to run the VI without any inputs wired to it.

E4.2   Open a new VI and switch to the block diagram window. Navigate to the **Functions≫Programming≫Timing** palette, select the Wait Until Next ms Multiple.vi and place it on the block diagram. Determine the errors that cause the **Broken Run** to appear. Determine a fix to the problem so that the VI can run properly.

E4.3   Open Slope.vi, which calculates the slope between two points, (X1, X2) and (Y1, Y2). The Slope.vi can be found in the Exercises&Problems folder in Chapter 4 of the Learning directory. The **Run** button is broken; hence, the VI is not executable. Click the **Run** button to access the **Error List** and you will find that the error is found within the subVI. You must open the subVI by double-clicking on it and correct the errors within it before the main VI will run. From within the SlopeSub.vi, go to the block diagram and wire the output of the Subtract terminal to the input of the Divide terminal. Save the subVI, return to the main VI, and notice that the **Run** button is not broken. Enter values into the four numeric controls in Slope.vi and run the VI to vary the slope between two points. Stop the VI.

Leave the SlopeSub.vi open and go to File≫New VI to open a new VI. Go to the block diagram of the new VI to practice the various ways to insert subVIs into a main VI. First, click on the icon of the SlopeSub.vi and drag it onto the block diagram of the new VI. Place the **Wiring** tool over the icon and see that all the input and output terminals are present. Second, from the block diagram, go to **Functions≫Select a VI** and browse to the Exercises&Problems folder in Chapter 4 of the Learning directory. Click the **Open** button and place the icon of the SlopeSub.vi on the block diagram. Remember that both methods of placing subVIs on block diagrams generate the same result.

E4.4   Open a new VI and go to Help≫Find Examples.... Click on the **Browse** tab and press the **Directory Structure** button. In the center panel, double-click on the express folder. This will locate examples that predominantly use Express VIs to accomplish various tasks. Open the Lissajous2.vi by double-clicking on it and run the VI changing the controls on the front panel to alter the curve. Now go to the block diagram and see that there are many Express VIs needed to perform this program.

From the block diagram, go to View≫VI Hierarchy. Use the tools you learned in this chapter to become familiar with the **VI Hierarchy Window** and explore the various subVIs used in Lissajous2.vi.

E4.5 Create a VI called Sign.vi which accepts a numeric input. The VI should have a Boolean output that returns True if the input is positive, and one that returns True if the input is negative. Make an icon and configure the connector pane so that the Sign VI can be used as a subVI with another VI. Use the icon in Figure E4.5 as a guide. The connector terminals should be assigned one float input and two Boolean outputs. Once the VI terminals have been assigned, add documentation so that the **Context Help** will display the following statement: "Tests if an input is positive or negative." Go to File≫VI Properties and then go to **Documentation** to enter the information.

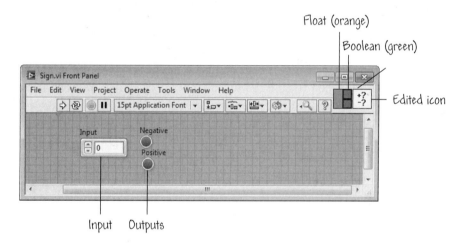

FIGURE E4.5
The front panel for the Sign VI illustrating the edited icon and the connector terminals.

E4.6 In this exercise you will modify a VI so that it can be employed as a subVI. Open the Find Greatest VI located in the Exercises&Problems folder in Chapter 4 of the Learning directory. Notice that the connector pane does not have the right number of inputs and outputs and the terminals have not been correctly linked to the connector. To fix this, take the following steps:

(a) Right-click on the connector pane and select **Disconnect All Terminals**.

(b) Right-click on the connector pane once again. On the **Patterns** menu, select the connector configuration with three terminals on the left and one terminal on the right.

(c) Assign the three controls to the terminals on the left. Assign the indicator to the terminal on the right.

(d) Open the **Icon Editor** by double-clicking on the icon and create an icon for the VI.

(e) Verify that you have set up the connector pane correctly by opening a new VI and placing Find Greatest within it as a subVI. Create controls and an indicator, and run the program several times with different inputs. The output should always be equal to the greatest value input to the subVI.

E4.7 An effective icon can convey information about the functionality of a subVI at a glance. The NI website has an icon art glossary containing a wide array of graphics which you may find useful to quickly design professional looking icons. Visit the Icon Art Gallery at www.ni.com/white-paper/6453/en/ and familiarize yourself with the available graphics. Follow the link titled **Guidelines for Creating Icons** to learn more about basic rules of thumb to keep in mind while creating your own icons.

The Calculator VI is a simple VI which performs four different operations on two numeric inputs. Open Calculator.vi found in the Exercises&Problems folder in Chapter 4 of the Learning directory. Right-click on the icon in the upper right hand corner of the front panel and select **Edit Icon** to create a new icon for this VI that is more descriptive. You can use the glyph library on the **Icon Editor** or create an icon by hand. Now right-click on the icon and select **Show Connector** to assign terminals to the controls and indicators.

Verify that you have set up the connector pane correctly by opening a new VI and placing Calculator within it as a subVI. Create controls and an indicator, and run the program several times with different inputs.

PROBLEMS

P4.1 Construct a VI that computes the average of three numbers input by the user. One computation in your program should be summing the three input numbers followed by a division by 3. The resulting average should be displayed on the front panel. Also, add a piece of code that multiplies the computed average by a random number in the range $[0, \ldots, 1]$. Create a subVI by grouping the parts of the code that compute the average. Remember to edit the icon so that it represents the function of the subVI, namely, the average of three numbers. Figure P4.1 can be used as a guide.

P4.2 Open Check Limit.vi, which generates a sine wave and plots it on a waveform graph. Check Limit.vi can be found in the Exercises&Problems folder in Chapter 4 of the Learning directory. The values of the sine wave are compared

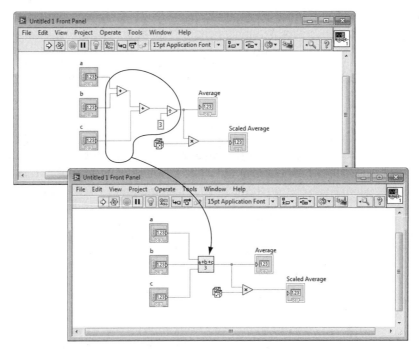

FIGURE P4.1
Computing the average of three numbers using a subVI to compute the average.

with a numeric control, therefore evaluating whether the sine wave exceeds the set limit. It also uses a While Loop that we will learn about it in the next chapter.

Edit the connector icon so that it only has the necessary number of terminals and connect them to the appropriate controls and indicators so that all of the front panel objects can be accessed if this VI were used in the future as a subVI. Also, change the VI so the toolbar, menu bar, and scroll bars are not visible when the VI executes. (**Hint:** Navigate to File≫VI Properties and choose **Window Appearance**, then choose **Customize**.)

P4.3 Create a subVI that multiplexes four inputs to a single output. The subVI should have four floating-point numeric controls (denoted In1 thru In4), one floating-point numeric indicator (denoted by Out), and one unsigned 8-bit integer control (denoted by Select). If Select = 1, then Out = In1; if Select = 2, then Out = In2; if Select = 3, then Out = In3; and if Select = 4, then Out = In4. (**Hint:** The Select function VI from the **Programming≫Comparison** palette may be useful.)

P4.4 Create a VI that executes the Quit LabVIEW VI from the **Functions≫Programming≫Application Control** palette. Open a new VI and place the Quit LabVIEW VI on the block diagram. To edit the VI properties, select File≫VI Properties and choose the **Execution** category. Check the box next to

Run when opened. Save the VI in Users Stuff. Close the VI and then open it again. What happens? Try to figure out how you can edit the VI. (**Hint:** A subVI may be useful.)

P4.5 Construct a VI that solves the quadratic formula to find the roots of the equation

$$ax^2 + bx + c = 0$$

for the variable x, where the constants a, b, and c are real numbers. In general, the roots of the quadratic equation can be either two real numbers or two complex numbers. For this problem, your VI only needs to find the real roots. The solution of the quadratic equation is given by the relations:

$$x_1 = \left(-b + \sqrt{b^2 - 4ac}\right)/2a \text{ and } x_2 = \left(-b - \sqrt{b^2 - 4ac}\right)/2a$$

Construct a VI that computes the roots according to the above relationships for x_1 and x_2. Use Figure P4.5 as a guide for the front panel. There should be two numeric outputs for the two roots and three inputs for the constants a, b, and c. Once you are finished building the block diagram, create an icon and configure the terminal connections so that your VI can be used as a subVI. Verify that you have done this correctly by placing your quadratic formula subVI in a new VI and connecting the terminals to controls and indicators.

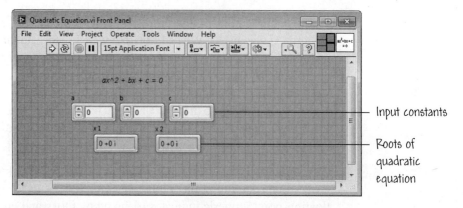

FIGURE P4.5
Front panel for the Quadratic Equation VI with edited icon.

P4.6 The error outputs of many mathematics functions indicate only an error number and do not use the Error Cluster (for optimizing memory and execution speed). Consider the VI shown in Figure P4.6. It is named Mean.vi and can be found in the Chapter 4 subfolder in the Exercises&Problems folder within the Learning directory. Its objective is to compute the mean of the input array (labeled **X** on the front panel) using the math function Mean found in **Probability & Statistics** on the **Mathematics** palette. Add the Simple Error Handler.vi to display an error code if an error occurs. Test the VI by passing an empty array to the Mean VI.

Save the VI with error handling as Mean.vi in the Users Stuff folder in the Learning directory.

 *The Simple Error Handler VI can be found in **Dialog & User Interface** on the **Programming** palette.*

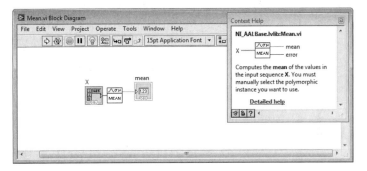

FIGURE P4.6
Using the Mean function with error handling.

DESIGN PROBLEMS

D4.1 Construct a subVI to calculate an age in years given a birth date and the current date as inputs. Use Figure D4.1 as a guide for the front panel. You should use three separate integer controls to enter each date: one for the day of the month, one for the month, and one for the year. Enter months as numbers. For example, use the number 1 for January, the number 2 for February, and so forth. Be sure to create an icon for the VI and connect the output and inputs to terminals on the icon. When you are finished, create documentation for your VI under **VI Properties**. Add the following information to the documentation: "Finds age in years of a person given the current date and the person's birth date as input."

D4.2 Create a subVI which will calculate the Body Mass Index (BMI) of an individual. The BMI is an internationally used measure of obesity. The subVI will have two numeric inputs: a weight in pounds and a height in inches. Calculate the BMI using the formula

$$\text{BMI} = (703 * W)/H^2$$

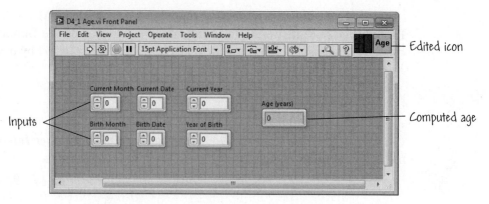

FIGURE D4.1
The front panel for the Age VI.

where W is the weight in pounds and H is the height in inches. Display the calculated BMI on a numeric indicator.

Classify the body weight according to Table D4.2 and output a string containing this information. There should also be a "warning" Boolean output that is True for any unhealthy input (i.e., underweight, overweight, or obese).

TABLE D4.2   BMI Classification

<18.5	Underweight
18.5–24.9	Healthy
25–29.9	Overweight
≥ 30	Obese

D4.3 Develop a subVI to compute the height of a trapezoid. The inputs to the VI are the lengths of the trapezoid top base, bottom base, left side, and right side, as shown in Figure D4.3. The VI's output is the height of the trapezoid. Edit the VI icon to show a trapezoid with the height labeled. Save the VI as Trap Height.vi in the Users Stuff folder in the Learning directory.

Verify that you have set up the connector pane correctly by opening a new VI and placing Trap Height.vi within it as a subVI. Create controls and an indicator, and run the program several times with different inputs. Test the algorithm with the following inputs: top base = 2, bottom base = 10, left side = 5, and right side = 5. The result should be height = 3.

D4.4 Develop a VI that generates a sinusoidal signal using the Sine Waveform.vi with the amplitude of the sine wave as a numeric control on the front panel. Then, using the Scaling & Mapping Express VI, amplify the sinusoidal signal

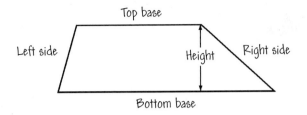

FIGURE D4.3
Trapezoid figure.

output of the **Sine Waveform VI** by a factor or 10 and plot the resulting signal on a waveform graph. Connect an error-cluster indicator to the **Scaling & Mapping Express VI** so that the error status, error code, and source of the error are displayed on the front panel.

*Hint: To wire the waveform graph to the **Scaling & Mapping Express VI**, right-click the **Scaled Signals** output of the express VI and select **Graph Indicator** from the **Create** menu. Similarly, to wire the error-cluster indicator, right-click the **error out** output of the express VI and select **Indicator** from the **Create** menu. Use **Context Help** for additional help.*

CHAPTER 5

Structures

Structures govern the execution flow in a VI. This chapter introduces the For Loop, the While Loop, the Case structure, and the Flat Sequence structure. Formula Nodes will be introduced to implement mathematical equations. The Diagram Disable structure is presented as a way to comment out code on the block diagram. We introduce the use of local variables and discuss software design architectures that exist within the graphical programming environment. We will also discuss controlling the execution timing of VIs.

GOALS

1. Study For Loops, While Loops, Case structures, and Flat Sequence structures, Formula Nodes, and Disable structures.

2. Understand how to use timing functions in LabVIEW.

3. Understand shift registers, feedback nodes, and local variables.

4. Gain an appreciation for common programming techniques.

5. Appreciate common wiring errors with structures.

228

5.1   THE FOR LOOP

For Loops and **While Loops** control repetitive operations in a VI, either until a specified number of iterations completes (i.e., the For Loop) or while a specified condition is True (i.e., the While Loop). A difference between the For Loop and the While Loop is that the For Loop executes a predetermined number of times, and a While Loop executes until a certain conditional input becomes True or False. For Loops and While Loops are found on the **Structures** palette of the **Functions≫Programming** subpalette—see Figure 5.1. In this section, we concentrate on For Loops. While Loops are discussed in the next section.

A For Loop executes the code inside its borders (known as the **subdiagram**) a total of N times, where the N equals the value in the count terminal, as illustrated in Figure 5.2.

The For Loop has two terminals: the **count terminal** (an input terminal) and the **iteration terminal** (an output terminal). You can set the count explicitly by wiring a value from outside the loop to the count terminal, or you can set the count implicitly with auto-indexing (a topic discussed in Chapter 6). The value of the count terminal is also exposed to the inside of the loop, allowing you access to the value of the count. The iteration terminal (see Figure 5.2) contains the current number of completed loop iterations, where 0 represents the first iteration, 1 represents the second iteration, and continuing up to $N - 1$ to represent the Nth iteration.

Both the count and iteration terminals are long integers with a range of 0 through $2^{32} - 1$. You can wire a floating-point number to the count terminal, but it will be rounded off and **coerced** to lie within the range 0 through $2^{32} - 1$. The For Loop does not execute if you wire 0 to the count terminal!

The For Loop is located on the **Functions≫Programming≫Structures** palette. Unlike many other **objects**, the For Loop is not simply dropped on the block diagram. Instead, a small icon representing the For Loop appears in the block diagram, giving you the opportunity to size and position the loop. To do so, first click in an area above and to the left of all the objects that you want to execute within the For Loop, as illustrated in Figure 5.3. While holding down the mouse button, drag out a rectangle that encompasses the objects you want to place inside the For Loop. A For Loop is created upon release of the mouse button. The For Loop is a resizable box—use the **Positioning** tool for resizing by grabbing a corner of the For Loop and stretching to the desired dimensions. You can add additional block diagram elements to the For Loop by dragging and dropping them inside the loop boundary. The For Loop border will highlight as objects move inside the boundaries of the

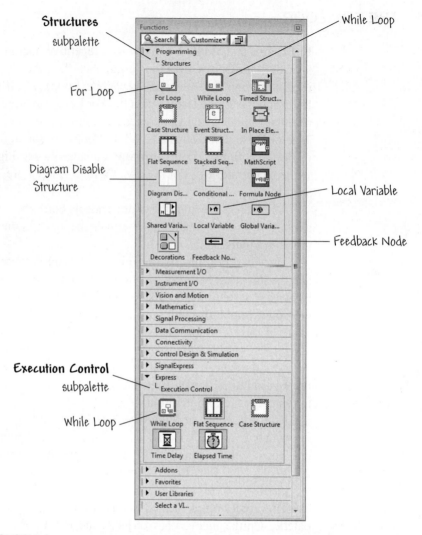

FIGURE 5.1
The For Loop and While Loop are found on the **Programming≫Structures** palette, the While Loop is also found on the **Express≫Execution Control** palette.

loop, and the block diagram border will highlight as you drag objects out of the loop.

If you move an existing structure (e.g., a For Loop) so that it overlaps another object on the block diagram, the partially covered object will be visible above one edge of the structure. If you drag an existing structure

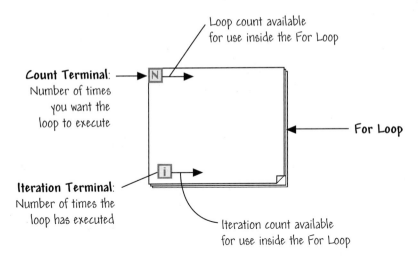

FIGURE 5.2
The For Loop.

completely over another object, the covered object will display a thick shadow
to warn you that the object is underneath.

5.1.1 Numeric Conversion

Most of the numeric controls and indicators you have used so far have been
double-precision, floating-point numbers. Numbers can be represented as integers (byte [I8], word [I16], or long [I32]) or floating-point numbers (single,
double, or extended precision). If you wire together two terminals that are of
different data types, LabVIEW converts one of the terminals to the same representation as the other terminal. As a reminder, a **coercion dot** is placed on the
terminal where the conversion takes place.

For example, consider the For Loop count terminal shown in Figure 5.4.
The terminal representation is long integer. If you wire a double-precision,
floating-point number to the count terminal, the number is converted to a long
integer. Notice the small dot in the count terminal of the first For Loop—that
is the coercion dot. To change the representation of the count terminal
input, right-click on the terminal and select **Representation**, as
illustrated in Figure 5.5. A palette will appear from which you can select the
desired representation. In the case of the For Loop, you can change
the count terminal input from double-precision, floating-point to long
integer.

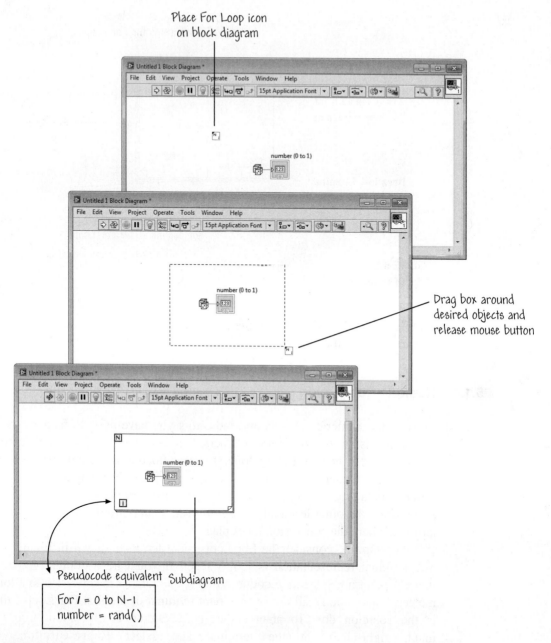

FIGURE 5.3
Placing a For Loop on the block diagram.

When the VI converts floating-point numbers to integers, it rounds to the nearest integer. If a number is exactly halfway between two integers, it is rounded to the nearest even integer. For example, the VI rounds 6.5 to 6, but rounds 7.5 to 8. This is an IEEE standard method for rounding numbers.

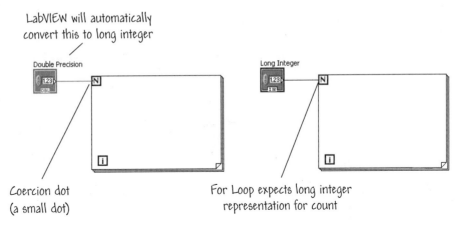

FIGURE 5.4
Converting double-precision, floating-point numbers at the count terminal.

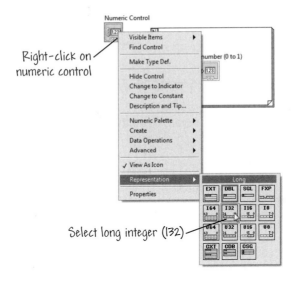

FIGURE 5.5
Changing the representation of a front panel numeric object.

 A For Loop Example

In this exercise we will place a random number object inside a For Loop and display the random numbers and For Loop counter on the front panel. The VI shown in Figure 5.6 can be constructed by following these steps:

1. Place a Random Number function on the block diagram. The Random Number function can be found in the **Functions≫Programming≫Numeric** palette. Create an indicator on the output of the Random Number function and label it number: 0 to 1.

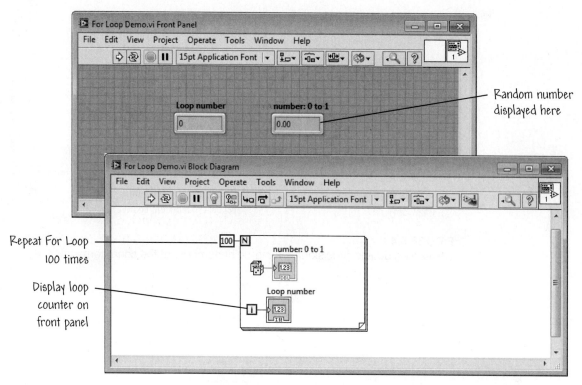

Repeat For Loop 100 times

Display loop counter on front panel

Random number displayed here

FIGURE 5.6
Displaying a series of random numbers using the For Loop.

2. Place the For Loop on the block diagram so that the Random Number function is enclosed within the loop, as shown in Figure 5.6.

3. Create a constant for the For Loop by right-clicking on the count terminal and selecting **Constant** from the **Create** menu. Set the value of the constant (which will be zero by default) to 100. This will let the For Loop execute 100 times.

4. Create an indicator on the iteration terminal and label it **Loop number**.

5. Debug and run the program. One suggestion is to run the program with **Highlight Execution** on, otherwise the program may run too fast to observe the loop execution.

6. On the front panel you will see the loop counter increment from 0 to 99 (that is, 100 iterations), and the random number between 0 and 1 should be displayed each iteration. Notice that the numeric indicator counts from 0 to 99, and *not* from 1 to 100!

7. Save the VI as **For Loop Demo.vi** in the **Users Stuff** folder in the **Learning** directory.

A working version of the VI called For Loop Demo.vi *can be found in the* Chapter 5 *folder in the* Learning *directory.*   ◆

5.1.2   **For Loops with Conditional Terminals**

If you need to be able to stop the iterations of a For Loop when a specified Boolean condition occurs or an error occurs, you can add a conditional terminal, as shown in Figure 5.7. A For Loop with a conditional terminal executes until the condition occurs or until all iterations complete. Also, you can wire an error cluster to the conditional terminal if you want the For Loop to execute a given number of times unless an error occurs, at which time the loop execution is terminated. A For Loop with conditional terminals is similar to the Break construct in textual programming.

To add or remove a conditional terminal in a For Loop, right-click the loop border and select **Conditional Terminal** from the shortcut menu, as shown in Figure 5.7. If you add a conditional terminal to a For Loop, it appears in the bottom right of the loop, and the count terminal appearance changes to include a red dot. Right-click the For Loop border and deselect **Conditional Terminal** to remove it (using the <Delete> key will not remove the conditional terminal). In a For Loop with a conditional terminal, you must wire the conditional terminal

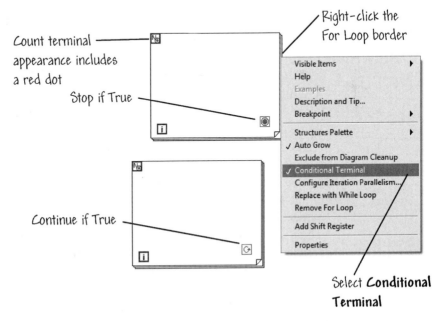

FIGURE 5.7
The For Loop with a conditional terminal.

and either wire a numeric value to the count terminal or autoindex an input array. (Arrays and autoindexing are discussed in Chapter 6.) The number wired to the count terminal determines the maximum number of loop iterations, assuming that the condition wired to the conditional terminal never occurs.

Use a While Loop if you want a loop to iterate until a condition occurs with no maximum number of iterations (see Section 5.2 for discussion on While Loops).

The default behavior and appearance of the conditional terminal is Stop if True, as shown in Figure 5.7. In this situation, the For Loop executes until the conditional terminal receives a TRUE value or until the maximum number of iterations is reached. You can change the behavior and appearance of the conditional terminal by right-clicking on it and selecting Continue if True from the shortcut menu. In this case, the For Loop executes until the conditional terminal receives a FALSE value. You also can use the **Operating** tool to click the conditional terminal to change the condition.

You cannot predetermine the number of iterations the For Loop executes when you wire the conditional terminal. You can determine the number of iterations the loop completes by wiring an indicator to the loop iteration terminal and checking the count after the loop executes.

5.2   THE WHILE LOOP

A While Loop is a structure that repeats a section of code until a certain condition is met. The While Loop in LabVIEW follows the behavior of the Do While construct in textual programming paradigms. The While Loop, shown in Figure 5.8, executes the subdiagram inside its borders until a certain condition is satisfied. The While Loop has two terminals: the **conditional terminal** (an input terminal) and the **iteration terminal** (an output terminal). The iteration terminal of the While Loop behaves exactly like the For Loop iteration terminal. It is an output numeric terminal that outputs the number of times the loop has executed, which is zero for the first iteration. The conditional terminal input is a Boolean variable: True or False. The While Loop executes until the Boolean value wired to its conditional terminal is either True of False, depending on whether the conditional terminal is set to **Stop if True** or **Continue if True**.

The VI checks the conditional terminal at the *end* of each iteration; therefore, the While Loop always executes at least once. If the value at the conditional terminal is False (and Stop if True is selected, which is default), another iteration is performed; otherwise the loop terminates.

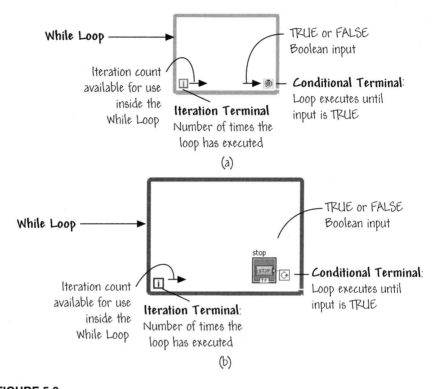

FIGURE 5.8
(a) The While Loop from the **Programming≫Structures** palette. (b) The While Loop from the **Express≫Execution Control** palette.

You place the While Loop in the block diagram by (a) selecting it from the **Functions≫Programming≫Structures** palette or (b) selecting it from the **Functions≫Express≫Execution Control** palette, as illustrated in Figure 5.1. Similar to the For Loop, the While Loop is not simply dropped on the block diagram. Instead, a small icon representing the While Loop appears in the block diagram, giving you the opportunity to size and position the loop, as shown in Figure 5.9. To do so, first click with the While Loop icon in an area above and to the left of all the objects that you want to execute within the While Loop, as shown in Figure 5.9. While holding down the mouse button, drag out a rectangle that encompasses the objects you want to place inside the While Loop. A While Loop is created upon release of the mouse button.

The completed While Loop is a resizable box—once the While Loop is placed on the block diagram, you can use the **Positioning** tool to resize the box by grabbing a corner and stretching the box to the desired dimensions. Additional block diagram elements can be added to the While Loop by dragging and dropping the desired objects inside the While Loop box. As with For Loops,

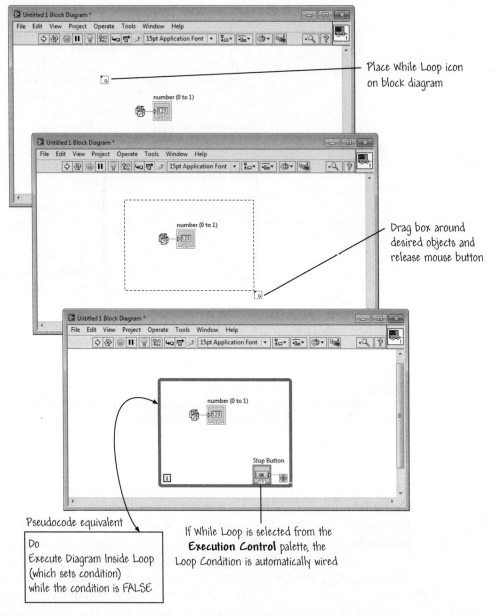

FIGURE 5.9
Placing a While Loop in the block diagram.

the While Loop border will highlight as the object moves inside, and the block diagram border will highlight when you drag an object out of the While Loop. Once the While Loop (or For Loop) is on the block diagram, you cannot place an object inside the structure by dragging the structure over the object! Doing this will simply cause the structure to be placed over the object (that is, the structure

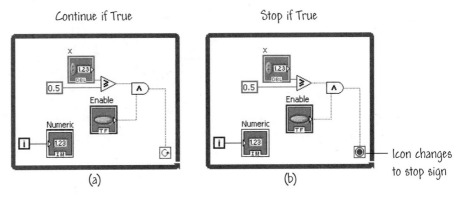

FIGURE 5.10
(a) Continue if True and (b) Stop if True.

will be the frontmost object). The correct procedure is to drag and drop objects inside existing While Loops (or other structures, such as For Loops).

You can change the way the conditional terminal functions by right-clicking on the While Loop and choosing **Continue if True**. Then, rather than having the While Loop **Stop if True**, it will now **Continue if True**. An example is shown in Figure 5.10. In Figure 5.10(a), the While Loop continues if the value of x is greater than or equal to 0.5 and the Enable Boolean is pushed (TRUE). Conversely, in Figure 5.10(b) the While Loop will stop if the value of x is greater than or equal to 0.5 and the Enable Boolean is pushed (TRUE).

The first time through a For Loop or a While Loop, the iteration count is zero. If you want to register how many times the loop has actually executed, you must add 1 to the count.

A While Loop Example

In this exercise we will place a random number object inside a While Loop and display the random numbers and While Loop counter on the front panel. This is very similar to the For Loop exercise in the previous section. The VI shown in Figure 5.11 can be constructed by following these steps:

1. Place a Random Number function on the block diagram (found in the **Functions≫Programming≫Numeric** palette). Create an indicator on the output of the Random Number function and label it number: 0 to 1.

2. Place the While Loop on the block diagram so that the Random Number function is enclosed within the loop, as shown in Figure 5.11. Use the **Functions≫Express≫Execution Control** palette to locate the While Loop.

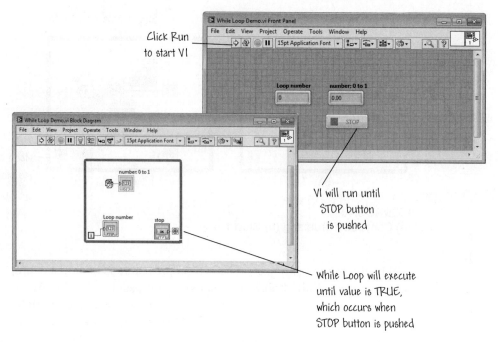

FIGURE 5.11
Displaying a series of random numbers using the While Loop.

3. When you place the While Loop from the **Express≫Execution Control** palette, a stop button will automatically be created. This will be used to stop the While Loop iterations while in Run mode.

4. Create an indicator on the output of the iteration terminal and label it Loop number.

5. Click on the **Run** button to start the program execution, and run the program with **Highlight Execution** to observe the program data flow.

6. On the front panel you will see the loop counter continue to increment until you push the Stop button. This causes the conditional terminal to change to True, and the While Loop iterations cease.

*If you want to run a VI multiple times, place all the code in a loop rather than using the **Run Continuously** button. It is better to stop a VI with a Boolean control than with the **Abort Execution** button.*

7. Save the VI as While Loop Demo.vi in the Users Stuff folder in the Learning directory.

A working version of the VI called **While Loop Demo.vi** *can be found in the* **Chapter 5** *folder in the* **Learning** *directory.*

If you place the terminal of the Boolean control outside the While Loop, as shown in Figure 5.12, you create either an infinite loop or a loop that executes only once, depending on the initial value. Why? Because the Boolean input data value is read before it enters the loop—remember data flow programming—and not within the loop or after completion of the loop.

To experiment with this, modify your VI by placing the Boolean terminal outside of the While Loop. Connect the Boolean terminal to the While Loop border, as illustrated in Figure 5.12. A green rectangle, called a tunnel, will appear on the loop border. Then wire the tunnel to the conditional terminal. You will learn more about using tunnels to pass data into and out of structures in later sections of this chapter.

Conduct the following two numerical experiments: First, set the button state on the front panel to "on" and then run the VI (use **Highlight Execution** to watch the data flow). After a reasonable period of time, press the Stop button. What happens? The VI does not stop running—it is in an infinite loop. Can

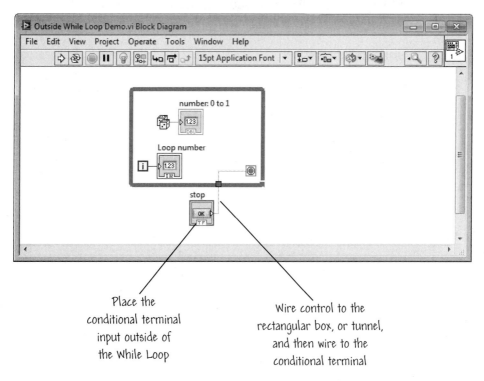

FIGURE 5.12
Placing the conditional terminal input outside of the While Loop.

you explain this behavior using the notion of data flow programming? Since LabVIEW operates under data flow principles, inputs to the While Loop must pass their data before the loop executes. A While Loop passes data out only after the loop completes all iterations. The infinite loop experiment can be stopped by clicking on the **Abort Execution** button on the front panel toolbar.

For a second experiment, push the Stop button on the front panel and then run the VI (use **Highlight Execution** to watch the data flow). What happens in this case? You should see that the VI will execute once through the While Loop.

Save the VI with the conditional terminal outside the While Loop as Outside While Loop Demo.vi in the Users Stuff folder in the Learning directory.

A working version of Outside While Loop Demo.vi can be found in the Chapter 5 *folder in the* Learning *directory.*   ◆

5.3   SHIFT REGISTERS AND FEEDBACK NODES

When programming with loops, you often need to access data from previous iterations of the loop. Two ways of accessing the data from previous iterations of the loop are by utilizing the shift register and the Feedback Node. Both methods are discussed in this section.

5.3.1   Shift Registers

Shift registers transfer values from one iteration of a For Loop or While Loop to the next. The shift register comprises of a pair of terminals directly opposite each other on the vertical sides of the loop border, as shown in Figure 5.13. You create a shift register by right-clicking on the left or right loop border and selecting **Add Shift Register** from the shortcut menu.

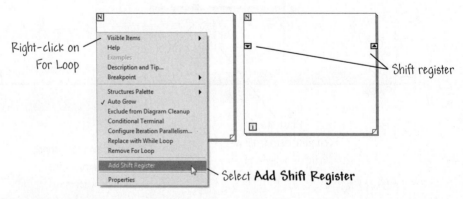

FIGURE 5.13
Shift register containing a pair of terminals.

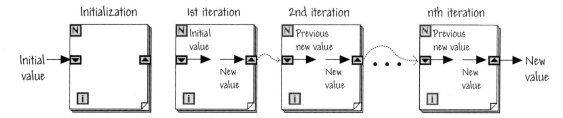

FIGURE 5.14
Passing data from one loop iteration to the next using shift registers.

The right terminal (the rectangle with the up arrow) stores the data as each iteration finishes. The stored data from the previous iteration is shifted and appears at the left terminal (the rectangle with the down arrow) at the beginning of the next iteration, as shown in Figure 5.14. A shift register can hold any data type, including numeric, Boolean, strings (see Chapter 9), and arrays (see Chapter 6), but the data wired to the terminals of each register must be of the same type. The shift register conforms to the data type of the first object that is wired to one of its terminals.

Consider a simple illustrative example. Suppose we have the two situations depicted in Figure 5.15. Both cases look very similar—but the code in

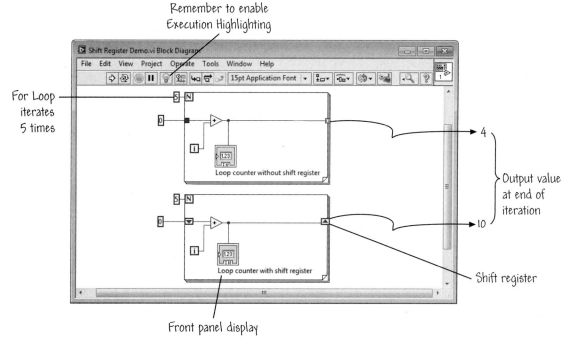

FIGURE 5.15
A simple example showing the effect of adding a shift register to a For Loop.

(b) contains a shift register. The code in (b) computes a running sum of the iteration count within the For Loop. Each time through the loop, the new sum is saved in the shift register. At the end of the loop, the total sum of 10 is passed out to the numeric indicator. Why 10? Because the sum of the iteration count is 10 ($= 0 + 1 + 2 + 3 + 4$). On the other hand, the code in (a) that does not contain the shift register does not save values between iterations. Instead, a zero is added to the current iteration count each time, and only the last value of the iteration counter ($= 4$) will be passed out of the loop.

> *You can run the simulation shown in Figure 5.15 by opening the VI titled* **Shift Register Demo.vi** *which is in* **Chapter 5** *of the* **Learning** *directory—enable* **Highlight Execution** *before running and watch the data flow through the code!*

5.3.2   Using Shift Registers to Remember Data Values from Previous Loop Iterations

You can configure the shift register to store data values from previous iterations. To prepare the loop to store previous values, first create additional terminals on the loop border, as illustrated in Figure 5.16. If, for example, you add four elements to the left terminal, you can store and access values from the previous

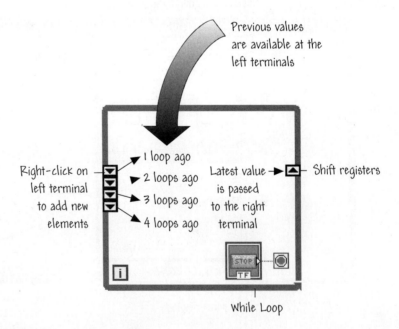

FIGURE 5.16
Adding elements to the shift register to access values from previous loop iterations.

four iterations. You add the additional elements by right-clicking on the left terminal of the shift register and choosing **Add Element** from the shortcut menu.

Using Shift Registers

In this example we will open an existing VI and use it to watch the data flow in a While Loop containing shift registers. Begin by opening Viewing Shift Registers.vi located in Chapter 5 of the Learning directory.

The front panel has four numeric indicators, as shown in Figure 5.17. The $X(i)$ indicator will display the current value, which will shift to the left terminal at the beginning of the next iteration. The $X(i-1)$ indicator will display the value one iteration ago, the $X(i-2)$ indicator will display the value two iterations ago, and the $X(i-3)$ indicator will display the value three iterations ago. The shift register is initialized to zero.

Before running the program, make sure that **Highlight Execution** is enabled. This will allow you to view the data flow and to watch the shift registers access data from previous iterations of the While Loop. Run the VI and watch the bubbles indicating the data flow. Notice that in each iteration of the While Loop, the VI "funnels" the previous values through the left terminals of the shift register. $X(i)$ shifts to the left terminal, $X(i-1)$, at the beginning of the next iteration. The values at the left terminal funnel downward through the terminals. In this

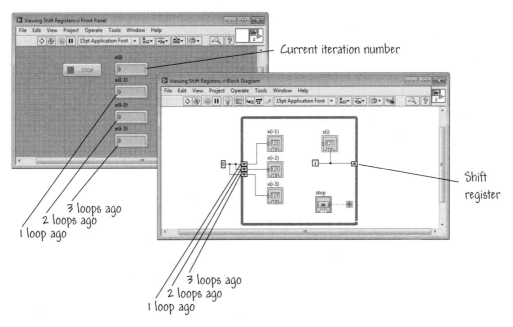

FIGURE 5.17
A demonstration of the use of shift registers to access data from previous iterations of a While Loop.

example, the VI retains the last three values. To retain more values, you would need to add more elements to the left terminal of the shift register. When you are finished observing the VI operation, close it and do not save any changes.   ◆

5.3.3   Initializing Shift Registers

Shift registers are initialized by wiring a constant or control to the left terminal of the shift register from outside the loop. Shift register initialization is demonstrated in Figure 5.18. On the first execution the final value of the shift register is 12. Why 12? Because the final value is a sum of the iteration count $(= 0 + 1 + 2 + 3 + 4)$ plus the initial value $(= 2)$. On the second execution of the code that contains the initialized shift register, the result is exactly the same as on the first execution. In fact, the results on all subsequent executions are identical to the first execution: the final value of the shift register is 12. That is, the final value of the shift register from the first execution does not play a role in the second run—this is as it should be!

Unless the shift register is explicitly initialized, the first time the VI is executed the initial value of the shift register will be the default value for the shift register data type—if the shift register data type is Boolean, the initial value will be FALSE. Similarly, if the shift register data type is numeric, the initial value will be zero. The bottom row of Figure 5.18 illustrates what happens if you execute code twice with uninitialized shift registers. On the first execution the final value of the shift register is 10. Why 10? Can you explain this result? Running the code again, without closing the VI first, results in a final value of the shift register equal to 20. This result is due to the fact that on the second execution

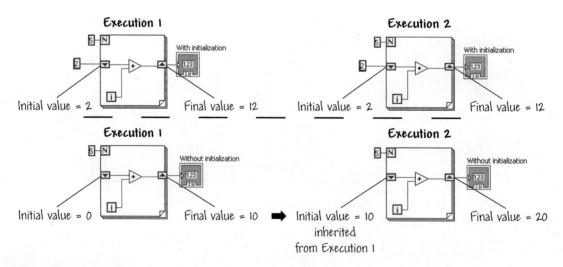

FIGURE 5.18
Initializing shift registers.

the initial value of the shift register is equal to 10 (left over from the previous run), and that is added to the sum of the iteration counter on the second run, which is equal to 10. When the shift register is not explicitly initialized, on the second run the shift register will take on the last value of the first run. Always use initialized shift registers for consistent results!

Data values stored in the shift register are stored until you close the VI and remove it from memory. That is, if you run a VI containing uninitialized shift registers, the initial values taken by the shift registers during subsequent executions will be the last values from previous executions. This can make debugging your VI a difficult process because this situation is hard to detect.

The shift register initialization demonstration depicted in Figure 5.18 can be found in **Chapter 5** *of the* **Learning** *directory and is called* **Shift Register Init Demo.vi**.

Computing a Running Average

In this example, we will use shift registers to compute the running average of a sequence of random numbers. Since the Random Number function in Lab-VIEW provides random numbers from 0 to 1, we expect that the average will be 0.5. How many random numbers will it take to obtain an average near 0.5?

Figure 5.19 shows a VI that computes the running average of several random numbers. The length of the random number sequence is input via the slide control, as depicted on the front panel in Figure 5.19. Using the block diagram in the figure as a model, develop your own VI to compute the running average of a sequence of random numbers. The formula that is coded to compute the average is

$$Ave_i = \frac{i}{i+1} Ave_{i-1} + \frac{1}{i+1} RN_i,$$

where $i = 0, 1, \cdots, N - 1$, Ave_i is the computed average at the ith iteration, and RN_i is the current random number from the Random Number function. If you aren't familiar with computing a running average, just concentrate on the programming aspects of the VI shown in Figure 5.19 and do not worry about the formula. The main point is to understand how to use the shift registers in conjunction with For Loops and While Loops!

In this example, the shift register is used to pass the value of the variable Ave_{i-1} from one iteration to the next. You should notice as you run the code that for small values of N (e.g., $N = 3$) the average is generally not close to the expected value of $Ave_N = 0.5$; however, as you increase N, the average gets closer and closer to the expected value. Try it out!

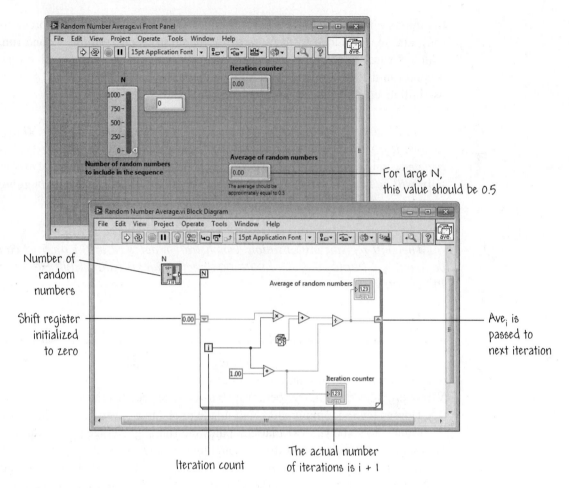

FIGURE 5.19
Computing a running average of a sequence of random numbers.

 *The VI depicted in Figure 5.19 can be found in **Chapter 5** of the **Learning** directory and is called **Random Number Average.vi**. Check it out if you can't get yours to work or if you want to compare your results.* ◆

5.3.4 Feedback Nodes

The Feedback Node can be utilized in a For Loop or While Loop if you wire the output of a subVI, function, or group of subVIs and functions to the input of that same VI, function, or group—that is, if you create a *feedback* path. Like a shift register, the Feedback Node stores data when the loop completes an iteration,

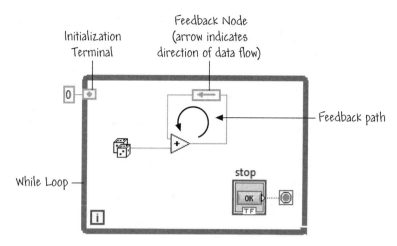

FIGURE 5.20
The Feedback Node.

sends that value to the next iteration of the loop, and transfers any data type. The Feedback Node is illustrated in Figure 5.20 inside of a While Loop.

Once the feedback is wired inside the loop, both the Feedback Node arrow and the Initializer Terminal will automatically appear. The Feedback Node arrow indicates the direction of data flow along the wire. Initializing a Feedback Node resets the initial value the Feedback Node passes the first time the loop executes when the VI runs. If you leave the input of the Initializer Terminal unwired, each time the VI runs, the initial input of the Feedback Node is the last value from the previous execution or the default value for the data type if the loop has never executed.

You also can select the Feedback Node on the **Programming≫Structures** palette and place it inside a For Loop or While Loop (see Figure 5.1 for location on the palette). If you place the Feedback Node on the wire before you branch the wire that connects the data to the tunnel, the Feedback Node passes each value to the tunnel. If you place the Feedback Node on the wire after you branch the wire that connects data to the tunnel, the Feedback Node passes each value back to the input of the VI or function and then passes the last value to the tunnel.

Use the Feedback Node to avoid unnecessarily long wires in loops.

You can configure the Feedback Node by right-clicking on the node and selecting **Properties**. In the **Object Properties** dialog box you can modify the appearance of the Feedback Node by selecting the **Appearance** tab, and you can configure the number of samples of data the Feedback Node stores before it begins to output those values by electing the **Configuration** tab. The appearance of the Feedback Node can include a label if selected, and you can select the header appearance and arrow direction.

By default, the Feedback Node stores data from only the previous execution or iteration. You can have it store *n* samples of data by delaying the output of the node for multiple executions. If you increase the delay on the **Configuration** page to more than one execution, the Feedback Node outputs the initializer value until the delay you specify is complete and then begins to output the stored values in subsequent order. The number on the Feedback Node represents the specified number of delays. Besides configuring the delay, you can also display the enable terminal, which if set to TRUE executes the Feedback Node as configured and if set to False outputs the last value from the previous execution until set to TRUE again. The Initialization box provides information on the initialization configuration of the Feedback Node and changes depending on whether the initializer terminal is wired.

Practice with Feedback Nodes

In this example we compare the use of the shift register with the Feedback Node and illustrate the differences in placing the Feedback Node in the feedback path and the feedforward path.

Open the Feedback Node Demo VI located in the **Chapter 5** folder of the **Learning** directory. The block diagram is shown in Figure 5.21. The block diagram has three parts:

1. In the top section of the code, the Feedback Node is placed after the branch.

2. In the middle section of the code, the Feedback Node is placed before the branch.

3. In the bottom section of the code, shift registers are used instead of Feedback Nodes.

Wiring a 0 to the Initialization Terminal or to the shift register initializes each diagram. The Time Delay function slows the operation of the code for better viewing. Run the VI and observe the outputs on the front panel.

The top section of the code reads the initialized Feedback Node and passes this value to the Add function.

The middle section of the code reads the initialized Feedback Node and passes this value to the indicator. This Add function does not execute until the next iteration of the loop. This section of code will output a number always one less than the first section of code.

The Feedback Node implementation in the top section of the code and the shift register implementation in the bottom section of the code have the same functionality.

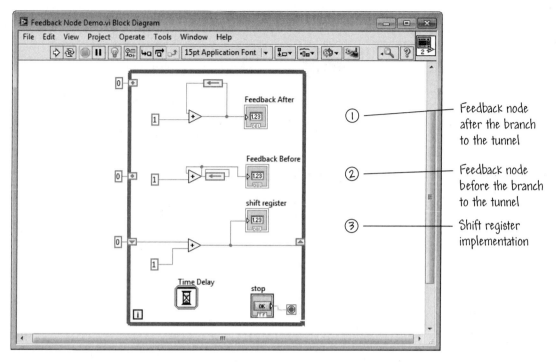

FIGURE 5.21
Different implementations of the Feedback Node.

*If you want to replace a Feedback Node with a shift register, right-click the Feedback Node and select **Replace with Shift Register**.* ◆

5.4   CASE STRUCTURES

A **Case structure** is a method of executing conditional text. This is analogous to the common If. . .Then. . .Else statements in conventional, text-based programming languages. You place the Case structure on the block diagram by selecting it from the **Structures** subpalette of the **Functions≫Programming** palette or from the **Execution Control** subpalette of the **Functions≫Express** palette, as shown in Figure 5.22. As with the For Loop and While Loop structures, you can either drag the Case structure icon to the block diagram and enclose the desired objects within its borders, or you can place the Case structure on the block diagram, resize it as necessary, and drag objects inside the structure.

The Case structure can have multiple subdiagrams. The subdiagrams are configured like a deck of cards of which only one card is visible at a time. At the top of the Case structure border is the selector label. The diagram identifier can

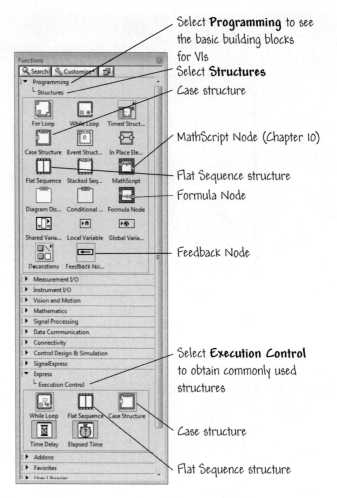

Select **Programming** to see the basic building blocks for VIs

Select **Structures**

Case structure

MathScript Node (Chapter 10)

Flat Sequence structure

Formula Node

Feedback Node

Select **Execution Control** to obtain commonly used structures

Case structure

Flat Sequence structure

FIGURE 5.22
(a) Select Case structure from the **Programming≫Structures** palette; (b) selecting the Case structure from the **Express≫Execution Control** palette.

be numeric, Boolean, string, or enumerated type control. More information on enumerated type controls can be found in the LabVIEW help. An enumerated type control is unsigned byte, unsigned word, or unsigned long and is selectable from the decrement and increment buttons, as depicted in Figure 5.23(a) and (b). The selector label displays the values that cause the corresponding subdiagrams to execute. The selector label value is followed by a selector label range, which shows the minimum and maximum values for which the structure contains a subdiagram. To view other subdiagrams (that is, to see the subdiagrams within the "stack of cards"), click the decrement (left) or increment (right) button to display the previous or next subdiagram, respectively. Decrementing from the first subdiagram displays the last, and incrementing from the last subdiagram displays the first subdiagram—it wraps around!

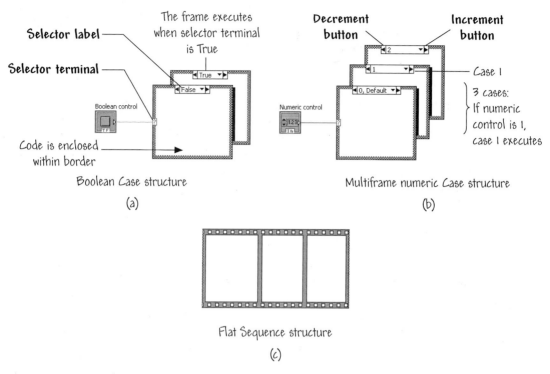

FIGURE 5.23
Overview of Case and Flat Sequence structures.

You can position the selector terminal anywhere on the Case structure along the left border. The selector label automatically adjusts to the input data type. For example, if you change the value wired to the selector terminal from a numeric to a Boolean, then cases 0 and 1 change to False and True, respectively (in the selector label). Here is a point to consider in changing the data type: if the Case structure selector originally received numeric input, then n cases $0, 1, 2, \cdots, n$ may exist in the code. Upon changing the selector input data from numeric to Boolean, cases 0 and 1 change automatically to False and True. However, cases $2, 3, \cdots, n$ are not discarded! You must explicitly delete these extra cases before the Case structure can execute.

You can also type and edit values directly into the selector label using the **Labeling** tool. The selector values are specified as a single value, as a list, or as a range of values. A list of values is separated by commas, such as $-1, 0, 5,$ 10. A range is typed in as 10..20, which indicates all numbers from 10 to 20, inclusively. You also can use open-ended ranges. For example, all numbers less than or equal to 0 are represented by ..0. Similarly, 100.. represents all numbers greater than or equal to 100.

The case selector can also use string values that display in quotes, such as "red," and "green." You don't need to type in the quotes when entering the values unless the string contains a comma or the symbol "..". In a case selector using strings, you can use special backslash codes (such as \r, \n, and \t) for nonalphanumeric characters (carriage return, line feed, tab, respectively).

If you type in a selector value that is not the same type as the object wired to the selector terminal, then the selector value displays in red, and your VI is broken. Also, because of the possible round-off error inherent in floating-point arithmetic, you cannot use floating-point numbers in case selector labels. If you wire a floating-point type to the case terminal, the type is rounded to the nearest integer, and a coercion dot appears. If you try to type in a selector value that is a floating-point number, then the selector value displays in red, and your VI is broken.

5.4.1 Adding and Deleting Cases

If you right-click on the Case structure border, the resulting menu gives you the many options shown in Figure 5.24. You choose **Add Case After** to add a case after the case that is currently visible, or **Add Case Before** to add a case before the currently visible case. You can also choose to copy the currently shown case by selecting **Duplicate Case** and to delete the current visible case by selecting

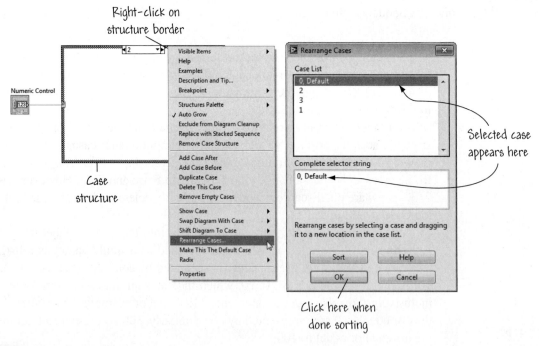

FIGURE 5.24
Adding and deleting cases in Case structures.

Delete This Case. When you add or remove cases (i.e., subdiagrams) in a Case structure, the diagram identifiers are automatically updated to reflect the inserted or deleted subdiagrams.

Sometimes you may want to rearrange the listed order of the cases in the structure. For example, rather than have the cases listed (0, 1, 2, 3, · · ·), you might want them listed in the order (0, 2, 1, 3, · · ·). Resorting the order in which the cases appear in the case structure on the block diagram does not affect the run-time behavior of the Case structure! It is merely a matter of programming preference. You can change the order in which the cases are listed in the structure by selecting **Rearrange Cases** from the shortcut menu. When you do so, the dialog box shown in Figure 5.24 appears. The **Sort** button sorts the case selector values based on the first selector value. To change the location of a selector, click the selector value you want to move (it will highlight when selected) and drag it to the desired location in the stack. In the **Rearrange Cases** dialog box, the section entitled "Complete selector string" shows the selected case selector name in its entirety in case it is too long to fit in the "Case List" box located in the top portion of the dialog box. Online assistance can be found in the context-sensitive help in the **Help** menu.

The **Make This The Default Case** item in the shortcut menu specifies the case to execute if the value that appears at the selector terminal is not listed as one of the possible choices in the selector label. Case statements in other programming languages generally do not execute any case if a case is out of range, but in LabVIEW you must either include a default case that handles out-of-range values or explicitly list every possible input value. For the default case that you define, the word "Default" will be listed in the selector label at the top of the Case structure.

**A Simple
Case
Structure
Example**

In this exercise you will build a VI that uses a Boolean Case structure, shown in Figure 5.25. The input numbers from the front panel will pass through **tunnels** to the Case structure, and there they are either added or subtracted, depending on the Boolean value wired to the selector terminal.

What is a tunnel? A tunnel is a data entry point or exit point on a structure. You can wire a terminal from outside the Case structure to a terminal within the structure. When you do this a rectangular box will appear on the structure border—this represents the tunnel. You can also wire an external terminal to the structure border to create the tunnel, and then wire the terminal to an internal terminal in a second step. Tunnels can be found on other structures, such as Sequence structures, While Loops, and For Loops. The data at all input tunnels is available to all cases.

In this exercise you will need to create two tunnels to get data into the Case structure and one tunnel to pass data out of the structure. Using Figure 5.25 as a guide, construct a VI that utilizes the Case structure to perform the following

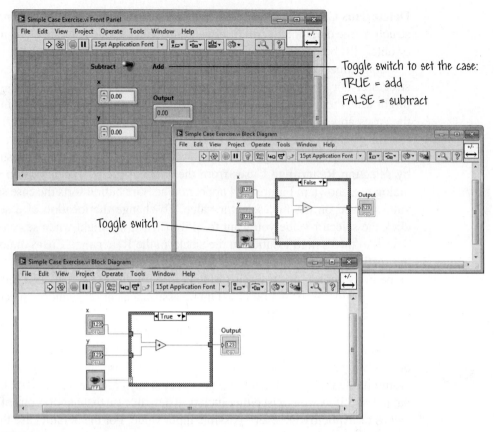

FIGURE 5.25
A Boolean Case structure example to add or subtract two input numbers.

task: if the Boolean wired to the selector terminal is True, the VI will add the numbers; otherwise, the VI will subtract the numbers. The Boolean value of the selector terminal is toggled using a horizontal toggle switch found on the front panel.

 The VI depicted in Figure 5.25 can be found in **Chapter 5** *of the* **Learning** *directory and is called* **Simple Case Exercise.vi.** ◆

5.4.2   Wiring Inputs and Outputs

As previously mentioned, the data at all input terminals (tunnels and selection terminal) is available to all cases. Cases are not required to use input data or to supply output data, but if any one case supplies output data, all must do so. Forgetting this fact can lead to coding errors. Correct and incorrect wire situations are depicted in Figure 5.26, where for the FALSE case two input numbers are

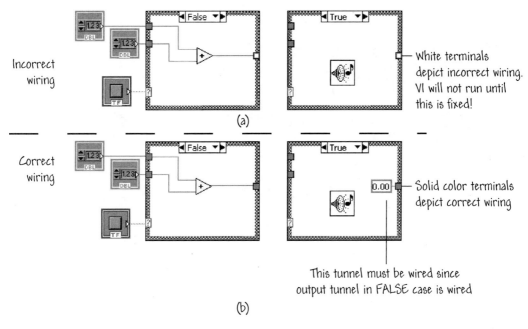

(a)

(b)

This tunnel must be wired since
output tunnel in FALSE case is wired

FIGURE 5.26
Wiring Case structure inputs and outputs.

added and sent (through the tunnel) out of the structure, and for the TRUE case the computer system beeps.

When you create an output tunnel in one case, tunnels appear at the same position on the structure border in the other cases. You must define the output tunnel for each case. Unwired tunnels look like white squares, and when they occur, the **Broken Run** button will also appear, so be sure to wire to the output tunnel for each unwired case. You can wire constants or controls to unwired cases by right-clicking on the white square and selecting **Constant** or **Control** from the **Create** menu. When all cases supply data to the tunnel, it takes on a solid color consistent with the data type of the supplied data, and the **Run** button appears.

**Using
Case
Structures**

In this exercise, you will build a VI that computes the ratio of two numbers. If the denominator is zero, the VI outputs ∞ to the front panel and causes the system to make a beep sound. If the denominator is not zero, the ratio is computed and displayed on the front panel.

Begin by opening a new VI. Build the simple front panel shown in Figure 5.27. The numeric control **x Numerator** supplies the numerator x, and the numeric control **y Denominator** supplies the denominator y. The **x/y** numeric indicator displays the ratio of the two input numbers. Switch to the block diagram and place the Case structure on the diagram.

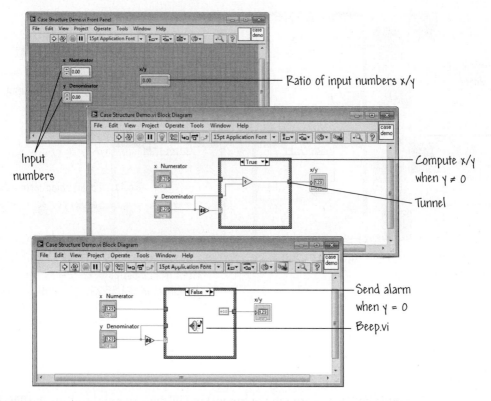

FIGURE 5.27
A VI that computes the ratio of two numbers and uses a Case structure to handle division by zero.

By default, the Case structure selector terminal is Boolean—this is the format we desire in this situation. You can display only one case at a time, and we will start by considering the True case. Remember that to change cases, you need to click on the increment and/or decrement arrows in the top border of the Case structure.

Place the other diagram objects and wire them as shown in Figure 5.27. The main objects are:

1. The Not Equal to 0? function (found in the **Functions≫Programming≫ Comparison** menu) checks whether the denominator is zero. The function returns a TRUE if the denominator is not equal to 0.

2. The Divide function (found in the **Functions≫Programming≫Numeric** menu) returns the ratio of the numerator and denominator numbers.

3. The Positive Infinity constant indicates that a divide by zero has been attempted (see the **Functions≫Programming≫Numeric** menu).

4. The Beep.vi function (found in the library **Function≫Programming≫ Graphics & Sound** menu) causes the system to issue an audible tone when

division by zero has been attempted. On Windows platforms, all input parameters to the Beep function are ignored, while on the Macintosh, you can specify the tone frequency in Hertz, the duration in milliseconds, and the intensity as a value from 0 to 255, with 255 being the loudest. Double click on the icon to see the VI front panel.

You must define the output tunnel for each case. When you create an output tunnel in the TRUE case, an output tunnel will appear at the same position in the FALSE case. Unwired tunnels look like white squares. In this and all other VI development, be sure to wire to the output tunnel for each unwired case. In Figure 5.27, the constant ∞ is wired to the output tunnel in the FALSE case.

The VI will execute either the TRUE case or the FALSE case. If the denominator number is not equal to zero, the VI will execute the TRUE case and return the ratio of the two input numbers. If the denominator number is equal to zero, the VI will execute the FALSE case and output ∞ to the numeric indicator and make the system beep.

Once all objects are in place and wired properly, return to the front panel and experiment with running the VI. Change the input numbers and compute the ratio. Make the denominator input equal to zero and listen for the system beep. When you are finished, save the VI as **Case Structure Demo.vi** in the **Users Stuff** folder in the **Learning** directory. Close the VI.

A working version of this VI called **Case Structure Demo.vi** *can be found in* **Chapter 5** *of the* **Learning** *directory.*   ◆

To conclude Section 5.4, we present an example that illustrates how Case structures can be used to construct a VI that computes factorials.

◆ **Using Case Structures to Compute Factorials**

The factorial $n!$ is defined for a positive integer n as

$$n! \equiv n * (n - 1) \ldots * 2 * 1.$$

The factorial of 0 is defined as 1. In this example, we will create a VI which performs this function, using a For Loop with a shift register.

1. Open a new blank VI and place a Case structure on the block diagram. The case structure can be found on the **Programming≫Structures** palette. We will use the Case structure to determine whether or not the user has entered a nonnegative integer.

2. Place a numeric control to the left of the Case structure and change the representation to I32. Label this control n. Use Figure 5.28 as a guide in the development of your VI.

3. Navigate to the **Programming≫Comparison** palette and find the Greater Than or Equal to 0? function. Connect the input of this function to n and connect the output to the selector terminal of the Case structure. The True frame of the case structure will execute when the input is a value greater than or equal to zero. The False frame will execute in all other cases.

4. Click either the increment or decrement button on the selector terminal to view the False case. Go to the **Programming≫Dialog & User Interface** palette and place a One Button Dialog function within the Case structure. Create a constant on the message input to this node with a message prompting the user to enter a nonnegative input.

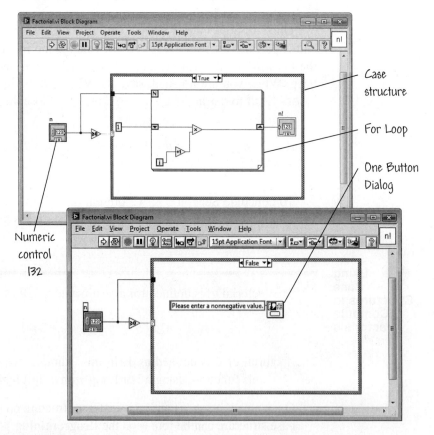

FIGURE 5.28
The block diagram for the Factorial VI.

5. Now switch to the True case by clicking either the increment or decrement button on the selector terminal. Place a For Loop within the Case structure. The For Loop is used because we know that we need to perform the multiplication operation exactly n times.

6. Wire the input n to the count terminal of the For loop so that the loop will execute n times. When you do this, a tunnel will automatically appear on the border of the Case structure.

7. Place a Multiply function and an Increment function within the For Loop. These can be found on the **Programming≫Numeric** palette. The Multiply function will be used to implement the operation $n*(n-1)$. Wire the output of the iteration terminal to the increment function and then connect the output of the Increment function to one of the inputs to the Multiply function. The output of the iteration terminal must be incremented since we want to use the values 1, 2, 3, $\dots n$ in our operations, rather than the values 0, 1, 2, $\dots n-1$.

8. Right-click on the border of the For Loop and select **Add Shift Register** from the menu. You will need to use a shift register so that the VI stores each product and saves it to be used in the next loop iteration. This can be accomplished by connecting the output of the Multiply function to the input of the shift register terminal on the right side of the loop. Wire the other input of the Multiply function to the output of the shift register terminal on the left side of the For Loop.

The shift registers turn blue to indicate that they are handling an integer data type.

9. Since the factorial of 0 is 1, we want to initialize the shift register with the value 1. Initialize the input of the shift registers by right-clicking on the input to the shift register terminal on the left side of the For Loop and selecting **Create Constant**. Change the value from 0 to 1.

10. Create an indicator labeled n! on the output of the shift register terminal on the right side of the For Loop.

You have now created a VI which calculates factorials. Experiment with different input values. Try running the VI with highlight execution turned on so that you can see what values appear on the output of the iteration terminal and shift register with each loop iteration. Compare your VI to the factorial VI found on the **Mathematics≫Elementary & Special Functions≫Discrete Math** palette.

A working version of this VI called **Factorial.vi** *can be found in* **Chapter 5** *of the* **Learning** *directory.*

◆

5.5   FLAT SEQUENCE STRUCTURES

The **Sequence structure** executes subdiagrams sequentially. The subdiagrams look like frames of film; hence they are known as **frames**. There are two classes of sequence structures: the Flat Sequence and the Stacked Sequence. Although the two classes of Sequence structures are similar, there are key differences between the two. For most purposes, the Flat Sequence structure is the better choice, for that reason we will only cover the Flat Sequence structure in this text. The Flat Sequence structure is depicted in Figure 5.23(c).

Determining the execution order of a program by arranging its elements in a certain sequence is called **control flow**. Most text-based programming languages (such as Fortran and C) have inherent control flow because statements execute in the order in which they appear in the program. In data flow programming, a node executes when data is available at all of the node inputs (this is known as **data dependency**), but sometimes you cannot connect one node to another. When data dependencies are not sufficient to control the data flow, the Sequence structure is a way of controlling the order in which nodes execute. You use the Sequence structure to control the order of execution of nodes that are not connected with wires. Within each frame, as in the rest of the block diagram, data dependency determines the execution order of nodes.

Data flow for the Flat Sequence structure differs from data flow for other structures. Frames in a Flat Sequence structure execute in order and when all data wired to the frame is available. New frames are created by right-clicking on the structure border and selecting **Add Frame After** or **Add Frame Before**, as illustrated in Figure 5.29. Data is passed between frames using tunnels in Flat Sequence structures. The data leaves each frame as the frame finishes executing.

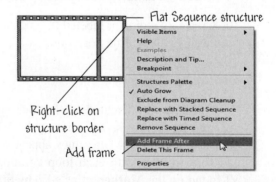

FIGURE 5.29
Removing and adding frames to the Flat Sequence structure.

5.5.1   Evaluate and Control Timing in a Sequence Structure

It is useful to be able to control and time the execution of VIs. You can use Express VIs or traditional VIs to accomplish the timing task. The Wait (ms) and Tick Count (ms) functions (located in the **Timing** palette of the **Functions≫ Programming** palette) in conjunction with a sequence structure let you accomplish these tasks using traditional VIs. The units of time are given in milliseconds.

You can also accomplish the timing task using Express VIs. The Time Delay Express VI and the Elapsed Time Express VI are located on the same **Timing** palette, as shown in Figure 5.30. When the Time Delay Express VI is placed on the block diagram, a dialog box appears in which you set the time delay value. A similar dialog box opens with the Elapsed Time Express VI. All units are given in seconds.

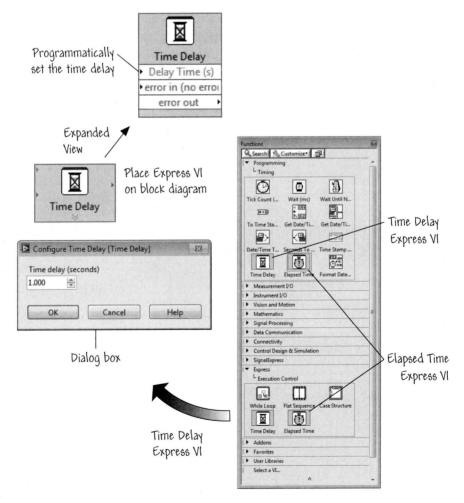

FIGURE 5.30
The Time Delay Express VI and the Elapsed Time Express VI.

The Wait (ms) function causes your VI to wait a specified number of milliseconds before it continues execution. The function waits the specified number of milliseconds and then returns the millisecond timer's end value. The Tick Count (ms) function returns the value of the millisecond timer and is commonly used to calculate elapsed time. The base reference time (that is, zero milliseconds) is undefined. Therefore, you cannot convert the millisecond timer output value to a real-world time or date. Note that the value of the millisecond timer wraps from $2^{32} - 1$ to 0.

The internal clock does not have high resolution—about 1 ms on Windows and Macintosh. The resolutions are driven by operating system limitations and not by LabVIEW.

A simple example illustrating the use of the Time Delay Express VI and the Elapsed Time Express VIs is shown in Figure 5.31. Open the VI Timing with Sound Demo located in **Chapter 5** of the **Learning** directory. Set the time between beeps on the slider control and run the VI in **Run Continuously** mode. You should hear a sound approximately every *n* seconds according to the setting on the slider control. The actual time between beeps (to the resolution of the clock) is displayed on the front panel to verify that the desired timing has been achieved. On Windows and Macintosh, if you reduce the time between beeps to 0.001 second (using the **Labeling** tool to type in the value in the digital

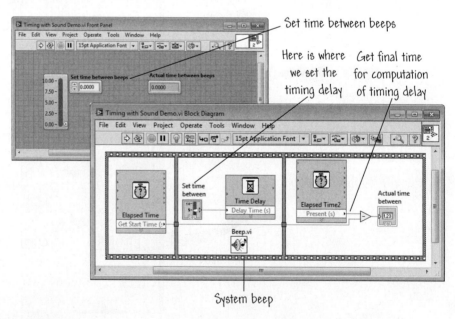

FIGURE 5.31
Controlling the timing of a VI using the Flat Sequence structure.

display of the slider control), you will find that the actual time between beeps is about 1 ms. However, if you continue to reduce the time below 0.001, the actual time between beeps will not reduce accordingly—you have reached the clock resolution of your system.

5.5.2   Avoid the Overuse of Sequence Structures

In general, VIs can operate with a great deal of inherent parallelism. Sequence structures tend to hide parts of the program and to interfere with the natural data flow. The use of Sequence structures prohibits parallel operations, but does guarantee the order of execution. Asynchronous tasks that use I/O devices can run concurrently with other operations if Sequence structures do not prevent them from doing so.

We will be discussing I/O devices (such as GPIB, serial ports, and data acquisition boards) more in subsequent chapters and you will understand better how the use of Sequence structures can inhibit the performance of such devices. The objective here is to alert you to the idea that you should avoid the overuse of Sequence structures. The use of Sequence structures does not negatively impact the computational performance of your program, but it does interrupt the data flow. You should write programs that take advantage of the concept of data flow programming!

5.6   THE FORMULA NODE

The **Formula Node** is a structure that allows you to program one or more algebraic formulas using a syntax similar to most text-based programming languages. It is useful when the equations have many variables or otherwise would require a complex block diagram model for implementation. The Formula Node itself is a resizable box (similar to the Sequence structure, Case structure, For Loop, and While Loop) in which you enter formulas directly in the code, in lieu of creating block diagram subsections.

Consider the equation

$$y = x - e \sin x,$$

where $0 \leq e \leq 1$. This is a famous equation in astrodynamics, known as Kepler's equation. If you implement this equation using regular LabVIEW arithmetic functions, the block diagram looks like the one in Figure 5.32. You can implement the same equation using a Formula Node as shown in the same figure.

The Formula Node operates in a similar fashion to the MathScript Node discussed in Chapter 10, but has a more limited set of functions that it can execute.

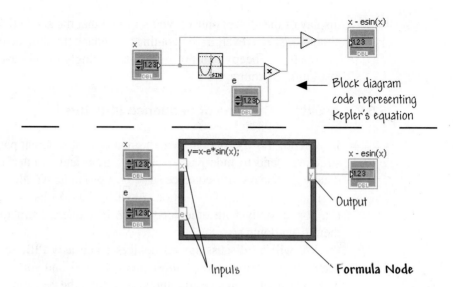

FIGURE 5.32
Implementing formulas in a Formula Node can often simplify the coding.

5.6.1   Formula Node Input and Output Variables

You place the Formula Node on the block diagram by selecting it from the **Structures** subpalette of the **Functions≫Programming** palette (see Figure 5.22). You create the input and output terminals of the Formula Node by right-clicking on the border of the node and choosing **Add Input** or **Add Output** from the shortcut menu, as shown in Figure 5.33. Output variables have a thicker border than input variables. You can change an input to an output by selecting **Change to Output** from the shortcut menu, and you can change an output to an input by selecting **Change to Input** from the shortcut menu.

Once the necessary input and/or output terminals are on the Formula Node, enter the input and output variable names in their respective boxes using the **Labeling** tool. Every variable used in the Formula Node must be declared as an input or an output with no two inputs and no two outputs possessing the same name. An output can have the same name as an input. Intermediate variables (that is, variables used in internal calculations) must be declared as outputs, although it is not necessary for them to be wired to external nodes. In Figure 5.33, the y and z variables are both declared as outputs, although the intermediate variable y does not have to be wired to an external node.

5.6.2   Formula Statements

Formula statements use a syntax similar to most text-based programming languages for arithmetic expressions. You can add comments by enclosing them inside a slash-asterisk pair (/*comment*/). Figure 5.34 shows the operators and

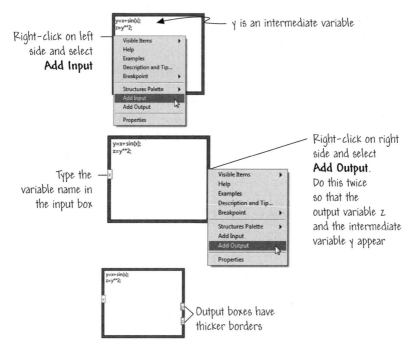

Right-click on left side and select **Add Input**

y is an intermediate variable

Right-click on right side and select **Add Output**. Do this twice so that the output variable z and the intermediate variable y appear

Type the variable name in the input box

Output boxes have thicker borders

FIGURE 5.33
Formula Node input and output variables.

functions that are available inside the Formula Node. You can access the same list of available operators and functions by choosing Help≫Show Context Help and moving the cursor over the Formula Node in the block diagram.

You enter the formulas inside the Formula Node using the **Labeling** tool. Each formula statement must terminate with a semicolon, and variable names are case-sensitive. There is no limit to the number of variables or formulas in a Formula Node. If you have a large number of formulas, you can either enlarge the Formula Node using the **Positioning** tool, or you can right-click in the For-mula Node (not on the border) and choose **Scrollbar**. The latter method will put a scrollbar in the Formula Node, and with the **Operating** tool you can scroll down through the list of formulas for viewing.

The following example shows how you can perform conditional branching inside a Formula Node. Consider the following code fragment that computes the ratio of two numbers, x/y:

if $(y \neq 0)$ then
$z = x/y$
else
$z = +\infty$
end if

Select **Show Context Help**
and move cursor
over Formula Node

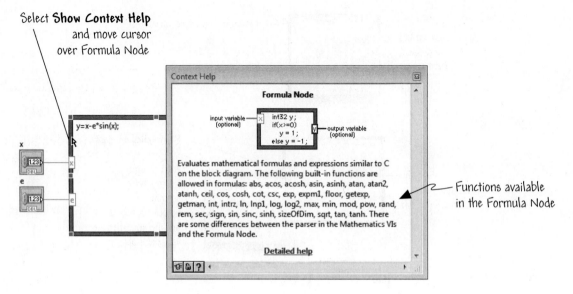

FIGURE 5.34
Functions available in Formula Nodes.

When $y = 0$, the result is set to ∞. You can implement the code fragment given above using a Formula Node, as illustrated in Figure 5.35.

The Formula Node demonstration depicted in Figure 5.35 can be found in **Chapter 5** *of the* **Learning** *directory and is called* **Formula Code.vi.** *Open the VI and try it out!*

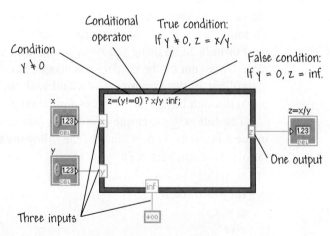

FIGURE 5.35
Implementing the formula $z = x/y$ in a Formula Node.

5.7   DIAGRAM DISABLE STRUCTURES

The Diagram Disable structure is used to disable specific sections of code on the block diagram. This is equivalent to commenting out code in a text-based programming language. The structure is found on the **Programming≫Structures** palette as shown in Figure 5.1.

To disable specific sections of code, place a Diagram Disable structure around the code on the block diagram that you want to disable. When placed on the block diagram, the selected code is automatically placed in the Disabled case of the Diagram Disable structure. The structure also has a second case, the Enabled diagram case, which is empty by default. The Enabled subdiagram is used to include the code that you want to enable in place of the code in the Disabled subdiagram. In order for the VI to execute, no coding errors can exist in the Enabled subdiagram. Coding errors can exist in the Disabled subdiagram and the VI will still compile and execute.

In Figure 5.36, a section of code in the Disabled subdiagram contains an error that prevents the VI from running. With that code encased within the

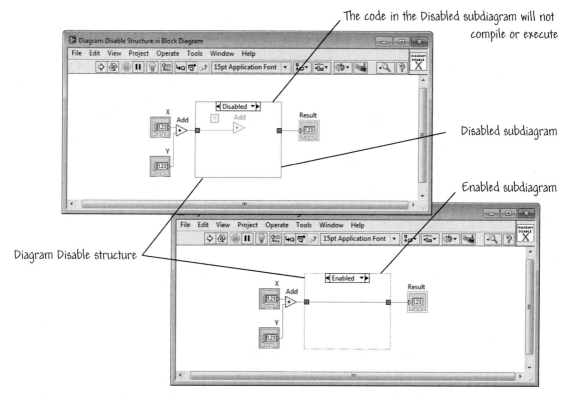

FIGURE 5.36
Using the Diagram Disable structure.

Diagram Disable structure and the Enabled subdiagram properly debugged, we can bypass the faulty code until it is debugged and fixed. This is a simple example and the error is readily fixed. Other circumstances beyond simple debugging, however, can benefit from the ability to bypass sections of code.

The VI in Figure 5.36 can be found in **Chapter 5** *of the* **Learning** *directory. It is named* **Diagram Disable Structure.vi.**

Open the Diagram Disable Structure VI shown in Figure 5.36 and run the VI. Notice that only the code in the Enabled subdiagram executes. Should you want to execute the disabled section of the block diagram, right-click on the structure while the Disabled subdiagram is visible and select **Enable This Subdiagram**. If you want to permanently remove the Diagram Disable structure and leave the subdiagram visible, right-click on the structure and select **Remove Diagram Disable Structure**.

In the VI shown in Figure 5.36, if you enable the Disabled subdiagram, the VI will show a **Broken Run** condition until you fix the coding error (by wiring the constant π to the Add function).

5.8   LOCAL VARIABLES

When you create VIs and subVIs, front panel objects utilize block diagram terminals to enable reading or writing of data to and from the front panel objects. Each front panel object has only one corresponding block diagram terminal. However, you may need to access the data in that terminal from more than one location in the code. Using local variables allows you to access front panel objects from more than one location in a single VI and to read or write to one of the controls or indicators on the front panel of a VI. In other words, local variables pass information between objects that you cannot connect with a wire. A common application of Local Variables is sharing data between two parallel While Loops.

Global variables are used to access and pass data among several VIs. This advanced LabVIEW topic is not covered here.

5.8.1   Creating Local Variables

When you create a local variable, an icon appears on the block diagram, as shown in Figure 5.37. The VI in Figure 5.37 uses a toggle switch on the front

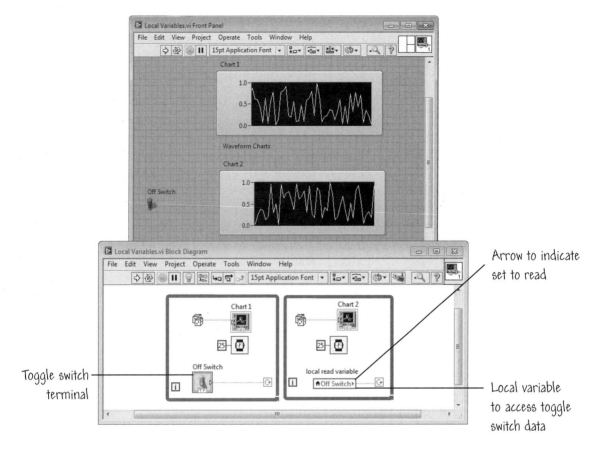

FIGURE 5.37
Using Local Variables.

panel. The toggle switch has the corresponding terminal in the While Loop on the left side of the block diagram. In this particular VI, the information from the toggle switch is also required in the While Loop on the right side of the block diagram. In this situation a local read variable can be used to pass the information from the toggle switch to two locations within the block diagram. Local variable nodes contain arrows to indicate whether the variable is set to read or write. An arrow appears on the right if the variable is set to read, and an arrow appears on the left if the variable is set to write. In Figure 5.37, the local variable named **local read variable** has an arrow on the right side to indicate it is set to read the **Off Switch** Boolean.

Writing to a local variable is similar to passing data to any other terminal. However, you can write to a local variable even if it is a control, or read from a local variable even if it is an indicator. In effect, with a local variable you can provide programmatic access to a front panel object as both an input and an output. When you create a local variable, it appears on the block diagram but not on the front panel.

The VI in Figure 5.37 can be found in **Chapter 5** *of the* **Learning** *directory. It is named* **Local Variables.vi.**

To create a local variable right-click on the desired front panel object or block diagram terminal of interest and select **Create≫Local Variable** from the shortcut menu. You can accomplish the same thing from the **Functions** palette by selecting Local Variable, as shown in Figure 5.1. Once the local variable node is placed on the block diagram, you must associate it with a control or indicator by right-clicking the node and choosing **Select Item** from the shortcut menu. The expanded shortcut menu lists all the front panel objects that have owned labels.

Since LabVIEW uses owned labels to associate local variables with front panel objects, be sure to label the front panel controls and indicators with descriptive owned labels.

5.8.2 Use Local Variables With Care

Since local variables are not inherently part of the LabVIEW dataflow execution model, overusing or misusing them can make block diagrams difficult to read or create unpredictable situations. A misuse of local variables, for example, would be to use them to access values in each frame of a Sequence structure. An overuse of local variables would be to use them to avoid long wires across the block diagram.

When factoring in issues associated with dataflow programming (discussed in Chapter 2), the outcome of the VI can depend on which actions occur first when two or more sections of code that execute in parallel change the value of a shared resource (as can occur with local variables). This *race condition* can cause unpredictable outcomes. In some programming languages, data flows from the top down, thereby ensuring the execution order. In LabVIEW, the dataflow concept is such that the execution order is not ensured beforehand. You can use wiring to perform multiple operations on a variable while avoiding race conditions and use other coding strategies such as the use of Sequence structures to guarantee the execution order. In the early stage of learning LabVIEW you will likely not need to use local variables—just make sure to search the on-line help for more detailed information if needed.

5.8.3   Initializing Local Variables

It is recommended that you initialize local variables with known data values before the VI runs. Otherwise the variable contains the default value of the associated front panel object. This might cause the VI to behave incorrectly. Also, if the variable relies on a computation result for the initial value, make sure LabVIEW writes the value to the variable before it attempts to access the variable for any other action (see the discussion of the race condition above).

5.8.4   Memory and Execution-Speed Considerations

Using local variables has memory and execution-speed implications. Local variables make copies of data buffers, so that when you read from a local variable, you create a new buffer for the data. If you use local variables to transfer large amounts of data from one place on the block diagram to another, you generally use more memory and have slower execution speed than if you used a wire. Also, if you need to store data during execution, consider using a shift register rather than local variables.

5.9   COMMON PROGRAMMING TECHNIQUES

As your understanding of LabVIEW basics develops and you are ready to build more sophisticated projects, you need to begin exploring the various software design architectures that exist within graphical programming environment. The most common of these are grouped into patterns recognized by other users and developers. The use of accepted design patterns will help you understand and improve existing projects. Many common design patterns are available to LabVIEW developers, and most VI projects use at least one.

Software design architectures are advanced topics, and in this introductory text we can only skim the surface. Here we consider sequential programming, state programming and state machines, and parallelism.

5.9.1   Sequential Programming

Many of your programs will need to perform tasks sequentially. For example, consider the VI code shown in Figure 5.38. Its purpose is to play a waveform at one frequency (say 200 Hz) and then display a screen message to the user, stating the frequency. Next, the VI should play a second waveform at a different frequency (say 300 Hz) and display a screen message announcing that frequency to the user. The idea is to allow the user to compare the sound made by different waveforms with given frequencies. This is a sequential task. However, the reality of LabVIEW dataflow programming is such that you cannot

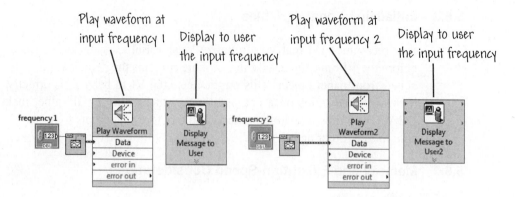

FIGURE 5.38
Unsequenced tasks.

predict the execution order of the four tasks (play waveform at the first frequency, show message one, play waveform at second frequency, and show message two). There is nothing in the block diagram to force the execution order, and in fact any one of these four events could happen first.

Using error-cluster wires (discussed in Section 4.7), you can force sequential execution of the two Play Waveform Express VIs as shown in Figure 5.39. Even then, however, the execution of the two Display Message to User Express VIs remains unpredictable, since they themselves are not wired. Although in this case the Display Message to User Express VIs could have been wired to the error clusters, often it is not possible to control the execution order of all subVIs or functions.

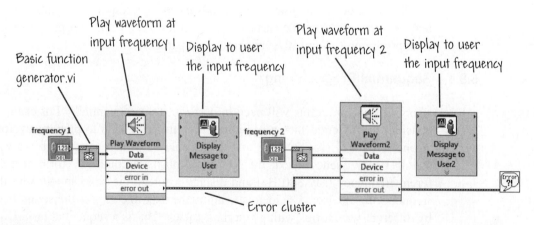

FIGURE 5.39
Partially sequenced tasks using error clusters.

An obvious choice is to utilize Sequence structures to force the execution order. As discussed in Section 5.5, a Sequence structure contains one or more subdiagrams that execute in sequential order. Since a frame cannot begin execution until everything in the previous frame has executed, this provides direct control of the execution order. Figure 5.40 shows an example of this VI using a Sequence structure to force execution order. There are valid reasons to avoid overusing Sequence Structures (see Section 5.5.2). While they ensure the order of execution, they hinder parallel operations and restrict the ability to halt execution of the sequence.

A recommended way to construct this VI is to enclose the Display Message to User Express VIs in Case structures and wire the error clusters to the case selector, as shown in Figure 5.41. This coding approach using Case structures and error clusters enables the sequential operations as desired, using Sequence structures sparingly while relying on data flow to control the execution order.

5.9.2   State Programming and State Machines

Software design based on sequential programming (as discussed in Section 5.9.1) may provide a good solution—but it may not. Even if the sequential-programming architecture is sufficient for the task at hand, it may be a good idea to consider other possibilities. The state programming architecture is a good choice, as it provides flexibility to address important questions. What if the problem you are solving requires that the sequence of the events be changed or that one or more events repeat more often than the others? Often, specific sections of code should execute only if certain conditions are satisfied. What if you need to

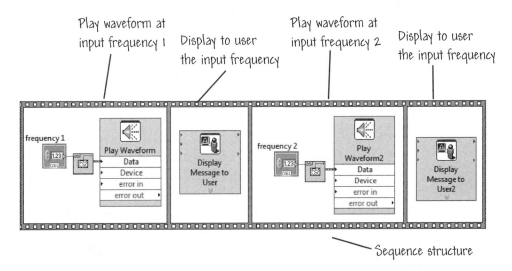

FIGURE 5.40
Sequenced tasks using Sequence structures.

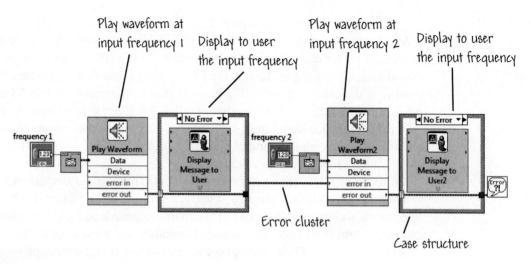

FIGURE 5.41
Sequenced tasks using Case structures and error clusters.

stop code execution of the code immediately rather than waiting for a sequence to end naturally? These questions can be addressed using a state programming architecture.

The common and useful **state machine design pattern** can be used to implement algorithms that can be described by a state diagram depicting the system's behavior. The **state diagram** describes the possible states of an object and can be used to graphically represent state machines. State machines have many variations, the most common being the Mealy machine and the Moore machine. A Mealy machine performs an action for each transition. A Moore machine performs a specific action for each state in the state transition diagram.

The state machine design pattern template in LabVIEW implements any algorithm described by a Moore machine.

What is a state machine? This is an advanced topic beyond the scope of this text. For our purposes we can imagine a state machine as a "model of behavior" consisting of states, state transitions, and actions. We can think of the state as an arrangement of information within the program (that is, within the machine). Since the current state of the system is reached by transitioning from previous states, the state is a record of past information (hence it supplies memory). A state transition designates a state change. Consider a lamp that can be either on or off (the two states). The actions are flipping the lamp switch up or down (two actions) and the two transitions are [lamp on to lamp off] and [lamp off to lamp on].

One advantage of using a state machine is that you can readily create a VI after you have created a state diagram. Translating the state diagram into a LabVIEW block diagram requires the following infrastructure components:

- While Loop—Continually executes the various states
- Case Structure—Contains a case for each state and the code to execute for each state
- Shift Register—Contains state transition information
- State Functionality Code—Implements the function of the state
- Transition Code—Determines the next state in the sequence

Figure 5.42 shows the basic structure of a state machine implemented in LabVIEW. The flow of the state transition diagram is implemented by the While Loop. The individual states are represented by cases in the Case structure. A shift register on the While Loop keeps track of the current state and communicates it to the Case structure input.

Enumerated type controls are often used to control initialization and transition of state machines. There are many ways to control what case a Case structure executes in the state functionality code and the state transition code. This is a software design issue, and you should implement an approach that matches the function and complexity of your state machine. The easiest approach is to use is the single Case structure transition code to transition between any number of states. This method provides for the most scalable, readable, and maintainable

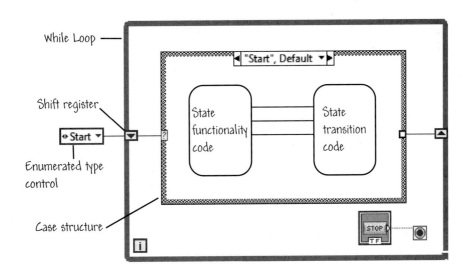

FIGURE 5.42
Basic infrastructure of a LabVIEW state machine.

state machine architecture. Other useful methods include the default transition, transition between two states, and transition between two or more states.

*LabVIEW provides a template for a standard state machine design pattern. On the **Getting Started Window**, select File≫New and choose the VI folder. Select the Standard State Machine VI in the **Design Patterns** subfolder within the **Frameworks** subfolder in the **From Templates** folder.*

An example VI that implements a test sequencer for automated execution of a set of VIs using a state machine is provided in the VI library **testseq.llb** in the **apps** subfolder within the **examples** folder in the **LabVIEW** directory. Open the VI library and select **State Machine Test Sequencer.vi** and navigate to the block diagram shown in Figure 5.43. From the documentation associated with the VI we learn that VIs to be tested are placed in the Case structure. The sequence ring constant (outside the loop) specifies the first test, and the size of that ring indicates the number of tests to be executed if all pass. The next test to be executed is determined by the value set in the ring in the current test case. Each VI passes out a Boolean value that is set to TRUE if the test passed and FALSE if the test failed. If the Stop on fail control is set to TRUE, the sequence will stop executing if any test fails. The Current Test Number indicator shows the number of the case that is currently executing. After the sequence finishes executing, the Result indicator is set to TRUE if the sequence passed (all VIs passed) or FALSE if the sequence failed (any VI failed).

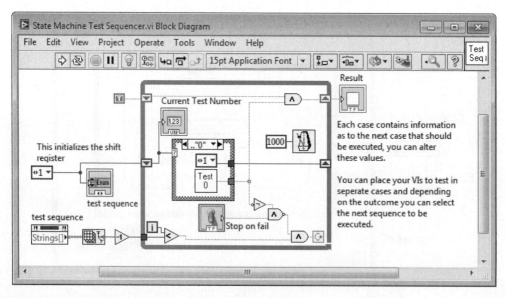

FIGURE 5.43
State machine example provided by LabVIEW.

5.9.3   Parallelism

Several important design patterns, such as parallel loop, master/slave, and producer/consumer, enable the execution of multiple tasks at the same time. These tasks can run in parallel if they do not have a data dependency between them and do not use the same shared resource.

More information on this advanced topic can be found at the National Instruments website, www.ni.com/white-paper/3023/en

The parallel loop design pattern is readily applicable to simple menu VIs, where the user selects from one of several buttons that perform different actions. The VI can handle simultaneous multiple independent tasks, such that responding to one action does not prevent responding to another action. For example, if a user clicks a button that displays a dialog box, parallel loops can continue to respond to I/O tasks.

The master/slave design pattern consists of multiple parallel loops. Each loop may execute at a different rate, where one loop serves as master and the others as slaves. The master loop controls the slave loops and communicates with them using messaging techniques. The master/slave design pattern is applicable when you need a VI to respond to user interface controls while simultaneously collecting data.

The producer/consumer design pattern is based on the master/slave pattern and improves data sharing among multiple loops running at different rates. Similar to the master/slave design pattern, the producer/consumer pattern separates tasks that execute at different rates. The parallel loops in the producer/consumer design pattern are separated into those that produce data and those that consume the data produced. Data queues communicate data among the loops. The producer/consumer design pattern is applicable when you need to acquire multiple sets of data that must be processed in order.

*LabVIEW provides a template for the master/slave and producer/consumer design patterns. On the **Getting Started Window**, select File≫ New and choose the VI folder. Select the Master/Slave Design Pattern VI or the Producer/ Consumer Design Pattern in the **Design Patterns** subfolder within the **Frameworks** subfolder in the From Template folder.*

LabVIEW has many functions and constructs to help tackle some of complexities of parallelism, such as Queues for data transfer, and Rendezvous, Semaphores, Notifiers, and Occurrences provide methods of synchronizing execution. Examples found in the **NI Example Finder** demonstrate the proper use of these techniques. The **NI Example Finder** is accessed by selecting Help≫ Find Examples. . . .

5.10   SOME COMMON PROBLEMS IN WIRING STRUCTURES

When wiring the structures presented in this chapter, you may encounter wiring problems. In this section, we discuss some of the more common wiring errors and present suggestions on how to avoid them. Some common problems with wiring structures are:

- Failing to wire a tunnel in all cases of a Case structure
- Overlapping tunnels
- Wiring underneath rather than through a structure.

5.10.1   Failing to Wire a Tunnel in All Cases of a Case Structure

When you wire from a Case structure to an object outside the structure, you must connect output data from all cases to the object. Failure to do so will result in a bad tunnel, as illustrated in Figure 5.44. This problem is a variation of the no source error. Why? Because at least one case would not provide data to the object outside the structure when that case executed. The problem is easily solved by wiring to the tunnel in all cases. Can you explain why this is not a multiple sources violation? It seems like one object is receiving data from mul-

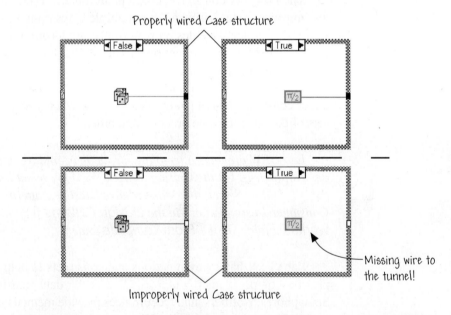

FIGURE 5.44
Failing to wire a tunnel in all cases of a Case structure leads to problems.

tiple sources. The answer is that only one case executes at a time and produces only one output value per execution of the Case structure. If each case did not output a value, then data flow execution would stop on the cases that did not output a value.

5.10.2   Overlapping Tunnels

Tunnels are created as you wire, resulting occasionally in tunnels that overlap each other. Overlapping tunnels do not affect the execution of the diagram, so this is not really an error condition. But overlapping tunnels make editing and debugging of the VI more difficult. You should avoid creating overlapping tunnels! The fix is easy—drag one tunnel away with the **Positioning** tool to expose the other tunnels. A problematic situation is illustrated in Figure 5.45.

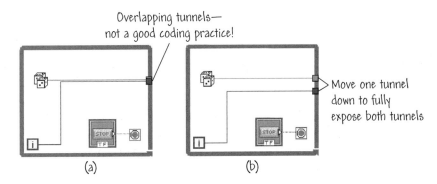

FIGURE 5.45
Sometimes overlapping tunnels occur when wiring structures.

5.10.3   Wiring Underneath Rather Than through a Structure

Suppose that you want to pass a variable through a structure. You accomplish this by clicking either in the interior or on the border of the structure as you are wiring. This process is illustrated in Figure 5.46. If you fail to click in the interior or on the border, the wire will pass underneath the structure, and some segments of the wire may be hidden. This condition is not an error per se, but hidden wires are visually confusing and should be avoided.

As you wire through the structure, you are provided with visual cues for guidance during the wiring process. For example, when the **Wiring** tool crosses the left border of the structure, a highlighted tunnel appears. This lets you know that a tunnel will be created at that location as soon as you click the mouse button. You should click the mouse button! However, if you continue to drag the tool through the structure (without clicking the mouse on the left border)

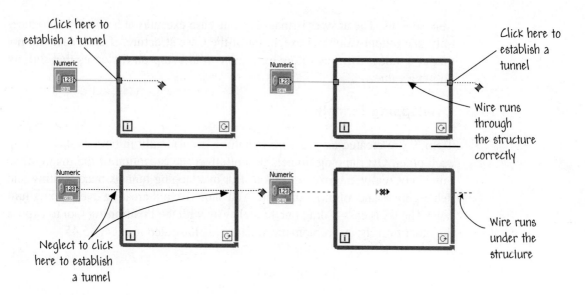

FIGURE 5.46
Wiring underneath rather than through a structure is a common problem.

until the tool touches the right border of the structure, a second highlighted tunnel appears. If you continue to drag the **Wiring** tool past the right border of the structure without clicking, both tunnels disappear. When this occurs, the wire passes underneath the structure rather than through it, as depicted in Figure 5.46.

BUILDING BLOCK

5.11   BUILDING BLOCKS: PULSE WIDTH MODULATION

In this Building Block you will create a Pulse Width Modulation VI using a While Loop, a Case structure, and Shift Registers. A new element of the design implementation will be the realization of the VI as a state machine. A state machine is an architecture employed by LabVIEW developers to build complex decision-making algorithms quickly as discussed in Section 5.9.2. State machines are useful in situations where distinguishable states exist, such as in the Pulse Width Modulation VI where the output is dependent upon the current output value as well as the input values from the Clock and the user controls Period and Duty Cycle.

The VI will have three states, defined by three different cases within a Case structure. To simplify the task, use the Standard State Machine VI Template. On the Getting Started screen, select File≫New and create a new VI that uses the **Standard State Machine** template found in VI≫From Template≫Frameworks≫ Design Patterns. Save the VI as PWM with LED.vi.

Your state machine will have three states: **Update**, **High**, and **Low**. The initial state will be called **Update**. In this state, the most recently entered values for Period and Duty Cycle will be read. These new values will then be stored in shift registers to be read by the other states. At the beginning of each cycle the state machine should return to this state to check for new values for Period and/or Duty Cycle. The VI is in the **High** state whenever the pulse output is high. Similarly, the VI is in the **Low** state whenever the output is low. In order to determine the transitions between the high and low state, you will need to use the Rising Edge and Falling Edge subVIs created in the previous Building Blocks.

Working versions of the Falling Edge subVI and the Rising Edge subVI can be found in the Building Blocks folder of the Learning directory.

Each state in the template block diagram is represented as a value of an enumeration. This enumeration is a customized control. Right-click on one of the enumerations (for example, **Initialize** on the left-hand side of the block diagram) and select **Open Type Def.** to modify the master copy of the custom control, as illustrated in Figure 5.47. Type definitions are used when you want to place a control of the same data type in many places.

Using the **Labeling** tool, change the values of the enumerations to reflect the three states of your state machine: **Update**, **High**, and **Low**. Edit **Initialize** and change it to **Update**. Similarly, edit **Stop** and change it to **High**. You will need to right-click on the enumeration and select **Add Item After** to add the **Low** state. Then save the control as PWM States.ctl. Note that now all instances of the type definition have been updated with your customized states.

An edited file named PWM States.ctl can be found in the Building Blocks folder of the Learning directory.

Using Figure 5.48, construct a front panel with Duty Cycle and Period as the numeric controls and an LED and Clock numeric indicator as outputs. You want to be able to stop your VI at any time, and since it does not have a final state, you should delete the current input to the **Stop Condition** terminal and create a stop button as the control.

The Case structure currently has only two cases, so you will need to add another case after **High**. Since you saved your new version of the PWM States control, LabVIEW will automatically name the case you add **Low**. Using the block diagrams in Figure 5.49 as a guide, add your subVIs to the Pulse Width

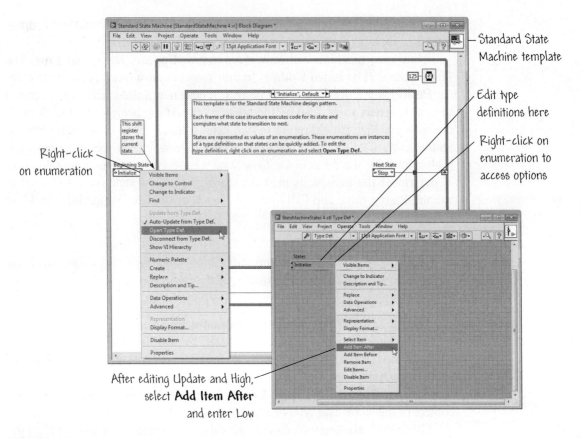

FIGURE 5.47
Editing the enumeration values to include Update, High, and Low.

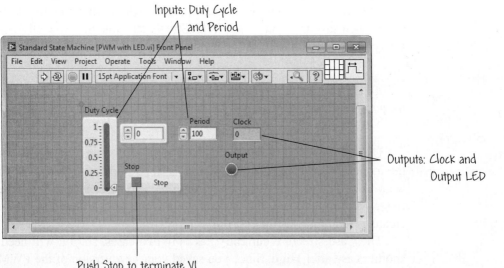

FIGURE 5.48
PWM with LED VI front panel.

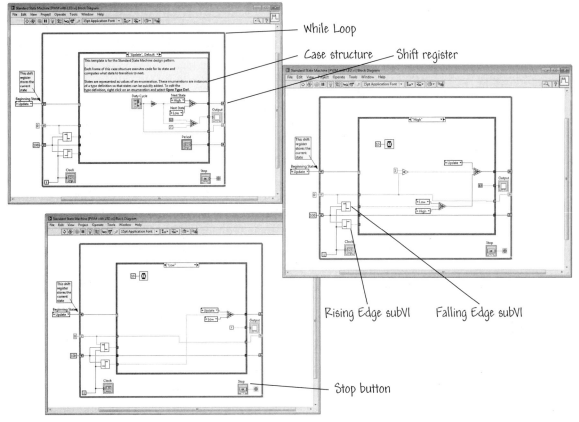

While Loop

Case structure   Shift register

Rising Edge subVI   Falling Edge subVI

Stop button

FIGURE 5.49
PWM with LED VI block diagram.

Modulation VI along with all other necessary coding to determine state transitions.

In later chapters we will learn how to graphically display the output waveform. For now, however, you will convert the output to a Boolean value and display it on the front panel using an LED. The LED should be on whenever the output waveform is high and off when it is low.

When you have completed your VI, experiment with different input values. The larger you make the **Period**, the easier it will be for you to observe the output pattern. You should find that the higher you make the **Duty Cycle**, the longer the LED remains lit for each cycle.

A working version of the PWM with LED VI can be found in the **Building Blocks** *folder in the* **Learning** *directory.*

5.12   RELAXED READING: REFINING THE PROCESS OF STEEL RECYCLING

In this reading, we learn about an automation system for a steel recycling facility that reduces the amount of energy consumed while improving safety for workers. Employing LabVIEW programmable automation controllers versus programmable logic controllers and ladder logic has led to a tenfold increase in efficiency and drastically reduced the costs of facility automation.

The most recycled material in North America is steel. In fact, two out of every three pounds of steel are produced from previously used steel. Steel companies are constantly refining their recycling operations to make them more efficient and environmentally friendly. Recycling steel consumes between 60% and 74% less energy than producing new steel from raw materials. The annual energy saving from using recycled rather than newly produced steel would power 18 million homes.

Nucor is the largest steel recycler in North America. In 2005, Nucor purchased the Marion Steel Company in Marion, Ohio, a location central to nearly 60% of the steel consumption in the United States. Nucor saw the need to implement a facility automation system to improve plant efficiency and safety.

A full line of rebar, sign supports, delineators, and cable barrier systems using recycled steel is manufactured at the Marion facility, shown in operation in Figure 5.50. During the steel recycling process scrap metal is heated in an electric arc furnace, and a combination of elements is added to the viscous steel to create the appropriate steel alloy. This process requires large amounts of energy, which vary significantly depending on the amount of scrap placed in the furnace. Originally operators relied on estimates, and consequently the metal often was overheated. The result was steel of unsatisfactory quality, which had to be recycled again at additional cost in time, money, and energy. To reduce the number of reheats, a low-cost scale and weighing system is now employed, using LabVIEW and NI Compact FieldPoint controllers that accurately calculate the amount of steel in each burn. Knowing the exact amount of scrap metal placed in the furnace allows a precise calculation of the amount of electricity required to heat it.

Drawing the large amount of electricity required to heat the furnace for recycling entails a risk of causing flicker on the power grid—an inconvenience to local residential communities. To limit electricity consumption, an online reactor was placed in series with the furnace, using the LabVIEW FPGA module and the CompactRIO platform to measure the amount of energy being drawn from the power grid. If the furnace approaches the prescribed limit, the system can quickly change control methods to reduce the power draw.

FIGURE 5.50
Using hardware and software from National Instruments, Nucor developed a variety of automation systems that have greatly reduced the amount of electricity used and eliminated potential safety issues. (Courtesy of National Instruments.)

Another key objective leading to facility improvements is employee safety. In rendering the steel recycling plant a safer place to work, for example, the method for turning the electric arc furnace on and off needed improvement. An operator had to manually pull the on/off switch, thus becoming vulnerable to injury if the fuse were to blow. The renovation used a Compact Field-Point programmable automation controller and a human-machine interface to create a remote power switch that does not put operators in potentially dangerous situations.

Using hardware and software from NI, a variety of automation systems have greatly reduced electricity usage and eliminated potential safety issues in the steel recycling plant at the Marion facility. Employing LabVIEW programmable automation controllers versus programmable logic controllers and ladder logic has led to a tenfold increase in efficiency and drastically reduced the costs of facility automation. Additionally, by development of a proactive approach for monitoring power intake using the National Instruments platform, the impact of power grid flicker has been significantly reduced.

For more information on steel at the Nucor Corporation, please visit the website http://www.nucor.com, and for more on steel recycling at Nucor visit http://sine.ni.com/cs/app/doc/p/id/cs-11227.

5.13   myDAQ BUILDING BLOCKS

In this myDAQ Building Block, you will continue to enhance the capabilities of the temperature monitoring system using a While Loop to enable the system to run continuously and a Case structure to respond to changes in temperature. Begin by opening Control System unedited.vi from the Chapter 5 folder in the myDAQ Building Blocks folder within the Learning directory.

Make sure that both Read Temp subVI.vi and LED Control subVI.vi are in the Support VIs folder in the myDAQ Building Blocks folder within the Learning directory. If the subVIs are not there, you can add them from the Chapter 4 Solution folder in the myDAQ Building Blocks» Solutions folder within the Learning directory.

Use the following steps as a guide to adding the While Loop and Case structure. The front panel and block diagram for the Control System unedited VI are shown in Figure 5.51.

1. Add the Read Temp subVI.vi to the block diagram by selecting **Select VI** ... on the **Functions** palette (or by dragging the subVI from the

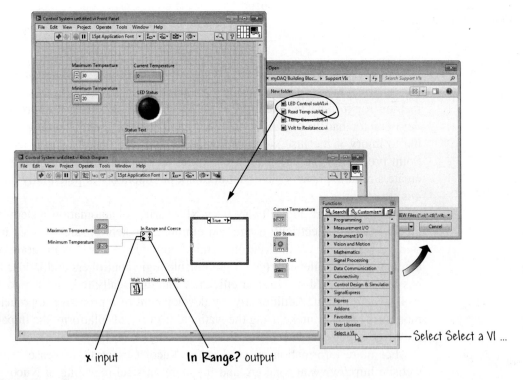

FIGURE 5.51
The Control System unedited.vi.

Support VIs folder). Wire the Temperature output of the Read Temp subVI.vi to the x input of the In Range and Coerce function. Wire the In Range? output of the In Range and Coerce function to the selector terminal of the Case structure.

2. Place a false constant and a string constant in the True case within the Case structure. In this case the temperature is in range, so you will not turn on the LED. In this case, let the string constant be shown as **In Range**.

*The Boolean constants can be found on the **Functions≫Programming≫Boolean** palette and the string constants can be found on the **Functions≫Programming≫String** palette.*

3. Switch to the False case of the Case structure. Place a true constant and a string constant in the False case within the Case structure. In this case the temperature is out of range, so the LED will turn on. In this case, let the string constant be shown as **Out of Range**.

4. Wire the constants from both cases to tunnels on the right side of the Case structure. You should have one Boolean tunnel (green) and one string tunnel (pink).

5. Add the LED Control subVI.vi to the block diagram from the Support VIs folder. Wire the output from the Boolean tunnel on the Case structure to the LED Control input of the LED Control subVI.vi. Wire the LED Status output of the LED Control subVI.vi to the LED Status indicator.

6. Wire the temperature and status text to the appropriate indicators.

7. Add a While Loop around all of the code on the block diagram to allow the program to run continuously. Right click on the loop condition terminal and select **Create Control** to create a stop button.

8. Create an appropriate input for the Wait Until Next ms Multiple function so that the loop runs at 10 Hz.

9. Run the VI. Adjust the Maximum Temperature and Minimum Temperature values to see what happens when the temperature is out of range.

10. Save the edited VI in the Chapter 5 folder in the myDAQ Building Blocks folder within the Learning directory and name it Control System.vi. Close the VI.

A working version of Control System.vi can be found in the Chapter 5 folder of the myDAQ Building Blocks≫Solutions folder in the Learning directory.

In this myDAQ Building Block, you enhanced the capabilities of the temperature monitoring system using a While Loop to enable the system to run continuously and a Case Structure to respond to changes in temperature. The VI is now ready to acquire multiple data points in Chapter 6.

5.14 SUMMARY

In this chapter we studied four structures (For Loops, While Loops, Flat Sequence structures, and Case structures). The While Loop and the For Loop are used to repeat execution of a subdiagram placed inside the border of the resizable loop structure. The While Loop executes as long as the value at the conditional terminal is either TRUE or FALSE. The For Loop executes a specified number of times. Shift registers are variables that transfer values from previous iterations to the beginning of the next iteration.

The Case structure and Sequence structures are utilized to control data flow. Case structures are used to branch to different subdiagrams depending on the values of the selection terminal of the Case structure. Sequence structures are used to execute diagram functions in a specific order. The portion of the diagram to be executed first is placed in the first frame of the Sequence structure, the diagram to be executed second is placed in the second frame, and so on. Use the Flat Sequence structure to avoid using sequence locals and to better document the block diagram.

You can directly enter formulas on the block diagram in the Formula Node. The Formula Node is useful when the function equation has many variables or otherwise would require a complex block diagram model for implementation.

The Diagram Disable structure was presented as a way to disable specific sections of code on the block diagram equivalent to commenting out code in a text-based programming language. We showed how to use local variables to access front panel objects from more than one location in a single VI and to read or write to one of the controls or indicators on the front panel of a VI. We briefly introduced software design architectures that exist within the graphical programming environment, including sequential programming, state programming and state machines, and parallelism.

We also discussed the matter of controlling and timing the execution of structures. The Wait (ms) and Tick Count (ms) functions in conjunction with a Sequence structure let you accomplish these tasks using traditional VIs. You can also accomplish the timing task using the Time Delay Express VI and the Elapsed Time Express VI.

KEY TERMS

Case structure: A conditional branching control structure that executes one and only one of its subdiagrams, based on specific inputs. It is similar to If-Then-Else and Case statements in conventional programming languages.

Coercion: The automatic conversion performed in LabVIEW to change the numeric representation of a data element.

Coercion dot: Dot that appears where LabVIEW is forced to convert a numeric representation of one terminal to match the numeric representation of another terminal.

Conditional terminal: The terminal of a While Loop that contains a Boolean value that determines whether the loop performs another iteration.

Control flow: Determining the execution order of a program by arranging its elements in a certain sequence.

Count terminal: The terminal of a For Loop whose value determines the number of times the For Loop executes its subdiagram.

Data dependency: The concept that block diagram nodes do not execute until data is available at all the node inputs.

Diagram Disable Structure: Structure used to disable specific sections of code.

For Loop: Iterative loop structure that executes its subdiagram a set number of times.

Formula Node: Node that executes formulas that you enter as text. Especially useful for lengthy formulas too cumbersome to build in the block diagram.

Frames: Diagrams associated with Sequence structures that look like frames of film and control the execution order of the program.

Iteration terminal: The terminal of a For Loop or While Loop that contains the current number of completed iterations.

Local Variables: Variables used to access front panel objects from more than one location in a single VI and to allow one to read or write to one of the controls or indicators on the front panel of a VI.

Object: Generic term for any item on the front panel or block diagram, including controls, nodes, and wires.

Parallism: A software design pattern such as parallel loop, master/slave, and producer/consumer that enables the execution of multiple tasks at the same time.

Sequence structure: Program control structure that executes its subdiagram in numeric order. Commonly used to force nodes that are not data-dependent to execute in a desired order.

Sequential Programming: A software design pattern in which tasks are performed sequentially.

Shift register: Optional mechanism in loop structures used to pass the value of a variable from one iteration of a loop to a subsequent iteration.

State Machines: A model of behavior consisting of states, state transitions, and actions.

State Programming: A common and useful software design pattern that can be used to implement algorithms that can be described by a state diagram depicting the behavior of the system.

Structure: Program control element, such as a Sequence, Case, For Loop, or While Loop.

Subdiagram: Block diagram within the border of a structure.

Tunnels: Relocatable connection points for wires from outside and inside the structures.

While Loop: Loop structure that repeats a section of code until a Boolean condition is False.

EXERCISES

E5.1 Open the block diagram of Rearrange Cases.vi found in the Exercises&Problems folder in Chapter 5 of the Learning directory. Analyze all the possible temperature values listed in the Case Structure. Notice that the case values are not in ascending numerical order. In order to change the order of case frames, right-click on the **Case** border and select **Rearrange Cases**. This will bring up the **Rearrange Cases** wizard, which displays the current order of the cases. Click the **Sort** button to arrange the cases in ascending numerical order. Click **OK** if you want to leave them in ascending order. Otherwise, if you want the case frames to display in any other order, click and drag the values in the Case List.

E5.2 Open the block diagram of Red Dots.vi found in the Exercises&Problems folder in Chapter 5 of the Learning directory. Notice there are two red dots to signify numeric conversion (see Figure E5.2). In general, you should try to eliminate as many red dots in your VIs as possible because they slow down execution and use more memory for mathematical operations. To eliminate red dots, right-click on the terminal with the red dot and select **Representation**. In this VI, right-click on the indicator terminal labeled Data Points and choose the same representation as the input value—which is DBL. To eliminate the other red dot on the N terminal of the For Loop, change the representation on the control terminal titled Number of Data Points to match the N terminal—which is I32.

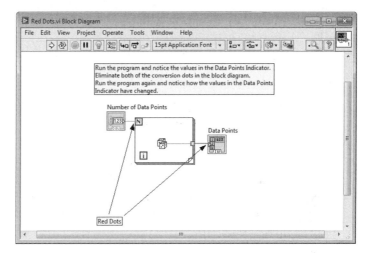

FIGURE E5.2
Eliminating red dots.

E5.3 Open Case Errors.vi found in the Exercises&Problems folder in Chapter 5 of the Learning directory and click on the **Broken Run** arrow. This will display a list of the errors in this VI (see Figure E5.3). Use the error list to determine how to fix the VI so you can run it.

E5.4 Using a single While Loop, construct a VI that executes a loop N times or until the user presses a stop button. Be sure to include the Time Delay Express VI so the user has time to press the **Stop** button.

E5.5 Open the block diagram of the Avoid Sequence.vi found in the Exercises& Problems folder in Chapter 5 of the Learning directory, and gain an

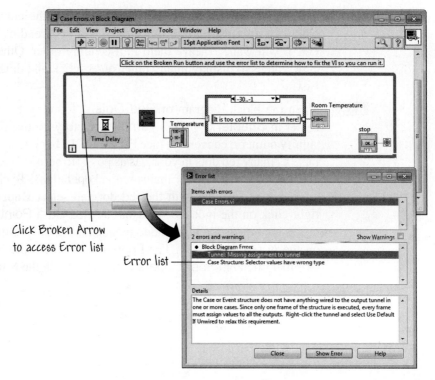

FIGURE E5.3
Eliminating Case structure errors.

understanding of the program. Rewrite this VI without using the Flat Sequence structure. The VI can execute in the correct order without the Sequence structure using data flow. Remember that it is important to use Sequence structures only when it is not possible to wire nodes together to ensure that they execute in the proper order.

E5.6   Open the block diagram of the Loan Calculator.vi found in the Exercises& Problems folder in Chapter 5 of the Learning directory. This program will compute monthly payments required to pay off a loan based on the inputs on the front panel. The user has three choices for how the interest is compounded and two choices for the payment frequency. Click on the **Broken Run** arrow to display the list of errors. You will need to program the following equations into the four Formula Express VIs:

Interest Compounded Monthly	$i = \text{APR}/\text{PymtFreq}$
Interest Compounded Daily	$i = (1 + (\text{APR}/365))^{\text{PymtFreq}} - 1$
Interest Compounded Continuously	$i = e^{(\text{APR}/\text{PymrFreq})} - 1$
Payment Amount	$A = P[i(1 + i)^N]/[(1 + i)^N - 1]$

E5.7 Open the block diagram of the **Compute Equilibrium.vi** found in the **Exercises&Problems** folder in **Chapter 5** of the **Learning** directory. This VI computes the equilibrium force required to keep a beam in equilibrium. Add a Case structure to the program so the user can choose whether the native LabVIEW functions or the Formula Express VIs will be used to calculate the result. The Formula Express VIs required are already programmed for you; you just need to add the Case structure, move them into the appropriate case, and wire them correctly.

Review: In order for a beam to be at equilibrium, the sum of the moments about any point must be zero. This example uses the sum of the moments about one end of the beam and the sum of the moments about the point where force A is located to determine the C and z values associated with the equilibrium force.

E5.8 Create a VI that uses a Flat Sequence structure and the Elapsed Time Express VI to measure how long it takes LabVIEW to execute **Open URL in Default Browser.vi**, found on the **Functions≫Programming≫Dialog & User Interface≫Help** palette. Display the execution time on a numeric indicator on the front panel. The elements of the block diagram that you will need to utilize in your VI are depicted in Figure E5.8.

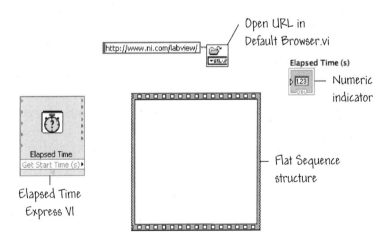

FIGURE E5.8
Elements needed on the block diagram.

E5.9 Open the While Loop Exercise VI found in the **Exercises&Problems** folder in **Chapter 5** of the **Learning** directory. This VI is supposed to run continuously until the random number (between 1 and 10) generated outside the While Loop is equal to the user input number (also ranging from 1 to 10).

The uncorrected block diagram is shown in Figure E5.9. Run the VI and verify that the operation of the VI is not as desired. There are two possible

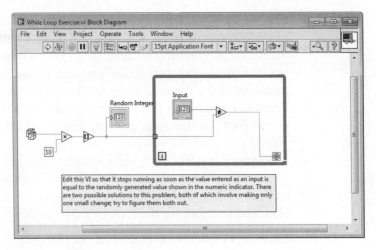

FIGURE E5.9
Block diagram for the While Loop Exercise VI.

modifications that would achieve the desired functionality—both minor. See if you can find both fixes to the VI.

E5.10 Create a front panel that has 8 LED indicators and a vertical slider control that is an 8-bit unsigned integer. Display a digital indicator for the slider, making sure that the LEDs are evenly spaced and aligned at the bottom. The problem is to turn the 8 LEDs into a binary (base 2) representation for the number in the slider. For example, if the slider is set to the number 10 (which in base 2 is $00001010 = 1 * (23) + 1 * (21)$), the LEDs 1 and 3 should be on. To test your solution, check the number 131. Since 131 is 10000011 in base 2, LEDs 0, 1, and 7 should be on.

What happens if you input a number greater than 255? Modify the VI using case structures to provide a check. If the number is greater than 255, send a warning message to the user.

This exercise is an extension of the design problem D3.2 in Chapter 3.

PROBLEMS

P5.1 Open the block diagram of the Statistics.vi found in the Exercises&Problems folder in Chapter 5 of the Learning directory. This program will compute the number of combinations possible if you are to select *r* out of *n* possible objects. Add Case structures to the code to utilize the inputs **Does Order Matter?** and

Can Samples Repeat? and display the appropriate result on the front panel. The table below outlines which equation should be used in which case:

	Order Matters	**Order Doesn't Matter**
Samples Repeat	n^r	$(n-1+r)!/(n-1)!r!$
Samples Cannot Repeat	$n!/(n-r)!$	$n!/r!(n-r)!$

P5.2 Use a For Loop to generate 100 random numbers. Determine the most current maximum and minimum number as the random numbers are being generated. This is sometimes referred to as a "running" maximum and minimum. Display the running maximum and minimum values as well as the current random number on the front panel.

Be sure to include the Time Delay Express VI so the user is able to watch the values update as the For Loop executes. Figure P5.2 shows the Time Delay Express VI configuration dialog box (accessed by double-clicking on the VI) with a recommended delay of 0.1 second.

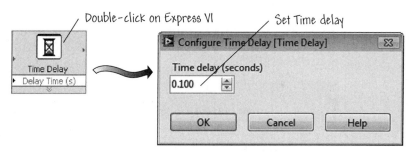

FIGURE P5.2
Time Delay Express VI with configuration dialog box showing 0.1 second delay.

Place the Time Delay Express VI within the For Loop.

P5.3 Construct a VI that displays a random number between 0 and 1 once every second. Also, compute and display the average of the last four random numbers generated. Display the average only after four numbers have been generated; otherwise display a 0. Each time the random number exceeds 0.5, generate a beep sound using the Beep.vi.

*The Beep VI can be found on the **Functions≫Programming≫Graphics& Sound** palette.*

P5.4 Construct a VI that has three Round LEDs on the front panel. When you run the program, the first LED should turn on and stay on. After one second, the second LED should turn on and stay on. After two more seconds, the third LED should turn on and stay on. All LEDs should be on for three seconds, and then the program should end.

Use the Flat Sequence structure with three frames and the Time Delay Express VI within each frame.

P5.5 Create a time trial program to compare the average execution times of the Formula Node and the native LabVIEW Math Functions. This program will require a For Loop, a Flat Sequence structure, and a Case structure. The For Loop is required to run the time trial *N* times and then the results can be averaged using the Statistics Express VI. The Sequence structure is required to sam-

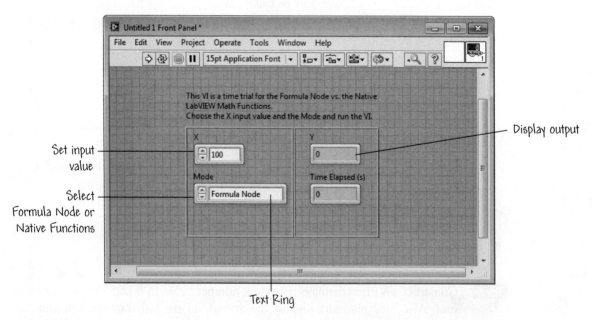

FIGURE P5.5
Suggested front panel for P5.5.

ple the Tick Count before and after the code executes. The Case structure is required to determine whether the user would like to execute the Formula Node or the native LabVIEW Math Functions. To test the timing, use the following formulas:

$$a = X^2/4;$$
$$b = (2 * X) + 1;$$
$$Y = \sin(a + b);$$

where X is the input and Y is the output. Run the time trial for each of the cases. Which method has the fastest execution time? Which method is the easiest to program? Which method is the easiest to understand if someone else were to look at your code?

P5.6 Using a While Loop which iterates every one-hundredth of a second, create a VI that calculates the percentage of the total run-time a push button on the front panel is depressed. The loop should run until a stop button is pressed by the user. There should be an indicator on the front panel displaying how many seconds the VI has been running. Update this indicator with each loop iteration. After the stop button is pressed, another indicator should display the ratio of time the button was depressed to total run-time. Your front panel should resemble the one shown in Figure P5.6.

P5.7 Consider the Magic Number Game VI shown in Figure P5.7. This VI is found in the **Exercises&Problems** folder in the **Chapter 5** folder of the **Learning**

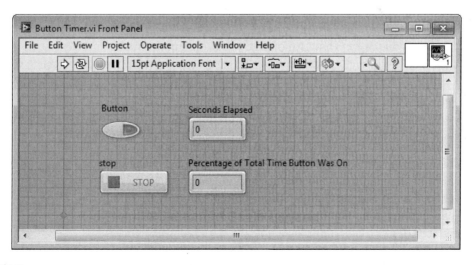

FIGURE P5.6
Front panel for the VI designed to compute the run-time a push button is depressed.

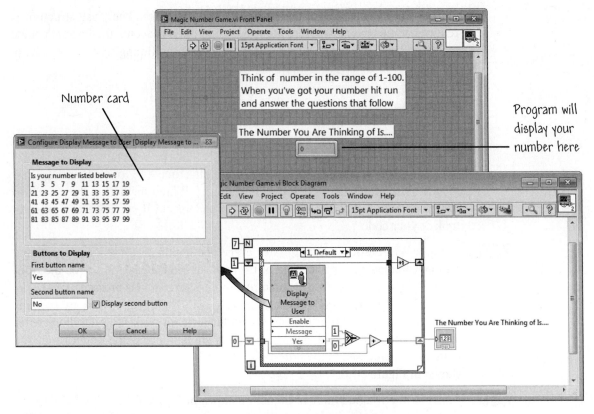

FIGURE P5.7
Magic number game.

directory. This program is based on simple binary math. First the user thinks of a number between 1 and 100. The VI is then run during which time a series of the question "Is your number listed below?" will be posed to the user and a number card will be displayed (see Figure P5.7) to the user. The user answers "yes" or "no" depending on the number in question. After seven iterations the VI will correctly guess the number. Open the VI and verify that it operates as described.

Navigate to the block diagram. List the main structures on the block diagram. Explore the code and describe the events that occur in order during the execution. Explain the "secret" of the VI. How does it always determine the number you were thinking of?

P5.8 Create a VI using a While Loop that continuously generates random numbers between 0 and 1000 until it generates a number that matches a number selected by the user. Determine how many random numbers the VI generated before the matching number.

DESIGN PROBLEMS

D5.1 Construct a VI that uses one Formula Node to calculate the following equations:

$$y_1 = 10\sin(a) + b$$
$$y_2 = a^3 + b^2 + 100$$

The inputs are the variables a and b. The outputs are y_1 and y_2.

D5.2 Use a Formula Node to calculate the Body Mass Index (BMI) of an individual (this design problem was originally presented in Chapter 4). The BMI is an internationally used measure of obesity. The subVI will have two numeric inputs: a weight in pounds and a height in inches. Calculate the BMI using the formula

$$BMI = (703 * W)/H^2$$

where W is the weight in pounds and H is the height in inches. Display the calculated BMI on a numeric indicator.

Classify the body weight according to the Table D5.2 and output a string containing this information. There should also be a "warning" Boolean output that is True for any unhealthy input (i.e., underweight, overweight, or obese).

TABLE D5.2 BMI Classification

<18.5	Underweight
18.5–24.9	Healthy
25–29.9	Overweight
≥ 30	Obese

D5.3 Construct a VI that calculates maximum heart rate using the following formulas:

Gender	Maximum Heart Rate
Male	214 − Age * 0.8
Female	209 − Age * 0.7

On the front panel, place a numeric control to input age and a slide switch to input gender. A numeric indicator should display the maximum heart rate. Use Formula Nodes to solve the heart rate formulas. The front panel should resemble the one depicted in Figure D5.3.

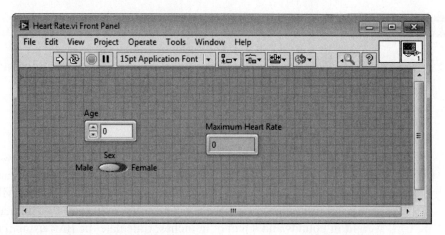

FIGURE D5.3
The Heart Rate VI front panel.

D5.4 Create a VI that calculates the hyperfactorial of any positive integer n, where the hyperfactorial is equivalent to the value obtained by the operation $1^1 * 2^2 * 3^{3*} \ldots n^n$. If the user inputs a value that is not a positive value, display a message informing the user that the input is not valid and request a new input.

Two important elements of the block diagram will be the One Button Dialog function and the Power of X subVI. The two functions are shown in Figure D5.4.

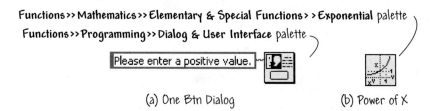

(a) One Btn Dialog   (b) Power of X

FIGURE D5.4
Use the One Button Dialog and the Power of X subVI.

D5.5 Re-consider the design problem D4.3 in Chapter 4 to develop a subVI to compute the height of a trapezoid. In this design, use a Formula Node to implement the computation of the height. The inputs to the VI are the lengths of the trapezoid top base, bottom base, left side, and right side, as shown in Figure D5.5. The output of the VI is the height of the trapezoid. Save the VI as Trap Height Formula.vi in the Users Stuff folder in the Learning directory.

Run the program several times with different inputs. Test the algorithm with the following inputs: top base = 2, bottom base = 10, left side = 5, and right side = 5. The result should be height = 3.

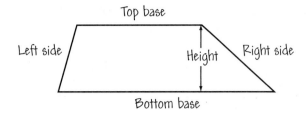

FIGURE D5.5
Trapezoid figure.

D5.6 Using Euclid's algorithm, find the greatest common factor between two numeric control inputs. Display a warning to the user if no common factor exists other than 1. As an example of Euclid's algorithm in action, find the greatest common factor of 48 and 15. First divide 48 by 15 (the greater divided by the lesser) yielding 3 with a remainder of 3. Then divide 15 by 3 (the previous remainder) yielding 5 with no remainder. The last non-zero remainder is the greatest common factor. Therefore the greatest common factor of 48 and 15 is 3.

D5.7 Using parallel Case structures, create a VI that calculates how many days are left until your next birthday (excluding leap days). The inputs to the VI are the current month (ranging from 1 to 12), current date (ranging from 1-31), and the month and date of your birthday. The output of the VI is the number of days until your next birthday.

D5.8 Calculate the binary representation of a number from -128 to 127 using eight-bit two's complement integers. Be sure to check in the VI that the input number is in range and clear all data if the number is not within range. Create a front panel that has 8 LED indicators and a vertical slider control that can accept integer inputs between -128 and 127. Display a digital indicator for the slider, and make sure that the LEDs are evenly spaced and aligned at the bottom. To check your solution, if the slider is set to the number 1, all eight LEDs should be on. As another test of your solution, check the number 127. In this case, LEDs 0, 1, 2, 3, 4, 5, and 6 should be on and LED 7 should be off (where LED 7 is the most-significant bit).

D5.9 Create a subVI to compute the square root of a real number that is input by the user. Using Case structures and error clusters (see Chapter 4 for more discussion on error handling) implement an error handling capability that displays the message "Square root of a negative number is not supported" anytime the user inputs a negative real number. The subVI should have as inputs error in and a real number, denoted by y, and as outputs it should have $\sqrt{y}$ (assuming y is nonnegative) and the error out. If the input is a negative number, set the output to the input. So if the input is y= 4, the output is 2 and the error cluster shows the status is FALSE (no error). If the input is y= -4, the output is -4 and the error cluster shows the status is TRUE (an error occurred), the error code is −1, and the error message listed above appears.

Save the VI as **Square Root.vi** in the **Users Stuff** folder in the **Learning** directory. Verify that you have set up the connector pane correctly by opening a new VI and placing Square Root within it as a subVI. Create controls and indicators, and run the program several times with different inputs, including a negative real number input.

CHAPTER 6

Arrays and Clusters

The array data type, the cluster data type, and the matrix data type are presented. An array is a variable-sized collection of data elements that are all the same type. A cluster is a fixed-sized collection of data elements of mixed types. The matrix data type stores rows or columns of real or complex scalar data for matrix operations and can be used in place of two-dimensional arrays. You will learn how to use built-in functions to manipulate arrays, clusters, and matrices. The important concept of polymorphism is introduced. Several rules that will lead to better memory usage are presented along with a short discussion on memory management.

GOALS

1. Understand how to create and use arrays and array functions.
2. Understand the concept of polymorphism.
3. Become familiar with using clusters and cluster functions.
4. Understand how to create and use matrix data types.
5. Be introduced to memory usage issues.

6.1 ARRAYS

An **array** is a *variable-sized* collection of data elements that are all the *same type*. In contrast, a **cluster** is a *fixed-sized* collection of data elements of *mixed types*. In what situations might you use arrays? One scenario that benefits from the use of arrays and which you will encounter frequently involves working with a collection of data for plotting purposes. Arrays are quite useful as a mechanism for organizing data for plotting. Arrays are also helpful when you perform repetitive computations or when you are solving problems that are naturally formulated in matrix-vector notation, such as solving systems of linear equations. Using arrays in your VIs leads to compact block diagram codes that are easier to develop because of the large number of built-in array functions and VIs.

Arrays can have one or more dimensions with up to $2^{31} - 1$ elements per dimension. The maximum number of elements depends on the available memory. The individual elements of an array can be any type with the exceptions that you cannot have an array of arrays, an array of charts, or an array of graphs. You access an individual array element by its index. The index is zero-based, implying that the array index is in the range 0 to $n - 1$, where n is the number of elements in the array. One-dimensional (1D) arrays are shown in Figure 6.1. The first element of a 1D array has index 0, the second element has index 1, and so on.

Another way to display an array of numbers is with a waveform graph. In Figure 6.1, the waveform graph is used to display the numeric array in which

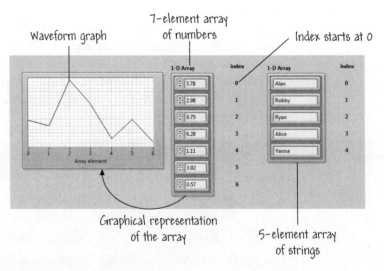

FIGURE 6.1
One-dimensional array examples.

each successive element of the array is plotted on the graph. To generate an *x* versus *y* graph, you could use a two-dimensional (2D) array with one column containing the *x* data points and the other column containing the *y* data points. To learn more about graphs, refer to Chapter 7.

6.1.1   Creating Array Controls and Indicators

It takes two steps to create an array control or indicator. The two steps involve combining an **array shell** from the **Array, Matrix & Cluster** subpalette of the **Controls≫Silver** palette, as shown in Figure 6.2, with a valid element, which can be a numeric, Boolean, or cluster. Actually, valid elements also include strings—we will discuss strings in Chapter 9. In any case, charts, graphs, or other arrays are not valid elements to combine with the array control or indicator.

The two steps that lead to the creation of an array control or indicator follow:

- Select an empty array shell from the **Array, Matrix & Cluster** subpalette of the **Controls≫Silver** palette and drop it onto the front panel, as illustrated in Figure 6.3.

- Drag a valid data object (such as a numeric, Boolean, or string) into the array shell, as shown in Figure 6.4. The array shell resizes to accommodate its new type.

To display more elements of the array, use the **Positioning** tool to grab a resizing handle on the corner of the array window and stretch the object to the desired number of visible array elements. The index value shown in the box on the left side of the array corresponds to the first visible element in the array. You

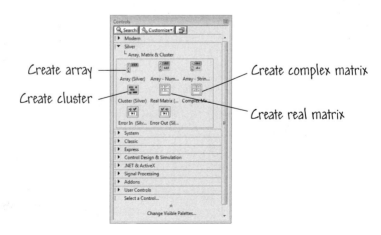

FIGURE 6.2
Creating an array control from the **Array, Matrix & Cluster** subpalette of the **Controls≫Silver** palette.

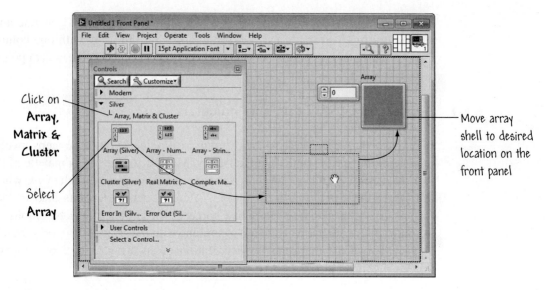

FIGURE 6.3
Placing an empty array shell on the front panel.

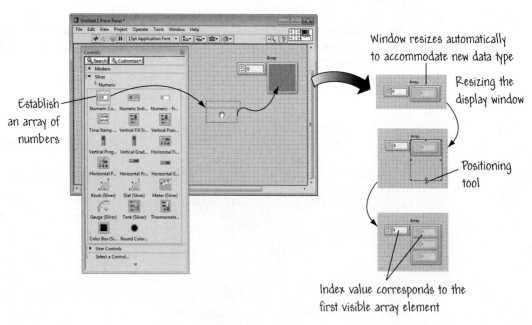

FIGURE 6.4
Drag a valid data object into the array shell to establish the array type.

can move through the array by clicking on the up and down arrows in the index display.

The block diagram terminal of an array shell is black when first dropped on the front panel. This indicates that the data type is undefined. When you drop a valid data object (such as a numeric, Boolean, or string) into the array shell, the array terminal on the block diagram turns from black to a color reflecting the data type. In Figure 6.5 the array block diagram terminal is pink, indicating that the array contains strings. When you wire arrays in the block diagram, you will find that array wires are thicker than wires carrying a single value.

Remember that you must assign a data object to the empty array shell before using the array on the block diagram. If you do not assign a data type, the array terminal will appear black with an empty bracket.

6.1.2 Multidimensional Arrays

A two-dimensional (2D) array requires two indices—a row index and a column index—to locate an element. A three-dimensional array requires three indices,

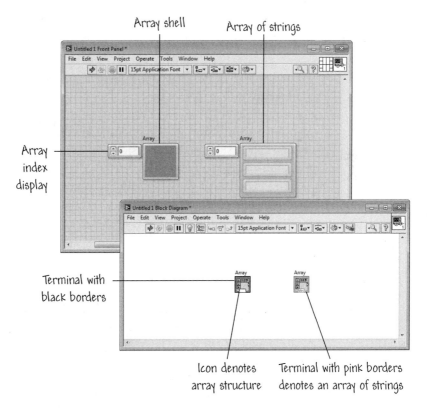

FIGURE 6.5
The terminal of an array shell is black, denoting an undefined data type.

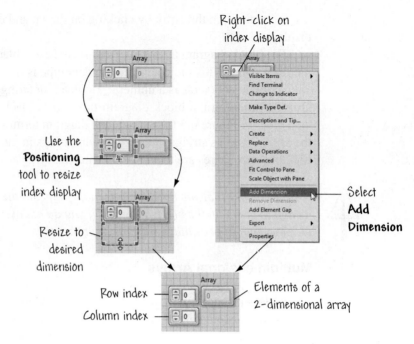

FIGURE 6.6
Adding dimensions to an array.

and in general, an *n*-dimensional array requires *n* indices. You add dimensions to the array in one of two ways: (a) by using the **Positioning** tool to resize the index display, or (b) by right-clicking on the array index display and choosing **Add Dimension** from the shortcut menu (see Figure 6.6). An additional index display appears for each dimension you add. The example in Figure 6.6 shows a 2D numeric control array.

You can reduce the dimension of an array by resizing the index display appropriately or by selecting **Remove Dimension** from the shortcut menu.

6.2 CREATING ARRAYS WITH LOOPS

With the For Loop and the While Loop you can create arrays automatically with a process known as **auto-indexing**. Figure 6.7(a) shows a For Loop creating a 10-element array using auto-indexing. On each iteration of the For Loop, the next element of the array is created. In this case the loop counter is set to 10; hence a 10-element array is created. If the loop counter was set to 20, then a 20-element array would be created. The array passes out of the loop to the indicator after the loop iterations are complete.

It is possible to pass a value out of a loop without creating an array. This requires that you disable auto-indexing by right-clicking on the tunnel (the square

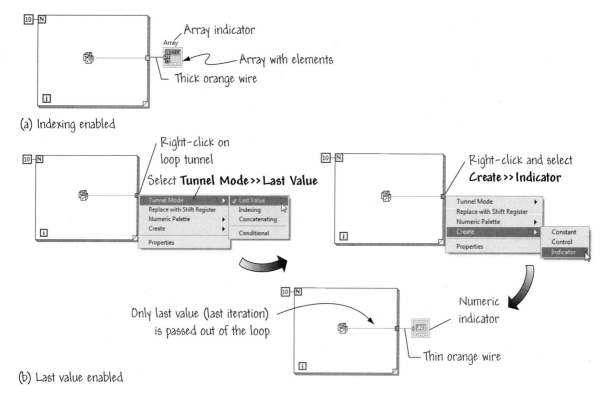

(a) Indexing enabled

(b) Last value enabled

FIGURE 6.7
Auto-indexing is the ability of For Loops and While Loops to automatically index and accumulate arrays at their boundaries.

on the loop border through which the data passes) and selecting **Tunnel Mode≫ Last Value** from the shortcut menu. In Figure 6.7(b), last value is enabled; thus when the VI is executed, only the last value returned from the Random Number function passes out of the loop.

*Open and run **Array Auto Index Demo.vi** shown in Figure 6.7. It can be found in the **Chapter 6** folder in the **Learning** directory. After running the program, you should find that with auto-indexing enabled, the indicator array will have 10 elements (indexed 0, 1, ⋯ , 9), and with auto-indexing disabled only the last value of the Random Number function is passed out of the For Loop.*

You can also pass arrays into a loop one element at a time or the entire array at once. With auto-indexing enabled, when you wire an array (of any dimension) from an external node to an input tunnel on the loop border, then elements of the array enter the loop one at a time, starting with the first component. This is illustrated in Figure 6.8(a). With auto-indexing disabled, the entire array passes into the loop at once. This is illustrated in Figure 6.8(b).

With For Loops auto-indexing is enabled by default. In contrast, with While Loops auto-indexing is disabled by default. If you desire auto-indexing, you need to right-click on the While Loop tunnel and choose **Enable Indexing** *from the shortcut menu.*

With a For Loop with auto-indexing enabled, an array entering the loop automatically sets the loop count to the number of elements in the array, thereby eliminating the need to wire a value to the loop count, N. What happens if you explicitly wire a value to the loop count that is different than an array size

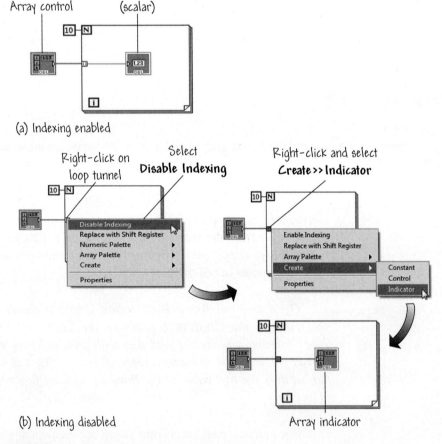

FIGURE 6.8
Auto-indexing applies when you are wiring arrays into loops.

entering the loop? Or what if you wire two arrays with different numbers of elements to a For Loop? The answer is that if you enable auto-indexing for more than one array entering the same For Loop, or if you set the loop count by wiring a value to N with auto-indexing enabled, the actual loop count becomes the smaller of the two. For example, in Figure 6.9, the array size ($= 4$), and not the loop count N ($= 5$), sets the For Loop count—because the array size is the smaller of the two. The Array Size function used in the VI shown in Figure 6.9 will be discussed in the next section along with other common built-in array functions.

*Open and run **Array Auto Count Set.vi** shown in Figure 6.9. It can be found in the **Chapter 6** folder in the **Learning** directory. After running the program, you should find that only four random numbers are generated, despite the fact that the loop count* N *has been set to five!*

6.2.1   Creating Two-Dimensional Arrays

You can use two nested For Loops (that is, one loop inside the other) to create a 2D array. The outer For Loop creates the row elements, and the inner For Loop

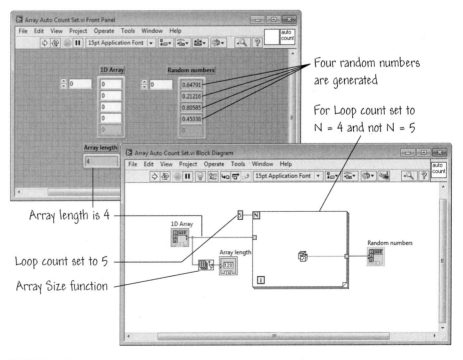

FIGURE 6.9
Automatically setting the For Loop count to the array size.

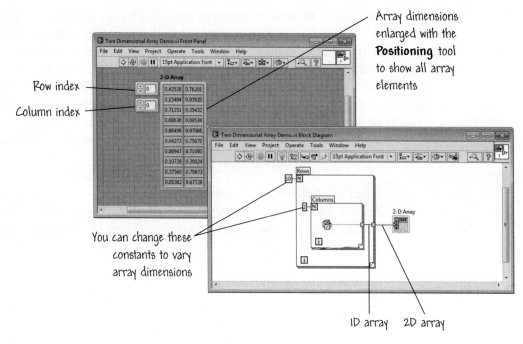

Array dimensions enlarged with the **Positioning** tool to show all array elements

Row index

Column index

You can change these constants to vary array dimensions

1D array   2D array

FIGURE 6.10
Creating a 2D array using two For Loops and auto-indexing.

creates the columns of the 2D array. Figure 6.10 shows two For Loops creating a 2D array of random numbers using auto-indexing.

Open and run Two Dimensional Array Demo.vi *shown in Figure 6.10. It can be found in the* **Chapter 6** *folder in the* **Learning** *directory. After running the program, you should find that a 2D array has been created.*

6.3   ARRAY FUNCTIONS

Many built-in functions are available to manipulate arrays. Most of the common array functions are found in the **Array** palette of the **Functions≫Programming** palette, as illustrated in Figure 6.11. Several of the heavily used functions are pointed out in Figure 6.11 and discussed in this section.

6.3.1   Array Size

The function Array Size returns the number of elements in the input array, as illustrated in Figure 6.12. If the input is an *n*-dimensional array, the output of

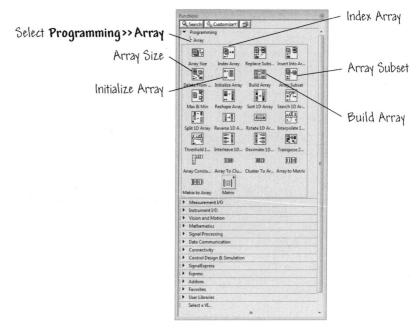

FIGURE 6.11
The array functions palette.

the array function Array Size is a one-dimensional array with n elements—each element containing a dimension size.

You can run the array size demonstration shown in Figure 6.12 by opening **Array Size Demo.vi** *in the* **Chapter 6** *folder in the* **Learning** *directory. After executing the VI once to verify that the 1D array has length 4 and the 2D array has 2 rows and 4 columns, add an element to the 1D array and run the VI again. What happens? The* **1D Array size** *value increments by 1.*

6.3.2   Initialize Array

The function Initialize Array creates an n-dimensional array with elements containing the values that you specify. All elements of the array are initialized to the same value. To create and initialize an array that has more than one dimension, right-click on the lower left side of the Initialize Array node and select **Add Dimension** or use the **Positioning** tool to grab a resizing handle to enlarge the node. You can remove dimensions by selecting **Remove Dimension** from the function shortcut menu or with the **Resizing** cursor. The Initialize Array is useful for allocating memory for arrays. For instance, if you are using shift registers to pass an array from one iteration to another, you can initialize the shift register using the Initialize Array function.

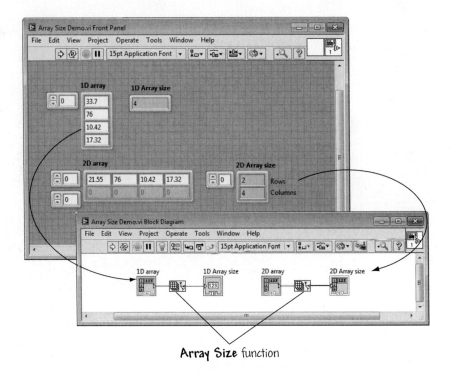

Array Size function

FIGURE 6.12
The function Array Size returns the number of elements in the input array.

The input determines the data type and the value of each element of the initialized array. The dimension size input determines the length of the array (see Figure 6.13(a)). For example, if the input value **element** is a double-precision floating-point number with the value of 5.7, and dimension size has a value of 100, the result is a 1D array of 100 double-precision floating-point numbers all set to 5.7, as illustrated in Figure 6.13(b). You can wire inputs to the Initialize Array function from front panel control terminals, from block diagram constants, or from calculations in other parts of the block diagram. One method of obtaining the output array is to right-click on the Initialize Array function and choose Create≫Indicator. Also, you can right-click and choose Create≫Control and Create≫Constant to create inputs for the element and dimension size. Figure 6.13(c) depicts a 2D array with 3 rows and 2 columns initialized with the long integer value of 0.

You can run the array initialization demonstration shown in Figure 6.13 by opening Array Initialization Demo.vi *located in the* **Chapter 6** *folder in the* **Learning** *directory. Change the initial value of the two arrays on the block diagram, run the VI, and examine the result.*

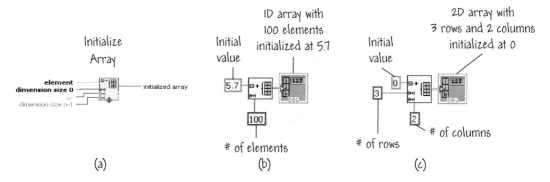

FIGURE 6.13
The function Initialize Array creates an array of a certain dimension containing the same value for each element.

6.3.3   Build Array

The function Build Array concatenates multiple arrays or adds extra elements to an array. The function accepts two types of inputs—scalars and arrays—therefore it can accommodate both arrays and single-valued elements.

The Build Array function appears with one scalar input when initially placed in the block diagram window, as shown in Figure 6.14.

Array inputs have brackets, while element inputs do not. Array inputs and scalar inputs are not interchangeable! Pay special attention to the inputs of a Build Array function, otherwise you may generate wiring errors that are difficult to detect.

You can add as many inputs as you need to the Build Array function. Each input to the function can be either a scalar or an array. To add more inputs, right-click on the left side of the function and select **Add Input**. You also can enlarge the Build Array node by placing the **Positioning** tool at one corner of the object to grab and drag the resizing handle. You can remove inputs by shrinking the node with the resizing handle or by selecting **Remove Input** from the shortcut menu.

The input type (element or array) is automatically configured when wired to the Build Array function. The Build Array function shown in Figure 6.14 is configured to concatenate an array and one scalar element into a new array.

The Build Array function concatenates the elements or arrays in the order in which they appear in the function, top to bottom. If the inputs are of the same dimension, you can right-click on the function and select **Concatenate Inputs** to concatenate the inputs into a longer array of the same dimension. If you leave the **Concatenate Inputs** option unselected, you will add a dimension to the array.

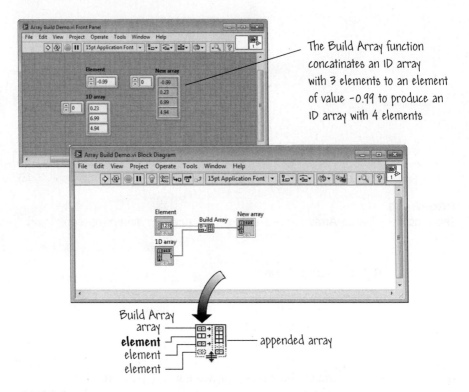

The Build Array function concatinates an 1D array with 3 elements to an element of value -0.99 to produce an 1D array with 4 elements

FIGURE 6.14
Build Array concatenates multiple arrays or appends elements to an array.

The Array Build demonstration shown in Figure 6.14 is located in the **Chapter 6** *folder in the* **Learning** *directory and is called* **Array Build Demo.vi.** *Open, run, and experiment with the VI.*

6.3.4 Array Subset

The function Array Subset returns a portion of an array starting at **index** and containing **length** elements. In the example shown in Figure 6.15 you will find that the array index begins with 0.

The array subset demonstration shown in Figure 6.15 can be found in the **Chapter 6** *folder in the* **Learning** *directory and is called* **Array Subset Demo.vi.** *Open and run the VI in the* **Run Continuously** *mode. Using the* **Operating** *tool, change the value of* **Index number** *and watch the Array Subset function return a portion of the array, starting at the index that you specify and including the three subsequent elements.*

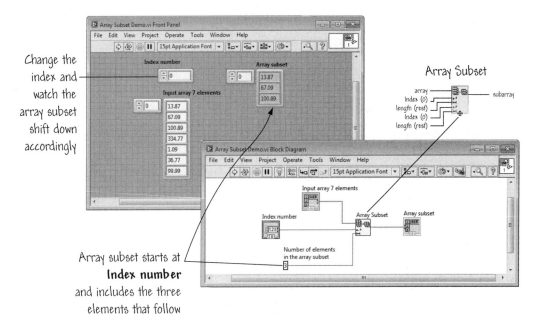

Change the index and watch the array subset shift down accordingly

Array subset starts at **Index number** and includes the three elements that follow

FIGURE 6.15
The function Array Subset returns a portion of an array starting at **index** and containing **length** elements.

6.3.5 Index Array

The function Index Array accesses an element of an array. The example shown in Figure 6.16 uses the input Index number to specify which element of the array to access. Remember that the index number of the first element is zero.

*The index array demonstration shown in Figure 6.16 can be found in the Chapter 6 folder of the Learning directory and is called Array Index Demo.vi. Open and run the VI in **Run Continuously** mode. Using the **Operating** tool, change the value of Index number and watch the Index Array function return a different element based on the index value.*

The Index Array function automatically resizes to match the dimensions of the wired input array. For example, if you wire a 1D array to the Index Array function, a single index input will show. Similarly, if you wire a 2D array to the Index Array function, two index inputs will show—one for the row and one for the column.

Once you have wired an input array to the Index Array function, you can access more than one element, or subarray (e.g., a row or a column) using the **Positioning** tool to manually resize the function once placed on the block diagram. When you expand the Index Array function it will expand in increments

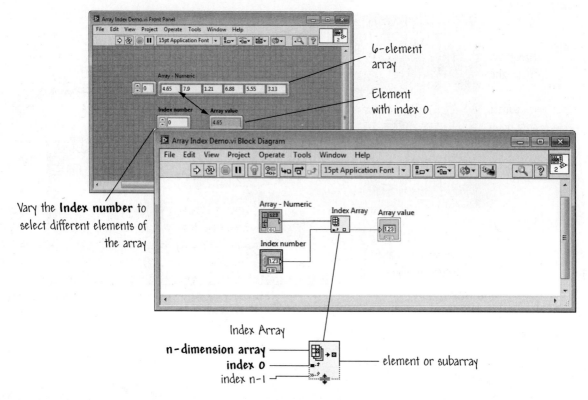

6-element array

Element with index 0

Vary the **Index number** to select different elements of the array

Index Array

n-dimension array
index 0
index n-1

element or subarray

FIGURE 6.16
The function Index Array accesses an element of an array.

determined by the dimensions of the array wired to the function. In Figure 6.17, the Index Array function is expanded so that three subarrays can be extracted: a row, a column, and a single element. The index inputs you wire determine the shape of the subarray you want to access or modify. For example, if the input to an Index Array function is a 2D array and you wire only the row input, you extract a complete 1D row of the array. If you wire only the column input, you extract a complete 1D column of the array. If you wire the row input and the column input, you extract a single element of the array. These three cases are illustrated in Figure 6.17. Each input group is independent and can access any portion of any dimension of the array.

*The 2D index array demonstration shown in Figure 6.17 can be found in the Chapter 6 folder of the Learning directory and is called 2D Array Index Demo.vi. Open and run the VI in **Run Continuously** mode. Using the **Operating** tool, change the **Column number** and **Row number** values and verify the numbers returned in **Column** and **Row**.*

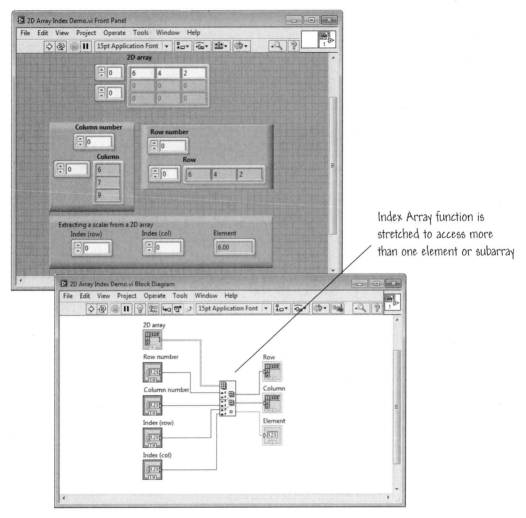

Index Array function is stretched to access more than one element or subarray

FIGURE 6.17
Using Index Array to extract a row or column of an array.

Practice with Arrays

In this exercise you will get more practice using array functions. Open Practice with Arrays.vi located in the Chapter 6 folder of the Learning directory. The front panel and block diagram of the incomplete VI are shown in Figure 6.18. This VI is not completed—you will finish wiring and debugging the VI for practice.

The front panel contains four arrays and one numeric control. The completed VI concatenates the input arrays and the numeric control values to form a new array. Using the Array Size function and the Array Initialize function, the VI creates a new array of appropriate dimension and with all elements of the array initialized to 1. In the final computation of the VI, the difference between

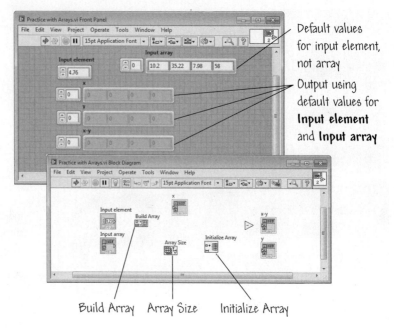

Default values for input element, not array

Output using default values for **Input element** and **Input array**

Build Array   Array Size   Initialize Array

FIGURE 6.18
Practice with array functions.

the two new arrays is computed, and the result is displayed on the front panel. When you compute the difference of two arrays with the same number of elements, the differencing operation subtracts the array values element by element.

The completed block diagram should resemble the one shown in Figure 6.19. Remember to save the VI in the Users Stuff folder and close it.

The completed VI for this exercise can be found in the **Chapter 6** *folder of the* **Learning** *directory and is called* **Practice with Arrays Done.vi.** ◆

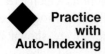

Practice with Auto-Indexing

In this example you will create a VI that computes the final grades in a course using three input arrays containing student scores on three different exams.

The objective of this example is to demonstrate the use of auto-indexing using Arrays and For Loops. To do this, you will compare two different approaches to constructing the VI block diagram. The first approach uses a For Loop with auto-indexing enabled on the input. When auto-indexing is enabled, one element of each array is passed through the input tunnel with each loop iteration (incrementing through the array). The second approach to constructing the

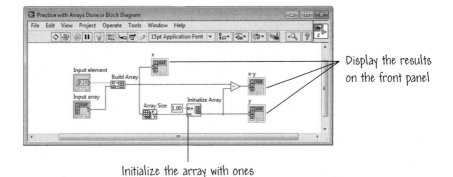

Display the results on the front panel

Initialize the array with ones

FIGURE 6.19
The completed block diagram for the array practice exercise.

VI is to disable auto-indexing on the For Loop and instead use the Index Array function. When auto-indexing is disabled, the entire array is passed through the input tunnel at once—the entire array is available on each iteration. This means that the Index Array function must be used to iterate through each element in the array.

The input to the VI consists of three 1D arrays: Exam 1 Scores (weighted as 30% of the final course grade), Exam 2 Scores (weighted as 30% of the final course grade), and Final Exam Scores (weighted as 40% of the final course grade). You can use the VI front panel shown in Figure 6.20 as a guide. The number of elements in each array is equal to the number of students in the class. The VI should output a 1D array containing the final course grade for each student.

Using the following steps as a guide, you can construct a VI that will compute the average student scores using the Index Array function rather than auto-indexing.

1. Open a new VI and save it as Array Auto-Indexing.vi in the Users Stuff folder in the Learning directory.

2. On the front panel create three numeric arrays and label them Exam 1 Scores, Exam 2 Scores, and Final Exam Scores. Use Figure 6.20 as a guide.

3. Switch to the block diagram and place a For Loop and wire the three array inputs to the border of the For Loop.

4. The Auto-Indexing is enabled by default for the For Loop. To disable it, right-click each tunnel and select **Disable Indexing**, as illustrated in Figure 6.21(a). This will make the entire array available at each iteration of the For Loop. The Index Array function will then be needed to access

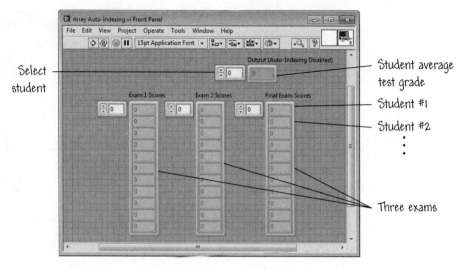

FIGURE 6.20
The Array Auto-Indexing VI front panel.

the individual scores within each array to be used in the weighted average calculations.

5. You need to specify how many times the For Loop has to iterate. The output array that you are creating should have the same number of elements as the input arrays, that is, one score for each student in the class. To do this, use the Array Size function and pass its output to N, as shown in Figure 6.21(b).

6. The Index Array function is used to extract a single scalar value from each array on every iteration of the For Loop. This way, each iteration of the For Loop will operate on the Exam 1 Scores, Exam 2 Scores, and Final Exam Scores of one specific student to output the final grade for that student. Wire the Loop Iteration terminal to the index input of the Index Array function so that the Index Array function will output the exam scores of the next student on each iteration.

7. Using the Compound Arithmetic and Multiply functions as shown in Figure 6.21(b), perform the necessary operations on the individual array elements to determine the course grade for each student.

8. Use the Build Array function to create a 1D array containing the course grades for the students. The entire output array will be built within the For Loop and then passed out after all calculations have been completed. You will need to use a Feedback Node to retain grades calculated on previous loop iterations. Continue appending newly calculated grades to the output array as the loop iterates for each student.

9. Create an indicator outside the loop to display the 1D output array. Name the indicator Output (Auto-Indexing Disabled). Note that if you do not

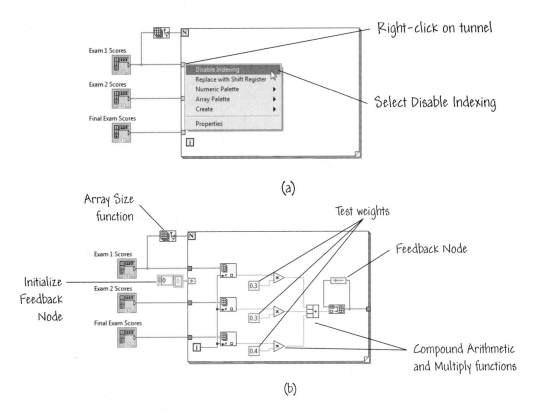

FIGURE 6.21
(a) Select Disable Indexing on the For Loop. (b) Completed block diagram showing Array Size function, Feedback Node, Compound Arithmetic and Multiply functions, and all necessary wiring to complete the diagram.

disable Auto-Indexing, the tunnel will output a 2D array and a broken wire appears when you connect the tunnel and indicator. Your block diagram should match the block diagram in Figure 6.21(b).

10. Complete the wiring of the block diagram and when ready, run the VI with different input values and verify that it computes the average grades correctly.

Using the following steps as a guide, you can construct a VI that will compute the average student scores using auto-indexing rather than the Index Array function. The objective is to create an identical output array, this time using the auto-indexing capability of the For Loop. This means that a new element from each input array will be passed into the loop upon each iteration, until all elements have been input.

1. Using the same front panel as above, add the Output (Auto-Indexing Enabled) indicator, as shown in Figure 6.22.

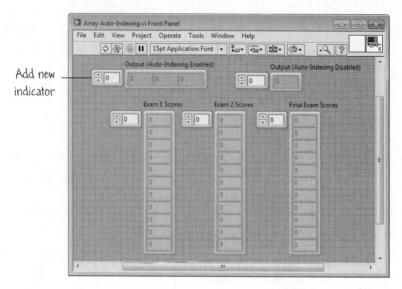

FIGURE 6.22
The Array Auto-Indexing front panel with the additional Output (Auto-Indexing Enabled) indicator.

2. On the same block diagram as in the Array Auto-Indexing VI constructed above, create a new section of the VI that uses the inputs Exam 1 Scores, Exam 2 Scores, and Final Exam Scores. Place the arithmetic functions necessary to calculate the weighted average for each student inside the For Loop. These will be the same as before, namely, the Compound Arithmetic and Multiply functions. Use Figure 6.23 as a guide.

3. Wire the exam score arrays directly to the inputs to the Multiply functions. Tunnels will automatically appear on the border of the For Loop with auto-indexing enabled by default.

4. Wire the array indicator Output (Auto-Indexing Enabled) to the output of the Compound Arithmetic function. Note that it is not necessary to provide an input to the Count terminal of the For Loop. Because the loop is now using auto-indexing, it will automatically iterate once for each element in the smallest array being input. In this case, the arrays are all the same size, and the loop will iterate once for each student in the class.

5. Your block diagram should resemble the one shown in Figure 6.23. Experiment with different input values and check that the output arrays contain identical values.

6. Remember to save the VI in the Users Stuff folder.

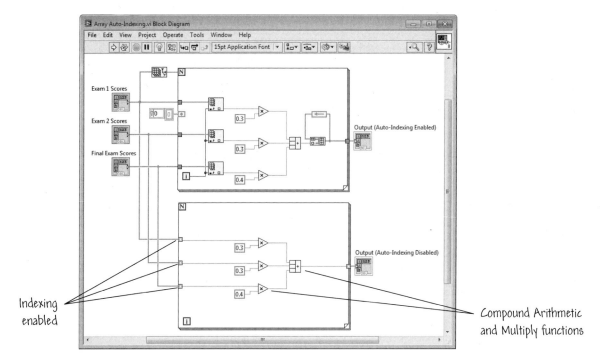

FIGURE 6.23
The completed Array Auto-Indexing block diagram.

*The completed VI for this exercise can be found in the **Chapter 6** folder of the **Learning** directory and is called **Array Auto-Indexing.vi.*** ◆

6.4   POLYMORPHISM

Polymorphism is the ability of certain LabVIEW functions (such as Add, Multiply, and Divide) to accept inputs of different dimensions and representations. Arithmetic functions that possess this capability are polymorphic functions. For example, you can add a scalar to an array or add together two arrays of different lengths. Figure 6.24 shows some of the polymorphic combinations of the Add function.

In the first combination, shown in Figure 6.24(a), the result of adding a scalar and a scalar is another scalar. In the second combination, shown in Figure 6.24(b), the result of adding a scalar to an array is another array. In this situation, the scalar input is added to each element of the input array. In the third combination, depicted in Figure 6.24(c), an array of length 2 is added to an array of length 3, resulting in an array of length 2 (the length of the shorter of the two input arrays). Array addition is performed component-wise; that is, each element of one array is added to the corresponding element of the other array.

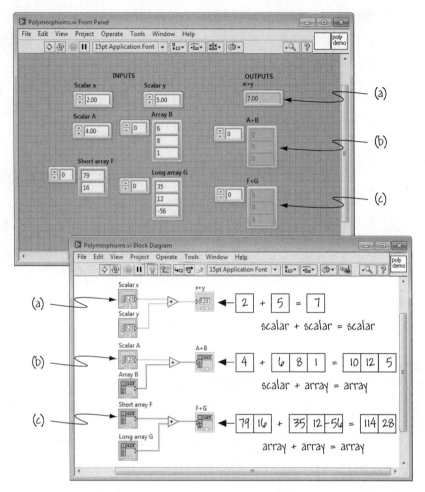

FIGURE 6.24
Polymorphic combinations of the Add function.

When two input arrays have different lengths, the output array resulting from some arithmetic operation (such as adding the two arrays) will be the same size as the smaller of the two input arrays. The arithmetic operation applies to corresponding elements in the two input arrays until the shorter array runs out of elements—then the remaining elements in the longer array are ignored!

Consider the VI depicted in Figure 6.25. Each iteration of the For Loop generates a random number that is stored in the array and readied for output once the loop iterations are finished. After the loop finishes execution (after 10 iterations), the Multiply function multiplies each element in the array by the scaling factor 2π. Notice that the Multiply function has two inputs: an array and a scalar. The polymorphic capability of the Multiply function allows the function

Array with 10 elements

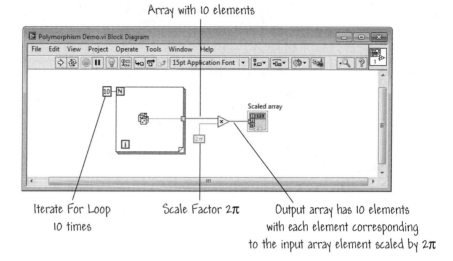

Iterate For Loop   Scale Factor 2π   Output array has 10 elements
  10 times   with each element corresponding
                                                 to the input array element scaled by 2π

FIGURE 6.25
Demonstrating the polymorphic capability of the Multiply function.

to take inputs of differing dimension (in this instance, an array of length 10 and a scalar) and produce a sensible output. What happens when you have two array inputs for a Multiply function? In that case, corresponding elements of each array are multiplied.

*The two demonstrations of polymorphism shown in Figures 6.24 and 6.25 can be found in the **Chapter 6** folder in the **Learning** directory and are called **Polymorphisms.vi** and **Polymorphism Demo.vi**, respectively. Open and run each VI. Vary the values of the array elements and examine the results. You can also edit the array lengths and test your knowledge of polymorphisms by first predicting the result and then verifying your prediction by running the VI.*

**Practice
Using
Polymorphism**

Open a new VI and recreate the front panel and block diagram shown in Figure 6.26. When you have finished the construction, experiment with the VI. A suggested way to run the VI is in **Run Continuously** mode. Increment and decrement the variable Array length and watch the array length change.

*The Generate Waveform VI in Figure 6.26 can be found in the folder **activity** of the **Learning** directory.*

This exercise demonstrates two applications of polymorphism. Referring to Figure 6.26, it can be seen that in one case the Multiply function operates on two inputs of different dimension: an array and a scalar $\pi/2$. In the second illustration of polymorphism, the Subtract function has two array inputs of

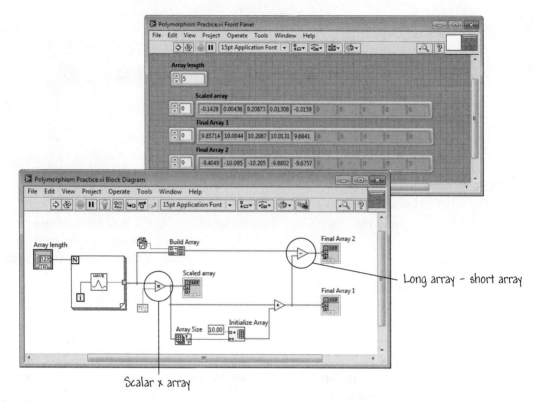

FIGURE 6.26
Practicing using polymorphism.

differing lengths. The resulting array **Final Array 2** has length equal to the length of the shorter array (the length is equal to the length of **Scaled array**). Three array functions are also utilized: Build Array, Array Size, and Initialize Array. Save the VI as **Polymorphism Practice.vi** and place it in the **Users Stuff** folder.

 The completed polymorphism practice VI shown in Figure 6.26 can be found in the Chapter 6 folder in the Learning directory and is called Polymorphism Practice.vi. ◆

6.5 CLUSTERS

A **cluster** is a data structure that, like arrays, groups data. However, clusters and arrays have important differences. One important difference is that clusters can group data of different types, whereas arrays can group only like data types. For example, an array might contain ten numeric indicators, whereas a cluster might contain a numeric control, a toggle switch, and a string control. And although cluster and array elements are both ordered, you access cluster elements by

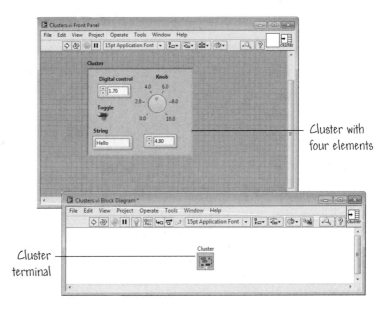

FIGURE 6.27
A cluster with four elements: numeric control, toggle switch, string control, and knob.

unbundling some or all the elements at once rather than indexing one element at a time. Clusters are also different from arrays in that they are of a fixed size.

One similarity between arrays and clusters is that both are made up of either controls or indicators. In other words, clusters cannot contain a mixture of controls and indicators. An example of a cluster is shown in Figure 6.27. The cluster shown has four elements: a numeric control, a horizontal toggle switch, a string control (more on strings in Chapter 9), and a knob. The cluster data type appears frequently when graphing data (as we will see in the next chapter).

Clusters are generally used to group related data elements that appear in multiple places on the block diagram. Because a cluster is represented by only one wire in the block diagram, its use has the positive effect of reducing wire clutter and the number of connector terminals needed by subVIs. A cluster may be thought of as a bundle of wires wherein each wire in the cable represents a different element of the cluster. On the block diagram, you can wire cluster terminals only if they have the same type, the same number of elements, and the same order of elements. Polymorphism applies to clusters as long as the clusters have the same number of elements in the same order.

6.6   CREATING CLUSTER CONTROLS AND INDICATORS

Cluster controls and indicators are created by placing a **cluster shell** on the front panel, as shown in Figure 6.28. A new cluster shell has a resizable border and an (optional) label. When you right-click in the empty element area of the

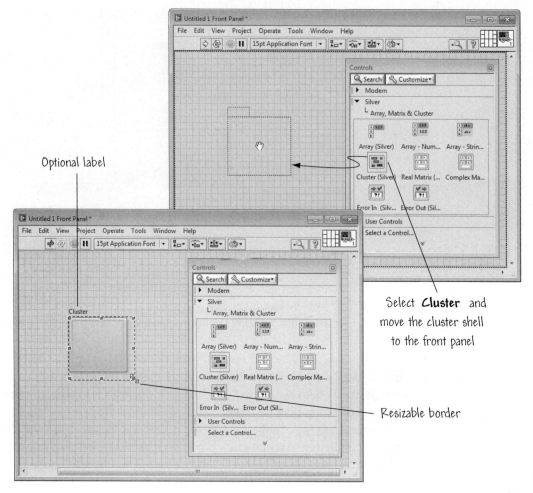

FIGURE 6.28
Creating and resizing a cluster.

cluster shell, the **Controls** palette appears, as illustrated in Figure 6.29. You create a cluster by placing any combination of numerics, Booleans, strings, charts, graphs, arrays, or even other clusters from the **Controls** palette into the cluster shell. Remember that a cluster can contain controls *or* indicators, but not both. The cluster becomes a control cluster or indicator cluster depending on the first element you place in the cluster. For example, if the first element placed in a cluster shell is a numeric control, then the cluster becomes a control cluster. Any objects added to the cluster afterwards become control objects. Selecting **Change To Control** or **Change To Indicator** from the shortcut menu of any cluster element changes the cluster and all its elements from indicators to controls or from controls to indicators, respectively. You can also drag existing objects from the front panel into the cluster shell.

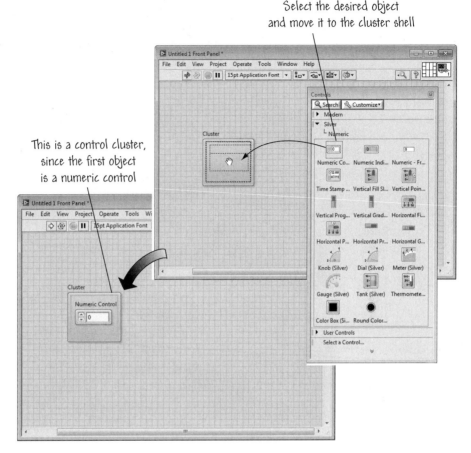

FIGURE 6.29
Creating a cluster by placing objects in the cluster shell.

6.6.1   Cluster Order

The elements of a cluster are ordered according to when they were placed in the cluster rather than according to their physical position within the cluster shell. The first object placed in the cluster shell is labeled element 0, the second object inserted is element 1, and so on. The order of the remaining elements is automatically adjusted when an element is removed from the cluster. The cluster order determines the order in which the elements appear as terminals on the Bundle and Unbundle functions in the block diagram (more on these cluster functions in the next sections). You must keep track of your cluster order if you want to access individual elements in the cluster, because individual elements in the cluster are accessed by order.

You can examine and change the order of the elements within a cluster by right-clicking on the cluster border on the front panel and choosing the item

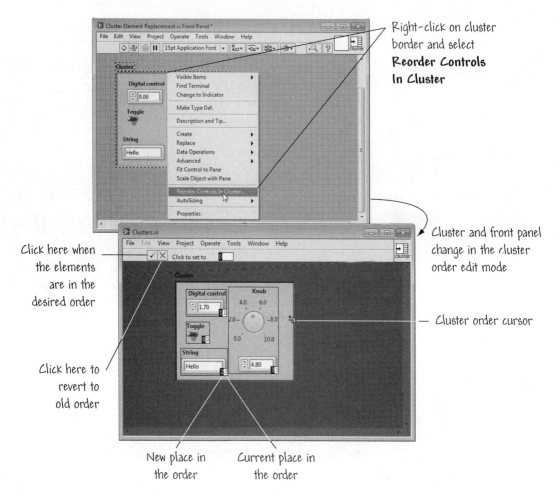

FIGURE 6.30
Examining and changing the cluster order.

Reorder Controls In Cluster from the shortcut menu, as shown in Figure 6.30. Notice in the figure that a new set of buttons replaces the toolbar, and the cluster appearance changes noticeably. Even the cursor changes to a hand with a # sign above—this is the cluster order cursor.

Two boxes appear side-by-side in the bottom right corner of each element in the cluster. The white boxes indicate the current place of that element in the cluster order. The black boxes indicate the new location in the order, in case you have changed the order. The numbers in the white and black boxes are identical until you make a change in the order. Clicking on an element with the cluster order cursor changes the current place of the element in the cluster order to the number displayed in the toolbar at the top of the front panel. When you have completed arranging the elements within the cluster to your satisfaction, click on the **OK** button, or you can revert to the old order by clicking on the **X** button.

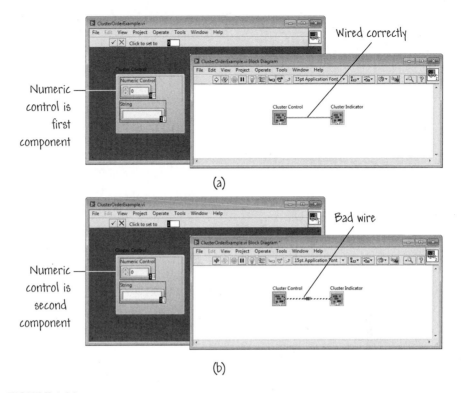

FIGURE 6.31
Example showing the importance of cluster order: (a) Wiring is correct; (b) wiring is incorrect.

A simple example shows the importance of the cluster order. Consider the situation depicted in Figure 6.31(a), where the front panel contains two clusters. In the first cluster, the first component is a numeric control, while in the second cluster, the first component is a numeric indicator. In the associated block diagram, the cluster control properly wires to the cluster indicator. Now, in Figure 6.31(b), the cluster order has been changed so that the string control is the first component of the cluster control. The numeric indicator is still the first component of the cluster indicator. The connecting wire is now broken, and if you attempted to run the VI you would get an error message stating that there is a type conflict.

6.6.2   Using Clusters to Pass Data to and from SubVIs

A subVI connector panel can have a maximum of 28 terminals (as discussed in Section 4.3). In general, you do not want to pass information individually to all 28 terminals when calling a VI or a subVI. A good rule-of-thumb is to use connector passes with 14 or fewer terminals. Otherwise, the terminals are

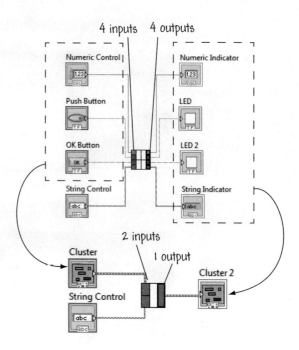

FIGURE 6.32
Using clusters to pass data to and from subVIs.

very small and may be difficult to wire. Using clusters, you can group related controls together. One cluster control uses one terminal on the connector, but it can contain many controls. Similarly, you can use a cluster to group indicators.

This allows the subVI to pass multiple outputs using only one terminal. Figure 6.32 illustrates the benefit of using clusters to pass data to and from subVIs.

6.7 CLUSTER FUNCTIONS

Two of the more important cluster functions are the Bundle and Unbundle functions. These functions (and others) are found in the **Cluster & Variant** sub-palette of the **Functions≫Programming** palette, as illustrated in Figure 6.33.

6.7.1 The Bundle Function

The Bundle function is used to assemble individual elements into a single new cluster or to replace elements in an existing cluster. When placed on the block diagram, the Bundle function appears with two element input terminals on the left side. You can increase the number of inputs by enlarging the function icon

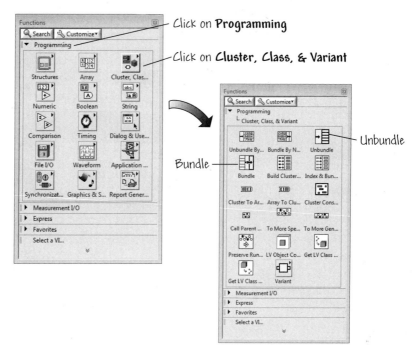

FIGURE 6.33
The **Cluster** palette.

(with the resizing handles) vertically to create as many terminals as you need. You can also increase the number of inputs by right-clicking on the left side of the function and choosing **Add Input** from the menu. Since you must wire all the inputs you create, only enlarge the icon to show the exact desired number of element inputs. When you wire to each input terminal, a symbol representing the data type of the wired element appears in the originally empty terminal, as illustrated in Figure 6.34. The order of the elements in the cluster is equal to the order of inputs to the Bundle function. The order is defined top to bottom, which means that the element you wire to the top terminal becomes element 0, the element you wire to the second terminal becomes element 1, and so on.

The example depicted in Figure 6.34 shows the Bundle function creating a cluster from three inputs: a floating-point real number, an integer, and an array of numbers generated by the For Loop. The output of the new cluster is wired to a waveform graph that displays the random numbers on a graph. Graphs (such as the waveform graph) are covered in Chapter 7.

*The example shown in Figure 6.34 can be found in the **Chapter 6** folder in the **Learning** directory and is called **Cluster Bundle Demo.vi**. When you open and run the VI you will see that varying the input x0 changes the x-axis origin.*

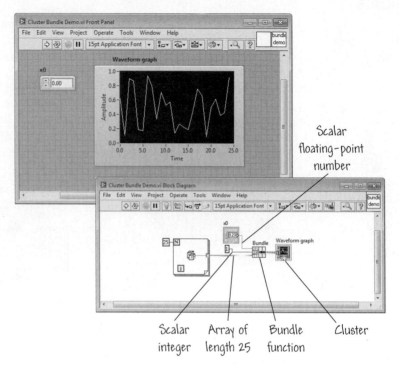

FIGURE 6.34
The Bundle function.

In addition to input terminals for elements on the left side, the Bundle function also has an input terminal for clusters in the middle, as shown in Figure 6.35. Sometimes you want to replace or change the value of one or two elements of a cluster without affecting the others. The example in Figure 6.35 shows a convenient way to change the value of two elements of a cluster. Because the cluster input terminal (that is, the middle terminal) of the Bundle function is wired to an existing cluster named **Cluster**, the only element input terminals that must be wired are those that are associated with cluster elements that you want to replace—in the illustration in Figure 6.35 the second and fourth element values are being replaced. To replace an element in the cluster using the Bundle function, you first place the Bundle function on the block diagram and then resize the function to show exactly the same number of input terminals as there are elements in the existing cluster that you want to modify. The next step is to wire the existing cluster to the middle input terminal of the Bundle function. Afterwards, any other input terminals on the left side that are wired will replace the corresponding elements in the existing cluster. Remember that if the objective is to create a new cluster rather than modify an existing one, you do not need to wire an input to the center cluster input of the Bundle function.

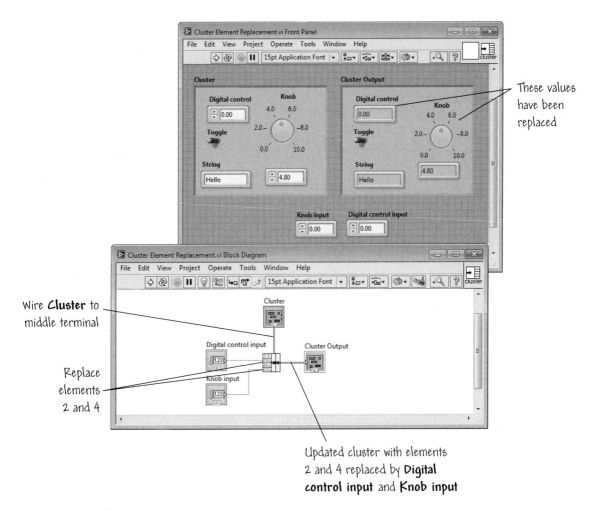

FIGURE 6.35
Using the Bundle function to replace elements of an existing cluster.

The example depicted in Figure 6.35 shows the Bundle function being used to change the value of the numeric control and the knob. The corresponding VI can be found in the Chapter 6 folder in the Learning directory and is called **Cluster Element Replacement.vi**. Open and run the VI in **Run Continuously** mode. Cluster is a control cluster and Cluster Output is an indicator cluster. Toggle the horizontal toggle switch (using the **Operating** tool). You should see that the toggle switch in Cluster Output also switches. Now vary the Knob value away from the default value of 4.8. Notice that the Knob in Cluster Output does not change. This phenomenon occurs because the Knob is the fourth element of the cluster and the value is being replaced by the value set in the numeric control Knob input. To change the value of the Knob in Cluster Output you change the value of Knob input. Try it out!

6.7.2   The Unbundle Function

The Unbundle function extracts the individual components of a cluster. The output components are arranged from top to bottom in the same order as in the cluster. When placed on the block diagram, the Unbundle function appears with two element output terminals on the right side. You adjust the number of terminals with the resizing handles following the same method as with the Bundle function, or you can right-click on the right side of the function and choose **Add Output** from the menu. Element 0 in the cluster order passes to the top output terminal, element 1 passes to the second terminal, and on down the line. The cluster wire remains broken until you create the correct number of output terminals, at which point the wire becomes solid. When you wire an input cluster to the correctly sized Unbundle, the previously blank output terminals will assume the symbols of the data types in the cluster.

While all elements are unbundled using the Unbundle function, the Unbundle By Name function can access one or more elements in a cluster.

The example depicted in Figure 6.36 shows the Unbundle function being used to unpack the elements of the data structure **Cluster**. The cluster has four elements, and each element is split from the cluster and wired to its own individual indicator for viewing on the front panel.

The example shown in Figure 6.36 can be found in the **Chapter 6** *folder in the* **Learning** *directory and is called* **Cluster Unbundle Demo.vi**. *Open and run the VI in* **Run Continuously** *mode and notice that varying any values in the control cluster immediately changes the values of the various indicators.*

6.7.3   Creating Cluster Constants on the Block Diagram

You can use the same technique you used on the front panel to create a cluster constant on the block diagram. On the block diagram, choose Cluster Constant from the **Functions≫Programming≫Cluster & Variant** palette to create the cluster shell as illustrated in Figure 6.37. Click in the block diagram to place the cluster shell, and place other constants of the appropriate data type within the cluster shell.

If you have a cluster control or indicator on the front panel and want to create a cluster constant containing the same components in the block diagram, you can either drag that cluster from the panel to the diagram or select **Create≫ Constant** from the shortcut menu, as depicted in Figure 6.38.

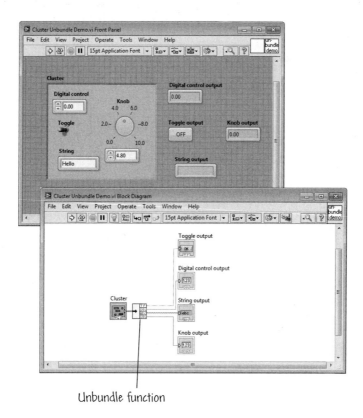

FIGURE 6.36
The Unbundle function.

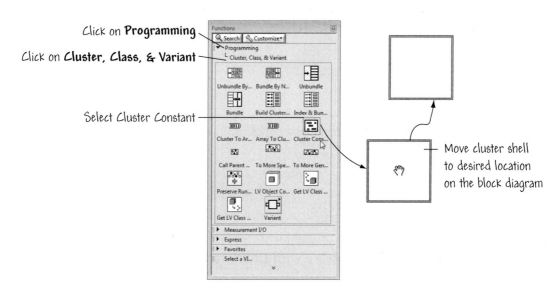

FIGURE 6.37
Creating a cluster constant on the block diagram.

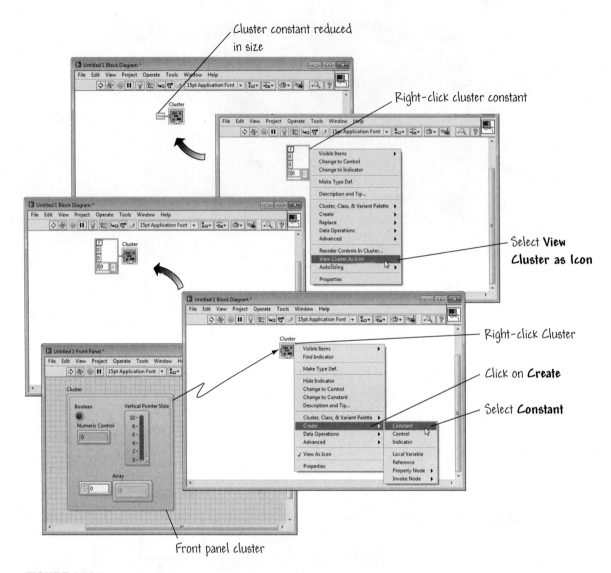

FIGURE 6.38
Create a cluster constant using a cluster on the front panel.

*You can right-click a cluster constant and select **View Cluster As Icon** from the shortcut menu to reduce the size of cluster constants on the block diagram, as shown in Figure 6.38.*

6.7.4 Using Polymorphism with Clusters

Since the arithmetic functions are polymorphic, they can be used to perform computations on clusters of numbers. As shown in Figure 6.39, you use the

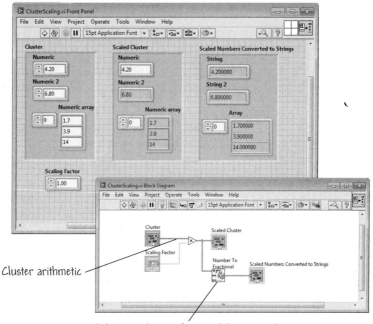

Cluster arithmetic

Select Number to Fractional String on the
String/Number Conversion subpalette found on the
Functions >> Programming >> String palette

FIGURE 6.39
Using polymorphism with clusters.

arithmetic functions with clusters in the same way you use them with arrays of numerics. You can also use the string-to-number functions to convert a cluster of numerics to a cluster of strings (strings are covered in Chapter 9).

The VI in Figure 6.39 can be found in **Chapter 6** *of the* **Learning** *directory. The VI is called* **ClusterScaling.vi.**

6.8   MATRIX DATA TYPE AND MATRIX FUNCTIONS

The **matrix data type** stores rows or columns of real or complex scalar data for matrix operations and can be used in place of two-dimensional arrays. A real matrix contains double-precision elements and a complex matrix contains complex elements with double-precision components. A matrix data type, since it represents a standard two-dimensional matrix, can have only two dimensions. Using this type to represent matrices (as opposed to using two-dimensional arrays) is advantageous for many linear algebra operations based on efficient matrix algorithms. You cannot create an array of matrices, but you can use the Bundle function to combine two or more matrices and create a cluster.

6.8.1   Creating Matrix Controls, Indicators, and Constants

To create a matrix on the front panel, navigate to **Modern≫Array, Matrix & Cluster** on the **Controls** palette and select RealMatrix or ComplexMatrix, as shown in Figure 6.2. Drag and drop the real or complex matrix on the front panel, as shown in Figure 6.40. Once on the front panel you can right-click on the matrix and select **Change to Indicator** if a matrix indicator is needed. The

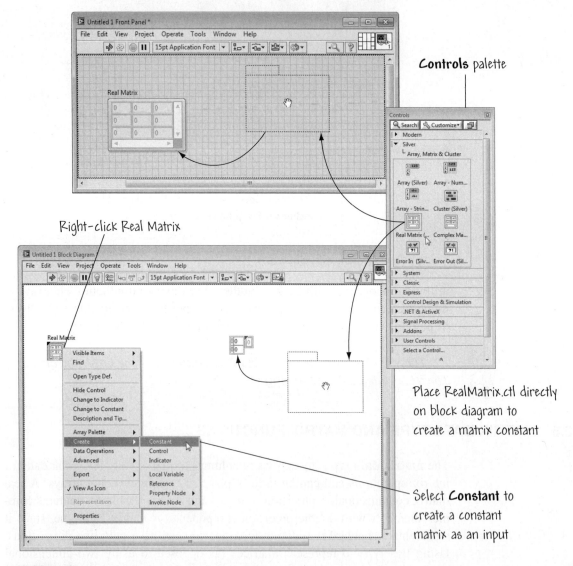

FIGURE 6.40
Creating matrix controls, indicators, and constants.

matrix data type resembles a real or complex two-dimensional array on the block diagram with a different wire pattern. By default, matrix controls and indicators show more than one element in two dimensions and display scrollbars for both dimensions. The control initially represents an empty matrix with each matrix element dimmed. Each dimension is initially 0. Although the size of a matrix cannot be limited to a fixed number of elements, you set the default matrix size when you set the values of a constant matrix. For matrix controls the default value is established by entering numbers and then right clicking on the control and selecting **Data Operations≫Make Current Value Default.**

Do not make the default size of a matrix larger than necessary. Making it unnecessarily large increases the size of the VI on disk.

A front panel matrix control has a matrix default value (which is either a floating-point or complex default value) and a scalar default value in each cell. The matrix default value is the value of the matrix when the VI is loaded. The scalar default value is the value used to pad the matrix when the matrix expands. For example, if you set the matrix index to a value beyond the defined portion of the matrix and enter a value in an element more than one row past the defined portion of the matrix, the elements between the previous end of the matrix and the element you added are set to the scalar default.

To create a matrix constant on the block diagram, select a matrix control from **Modern≫Array**, **Matrix & Cluster** on the **Controls** palette and place it directly on the block diagram, or create a constant from any matrix terminal, as shown in Figure 6.40. Remove matrix values by right clicking and selecting **Data Operations≫Delete Row** or **Delete Column**. The matrix can also be reinitialized to its default values.

6.8.2   Matrix Functions

The **matrix functions** can found on the **Programming≫Array≫Matrix** palette and on the **Mathematics≫Linear Algebra** palette, as shown in Figure 6.41. The matrix functions provide a convenient means to manipulate matrices. Many of them are similar to existing array functions (on the **Array** palette) but offer functionality for math algorithms based on matrices. Matrix functions accept the matrix data type and return a matrix data type.

VIs and functions that accept the matrix data type automatically support matrix operations when you wire a matrix data type as an input. In fact, most Numeric functions support the matrix data type and will perform matrix operations. For example, you can use the Multiply function to multiply a matrix by another matrix. If a function does not perform matrix operations but accepts a matrix data type, it automatically converts the matrix data type to a

FIGURE 6.41
Matrix functions.

two-dimensional array. Similarly, if you wire a two-dimensional array to a function that performs matrix operations, the function automatically converts the array to a real or complex matrix, depending on the array data type. For example, if you wire a matrix data type to an Array Size function, it will return a one-dimensional array with two elements representing the number of rows and columns of the matrix data type. If you wire the same matrix data type to the Matrix Size function, it will return two integers representing the number of rows and columns of the matrix data type, respectively.

Coercion dots appear on VIs and functions when data is converted to or from a matrix or two-dimensional array. This data conversion does not affect performance because matrices are stored in the same way as two-dimensional arrays.

Many polymorphic functions that accept the matrix data type return outputs in the same matrix data type, even if the function performs array operations internally. If a function or subVI in a block diagram converts a matrix data type to a two-dimensional array and subsequent operations in the data flow are array operations, you should permit the conversion from matrix to array to occur and work with the two-dimensional array. If you need an array in matrix format for saving or other purposes, you can always convert it back to matrix with the Array to Matrix function found on the **Programming≫Array** palette.

If your block diagram includes subVIs that accept matrix data types but return two-dimensional arrays, you do not need to convert the resulting two-dimensional arrays back into matrices before you wire the arrays to polymorphic VIs or functions that accept the matrix data type.

When you wire a matrix data type as an input to many programming functions typically associated with scalar operations (such as the Subtract or Square Root functions), a VI that contains matrix algorithms replaces the function. The resulting VI has the same icon, so it may not be readily apparent that the function is using matrix algorithms. You can restore the original function by disconnecting the matrix from the input and wiring other data types as inputs. This applies to the functions Equal?, Not Equal?, Absolute Value, Add, Multiply, Square Root, Subtract, Exponential, Natural Logarithm, Power Of X, Re/Im To Complex, Complex To Re/Im, Polar To Complex, and Complex To Polar.

The Equal? and Not Equal? functions are very useful for comparing matrices.

If you wire a data type to a function and that data type causes a basic math operation to fail, the function returns an empty matrix or NaN. For example, if you wire a matrix with two rows and three columns to one input of the Add function and wire another matrix with three rows and two columns to the other input of the Add function, the result is an empty matrix.

6.9   VI MEMORY USAGE

Unlike text-based programming languages, the LabVIEW dataflow paradigm eliminates the complexity associated with memory management. As you worked your way through the first five chapters, you spent no time considering memory allocation and usage. You created a VI by building a block diagram with connections representing the transition of data, and (probably unbeknown to you) the functions that generate the data managed the memory allocation for that data. The associated memory is deallocated when the data is no longer required. When you add information to arrays or strings (the topic of Chapter 9), sufficient memory is automatically allocated to manage the new information.

While automatic memory handling is a wonderful benefit of LabVIEW, it gives you no insight into the process. As a beginning LabVIEW programmer you will probably not work with large sets of data, and memory allocation will not be an overriding issue. As your VIs become more complex, however, you will need a deeper understanding of memory allocation. A grasp of the principles involved can help you create programs with significantly smaller memory requirements, leading to faster VI execution, because allocating memory and copying data can take a considerable amount of time.

Memory management is beyond the scope of this introductory text. For more information you can search the on-line help (use "memory" as the keyword for your search). Here we discuss a few rules that lead to better memory usage.

In practice, the following rules (listed in no particular order) can help you create VIs that use memory efficiently:

1. Initialize large data sets, later replacing elements during execution. Replacing memory elements takes less time than dynamically creating them on the fly.

2. You can improve memory usage by breaking a VI into subVIs, because the system can reclaim subVI data memory when the subVI is not executing.

3. Do not overuse local variables. Reading a local variable causes LabVIEW to generate a copy of the data.

4. Unless you need to, do not display large arrays and strings on open front panels. Controls and indicators on open front panels retain a copy of the data they display.

5. Watch for places on the block diagram where input size is different from output size. For example, if you are frequently increasing the size of an array using the Build Array function, you are generating copies of data.

6. Use consistent data types for arrays and watch for coercion dots when passing data to subVIs and functions. Changing data types results in copies being made of the data.

7. Do not use complicated hierarchical data types, such as clusters or arrays of clusters containing large arrays or strings.

8. Unless they are necessary, do not use transparent and overlapped front panel objects.

As an example, it is not recommended that you create an array in a loop by constantly calling Build Array to concatenate a new element. The Build Array function reuses the input array, continually resizing the buffer in each iteration to make room for the new array, and appends the new element. Consequently, execution is slow. Instead, initialize the array to establish the size and then replace the elements programmatically, or if you want to add a value to the array with every iteration of the loop, you can use the autoindexing on the edge of a loop. With For Loops, the VI can predetermine the size of the array (based on the value wired to N) and resize the buffer only once. Autoindexing is not quite as efficient with While Loops because the end size of the array is not known. However, by increasing the output array size in large increments, While Loop autoindexing avoids resizing the output array with every iteration. When the loop is finished, the output array is resized to the correct size. The performance of While Loop autoindexing is nearly identical to that of For Loop autoindexing.

BUILDING BLOCK

6.10   BUILDING BLOCKS: PULSE WIDTH MODULATION

The goal in this Building Block is to analyze the PWM being produced. In order to do so, you will use the arrays introduced in the chapter, then use the Pulse Measurement VI to display the duty cycle, period, and the duration of the pulse.

To begin, open the Building Blocks VI that you constructed in Chapter 5. It should be called PWM with LED.vi and located in the Users Stuff folder in the Learning directory. Once the VI is open, rename it as PWM with Measurements.vi and save the VI in the Users Stuff folder.

In case you do not have a working version of the Building Blocks VI from Chapter 5, a working version of the PWM with LED VI can be found in the Building Blocks *folder in the* Learning *directory.*

Switch to the block diagram of the PWM with Measurements VI. The changes that you will make are illustrated in Figure 6.42. To begin, add the Boolean To (0,1) function inside the While Loop but outside of the Case structure, as shown in Figure 6.42. The Boolean To (0,1) function is found on the **Functions≫Programming≫Boolean** palette. Wire the input to the Output indicator to the Boolean To (0,1) and then wire the output of the Boolean To (0,1) to the While Loop.

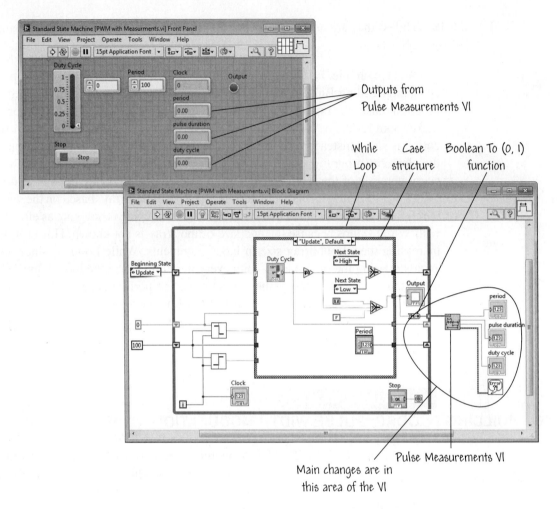

FIGURE 6.42
The PWM with Measurements VI.

You will need to convert the Boolean outputs to numeric values before creating the array. Recall that by default, auto-indexing is disabled on While Loops and so you will need to enable it before you can output an array from the loop. If you do not use auto-indexing with this While Loop, on each iteration of the While Loop, a single scalar value representing the value of the waveform on the last clock cycle will be output. When auto-indexing is enabled, each iteration of the loop will append a new output value to the end of an array containing all output values. To enable the auto-indexing, right-click on the tunnel and select **Tunnel Mode≫Indexing**, as shown in Figure 6.43.

Place the Pulse Measurements VI on your block diagram outside of the While Loop as shown in Figure 6.42. Use **Quick Drop** <Ctrl-Space> to find the Pulse Measurement VI. You can use **Context Help** to find out more about this

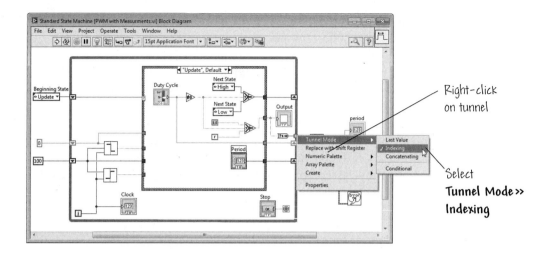

FIGURE 6.43
Enable indexing on the Boolean outputs.

VI, including the outputs, which include pulse, pulse duration, and duty cycle. Right-click on the appropriate terminal of the Pulse Measurements VI and add indicators for **Period**, **Pulse Duration**, and **Duty Cycle** that can be displayed on the front panel.

Again, use **Quick Drop** to add a Simple Error Handler VI and wire the appropriate output of the Pulse Measurements VI. Save your modified VI as PWM with Measurements.vi in the Users Stuff folder in the Learning directory.

What output values would you expect for inputs of Duty Cycle $= 0.75$ and Period $= 100$? Input these values on the front panel controls and run the VI. Check the clock before you stop the VI and confirm that it has been running for at least two output cycles so that the Pulse Measurements VI has enough data to make the appropriate calculations. Do your measurements confirm the input values you provided?

The Pulse Measurements VI, like many VIs in LabVIEW, has built-in error handling capabilities. Error handling in LabVIEW follows the data-flow model. Just as data flow through a VI, so can error information. Many nodes in Lab-VIEW have Error In and Error Out terminals which pass error information through the node. This information is passed in a cluster containing a numeric code identifying the error, a Boolean value reporting whether or not an error occurred, and a string identifying the node in which the error occurred. As the VI runs, LabVIEW tests for errors at each execution node. If LabVIEW does not find any errors, the node executes normally. If LabVIEW detects an error, the node passes the error to the next node without executing. The next node does the same thing, and so on.

*The numeric error code displayed by the error cluster is not always self-explanatory. For clarification, right-click on the error code and select **Explain Error** from the menu.*

Now what happens when you run the VI with Duty Cycle = 1? Try running the VI with a different value of Duty Cycle, but this time, stop execution before two output cycles complete. Observe how these errors in the Pulse Measurements VI are handled.

*A working version of the PWM with Measurements VI can be found in the **Building Blocks** folder of the **Learning** directory.*

6.11   RELAXED READING: AUTOMATIC LASER-ASSISTED NEURON GROWTH

In this reading we explore medical and biological applications of lasers and other photonic devices in a developing field known as biophotonics. In particular, we consider detection of the structural details of neuronal growth cones in vitro and using this information to correctly position a laser beam to grow the neuron toward a user-selected target. LabVIEW is used to determine and recognize the shape of a nerve ending from a charge-coupled device image in real time and to calculate the coordinates to position the laser.

A growth cone is a part of the neuron cell that guides the growth of the neuron's nerve endings to meet the nerve endings of another neuron cell. Optically guided neuronal growth is an emerging biophotonic field in which focus infrared laser beams are used to artificially guide neuronal growth cones. The ultimate goal is to employ a laser to control the direction of developing neuronal growth cones to rebuild broken spinal cords and damaged networks. A sufficiently powerful laser beam aimed onto the leading edge of the growth cone manipulates the cytoskeleton to grow toward the center of the beam. By continually adjusting the position of the beam, an operator manipulates the neuron to grow in an entirely artificial direction. Manipulating these structures could have far reaching medical and biotechnological effects. To date, the exact mechanism by which this phenomenon operates has not been explained.

The infrared laser is coupled into a microscope via a spatial light modulator. An image acquisition board is used to capture the CCD image and a custom-built LabVIEW program uses the CCD images to provide beam-steering information that is sent to the spatial light modulator to manipulate the beam. The shape detection algorithm is the most complex and crucial part of the program since mistakes can manifest as bad coordinates for the laser. Figure 6.44 shows a detailed step-by-step diagram of the shape detection process. Figure 6.44-1 shows the raw image from the CCD. The remaining images in Figure 6.44-2 to Figure 6.44-7 show the progression of the shape detection process that includes edge detection, dilation and fill operations, erosion, object detection with particle filtering, and smoothing. Figure 6.44-8 shows the final shape information overlaid on the original image, including the center of mass, a user-input target point, the trajectory between the two and the point where this line intersects the edge.

With the shape detected, the program extracts the data needed for the hologram. LabVIEW is employed to extract the center of mass of the detected shape

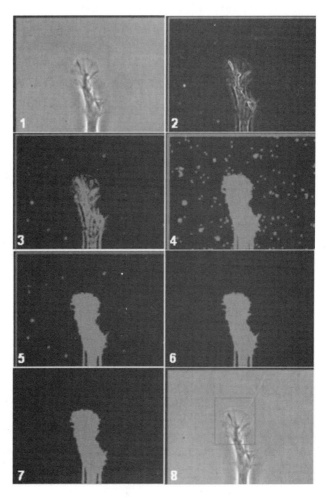

FIGURE 6.44
Image processing.

as a coordinate by means of a pixel weighting average. A straight line is plotted between the center of mass and the user-input target points and the point where this line intersects the edge of the shape is extracted, as illustrated in Figure 6.44-8. This is the point where the laser focus is directed. We use the coordinates from the edge intersection to direct the laser to the required image spot. As the edge moves, the laser spot tracks the edge.

A growth cone of an NG-108 cell guided by a laser of approximately 40 mW of power and a spot size of 1 μm is depicted in Figure 6.45. The growth cone is initially growing in an upper left direction, but a clear direction change is evident as the sequence progresses. Figure 6.45 shows the chronological progression of a growth cone guided by a line profile approximately every 2 minutes with a scale bar of 10 μm.

In summary, LabVIEW and NI hardware was successfully employed to develop a data acquisition system with complex process control creating a feedback control system that can automatically detect the shape and characteristics of a neuronal growth, then steer a laser beam to an appropriate point to induce artificial guidance toward a user-input target point. For further information, visit http://sine.ni.com/cs/app/doc/p/id/cs-13033 and http://research-repository.st-andrews.ac.uk/handle/10023/1860.

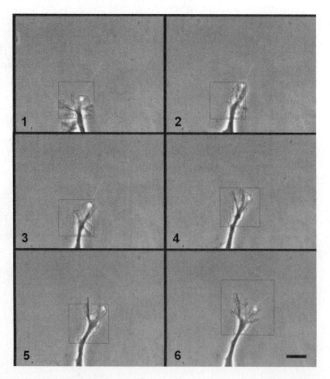

FIGURE 6.45
Chronological order of growth.

6.12   myDAQ BUILDING BLOCKS

In this myDAQ Building Block, you will add the capability to acquire multiple points from the temperature monitoring system. Additionally, you will add data smoothing using an averaging process to create a smoother, more consistent result.

Begin by opening Control System Multiple unedited.vi from the Chapter 6 folder in the myDAQ Building Blocks folder within the Learning directory. Display the block diagram, as illustrated in Figure 6.46, and use the following steps as a guide.

1. Add a For Loop around Read Temp.vi, the Current Temperature terminal and the Wait Until Next ms Multiple function. You should see an auto-indexing tunnel where the Current Temperature wire exits the loop. Notice that the wire is thicker outside the loop, indicating a change in the dimension of the data type. The indexing tunnel accumulates the temperature value for each iteration and builds an array of the results.

2. Create a constant for the count terminal of the For Loop and set the value to 10. This allows the (outer) While Loop to run at 1 Hz without changing the input to the Wait Until Next ms Multiple function (currently set to 100).

3. The Add Array Elements function will sum the elements of a numeric array. Use this function along with the For Loop counter and a Divide function to calculate the average temperature for one iteration of the While Loop. Wire this output to the Average Temperature terminal.

4. You may have noticed when you added the For Loop that the Boolean input to the selector terminal of the Case structure became a broken wire. This happens because the In Range and Coerce function is polymorphic. Since it can accept both scalar and array inputs, when the **x** input became an array, the In Range? output became an array as well. The broken wire occurs because the Case structure cannot accept an array input. Place an And Array Elements function on the block diagram and wire it between the In Range? Output of the In Range and Coerce function and the Boolean input to the selector terminal of the Case structure.

*The And Array Elements function, found on the **Functions≫Programming≫Boolean** palette, returns TRUE if all the elements in **Boolean array** are TRUE or if **Boolean array** is empty. Otherwise, the function returns FALSE. This function accepts an array of any size but returns only a single value based on all values in **Boolean array**. See Figure 6.46 for the definition of **Boolean array**.*

5. Switch to the front panel and run the VI.

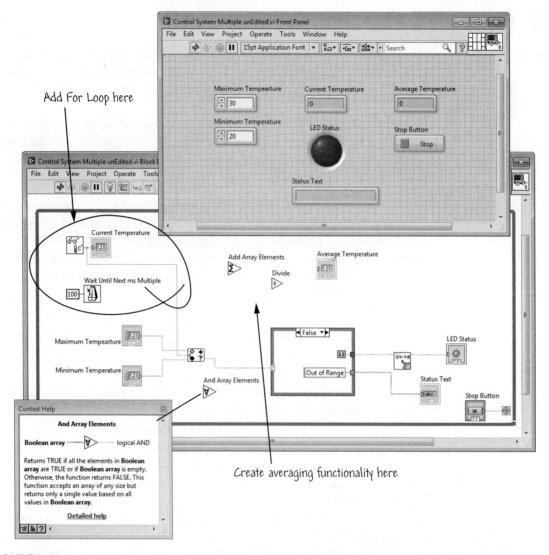

FIGURE 6.46
Acquiring multiple temperature data points and averaging to obtain a smoother output.

6. Save the edited VI in the Chapter 6 folder in the myDAQ Building Blocks folder within the Learning directory and name it Control System Multiple.vi. Close the VI.

A working version of Control System Multiple.vi can be found in the Chapter 6 folder of the myDAQ Building Blocks≫Solutions folder in the Learning directory.

In this myDAQ Building Block, you enhanced the capabilities of the temperature monitoring system using a For Loop to enable the system to acquire multiple temperature data points and to smooth the data using a simple averaging functionality. The VI is now ready to output the data in graphical formats in Chapter 7.

6.13   SUMMARY

The main topics of the chapter were arrays and clusters. Arrays are a variable-sized collection (or grouping) of data elements that are all the same type, such as a group of floating-point numbers or a group of strings. LabVIEW offers many functions to help you manipulate arrays. In this chapter we discussed the following array functions:

- Array Size
- Initialize Array
- Build Array
- Array Subset
- Index Array

A cluster is a fixed-sized collection of data elements of mixed types, such as a group containing floating-point numbers and strings. As with arrays, LabVIEW offers built-in functions to help you manipulate clusters. In this chapter we discussed the following cluster functions:

- Bundle
- Unbundle

The important concept of polymorphism was introduced. Polymorphism is the ability of a function to adjust to input data of different types, dimensions, or representations. For example, you can add a scalar to an array or add two arrays of differing lengths together.

We discussed the matrix data type that stores rows or columns of real or complex scalar data for matrix operations and can be used in place of two-dimensional arrays. Using this data type is advantageous for many linear algebra operations based on efficient matrix algorithms. We also discussed matrix functions that are convenient for manipulating matrices. Many of these are similar to existing array functions, but they offer functionality for math algorithms based on matrices.

The LabVIEW dataflow paradigm eliminates the complexity associated with memory management. While automatic memory handling is a wonderful benefit of LabVIEW, it allows you less insight into the process precisely because it is automatic. Memory management is beyond the scope of this introductory text. We did, however, present a few rules that will lead to better memory usage.

KEY TERMS

Array: Ordered and indexed list of elements of the same type.

Array shell: Front panel object that contains the elements of an array. It consists of an index display, a data object window, and an optional label.

Array Size: Array function that returns the number of elements in the input array.

Array Subset: Array function that returns a portion of an array.

Auto-indexing: Capability of loop structures to assemble and disassemble arrays at their borders.

Build Array: Array function that concatenates multiple arrays or adds extra elements to an array.

Bundle: A cluster function that assembles all the individual input components into a single cluster.

Cluster: A set of ordered, unindexed data elements of any data type, including numeric, Boolean, string, array, or cluster.

Index Array: Array function that accesses elements of arrays.

Initialize Array: Array function that creates an n-dimensional array with elements containing the values that you specify.

Matrix: A data type that stores rows or columns of real or complex scalar data for matrix operations and can be used in place of two-dimensional arrays.

Matrix functions: A convenient means to manipulate matrices, similar to existing array functions but offering functionality for math algorithms based on matrices.

Polymorphism: The ability of a function to adjust to input data of different dimension or representation.

Unbundle: A cluster function that splits a cluster into each of its individual components.

E6.1   Open While Loop Indexing.vi found in the Exercises&Problems folder in Chapter 6 of the Learning directory and click on the **Broken Run** button to access the error list. Highlight the error and press the **Show Error** button. This error is a result of an array wired to a terminal that is a scalar value. Recall that While Loops disable indexing on tunnels by default. To eliminate the error, right-click on the loop tunnel and choose **Tunnel Mode»Indexing**. Run the VI. After a few moments, press the **Stop** button on the front panel. An array will appear in the numeric indicator titled Temperature Measurements.

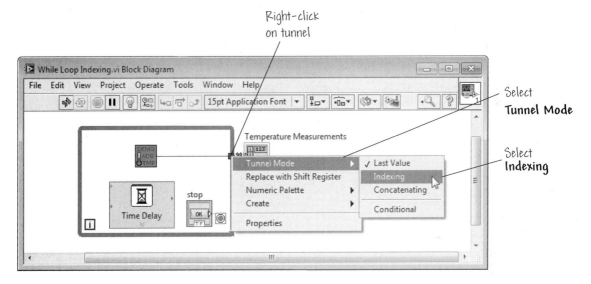

FIGURE E6.1
While Loop Indexing block diagram.

E6.2   Open Cluster Order.vi found in the Exercises&Problems folder in Chapter 6 of the Learning directory and click on the **Broken Run** button to access the error list. Highlight the error and press the **Show Error** button. This error is a result of the cluster elements not being in the same order as the elements that are bundled. To eliminate the error, right-click on the cluster border on the front panel and select **Reorder Controls in Cluster**. Change the order so Sinc Data Points is the first element, Over Limit? is the second element, and Limit is the third element. Now you can run the VI.

E6.3   Open **Unbundle By Name.vi** found in the **Exercises&Problems** folder in **Chapter 6** of the **Learning** directory and view the block diagram. This VI is the same as the **Cluster Unbundle Demo.vi** discussed earlier in this chapter. However, in this VI the Unbundle By Name function is used instead of the Unbundle function. The advantage to using the Unbundle By Name is that you can easily identify the cluster values that are unbundled. You can use Unbundle By Name only if every object in the cluster has an owned label associated with it. Using the LabVIEW Help as a source of information, write a short description of the cluster function Unbundle By Name.

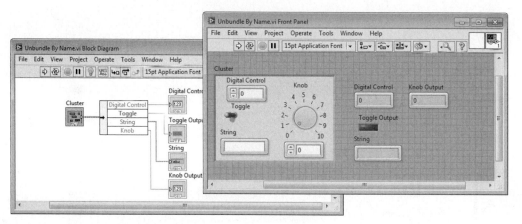

FIGURE E6.3
Unbundle By Name VI.

E6.4   Open **Reversed Array.vi** found in the **Exercises&Problems** folder in **Chapter 6** of the **Learning** directory and view the block diagram. Add code to this VI that will display the data that is in the Array indicator in reversed order in the Reversed Array indicator.

E6.5   Open **Rolling Dice.vi** found in the **Exercises&Problems** folder in **Chapter 6** of the **Learning** directory and view the block diagram shown in Figure E6.5. This program uses the random number generator to simulate rolling a six-sided die. The program should keep track of the number of times each value is rolled by incrementing the appropriate value in a 1D array. You will need to wire the Index Array, Increment, and Replace Array Subset nodes appropriately to run this VI.

E6.6   Open **Mean.vi** found in the **Exercises&Problems** folder in **Chapter 6** of the **Learning** directory. This VI computes the mean of the values in the input sequence **Array**. The block diagram is shown in Figure E6.6. Find a function on

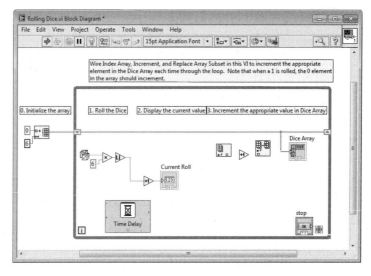

FIGURE E6.5
Incomplete Rolling Dice VI.

the **Programming≫Numeric** palette which will allow you to significantly simplify the VI. Then edit the VI so that it uses this function instead of a For Loop.

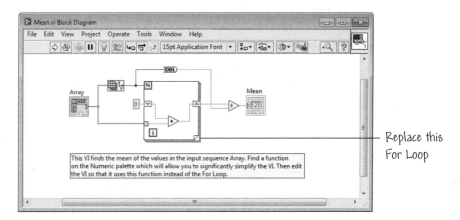

FIGURE E6.6
The block diagram for the Mean.vi.

E6.7   Construct a VI that outputs a 2D array with 10 rows and 5 columns. The first row should contain five random numbers in the range 0–1, the second row numbers in the range 1–2, and so forth to the tenth row, which should contain five random numbers in the range 9–10.

E6.8 Clusters are a good way to combine data. Open up the Product in Stock VI in **Chapter 6** of the **Exercises&Problems** folder in the **Learning** directory. Switch to the block diagram. Describe in your own words what the VI is doing. Click on the **Broken Run** to discover the errors in the VI. Correctly wire the Unbundle by Name functions to complete the VI.

Run the VI and describe the outcome when you input **Hammer** as the part name. What happens if you input **100** for the quantity of hammers?

PROBLEMS

P6.1 Open the VI called **Age.vi** that you created in Chapter 4 in Design Problem 4.1. Currently, the VI has 6 inputs: **Current Year, Current Month, Current Date, Birth Month, Birth Date,** and **Year of Birth**. Create a cluster so that all of the information about an individual date can be carried in one single wire on the block diagram. Then update the Age VI so that it has only 2 inputs: **Current Date** and **Birth Date**.

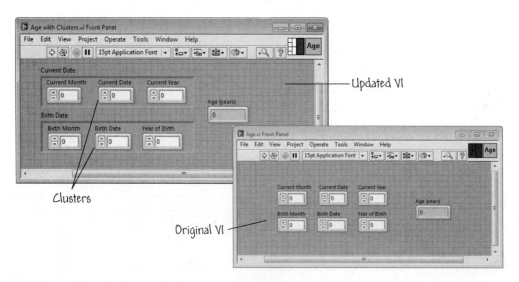

FIGURE P6.1
The original Age VI and the updated Age with Clusters VI front panels.

 If you did not complete Design Problem 4.1, a working version of the Age VI can be found in the Exercises&Problems folder in Chapter 6 of the Learning directory.

P6.2 Create a VI that calculates the dot (inner) product of two n-dimensional vectors. Double check your math by comparing your calculation using arrays and math functions with the results of the Dot Product.vi that can be found in the **Functions≫Mathematics≫Linear Algebra** palette.

Review: If the two vectors are denoted by v_1 and v_2, where

$$
v_1 = \begin{bmatrix} v_1(0) \\ v_1(1) \\ \vdots \\ v_1(n) \end{bmatrix} \quad \text{and} \quad v_2 = \begin{bmatrix} v_2(0) \\ v_2(1) \\ \vdots \\ v_2(n) \end{bmatrix},
$$

then the dot product is given by

$$
v_1 \cdot v_2 = v_1(0)v_2(0) + v_1(1)v_2(1) + \cdots + v_1(n)v_2(n).
$$

P6.3 Create a VI that calculates the cross product of two 3-dimensional vectors.

Review: If the two vectors are denoted by v_1 and v_2, where

$$
v_1 = \begin{bmatrix} v_1(0) \\ v_1(1) \\ v_1(2) \end{bmatrix} \quad \text{and} \quad v_2 = \begin{bmatrix} v_2(0) \\ v_2(1) \\ v_2(2) \end{bmatrix},
$$

then the cross product is given by

$$
v_1 \times v_2 = \begin{bmatrix} v_1(1)v_2(2) - v_1(2)v_2(1) \\ v_1(2)v_2(0) - v_1(0)v_2(2) \\ v_1(0)v_2(1) - v_1(1)v_2(0) \end{bmatrix}.
$$

P6.4 Create a VI that performs matrix multiplication for two input matrices **A** and **B**. The matrix **A** is an $n \times m$ matrix, and the matrix **B** is an $m \times p$ matrix. The resulting matrix **C** is an $n \times p$ matrix, where $\mathbf{C} = \mathbf{AB}$. Double-check your math by comparing your calculation using matrix data types and matrix functions with the results of the AxB.vi found in the **Mathematics≫Linear Algebra** palette.

P6.5 Create a VI that reads 12 temperature measurements using the Demo Temp.vi found in the Exercises&Problems folder in Chapter 6 of the Learning directory. With each temperature measurement, bundle the time (including seconds) and date of the measurement. Include a Time Delay Express VI to slow the loop so it executes twice per second. Run the program and review the time stamps in the output array to be sure there are two samples per second.

P6.6   Build a VI that generates and plots 500 random numbers on a waveform graph indicator. Compute the average of the random numbers and display the result on the front panel. Use the Statistics Express VI found in the **Mathematics≫ Probability and Statistics** palette to compute of the average of the random numbers.

P6.7   Create a VI that computes and plots the second order polynomial $y = Ax^2 + Bx + C$. The VI should use controls on the front panel to input the coefficients A, B, and C, and it should also use front panel controls to enter the number of points N to evaluate the polynomial over the interval x_0 to x_{N-1}. Plot y versus x on a waveform graph indicator.

P6.8   Create a VI with a cluster of six buttons labeled **Option 1** through **Option 6**. When executing, the VI should wait for the user to press one of the buttons. When a button is pressed, use the Display Message To User Express VI to indicate which option was selected. Repeat this process until the user presses the **Stop** button. Be sure to include the Time Delay Express VI to give the user time to press the buttons.

  Hint 1: This program will require a simplified version of the state machine architecture along with a user menu. A typical state machine in LabVIEW consists of a While Loop, a Case structure, and a Shift register (see Section 5.9). Each state of the state machine is a separate case in the Case structure. You place VIs and other code that the state should execute within the appropriate case. A Shift register stores the state to be executed upon the next iteration of the loop. This is required in a typical state machine because there are times when the results of the state you are currently in control the state you will go to next. In the program described above, no two states are dependent upon one another, so the shift register is optional.

  Hint 2: You can use latched Boolean buttons in a cluster to build a menu for a state machine application. The Cluster To Array function converts the Boolean cluster to a Boolean array, where each button in the cluster represents an element in the array. The Search 1D Array function searches the 1D array of Boolean values created by the Cluster To Array function for a value of True. A True value for any element in the array indicates that the user clicked a button in the cluster. The Search 1D Array function returns the index of the first True value it finds in the array and passes that index value to the selector terminal of a Case structure. If no button is pressed, Search 1D Array returns an index value of -1 and the -1 case executes, which does nothing. The While Loop repeatedly checks the state of the Boolean cluster control until you click the **Stop** button.

P6.9   Create a VI that determines all the factors of an integer that is input by the user. The results should be displayed in an array. For this problem it is okay if there are zeros in the answer array.

P6.10 As in P6.9, create a VI that determines all the factors of an integer that is input by the user. The results should be displayed in an array. However, unlike in P6.9, display only the factors on the front panel (no zeros). Also, create a subVI named Factoring without 0s in Array.vi. Save the VI in the Users Stuff folder in the Learning directory.

P6.11 Create a VI that determines whether an integer input by the user is prime. Also, create a subVI named Prime Number.vi. Save the VI in the Users Stuff folder in the Learning directory.

It might be helpful to use Factoring without 0s in Array.vi *that you created in P6.10.*

DESIGN PROBLEMS

D6.1 Construct a VI that searches the values of an input array of numbers for a specified value. If the specified value is found in the array, the VI should indicate that by turning on an LED and by indicating the corresponding index of the array. Use Figure D6.1 as a guide.

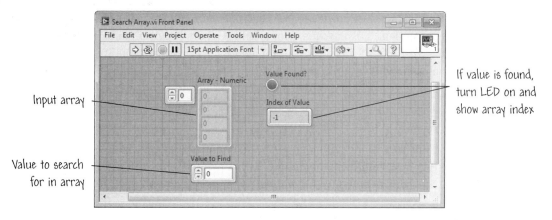

FIGURE D6.1
The Search Array VI front panel.

D6.2 Design a VI that produces an array of clusters containing the course average, letter grade, and class rank for each student in a class, given three arrays with the examination scores as input. There is one array of scores for each examination. The course average is determined as follows: 30% from the first exam, 30% from the second exam, and 40% from the final exam. The letter grade is determined from the course average by the criteria listed in Table D6.2.

A Case structure will be useful in determining letter grades.

TABLE D6.2   Course Average Criteria.

Course Average	Letter Grade
0%–59%	F
60%–69%	D
70%–79%	C
80%–89%	B
90%–100%	A

Rank the students so that the student with the highest grade has a rank of 1 and the student with the lowest grade in the class has a rank equal to the number of students in the class. You may wish to use the Sort 1D Array and Search 1D Array functions within a loop to establish the student ranks.

Bundle all of the information (course average, letter grade, and class rank) into an array of clusters where each cluster contains all the data for one individual student. Display this array of clusters on the front panel.

Test the VI with different inputs to verify its functionality. Since the array of clusters will become very large for a class with more than a few students, space can be conserved on the front panel by showing only one element of the output array at a time and then using the index display to scroll through the different elements of the cluster array.

D6.3   Construct a VI with one numeric input n that builds an array containing n Fibonacci numbers, beginning with F_1. Fibonacci numbers are the sequence of numbers, denoted by F_n, defined by the relation

$$F_n \equiv F_{n-2} + F_{n-1}$$

with $F_1 = 1$ and $F_2 = 1$. Table D6.3 shows the values of F_n you should obtain for the first few values of n (all inputs are positive integers).

Before deciding what type of loop to use in your VI, you may want to consider an important difference between For Loops and While Loops. A While Loop always executes at least once because the VI checks the conditional terminal at the end of each iteration. However, a For Loop does not possess this attribute; if a value less than one is wired to the Count terminal, the For Loop will not execute. Keep this in mind when initializing the output array. Consider carefully

TABLE D6.3   First Ten Fibonacci Numbers.

n	F_n
1	1
2	1
3	2
4	3
5	5
6	8
7	13
8	21
9	34
10	55

how to construct your VI so that the output array is accurate for $n = 1$ and $n = 2$, as well as the values computed using the formula.

D6.4 Amicable numbers are two different numbers where the sum of the factors (except itself) of one of the numbers is equal to the other number. Write a VI that outputs all amicable number pairs less than 10,000.

It might be helpful to use **Factoring without 0s in Array.vi** *that you created in P6.10.*

D6.5 A perfect number is a positive integer that is equal to the sum of its factors. For example, 6 is a perfect number, because $6 = 1 + 2 + 3$. Create a VI that finds all perfect numbers less than 10,000.

There are four perfect numbers less than 10,000.

D6.6 Create a VI that generates n random numbers, where n is a user input. Display the n random numbers in an array, and in a second array display the same random numbers in ascending order.

D6.7 Create a VI that computes all prime numbers less than n, where n is a user input. Display the result in an array. For example, if $n = 10$, then the VI should compute and display 1, 2, 3, 5, and 7.

It might be helpful to use **Prime Number.vi** *that you created in P6.11.*

D6.8   Design a VI that calculates how old you are on other planets based on a birthday that is input to the VI. The input should be the birth month, date, and year. Use a cluster to output the results showing the user age on all nine planets. Assume that the length of a year on each of the nine planets is as given in Table D6.8. We assume here that Pluto is still a planet!

TABLE D6.8   Length of year on the nine planets.

Planet	Length of year (days)
Mercury	88
Venus	225
Earth	365
Mars	687
Jupiter	4331
Saturn	10756
Uranus	30687
Neptune	60190
Pluto	90553

Data taken from http://solarsystem.nasa.gov/planets/index.cfm

Use the subVIs created in D4.1 and D5.7 to compute the age in years and the number of days, respectively, until your next birthday.

CHAPTER 7

Charts and Graphs

Graphs and charts are used to display data in a graphical form. Three types of charts are discussed—strip, scope, and sweep charts—two types of graphs—waveform and XY graphs—and two types of math plots—2D Graphs and 3D graphs. The math plots are very helpful in performing mathematical analysis and displaying results graphically. Data is presented on charts by appending new data to the existing plot as it becomes available, much in the same manner as on a strip chart you might find in a laboratory. On the other hand, graphs are used to display pregenerated arrays of data in a more traditional fashion, such as the typical *x-y* graph. The subject of customizing charts and graphs is also discussed in the chapter.

GOALS

1. Learn about charts and graphs and recognize their similarities and differences.

2. Understand the three modes of a chart: strip, scope, and sweep.

3. Learn about math plots as a way to present graphically the results of mathematical analysis.

4. Learn to customize the appearance of charts and graphs.

369

7.1 WAVEFORM CHARTS

A **waveform chart** is a special kind of indicator—located in the **Graph** sub-palette of the **Controls≫Modern** palette, as illustrated in Figure 7.1. The wave-form chart can also be found on the **Graph Indicators** subpalette of the **Controls≫Silver** palette.

There is only one type of waveform chart, but it has three different update modes for interactive data display—**strip chart**, **scope chart**, and **sweep chart**, as shown in Figure 7.2. You select the update mode that you want to use by right-clicking on the waveform chart and choosing one of the options from the Advanced≫Update Mode menu. This process is illustrated in Figure 7.3. This can also be selected from the **Appearance** menu of the **Properties** dialog box. The strip chart, scope chart, and sweep chart all handle incoming data a little differently. The strip chart has a display that scrolls so that as each new data point arrives, the entire plot moves to the left—the oldest data point falls off the chart, and the latest data point is added at the rightmost part of the plot. This action is very similar to a paper strip chart commonly found in laboratories.

The scope and sweep charts more closely resemble the action of an oscillo-scope. When the number of data points is sufficient so that the plot reaches the

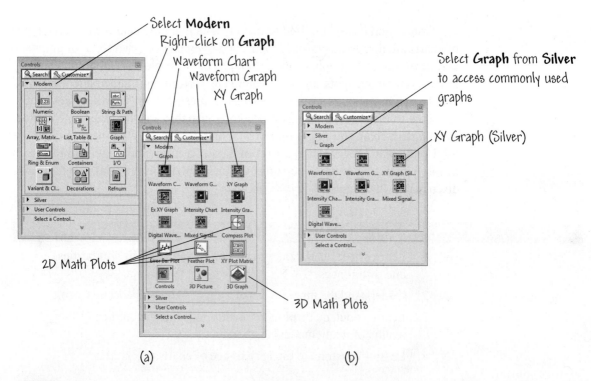

(a)   (b)

FIGURE 7.1
(a) The graphs and chart are located in the **Graph** subpalette of the **Modern** palette; (b) commonly used graphs are located on the **Graph** subpalette of the **Silver** palette.

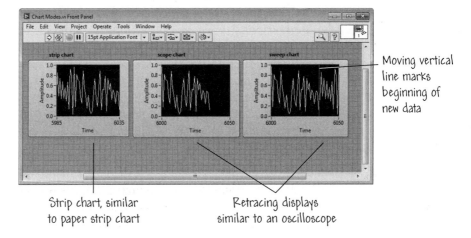

FIGURE 7.2
The waveform chart has three update modes.

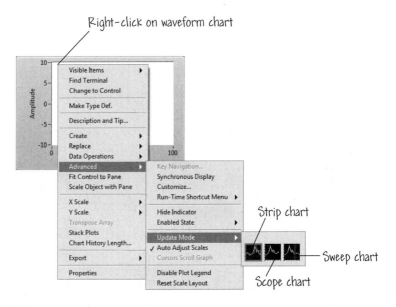

FIGURE 7.3
Changing waveform chart modes: strip chart, scope chart, and sweep chart.

right border of the plotting area of the scope chart, the entire plot is erased, and the plotting begins again from the left side. The sweep chart acts much like the scope chart except that, rather than erasing the plot when the data reaches the right border, a moving vertical line marks the beginning of new data and moves across the display as new data is added. The scope chart and the sweep chart run faster than the strip chart.

You can run the waveform chart demonstration shown in Figure 7.2 by opening Chart Modes.vi *located in* Chapter 7 *of the* Learning *directory. Run the VI in* **Run Continuously** *mode and examine the three different waveform charts: the strip chart, the scope chart, and the sweep chart. Select* **Highlight Execution** *if you want to observe the data as it plots in slow motion.*

Waveform charts may display single or multiple traces. An example of a multiple-plot waveform chart is shown in Figure 7.4.

To generate a single-plot waveform chart, you wire a scalar output directly to the waveform chart. The data type displayed in the waveform chart will match the input. In the example shown in Figure 7.5, a new temperature value is plotted on the chart each time the While Loop iterates.

As illustrated in Figure 7.4, waveform charts can display multiple plots. To generate a multi-plot waveform chart, you bundle the data together using the Bundle function or use the Merge Signal function located on the **Functions≫ Express Signal Manipulation** palette (see Chapter 6 for a review). In the example in Figure 7.5, the Bundle function groups the output of the two different data sources—Digital Thermometer.vi and the Random Number (0–1) function—that acquire temperature and random number data for plotting on the waveform

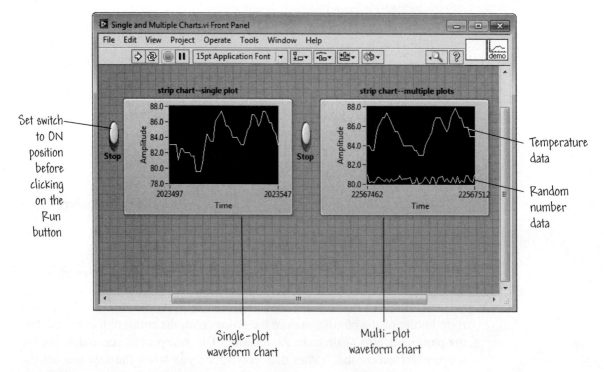

FIGURE 7.4
Examples of single- and multiple-plot waveform charts.

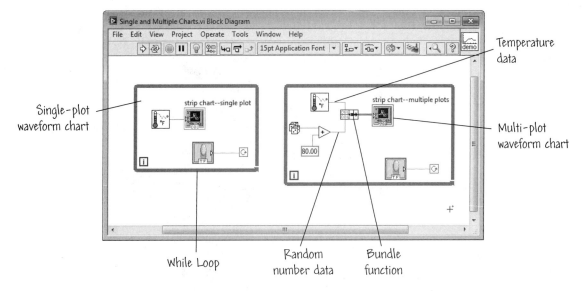

FIGURE 7.5
Block diagrams associated with single- and multiple-plot waveform charts.

chart. The random number is biased by the constant 80 so that the plots are easily displayed on the same *y*-axis scale. Remember that Digital Thermometer.vi simulates acquiring temperature data presented in degrees Fahrenheit in the general range of 80°. You can add more plots to the waveform chart by increasing the number of Bundle function input terminals—that is, by resizing the Bundle function.

You can run the single- and multi-plot waveform chart demonstration shown in Figure 7.4 by opening Single and Multiple Charts.vi located in Chapter 7 of the Learning directory. Run the VI and examine the single- and multi-plot waveform chart. You can stop the plotting of each chart using the vertical switch button. After stopping the plotting with the vertical switch, remember to reset the switch to the up position before the next execution of the VI.

Practice with Waveform Charts

In this exercise you will construct a VI to compute and display the circle measurements of circumference, area, and diameter on waveform strip charts. This VI utilizes the Circle Measurements subVI.vi you developed in Chapter 4. Figure 7.6 shows the front panel that you can use as a guide as you construct your VI. The front panel has five items: a numeric control defining the radius of the circle, a push button to stop execution of the VI, and three waveform charts.

Open a new VI front panel and place a numeric control in the window and a push-button Boolean. Label the numeric control r, and add a free label near the

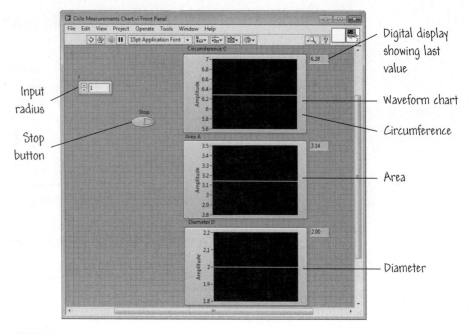

FIGURE 7.6
The front panel for waveform charts to plot circle measurements of circumference, area, and diameter.

push button indicating that pushing the button stops the execution. Place three waveform charts in the front panel window and locate the front panel objects in approximately the same locations as seen in Figure 7.6.

Enter the data r = 1 in the numeric control. In the **Operate** pull-down menu, select **Make Current Values Default**. In future sessions, when the VI is used it will not be necessary to input the default value.

Figure 7.7 shows the block diagram that you can use as a guide for building your VI. The numeric control terminal, waveform charts, and push-button Boolean will automatically appear on the block diagram after you have placed their associated objects on the front panel. The block diagram has two additional items:

- While Loop
- Circle Measurements subVI.vi

Switch to the block diagram window of your VI. First enclose the numeric control, waveform charts, and push-button terminals inside a While Loop. The only other object that is needed is the subVI to compute the circle measurements. The Circle Measurements subVI.vi is located in Chapter 4 of the Learning directory.

You can access the subVI by right-clicking on the block diagram and from the **Functions** palette choosing **Select a VI** as illustrated in Figure 7.8. You will

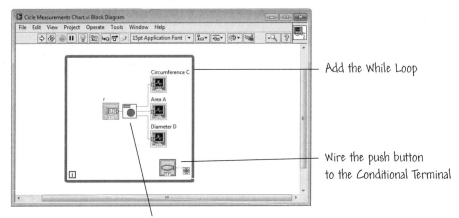

Add the While Loop

Wire the push button
to the Conditional Terminal

Add the Circle Measurements subVI and wire the terminals

FIGURE 7.7
The block diagram for waveform charts to plot circle measurements of circumference, area, and diameter.

need to navigate through the file structure to reach the Chapter 4 directory. Once there, you select the desired subVI—this is demonstrated in Figure 7.8.

Once all the required objects are on the block diagram, wire the diagram as shown in Figure 7.7. As an optional step, you can remove the *x*-axis scale. Right-click on each waveform chart on the front panel and under the **Visible Items** category choose **X Scale** to deselect the *x*-axis scale. In this example, the values of the circle measurements are being computed and plotted on each iteration of the While Loop—the **X Scale** does not have any physical meaning, other than as a count of the total loop iterations.

Once everything is ready for execution, run the VI. With the default value of r = 1, you should find that the computed circumference is C = 6.28, the area is A = 3.14, and the diameter is D = 2.0. Increase the radius and observe how it affects the circle measurements. As you vary the radius you should see a line across the waveform chart move up when the radius increases and back down as the radius decreases.

Save your VI in the Users Stuff folder and name it Circle Measurements Chart.vi.

You can find a completed VI for the waveform chart demonstration shown in Figure 7.6 in Chapter 7 *of the* Learning *directory. The VI is called* Circle Measurements Chart.vi.   ◆

In the previous example, the While Loop executed as quickly as the computer system would allow. In many situations you may want to acquire and plot data at specified intervals. This can be accomplished by controlling the loop

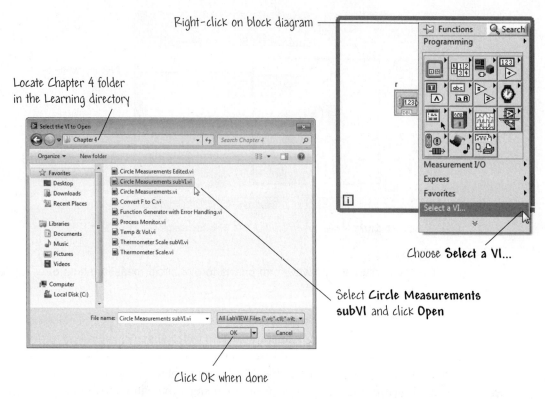

Right-click on block diagram

Locate Chapter 4 folder
in the Learning directory

Choose **Select a VI...**

Select **Circle Measurements
subVI** and click **Open**

Click OK when done

FIGURE 7.8
Placing the Circle Measurements subVI.vi on the block diagram.

timing using the Wait Until Next ms Multiple function. This timing control function, located in the **Functions≫Programming≫Timing** palette, as shown in Figure 7.9, waits until the millisecond timer is a multiple of the specified input value before returning the (optional) millisecond timer value (the output of the function need not be wired!). Therefore, you can force the loop iteration to execute only once every specified number of milliseconds. As with the Wait (ms) function, the timer resolution is system dependent and may be less accurate than 1 millisecond.

The example shown in Figure 7.10 shows how to control the timing of a For Loop to execute once every second (that is, once per 1000 ms). Placing the Wait Until Next ms Multiple function within the For Loop, each iteration will take approximately 1 second, and the random number will be plotted at 1-second intervals. Can you predict how much time it will take to complete the execution of the For Loop shown in Figure 7.10? It should take around 10 seconds. You can test it out by opening and running the VI shown in Figure 7.10. It is called Loop Timing Demo.vi and can be found in Chapter 7 of the Learning directory.

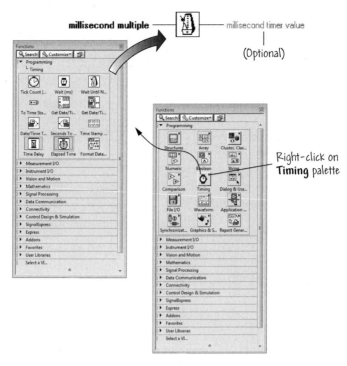

FIGURE 7.9
The Wait Until Next ms Multiple function.

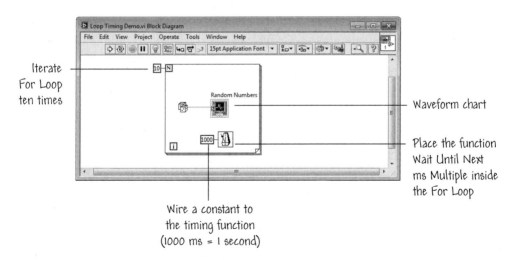

FIGURE 7.10
Controlling loop timing and the rate at which data is plotted on a waveform chart.

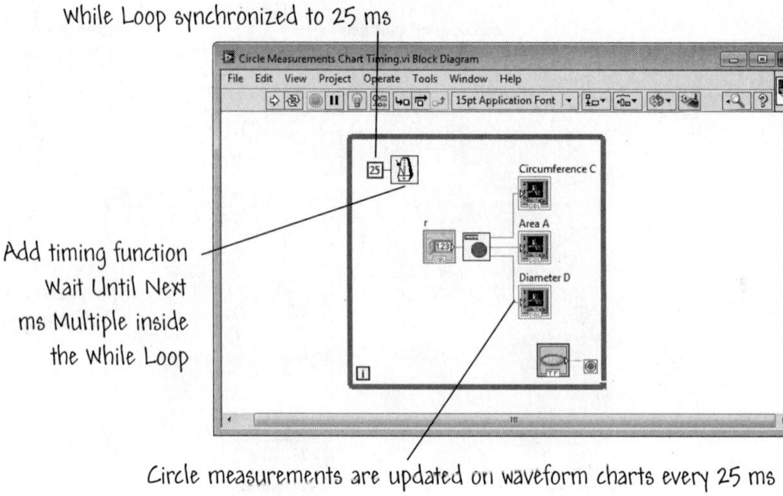

While Loop synchronized to 25 ms

Add timing function
Wait Until Next
ms Multiple inside
the While Loop

Circle measurements are updated on waveform charts every 25 ms

FIGURE 7.11
Controlling the rate at which the waveform charts display the computed circle measurements.

Practice
with
Timing

Open the VI called Circle Measurements Chart.vi that you constructed in the previous example (it should be located in the Users Stuff folder). Add the capability to control the loop timing by introducing a Wait Until Next ms Multiple function within the While Loop, as shown in Figure 7.11.

Set the loop timing for 25 ms. When the VI is now executing the circle measurements are computed and the waveform charts are updated at 25-millisecond intervals. The effect is that you should be able to watch the computed circle measurements change more clearly than when the While Loop was executing as fast as possible. Vary the loop-timing parameter and observe the effect on the waveform charts.

Save your updated VI in the Users Stuff folder and name it Circle Measurements Chart Timing.vi.

You can find a completed VI for the waveform chart timing demonstration in Chapter 7 *of the* Learning *directory. The VI is called* Circle Measurements Chart Timing.vi.

7.2 WAVEFORM GRAPHS

A **waveform graph** is an indicator that displays one or more data arrays. This is equivalent to the familiar 2D plot with horizontal and vertical axes. There are two types of graphs: waveform graphs and XY graphs. In this section the focus is on waveform graphs—XY graphs are the subject of the next section.

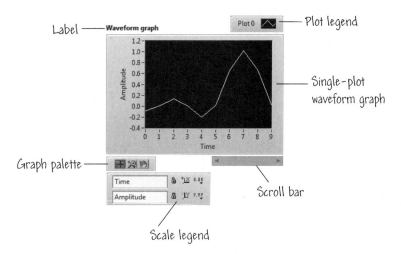

FIGURE 7.12
Example of a waveform graph.

Waveform graphs and XY graphs are functionally different, but they look the same on the front panel. Both waveform graphs and XY graphs plot existing arrays of data all at once, unlike waveform charts, which plot new data as they become available. An example of a waveform graph is shown in Figure 7.12. The waveform graph plots only single-valued functions with uniformly spaced points and is ideal for plotting arrays of data in which the points are evenly distributed. Conversely, XY graphs are general-purpose Cartesian graphs suitable for plotting data available at irregular intervals or plotting two dependent variables against each other. Waveform graphs and XY graphs accept different types of input data!

Waveform graphs are located on the **Graph** subpalette of the **Controls»** **Modern** palette (see Figure 7.1). For basic, single-plot graphs, an array of Y values (along the vertical axis) can pass directly to a waveform graph. This method assumes the initial X value (along the horizontal axis) is $X_0 = 0$ and the $\Delta X = 1$. The value of ΔX determines the X marker spacing. When passing only the Y values to the waveform graph, the graph icon will appear as an array indicator, as seen in Figure 7.13(a).

If you want to start plotting at an initial X value other than $X_0 = 0$, or if your data points are spaced differently than $\Delta X = 1$, you can wire a cluster consisting of the initial value X_0, ΔX, and a data array to the waveform graph. The graph terminal will then appear as a cluster indicator, as seen in Figure 7.13(b).

The waveform graph also accepts the waveform data type, as illustrated in Figure 7.13(c). When you wire waveform data type to a waveform graph (or chart), the graph (or chart) automatically plots the waveform. Similarly, wiring an array of waveform data to a waveform graph (or chart) will produce a graph (or chart) with all waveforms. The waveform data type contains the data, the

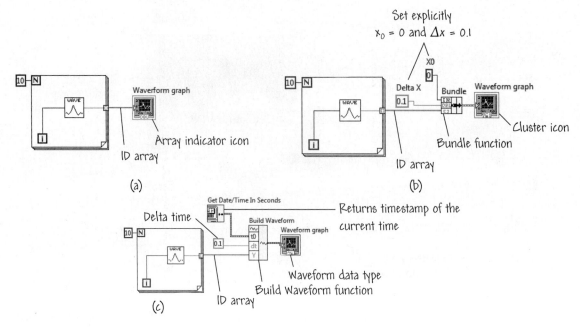

FIGURE 7.13
Single-plot waveform graphs.

start time, and the Δt of the waveform. The start time is the timestamp of the first measurement point in the waveform and can be used to synchronize plots on a multiplot waveform graph or to determine delays between waveforms. The Δt is the time interval in seconds between any two points in the waveform. The waveform data type is created using the Build Waveform function found on the **Programming≫Waveform** palette. In Figure 7.13(c) the Get Date/Time In Seconds function found on the **Programming≫Timing** palette returns the timestamp of the current time.

*Many VIs and functions used to acquire or analyze waveforms accept and return waveform data by default. For example, the Get Final Time Value function found on the **Programming≫Waveform** palette accepts the waveform data type as an input and outputs the waveform's ending time.*

If you want to plot more than one curve on a single graph, you can pass a 2D array of data to create a multiple-plot waveform graph. Two methods for wiring multiple-plot waveform graphs are illustrated in Figure 7.14. The graph icon assumes the data type to which it is wired.

Graphs always plot the *rows* of a two-dimensional array. The two-dimensional array in Figure 7.14 has two rows with 10 columns per row—a 2×10 array. If your data is given in columns, you must transpose the array

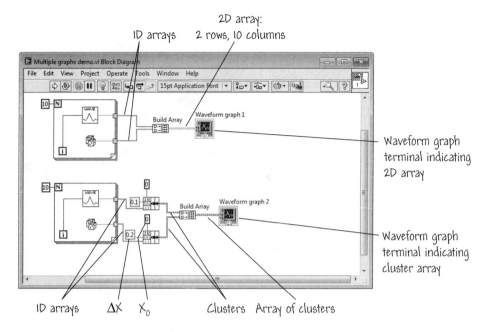

FIGURE 7.14
Two ways to wire multiple-plot waveform graphs.

before graphing. Once you have wired your 2D array of data to the graph terminal, go to the front panel, right-click on the graph, and select **Transpose Array**.

The single-plot waveform graph example shown in Figure 7.13 is called Waveform Graphs demo.vi. *The multiple-plot waveform graph example shown in Figure 7.14 is called* Multiple graphs demo.vi. *Both VIs can be found in* Chapter 7 *of the* Learning *directory.*

Practice with Waveform Graphs

Open a new VI and build the front panel shown in Figure 7.15. Place three objects on the front panel:

- Numeric control labeled No. of points
- Numeric control labeled Rate of growth, r
- Waveform graph labeled Population

We would like the waveform graph *y*-axis scale to be driven by the data values, rather than remaining at some predetermined fixed values. To modify the *y*-axis limits of the waveform graph, right-click on the waveform graph and change the **Y Scale** to **AutoScale Y**, as illustrated in Figure 7.16.

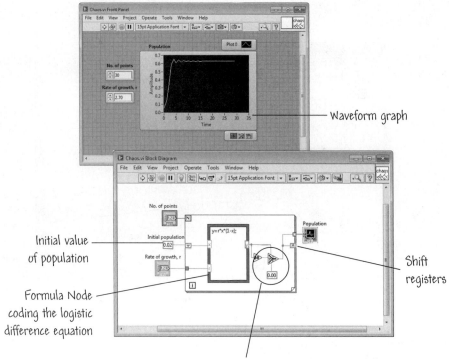

Waveform graph

Initial value of population

Shift registers

Formula Node coding the logistic difference equation

If the population falls below zero the population is extinct

FIGURE 7.15
Front panel and block diagram arrangement to experiment with chaos.

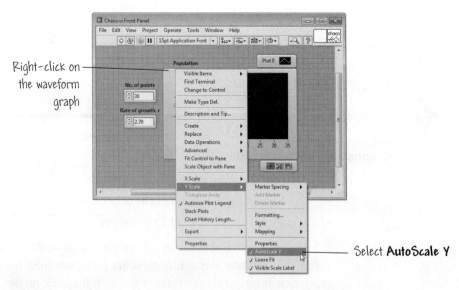

Right-click on the waveform graph

Select **AutoScale Y**

FIGURE 7.16
Changing the waveform graph y-axis limits to allow autoscaling.

The block diagram that you need to construct is shown in Figure 7.15. You will need to add the following objects:

- For Loop with the loop counter wired to the numeric control **No. of points**
- Shift registers on the For Loop to transfer data from one iteration to the next
- A MathScript Node containing the equation

$$y = rx(1 - x),$$

where the source of the parameter r is the digital control **Rate of growth, r**.

The equation in the MathScript Node is known as the *logistic difference equation*, and is given more formally as

$$x_{k+1} = rx_k(1 - x_k)$$

where $k = 0, 1, 2, \cdots$, and x_0 is a given value. In the block diagram (see Figure 7.15), the initial value is wired as $x_0 = 0.02$. The logistic difference equation has been used as a model to study population growth patterns. The model has been scaled so that the values of the population vary between 0 and 1, where 0 represents extinction and 1 represents the maximum conceivable population.

Wire the block diagram as shown in Figure 7.15 and prepare the VI for execution. Set the variable **No. of points** initially to 30 and the variable **Rate of growth, r** to 2.7. Execute the VI by clicking on the **Run** button. The value of the population is shown graphically in the waveform graph. You should observe the population reach a steady-state value around 0.63.

Experiment with the VI by changing the values of the **Rate of growth, r**. For values of $1 < r < 3$, the population will reach a steady-state value. You will find that as $r \to 3$, the population begins to oscillate, and in fact, the "steady-state" value oscillates between two values. When the parameter $r > 4$, the behavior appears erratic. As the parameter increases, the behavior becomes chaotic.[1] Verify that when the parameter $r = 4.1$, the population exceeds the maximum conceivable value (that is, it exceeds 1.0 on the waveform graph) and subsequently becomes extinct! Save your updated VI in the **Users Stuff** directory and name it **Chaos.vi**.

The working version of the chaos example shown in Figure 7.15 is called **Chaos.vi** *and is located in* **Chapter 7** *of the* **Learning** *directory.* ◆

---

1. For more on chaos, see *Chaos: Making a New Science*, by James Gleick, Penguin Books (New York, 1987).

Waveform Charts and Graphs

In this example, we will illustrate the difference between charts and graphs in LabVIEW by constructing a VI that plots multiple signals. The VI will plot multiple signals on a Waveform Chart and plot those same signals on a Waveform Graph.

To get started, open a new VI and place Triangle Wave PtByPt.vi and Sine Wave PtByPt.vi on the block diagram. The signals will be generated using these two VIs found on the **Signal Processing≫Point By Point≫Signal Generation PtByPt** palette, as illustrated in Figure 7.17. The inputs to the sine and triangle wave point-by-point VIs are amplitude, frequency (f), phase, and time. The output is the corresponding output signal (sine or triangle wave) at the requested input time. The use of these signals in a While Loop simulates point-by-point sampling of an actual input signal.

Use Figure 7.18 as a guide to constructing the VI. Add six numeric controls to the block diagram and label them amplitude, amplitude 2, frequency, frequency 2, phase, and phase 2. Wire three controls (amplitude, frequency, and phase) to the Triangle Wave PtByPt VI and the remaining three to the Sine Wave PtByPt VI.

Notice that the iteration terminal of the While Loop is scaled (by 500) and used to provide a time input signal to the point by point signal generation VIs. Adding the delay of 10 ms to the loop using the Wait Until Next ms Multiple

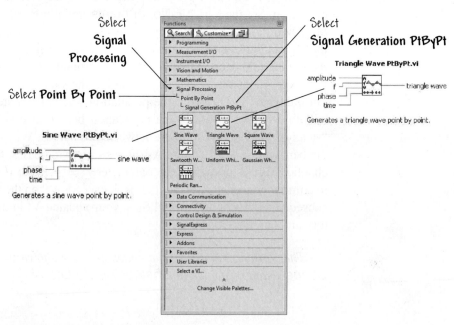

FIGURE 7.17
Accessing the SineWave PtByPt VI and Triangle Wave PtByPt VI.

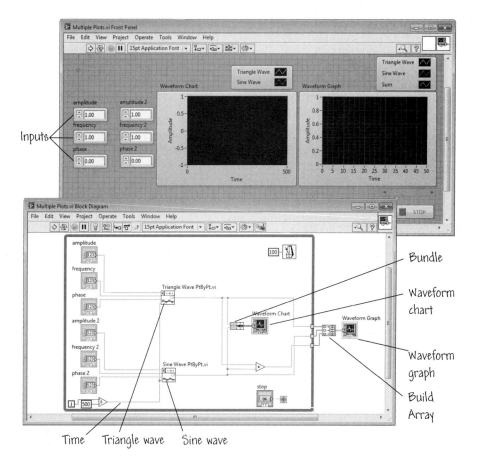

FIGURE 7.18
Generating and plotting a sine wave and a triangle wave on a waveform graph and a waveform chart.

function slows down the data generation and makes the display in the waveform chart easier to follow.

To display the data as it is generated, place a waveform chart within the While Loop. The chart is updated on each loop iteration with the most recently generated data point from each point by point signal generation VI. In order to display both signals as separate plots on the same chart, you need to bundle the signals into one cluster to input to the chart. The Bundle function is shown in Figure 7.18. The waveform chart will be updated with two points (one from each signal generation VI) upon each loop iteration. On the front panel, right-click on the waveform chart and select Visible Items≫Plot Legend. Rename the two plots Triangle Wave and Sine Wave.

Place a waveform graph on the block diagram outside of the While Loop. The graph will be used to display both the sine wave and the triangle wave, as well as the sum of these two signals (refer to the block diagram in Figure 7.18).

The waveform graph will display the acquired signals after all data points have been generated.

When displaying multiple plots on a waveform graph you must use the Build Array function to combine the signals to be displayed on the graph. Place the Build Array function on the block diagram outside of the While Loop. Right-click on the Build Array function and verify that **Concatenate Inputs** is not selected. With **Concatenate Inputs** deselected, the function will produce a 2D array with a row for each of the input arrays, as desired. Otherwise, the function will build a 1D array by concatenating all of the input arrays, leading to a single signal on the graph.

In order to display an additional signal, use the Add function to sum the sine and triangle signals. Then create tunnels on the While Loop for the sine wave signal, the triangle wave signal, and the sum of the two signals. Wire the three signals appropriately to the Build Array function, as depicted in Figure 7.18. On the front panel, right-click on the waveform graph and select **Visible Items≫ Plot Legend**. Resize the **plot legend** and rename each plot to accurately reflect the signal it represents.

Complete the wiring of the block diagram. Then run the VI and check that your graph plots three different waveforms. You should find that as the While Loop iterates, the chart indicator is continuously updated with new data points. The data generated appears all at once on the graph indicator only when the loop is stopped and data acquisition is complete.

Vary the inputs and observe the effects on the sine and triangle signals. Save the VI as **Multiple Plots.vi** in the **Users Stuff** folder when you are finished.

You can find a completed VI for the waveform chart and graph demonstration shown in Figure 7.18 in **Chapter 7** *of the* **Learning** *directory. The VI is called* **Multiple Plots.vi.** ◆

7.3 XY GRAPHS

As previously mentioned, waveform graphs are ideal for plotting evenly sampled waveforms. But what if your data is available at irregular intervals, or what if you want to plot two dependent variables against each other (e.g., x versus y)? The **XY graph** is well-suited for use in situations where you want to specify points using their (x, y) coordinates. The XY graph is a general-purpose Cartesian graph that can also plot multivalued functions—such as circles and ellipsoids. The XY Graph VI is located on the **Graph** subpalette of the **Controls≫Modern** palette.

A single-plot XY graph and its corresponding block diagram are shown in Figure 7.19. The Merge Signal function is used to combine the **X** and **Y** arrays into a single output for use with the XY Graph. The Express XY Graph VI

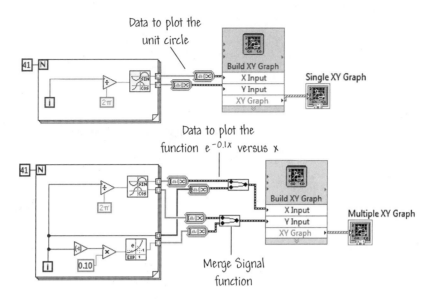

FIGURE 7.19
Examples of the XY graphs: single and multiple plots.

makes creating XY graphs simple. The Express XY Graph VI is also located on the **Graph Indicators** subpalette of the **Controls≫Express** palette, as shown in Figure 7.1. When placed on the front panel, the Express XY Graph VI places a corresponding terminal on the block diagram and an Express VI for quick configuration of your X and Y inputs. Wire the data arrays into the **X Input** and **Y Input** and they will automatically be configured for the graph, as shown in Figure 7.19.

The XY graph example shown in Figure 7.19 is called XY Graphs demo.vi *and is located in* Chapter 7 *of the* Learning *directory. The single XY graph produces a unit circle, and the multiple XY graph produces a unit circle and a plot of $e^{-0.1x}$ for $0 \leq x \leq 41$.*

Practice with XY Graphs

Suppose that you need to borrow $1000 from a bank and are given the option of choosing between simple interest and compound interest. The annual interest rate is 10% and, in the case of compound interest, the interest accrues annually. In either case, you pay the entire loan off in one lump sum at the end of the loan period. Construct the VI shown in Figure 7.20 to compute and graph the amount due at the end of each loan period, where the loan period varies from $N = 1$ year to $N = 20$ years in increments of 1 year.

Create a multiple-plot XY graph showing the final payments due for both the simple interest and the compound interest situations.

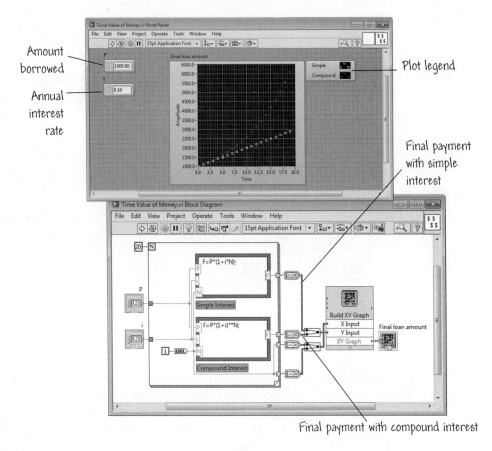

FIGURE 7.20
A VI to compute and plot the time value of money.

Define the following variables: F = final payment due, P = amount borrowed, i = interest rate, N = number of interest periods. The relevant formulas are:

- Simple interest

$$F = P(1 + iN)$$

- Compound interest

$$F = P(1 + i)^{N}$$

Open a new front panel and place the following three items on it:

- Numeric control labeled P
- Numeric control labeled i
- XY Graph

The XY graph in Figure 7.20 has been edited somewhat. Right-click on the first plot in the legend and select **Common Plots** and choose the scatter plot (top row, center). Then right-click in the same place again and select a **Point Style**. Repeat this process for the second plot listed in the legend.

One solution to coding the VI is shown in Figure 7.20. In this block diagram, the formulas are coded using MathScript Nodes. Can you think of other ways to code the same equations?

Once your VI is working, you can use it to experiment with different values of annual interest and different initial loan amounts. Based on your investigations, would you prefer to have simple interest or compound interest?

The XY graph example shown in Figure 7.20 is called Time Value of Money.vi *and is located in* Chapter 7 *of the* Learning *directory.* ◆

7.4 CUSTOMIZING CHARTS AND GRAPHS

Charts and graphs have editing features that allow you to customize your plots. This section covers how to configure some of the more important customization features. In particular, the following items are discussed:

- Autoscaling the chart and graph *x*- and *y*-axes
- Using the Plot Legend
- Using the Graph Palette
- Using the Scale Legend
- Chart customizing features, including the scrollbar and the digital display

By default, charts and graphs have the **plot legend** showing when first placed on the front panel. Using the **Positioning** tool, you can move the scale and plot legends and the graph palette anywhere relative to the chart or graph. Figure 7.21 shows some of the more important components of the chart and graph customization objects.

7.4.1 Axes Scaling

The *x*- and *y*-axes of both charts and graphs can be set to automatically adjust their scales to reflect the range of the plot data. The **autoscaling** feature is enabled or disabled using the **AutoScale X** and **AutoScale Y** options from the **X Scale** and **Y Scale** submenus of the shortcut menu, as shown in Figure 7.22.

The use of autoscaling may cause the chart or graph to update more slowly.

The *X* and *Y* scales can be varied manually using the **Operating** or **Labeling** tools—just in case you do not want to use the autoscale feature. For instance,

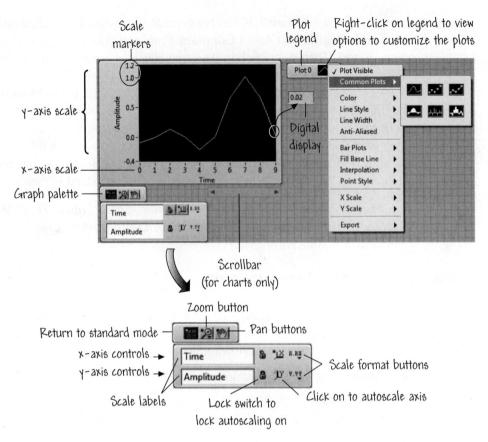

FIGURE 7.21
The chart and graph customization objects.

if the graph x-axis end marker has the value 10, you can use the **Operating** tool to change that value to 1 (or whatever other value you want!). The graph x-axis marker spacing will then change to reflect the new maximum value. In typical situations, the scales are set to fit the exact range of the data. You can use the **Loose Fit** option (see Figure 7.22) if you want to round the scale to multiple values of the scale increment—in other words, with a loose fit the scale marker numbers are rounded to a multiple of the increment used for the scale. For example, if the scale markers increment by 5, then with a loose fit, the minimum and maximum scale values are set to multiples of 5.

On both the **X Scale** or **Y Scale** submenus, you will find the **Formatting** option. Choosing the option opens up a dialog box, as shown in Figure 7.23, that allows you to configure many components of the chart or graph. This dialog can also be opened by right-clicking and selecting **Properties** to change the properties of the graph. You can modify the x- and y-axis characteristics individually. The following choices are available:

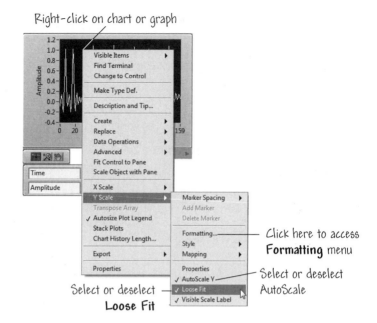

FIGURE 7.22
Using the **Autoscaling** feature of charts and graphs.

- The **Appearance** tab is used to specify which elements of the object are visible. This provides a way to choose to show the graph palette, plot legend, scrollbar, scale legend, and cursor legend. You can also select these items on the **Visible Items** pull-down menu by right-clicking on the graph.

- Within the **Format and Precision** area, you change the format and precision of numeric objects. You can choose the number of digits of precision for the scale marker numbers, as well as the notation. You can choose among floating point, scientific, SI notation, hexadecimal, octal, binary, absolute time, and relative time. The choices depend on the format choice.

- The **Plots** tab is used to configure the appearance of plots on a graph or chart. You can select the line type (e.g., solid versus dashed), point markers, and scale titles.

- Under the **Scales** tab you format scales and grids on graphs and charts. For example, you can set the axis origin $X0$ (or $Y0$) and the axis marker increment ΔX (or ΔY). The grid style and colors provide control over the gridlines: no gridlines, gridlines only at major tick mark locations, or gridlines at both major and minor tick marks. This option also allows you to change the color of the gridlines. Here you can also specify the scale as either a linear or a logarithmic scale. The **Scales** tab lets you select major and minor tick marks for the scale, or none at all. A major tick mark corresponds to a scale label, while a minor tick mark denotes an interior point between labels. You

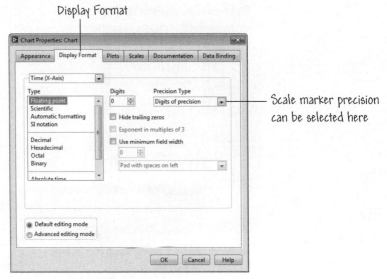

Precision Formatting

(a)

Scale Formatting

(b)

FIGURE 7.23
Scale and precision formatting.

have to click on the **Scale Style** icon to view the choices. With this menu, you can make the axes markers invisible.

- The **Cursors** tab is used to add cursors to a graph or chart and to configure the appearance of the cursors.

- Use the **Documentation** tab to describe the purpose of the object and to give users instructions for using the object.

7.4.2   The Plot Legend

The plot legend provides a way to customize the plots on your charts and graphs. This is where you choose the data point style, the plot style, the line style, width and color, and other characteristics of the appearance of the chart or graph. For example, on multiple-plot graphs you may want one curve to be a solid line and the other curve to be a dashed line or one curve to be red and the other blue. An example of a plot legend is shown in Figure 7.24. When you move the chart or graph around on the front panel, the plot legend moves with it. You can change the position of the plot legend relative to the graph by dragging the plot legend with the **Positioning** tool to the desired location.

The plot legend can be visible or hidden—and you use the **Visible Items** submenu of the chart or graph shortcut menu (see Figure 7.24) to choose. After you customize the plot characteristics, the plot retains those settings regardless

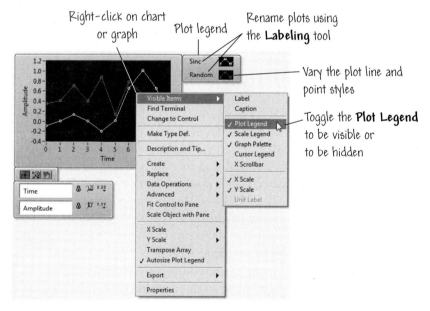

FIGURE 7.24
The plot legend for charts and graphs.

of whether the plot legend is visible or not. If your chart or graph receives more plots than are defined in the plot legend, they will have the default characteristics.

When the plot legend first appears, it is sized to accommodate only a single plot (named Plot 0, by default). If you have a multiple-plot chart or graph, you will need to show more plot labels by dragging down a corner of the legend with the cursor to accommodate the total number of curves. Each curve in a multiple-plot chart or graph is labeled as Plot 0, Plot 1, Plot 2, and so forth. You can modify the default labels by assigning a name to each plot in the legend with the **Labeling** tool. Choosing names that reflect the physical value of the data depicted in the chart makes good sense—for instance, if the curve represents the velocity of an automobile you might use the label Automobile velocity in km/hr. The legend can be resized on the left to give the plot labels more room or on the right to give the plot samples more room. This is useful when you assign long names to the various curves on a multiple-plot chart or graph. You can right-click on each plot in the legend and change the plot style, line, color, and point styles of the plot. The shortcut menu is shown in Figure 7.21.

If the graph or chart contains multiple plots, there are two ways to choose whether to display a particular plot on a graph or chart. First, you can right-click the glyph in the plot legend and select **Plot Visible** from the shortcut menu, as shown in Figure 7.25(a). You can also right-click the plot legend and select **Visible Items≫Plot Visibility Checkbox** from the shortcut menu to display a checkbox next to each plot, as shown in Figure 7.25(b). You can then use the checkboxes to select which plots you want to display.

Using Property Nodes

Property Nodes allow you to set the properties and to get the properties of objects. (Refer to Chapter 3 for more details on Property Nodes.) One of their applications is to programmatically modify the appearance of graphs and charts. For example, you might want to change the color of a trace on a chart when data points are above a certain value. The block diagram in Figure 7.26 shows a waveform graph with an associated Property Node configured to set the Plot Color property.

When the output of the Generate Waveform VI exceeds 0, the plot line is red, otherwise it is green. The RGB to Color VI converts a red, green, and blue value from 0 to 255 to the corresponding RGB value required by the Property Node. The RGB to Color VI can be found on the **Programming≫Numeric≫ Conversion** palette.

Recall that you create a property from a front panel object by right-clicking the object, selecting **Create≫Property Node**, and selecting a property from the shortcut menu. A Property Node is created on the block diagram that is implicitly linked to the front panel object. Besides the plot color, many other graph and chart properties can be programmatically changed, including line style, line width, plot name, point style, axes scales, visibility, legend parameters, and update mode.

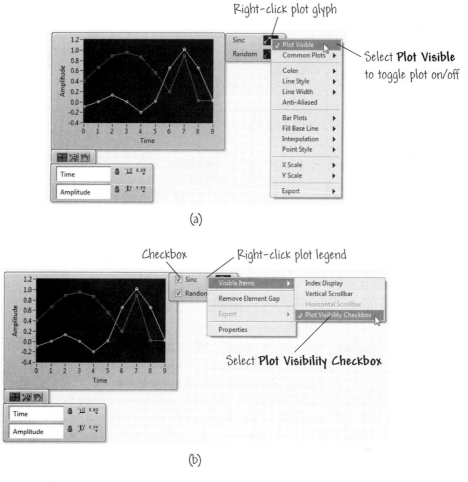

(a)

(b)

FIGURE 7.25
(a) Using the **Plot Visible** to choose to display a plot; (b) using the **Plot Visibility Checkbox** to choose to display a plot.

The VI shown in Figure 7.26 can be found in **Chapter 7** *of the* **Learning** *directory. Open the VI called* **Waveform Graph with Property Node** *and run the VI to watch the plot color change as the value of the Generate Waveform VI varies.* ◆

7.4.3 The Graph Palette and Scale Legend

The graph palette and scale legend allow you access to various aspects of the chart and graph appearances while the VI is executing. You can autoscale the axes and control the display format of the axes scales. To aid in the analysis of the data presented on the charts and graphs, buttons allow you to pan and zoom

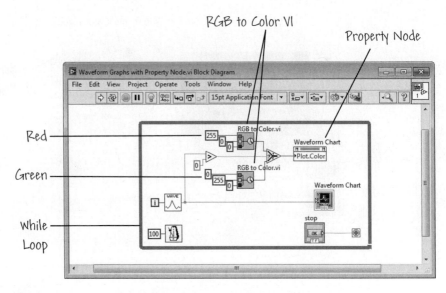

FIGURE 7.26
Modifying the properties of graphs and charts using Property Nodes.

in and out. The graph palette and scale legend shown in Figure 7.27 can be displayed using the **Visible Items** submenu of the chart or graph shortcut menu.

On the right of the scale legend are two buttons for each axis that control the autoscaling of the axes. Clicking on the autoscaling button (either the *x*- or *y*-axis buttons) will cause the chart or graph to autoscale the axis. When you press the lock button, the chart or graph will autoscale either of the axes scales continuously.

The *x*-axis and *y*-axis labels can also be modified by entering the desired labels (using the **Labeling** tool) into the text area provided on the left side of the scale legend. The upper text is for the *x*-axis.

The scale format buttons (on the right of the scale legend) provide run-time control over the format and precision of the *x* and *y* scale markers, respectively. As shown in Figure 7.27, you can control the **Format** of the axes markers, the **Precision** of the marker numbers (from 0 to 6 digits after the decimal point), and the **Mapping Mode**. Remember that in the Edit Mode, you can control the format, precision, and mapping mode through shortcut menus on the chart or graph. The chart or graph scale legend gives you the added feature of being able to control the format and precision of the *x* and *y* scale markers while the VI is executing.

The three buttons on the graph palette allow you to control the operational mode of the chart or graph. You are in the standard mode under normal circumstances where the button with the crosshair (left side of the palette) is selected. This indicates that you can click in the chart or graph area to move cursors around. Other operational modes include panning and zooming. When you press

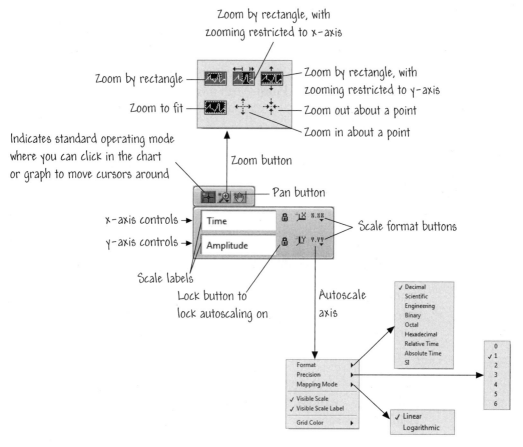

FIGURE 7.27
The graph palette and scale legend.

the pan button (depicted in Figure 7.27), you can scroll through the visible data by clicking and dragging sections of the graph with the pan cursor. Selecting the zoom button gives you access to a shortcut menu that lets you choose from several methods of zooming.

As depicted in Figure 7.27, you have several zoom options:

- **Zoom by Rectangle**. In this mode, you use the cursor to draw a rectangle around the area you want to zoom in and when you release the mouse button, the axes will rescale to zoom in on the desired area.

- **Zoom by Rectangle**—with zooming restricted to *X* data. This is similar in function to the **Zoom by Rectangle** mode, but the *Y* scale remains unchanged.

- **Zoom by Rectangle**—with zooming restricted to *Y* data. This is similar in function to the **Zoom by Rectangle** mode, but the *X* scale remains unchanged.

- **Zoom to Fit**. This autoscales all *x*- and *y*-scales on a graph or chart.

- **Zoom in about a Point**. With this option you hold down the mouse on a specific point on the chart or graph to continuously zoom in until you release the mouse button.

- **Zoom out about a Point**. With this option you hold down the mouse on a specific point and the graph will continuously zoom out until you release the mouse button.

For the zoom in and zoom out about a point modes, <Shift>-clicking will zoom in the other direction.

7.4.4   Special Chart Customization Features

Charts have the same customizing features as graphs—with two additional options: **Scrollbars** and **Digital Displays**. The **Visible Items** submenu (of the chart shortcut menu) is used to show or hide the optional digital display(s) and a scrollbar.

Charts have scrollbars that can be used to display older data that has scrolled off the chart. The chart scrollbar is depicted in Figure 7.21. The scrollbar can be made visible by right-clicking on the chart and selecting the scrollbar from the **Visible Items** submenu, as illustrated in Figure 7.28.

There is one digital display per plot that displays the latest value being plotted on the chart. You can place a digital display on the chart by selecting **Digital Display** from the **Visible Items** submenu, as illustrated in Figure 7.28. The digital display can be moved around relative to the chart using the **Positioning** tool. The last value passed to the chart from the block diagram is the latest value for the chart and the value that is displayed on the digital display. If you want to view past values contained in the chart data you can navigate through the older data using the scrollbar. There is a limit to the amount of data the chart can hold in the buffer. By default, a chart can store up to 1024 data points. When the chart reaches the limit, the oldest point(s) are discarded to make room for new data. If you want to change the size of the chart data buffer, select **Chart History Length** from the shortcut menu (as shown in Figure 7.2) and specify a new value of up to 100,000 points.

When you are not running the VI, you can clear the chart by right-clicking on the chart and selecting **Clear Chart** from the **Data Operations** menu, as shown in Figure 7.28. However, if you are running the VI (you are in the Run Mode), then **Clear Chart** becomes a shortcut menu option and it will automatically appear in the chart shortcut menu after clicking on the **Run** button. This allows you to clear the chart while the VI is executing.

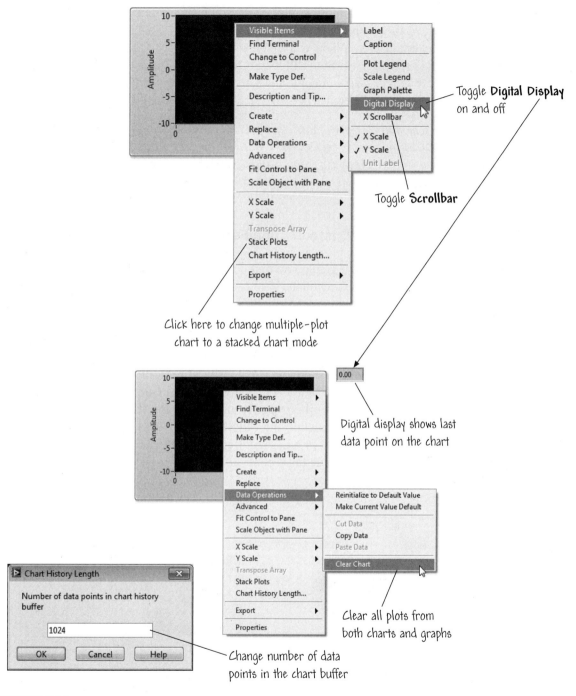

FIGURE 7.28
The scrollbar and the digital display for charts.

If you have a multiple-plot chart, you can choose between overlaid plot or stacked plot. When you display all plots on the same set of scales, you have an overlaid plot. When you give each plot its own scale, you have a stacked plot. You can select **Stack Plots** or **Overlay Plots** from the chart shortcut menu (see Figure 7.28) to toggle between stacked and overlaid modes. Figure 7.29 illustrates the difference between stacked and overlaid plots.

The multiple-plot chart example shown in Figure 7.29 is called Chart Cus-tomizing.vi *and is located in* Chapter 7 *of the* Learning *directory. The two charts on the block diagram display the same data, except that the chart on the left uses the stacked chart, and the chart on the right uses the overlaid chart.*

7.4.5   Special Graph Customization Features: Cursor Legend

To add cursors to a graph, you first bring the **Cursor Legend** to the front panel graph by right-clicking on the graph and selecting Visible Items≫Cursor Leg-end from the shortcut menu, as illustrated in Figure 7.30. The cursor legend is shown in Figure 7.31. To add a cursor to the graph, right click in the **Cursor Legend** and select **Create** Cursor. You will be given three options to choose from: **Free**, **Single-Plot**, and **Multi-Plot**. Choosing **Free** creates a cursor that moves freely within the plot area. Choosing **Single-Plot** creates a cursor that is positioned on a single plot associated with that cursor. If you have one plot on the graph, the cursor will move only along that plot. If you have multiple plots on a single graph, then choosing **Single-Plot** will associate the cursor with the first plot listed in the bf Plot Legend. Repeating the process and selecting **Single-Plot** again will associate the second cursor with the second plot on the graph as

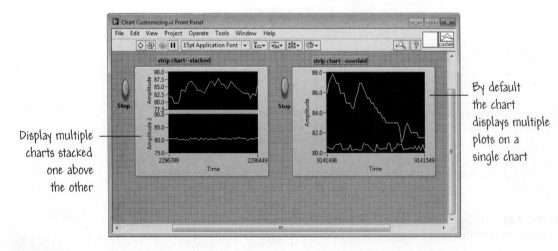

Display multiple charts stacked one above the other

By default the chart displays multiple plots on a single chart

FIGURE 7.29
Stacked and overlaid charts.

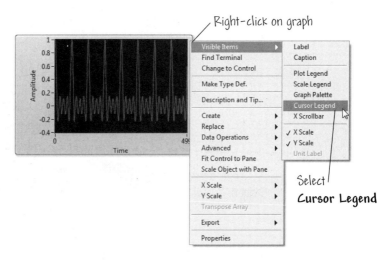

FIGURE 7.30
Placing the **Cursor Legend** on the front panel.

listed in the **Plot Legend**. To modify the association of the cursors with their respective plots, right-click the **Cursor Legend** on the row of the desired cursor and select **Snap To** from the shortcut menu to associate one or all plots with the cursor. The third option, **Multi-Plot**, positions the cursor only on a specific data point in the plot area. This mode is only valid for mixed-signal graphs.

Various attributes of the cursors can be modified and personalized. As illustrated in Figure 7.31, right-click on the **Cursor Legend** on the row of the cursor that you wish to modify. For example, in Figure 7.31 the attributes of Cursor 1 are being modified. The attributes that can be modified include: **Color**, **Cursor Style**, **Point Style**, **Line Style**, **Line Width**, **Show Name**, and **Allow Drag**. Selecting **Color** displays the color picker so you can select the color of the cursor. **Cursor Style** provides various cursor styles, including combinations of dots and dashes in an assortment of configurations. **Point Style** provides various point styles for the intersection of the cursor, including circles, filled circles, squares, filled squares, crosses, and a host of other choices. **Line Style** and **Line Width** offer various solid- and dotted-line styles and line widths, respectively.

Two other important options appearing in the **Attributes** menu include **Show Name** and **Allow Drag**. **Show Name** displays the name of the cursor on the graph and **Allow Drag** lets the mouse drag the cursor.

You can control the location of the cursor using the **Bring to Center** and **Go to Cursor** options found on the **Cursor Legend** pull-down menu. Selecting **Bring to Center** results in centering the cursor on the graph without changing the x- and y-scales. When the cursor mode is **Single-Plot** or **Multi-Plot**, this option centers the cursor on the plot on which the cursor is currently positioned and updates the cursor coordinates in the cursor legend. When the cursor mode is **Free**, this option centers the cursor in the plot area and updates the cursor

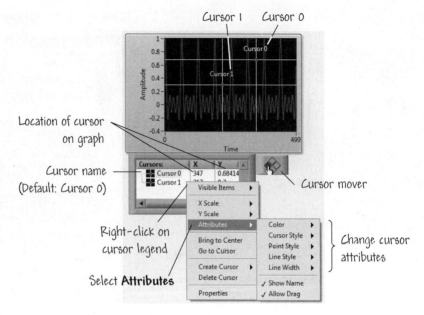

FIGURE 7.31
The **Cursor Legend** is associated with graphs.

coordinates in the cursor legend. The **Go to Cursor** option changes the *x*- and *y*-scales to show the cursor at the center of the graph.

You cannot change the mode of a cursor after you create it. You must delete the cursor and create another.

7.4.6 Using Graph Annotations

It is often very useful to add annotations to graphs, especially when the graphs are in their final form and are being used in reports or presentations. In Lab-VIEW you can use annotations to highlight important data points in the plot area. The annotation includes a label and an arrow that identifies the annotation and data point. A graph can have any number of annotations. Figure 7.32 shows an example of a graph using annotations.

To add an annotation to a graph, right-click the graph and select Data Operations≫Create Annotation from the shortcut menu to display the **Create Annotation** dialog box, as illustrated in Figure 7.32. You can assign the annotation name and you can specify how the annotation snaps to plots in the plot area. To specify the annotation name, just type the desired name or phrase in the **Annotation Name** field.

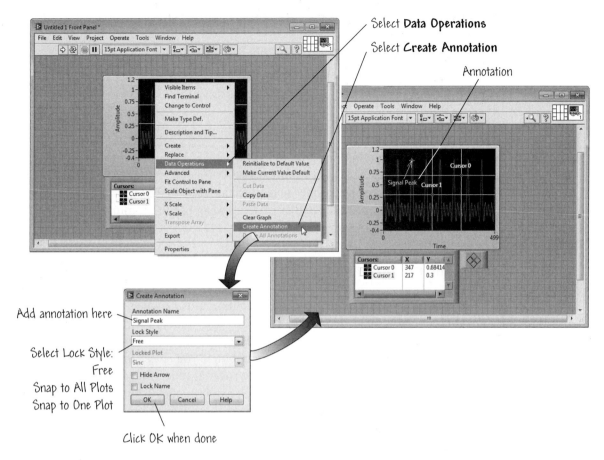

FIGURE 7.32
Annotating a graph.

To specify how the annotation snaps to plots in the plot area, use the **Lock Style** pull-down menu. The **Lock Style** component includes the following options:

- **Free**—Allows you to move the annotation anywhere in the plot area.

- **Snap to All Plots**—Allows you to move the annotation to the nearest data point along any plot in the plot area.

- **Snap to One Plot**—Allows you to move the annotation only along the specified plot.

You can customize the behavior and appearance of the annotation. You can hide or show the annotation name or arrow in the plot area, specify the color of the annotation, and specify line, point, and annotation style. To accomplish this, right-click the annotation and select options from the shortcut menu to customize the annotation.

7.4.7   Exporting Images of Graphs, Charts, and Tables

You can export images of graphs, charts, tables, digital data, and digital waveform controls and indicators into the following formats:

- Windows: .emf, .bmp, and .eps files
- Mac OS: .pict, .bmp, and .eps files

To export an image, right-click the control and select Data Operations≫ Export Simplified Image from the shortcut menu. In the **Export Simplified Image** dialog box, select the image format and whether you want to save the image to the clipboard or to disk. You also can use the **Export Image** method to export an image programmatically.

7.4.8   Using Context Help

If you are new to LabVIEW, it can often be confusing when you try to wire data to charts and graphs. Do you use a Build Array function, a Bundle function, or both? What order do the input terminals use? You can find valuable information in the **Context Help** window. For example, if you select **Show Context Help** from the **Help** menu and put your cursor over a waveform graph terminal in the diagram, you will see the information shown in Figure 7.33. The **Context Help** window shows you what data types to wire to the waveform graph, how to specify point spacing with the Bundle function, and which example to use when you want to see the different ways you can use a waveform graph. The **Context Help** window also shows similar information for XY graphs and waveform charts.

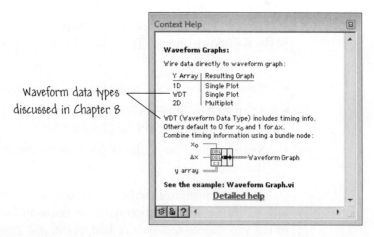

FIGURE 7.33
Using Context Help to determine which data types to wire to the waveform graph.

7.5 USING MATH PLOTS FOR 2D AND 3D GRAPHS

The LabVIEW math plots are very helpful in performing mathematical analysis and displaying results graphically. They offer many customization options, making them particularly useful in presenting data for analysis. There are two types of math plots: 2D Graphs and 3D Graphs. 2D Graphs are used to display 2D data in one of four different ways on a 2D plot on the front panel. 3D Graphs are used to display 3D data in one of eleven ways on a 3D plot on the front panel.

The available 2D graphs found on the **Modern≫Graph** subpalette on the **Controls** palette include the Compass Plot, Error Plot, Feather Plot, and XY Plot Matrix. The 3D graphs found on the **Modern≫Graph≫3D Graph** subpalette on the **Controls** palette include the Bar, Comet, Contour, Mesh, Pie, Quiver, Ribbon, Scatter, Stem, Surface, and Waterfall.

*For more detailed information on math plots, see **Property and Method Reference≫Math Plots** on the **Contents** tab of the **LabVIEW Help**.*

7.5.1 2D Graphs

There are four 2D math plots that use x and y data to mark and connect points to form a two-dimensional view of the data. All 2D graphs are XY Graphs with distinctive characteristics, as shown in Figure 7.34. The Compass Plot shown in Figure 7.34(a) graphs vectors that emanate from the center of a compass graph. The Error Bar Plot shown in Figure 7.34(b) graphs a user-specified error bar at each point above and below the line graph. The Feather Plot shown in Figure 7.34(c) graphs vectors that emanate from equally spaced points along a horizontal axis. The XY Plot Matrix shown in Figure 7.34(d) graphs rows and columns of scatter graphs.

When you add a 2D graph to the front panel, the graph on the block diagram is automatically wired to a helper VI, as illustrated in Figure 7.35. The helper VI converts the input data into the data type its associated 2D graph accepts. You can use **Context Help** to discover what inputs are needed to create the desired 2D graph. For example, in Figure 7.35 we see that the 2D Compass graph needs the theta vector and the radius vector as inputs. The theta vector specifies the angles on the compass graph you want to plot, and the radius vector specifies the vector length as it extends from the center.

The helper VIs for the 2D Error Bar graph and the 2D Feather graph require as inputs x vector and y vector specifying the x- and y-coordinates to plot, respectively. The length of x vector must equal the length of y vector. The 2D Error Bar graph also requires two key additional inputs specifying the error percentages that define the error bars that appear at each point above and below

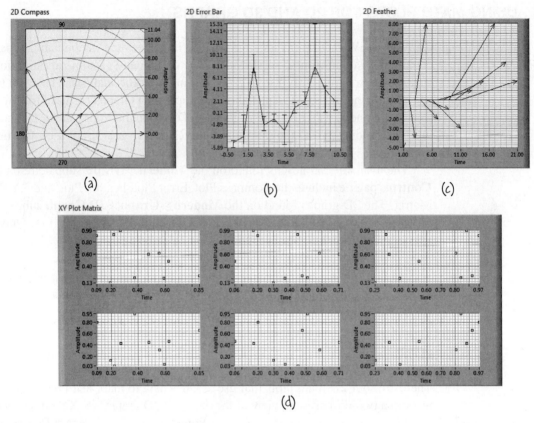

FIGURE 7.34
2D math plots: (a) 2D Compass; (b) 2D Error Bar; (c) 2D Feather; and (d) XY Plot Matrix.

the graphed line. The helper VI for the XY Plot Matrix graph requires **X matrix** and **Y matrix** as inputs. The number of columns of **X matrix** specifies the number of columns of XY Plot Matrix graphs you want to generate. The number of columns of **Y matrix** specifies the number of rows of XY Plot Matrix graphs you want to generate. The number of rows of **X matrix** must equal the number of rows of **Y matrix**. We see in Figure 7.34(d) that the input **X matrix** has 3 columns and the input **Y matrix** 2 columns.

*In the NI Example Finder window, select the **Browse** tab and then browse according to **Directory Structure** to the 2D Math Plots subfolder within the Math Plots folder. Here you will find examples of plotting data using all four 2D graph types.*

You can edit and customize the appearance of any of the four 2D graphs by right-clicking on the graph and selecting **Properties** from the shortcut menu to display the dialog box, as illustrated in Figure 7.36. You can format the

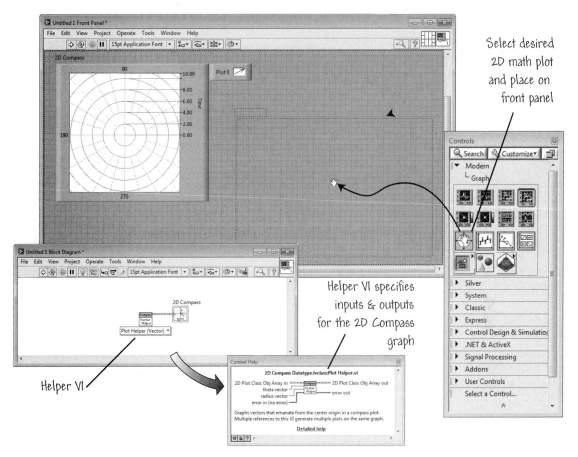

FIGURE 7.35
Creating 2D math plots and using the helper VI to assist in wiring the graph.

properties and change the appearance of the 2D graph, including the display format, plot styles, scales, and cursors. Alternatively, you can click the plot name on the plot legend to the right of the Compass, Error Bar, and Feather graphs to customize the line style (set the line style of the plot), line width (sets the line width of the plot), size (sets the arrow size for the Compass and Feather graphs and the error bar size for the Error Bar graph), and color (sets the color of the plot line).

7.5.2   3D Graphs

Often you will need to visualize data in three dimensions. The 3D Graphs allow you to visualize three-dimensional data and readily adjust the appearance of the graphs by modifying the 3D graph properties. Eleven types of 3D graphs are

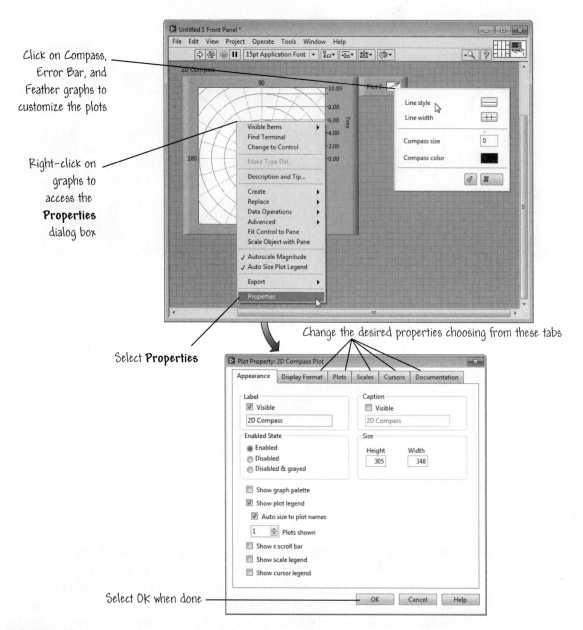

FIGURE 7.36
Modifying the properties of 2D math plots.

available, including the Bar, Comet, Stem, Pie, Scatter, Surface, Contour, Mesh, Waterfall, Quiver, and Ribbon graphs.

As with the 2D graphs, when you add a 3D graph to the front panel, LabVIEW wires the graph on the block diagram to one of the helper VIs, depending on which 3D graph you select. The helper VIs convert the input data types into the generic data type the 3D graph accepts.

The Bar, Comet, Stem, and Pie graphs are illustrated in Figure 7.37. They are grouped together here, as they present data without using 3D surfaces. The Bar graph generates a plot of vertical rectangular bars with lengths proportional to the data, as shown in Figure 7.37(a). Bar charts are generally used when the amount of data is relatively small. The Comet graph creates an animated graph with a circle that follows the data points. The Comet plot, shown in Figure 7.37(b), is probably the least known among this group of charts but provides a unique way to visualize data. The comet head is the circle and the comet tail is the straggling line that follows the comet head. The Stem graph displays an impulse response and organizes the data by its distribution, as shown in Figure 7.37(c). The Stem chart is very commonly used to visualize the shape of a distribution. The Pie graph is circular and divided into segments indicating the relative frequencies or magnitudes of the data, as shown in Figure 7.37(d). The Bar and Pie graphs should be very familiar to students, as they are used a great deal in presentations.

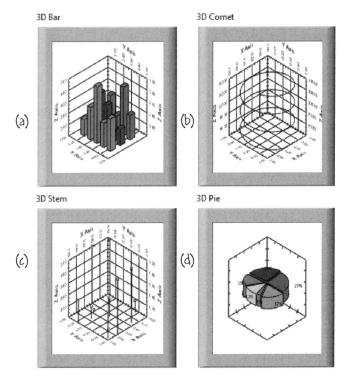

FIGURE 7.37
3D math plots: (a) Bar; (b) Comet; (c) Stem; and (d) Pie.

*Graphical hardware acceleration is used in the 3D graphs in the render window, offering performance benefits. Right-click the 3D graph and select **Render Window** from the shortcut menu to view the 3D graph in the render window.*

The Contour, Mesh, Ribbon, Quiver, Scatter, Waterfall and Surface graphs are illustrated in Figure 7.38. This group of graphs shows the data in some form of surface plot. The Contour graph shown in Figure 7.38(a) displays the data on a

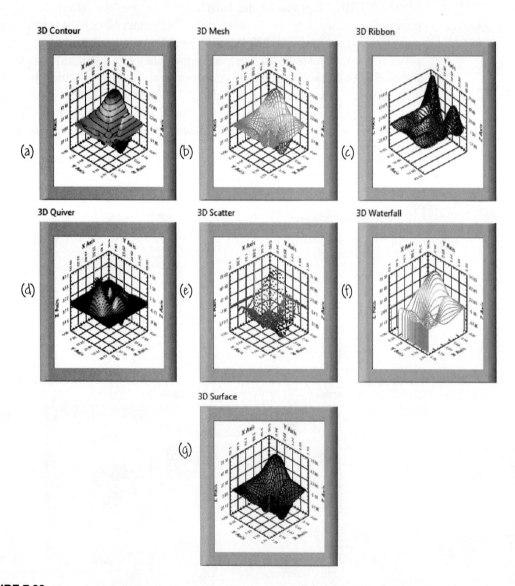

FIGURE 7.38
3D math plots: (a) Contour; (b) Mesh; (c) Ribbon; (d) Quiver; (e) Scatter; (f) Waterfall; and (g) Surface.

surface using contour lines. The Mesh graph is a representation of the data using a mesh surface with open spaces, as shown in Figure 7.38(b). The Ribbon graph generates a plot of parallel lines, as shown in Figure 7.38(c). The Quiver graph, shown in Figure 7.38(d), generates a velocity plot. The Scatter graph shows trends in statistics and the relationship between two sets of data, as shown in

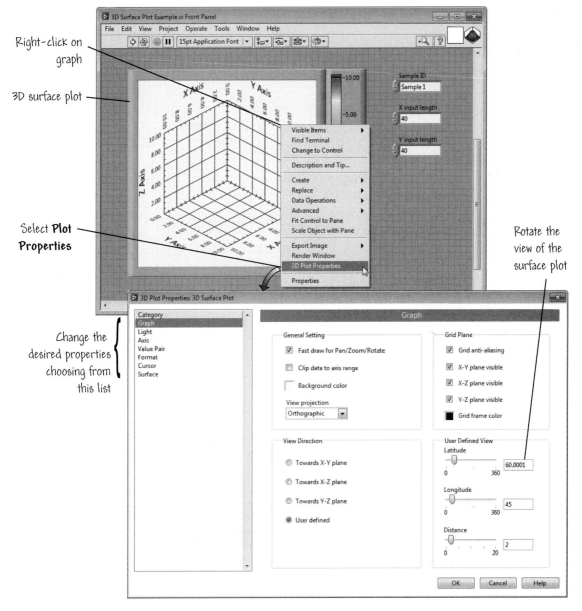

FIGURE 7.39
Modifying the properties of 3D math plots using the **Plot Properties** dialog box.

Figure 7.38(e). The Waterfall graph depicts the surface of the data and the area on the *y*-axis below the data points, as illustrated in Figure 7.38(f). The final 3D graph is the Surface graph, which presents the data with a connecting surface. A surface plot uses 3D data to plot points on the graph. The surface plot then connects these points, forming a three-dimensional surface view of the data, as shown in Figure 7.38(g).

*In the NI Example Finder window, select the **Browse** tab and then browse according to **Directory Structure** to the 3D Math Plots subfolder within the Math Plots folder. Here you will find examples of plotting data using all eleven 3D graph types.*

You can customize the appearance of any of the eleven 3D graphs by right-clicking on the graph and selecting **Plot Properties** from the shortcut menu to display the dialog box, as illustrated in Figure 7.39. You can modify the appearance of the 3D graph including the view direction, lighting, axes, and cursors. On Bar and Pie graphs you can use the color palette to the right of the graphs to customize the color for each bar or pie slice. Place the cursor over the name of each bar or pie slice and click to change the name. You can also click the color to the left of the name to select a new color. Use the projection palette near the bottom-right corner of the graph to customize the plane projection view as well as use the **Plot Properties** dialog box. On the projection palette you can display the plot projection on the XY plane by clicking on the **XY plane projection** button. Similarly, you can display the plot projection on the XZ plane or YZ plane by clicking on the **XZ plane projection** button or **YZ plane projection** button, respectively. The **Default Projection** button returns the display to the 3D plane.

   BUILDING BLOCK

7.6   BUILDING BLOCKS: PULSE WIDTH MODULATION

The goal of this Building Block is to plot the waveform output of a pulse width modulation VI using a waveform chart. To begin, open the Building Blocks VI that you constructed in Chapter 6. It should be called **PWM with Measurements.vi** and located in the **Users Stuff** folder in the **Learning** directory. Once the VI is open, rename it **PWM with Chart.vi** and save the VI in the **Users Stuff** folder.

In case you do not have a working version of the Building Blocks VI from Chapter 6, a working version of the PWM with Measurements VI can be found in the Building Blocks folder in the Learning directory.

In the PWM with Measurements VI, the output was displayed using an LED indicator. You should replace that LED with a waveform chart. To accomplish this, in the block diagram of the PWM with Chart VI, select and delete the Boolean LED named Output. Then add a waveform chart and wire the chart to the output of the Boolean to (0,1) function, as illustrated in Figure 7.40. Clear all broken wires. Rename the waveform chart Pulse Width Modulated Signal, as illustrated in Figure 7.40.

When you have added and named the waveform chart on your VI, experiment with different input values. By default, the **Chart History Buffer** keeps the most recent 1024 data points in memory. Thus you may find that, for large periods, you will need to increase this value in order to view a significant portion of the output signal. This can be accomplished by right-clicking on the plot, selecting **Chart History Length**, and replacing the default **Chart History Length** with a larger number—preferably one that is several times the size of the period. For example, you might choose to keep 5000 data points in memory.

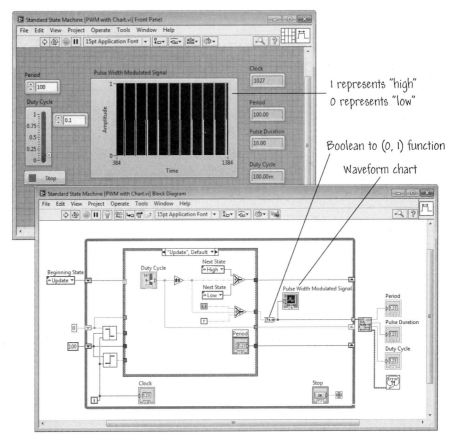

FIGURE 7.40
The PWM with Chart VI showing the use of the waveform chart.

We have not specifically stated units for either the x-axis or the y-axis of the chart. Since this signal is not being used to communicate with any physical devices, explicit units of measurement are not critical. The important information that the signal carries is how the magnitude varies with respect to time. As long as the units of magnitude and time remain consistent in this simulation, it is not necessary to expressly relate them to physical units. We denote the high and low values of the signal 1 and 0.

Observe what the output looks like for various input values. Change the update mode and watch how the data is plotted for all three update modes. To change the update mode, right-click on the chart and select Advanced≫Update Mode. The three options are **Strip Chart**, **Scope Chart**, and **Sweep Chart**. When you are finished, save the VI as PWM with Chart.vi.

A working version of the PWM with Chart VI can be found in the Building Blocks *folder of the* Learning *directory.*

7.7   RELAXED READING: ENVIRONMENTAL MONITORING IN THE COSTA RICAN RAIN FOREST

In this reading, we discuss a new strategy to investigate the emission of greenhouse gases. Employing a wide range of wireless environmental measurements using a single device that provides robotic control, remote configuration, and data sharing over the Web, researchers in Costa Rica are testing a hypothetical explanation of the exchange of CO_2 and other materials between the forest floor and the atmosphere—the so-called "gap theory."

Approximately 70% of solar energy is absorbed by the Earth's atmosphere. As the Earth's surface emits this energy in the form of thermal radiation, the atmosphere naturally captures and recycles a large portion of it, keeping the planet warm. This process is known as the greenhouse effect. Recently, the greenhouse effect has been artificially enhanced by the increased emission of gases that absorb infrared radiation, such as carbon dioxide (CO_2), methane, and nitrous oxide. The increased absorption of thermal radiation may contribute to the climate change known as global warming.

To better understand the impact of the emission of greenhouse gases on the environment, researchers are conducting a study at La Selva Biological Station in the Costa Rican rainforest to measure the exchange of CO_2 (also known as the carbon flux) and other materials between the forest floor and the atmosphere. The area under observation lies within a 3,900-acre tropical rainforest that averages 13 feet of rainfall per year and is located at the confluence of two major rivers in the lowlands of northeastern Costa Rica.

This area was chosen for observation because rainforests are naturally rich in biodiversity and are carbon sinks, meaning their function is opposite to that of a human lung–absorbing CO_2 and releasing oxygen into the environment. Tropical rainforests absorb more CO_2 than any other terrestrial ecosystem and affect the climate locally and globally. However, in rainforests, carbon flux is unusually complex because of the multilayered, diverse forest structure.

The "gap theory" is a hypothetical explanation for the complexity of carbon fluxes. It hypothesizes that small, open areas in the forest canopy caused by natural processes (such as tree falls) function as a chimneys, pulling out CO_2 produced by soil respiration and leaking it into the atmosphere at local points. Due to the difficulty in making measurements from multiple points on the forest floor and corresponding points in the canopy, a balanced budget for CO_2 fluxes has been historically difficult to measure.

To increase the accuracy of the measurements and to determine the effects of uneven carbon flux, a mobile, wireless, aerially suspended robotic sensor system was deployed, as shown in Figure 7.41(a). The technology deployed in Costa Rica is a networked infomechanical system (NIMS) based on LabVIEW software and CompactRIO hardware. The NIMS application was developed at the University of California Los Angeles by the Center for Embedded Networked Sensing. The system is capable of measuring the transfer of carbon and other materials between the atmosphere and the Earth.

A wide range of measurements are necessary to characterize the carbon flux, including temperature, CO_2, humidity, precise wind movement, heat flux, solar radiation, and photosynthetic active radiation. In the past, acquiring this breadth of measurements required the use of multiple data loggers from different vendors. The deployed system in Costa Rica employs the CompactRIO platform to support a wide range of measurements using C Series modules from National Instruments and third-party vendors. The NI Wireless Access Point was chosen to transfer data between the distributed sensors, the towers, and the canopy floor.

The flexibility of CompactRIO addressed the current measurement needs with a single platform while leaving room to easily add new measurement modules in the future. The wireless sensor systems are arranged at points on the forest floor and on aerially suspended robotic shuttles, creating the first environmental monitoring system capable of taking measurements three-dimensionally.

The wireless mobile sensing platforms traverse cables along three separate transects of the forest understory. During the deployment, the shuttle stops at 1-m intervals along each transect for 30 s to allow sensors to equilibrate and take the required measurements. Each transect pass requires 30 minutes, and each transect runs for 24 hours. LabVIEW provides advanced analysis tools for real-time embedded processing to perform local mass flux analysis and post-processing for remote researchers. In addition, LabVIEW is equipped with a

(a)

(b)

FIGURE 7.41
(a) The NIMS measurement unit using CompactRIO and LabVIEW traverses on a cable between towers at the La Selva Biological Station. (b) This diagram shows how researchers can collect measurements using the wireless measurement systems connected between towers in the forest canopy. (Courtesy of National Instruments.)

human-machine interface for observation of the real-time measurements. Before the development of this real-time analysis system, researchers typically spent significant time on-site collecting large amounts of data to bring back to their respective labs for further analysis.

The plan is to expand the system by adding high towers approximately 45 m above the forest floor with canopy walkways, as illustrated in Figure 7.41(b). Students from around the world can access the canopy walkways to experience the unique atmosphere and biodiversity of the rainforest canopy. Additionally, remote data access will be provided through the Web to researchers and students who are not on-site. Using a Web browser and the Web capabilities of LabVIEW, researchers everywhere will be able to access and download live and archived data for their own analysis.

Additional measurements will provide the data needed to validate the "gap theory" hypothesis that carbon transfer occurs unevenly across the rainforest. With this research, scientists will better understand the carbon absorption impact of rainforests and potentially calculate the carbon absorption value of an acre of forest, ultimately providing a method of quantifying carbon credits. For more information on the Center for Embedded Networked Sensing, please visit the website http://research.cens.ucla.edu/ or the NI website at http://sine.ni.com/cs/app/doc/p/id/cs-11143.

7.8   myDAQ BUILDING BLOCKS

In this myDAQ Building Block, you will chart the averaged temperature and graph the actual temperature. Recall that a chart appends new data points to those points already in the display to create a history. A graph plots an entire set of data, usually stored as an array, and overwrites old data to display new data as it arrives.

Begin by opening the Control System Multiple.vi that you worked on in Chapter 6 and saved in the Chapter 6 folder in the myDAQ Building Blocks folder within the Learning directory. Before making any changes, save the VI as Control System Graph.vi in the Chapter 7 folder in the myDAQ Building Blocks folder.

A working version of Control System Multiple.vi *can be found in the* Chapter 6 *folder of the* myDAQ Building Blocks≫Solutions *folder in the* Learning *directory.*

The goal is to create a VI with a front panel as illustrated in Figure 7.42. Use the following steps as a guide.

1. Both graphs and charts can be found on the **Controls≫Silver≫Graph** palette. Using Figure 7.42 as a guide, add a Waveform Graph and a Waveform Chart to the front panel. Label the graph Current Temperature Graph and a label the chart Average Temperature Chart.

2. Rename the y-axis of both the chart and graph to be Temperature and make sure that auto-scaling is enabled for both.

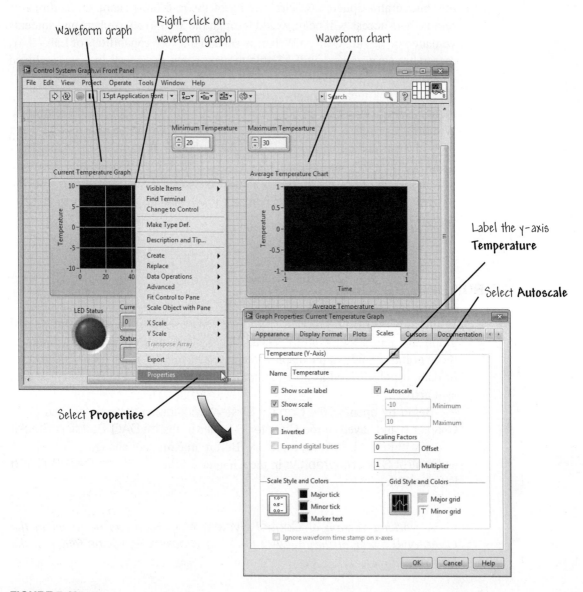

FIGURE 7.42
Selecting **Properties** for the waveform graph and waveform charts on the front panel of the Control System Graph.vi.

*Right-click on the chart or graph to access the **Properties**, as illustrated in Figure 7.42*

3. Switch to the block diagram. Connect the graph to the array output of the For Loop, and connect the chart to the same input as the **Average Temperature** indicator.

4. Switch back to the front panel and run the VI. Observe the differences between the chart and graph displays. Can you think of a situation where both would be useful?

5. Save and close the VI.

*A working version of **Control System Graph.vi** can be found in the **Chapter 7** folder of the **myDAQ Building Blocks ≫ Solutions** folder in the **Learning** directory.*

In this myDAQ Building Block, you enhanced the capabilities of the temperature monitoring system to graph the data on both a Waveform Graph and a Waveform Chart. Graphical output of acquired data is an important capability of LabVIEW.

7.9   SUMMARY

You can display plots of data using charts or graphs. Three types of charts were discussed—strip, scope, and sweep—two types of graphs—waveform and XY—and two types of math plots—2D Graphs and 3D graphs. Both charts and graphs can display multiple plots at a time. Charts append new data to old data by plotting one point at a time. Graphs display a full block of data. The waveform graph plots only single-valued functions with points that are evenly distributed. Conversely, the XY graph is a general-purpose Cartesian graph that lets you plot unevenly spaced, multivalued functions. You can customize the appearance of your charts and graphs using the legend and the palette. The format and precision of the X and Y scale markers and zoom in and out are controlled with the palette even while the VI is executing. Using Property Nodes, it is also possible to programmatically customize graphs and charts.

Math plots are very helpful in performing mathematical analysis and displaying results graphically. The math plots offer many customization options, making them particularly useful in presenting data for analysis. We found that there are two types of math plots: 2D Graphs and 3D Graphs. The 2D Graphs are used to display 2D data in one of four different ways on a 2D plot on the front panel. The 3D Graphs are used to display 3D data in one of eleven ways on a 3D plot on the front panel.

KEY TERMS

Autoscaling: The ability of graphs and charts to adjust automatically to the range of plotted values.

Legend: An object owned by a chart or graph that displays the names and styles of the plots.

Math Plots: Helpful 2D and 3D graphs for displaying graphically the results of mathematical analysis.

Palette: An object owned by a chart or graph from which you can change the scaling and format options while the VI is running.

Plot: A graphical representation of data shown on a chart or graph.

Scope chart: A chart modeled after the operation of an oscilloscope.

Strip chart: A chart modeled after a paper strip chart recorder, which scrolls as it plots data.

Sweep chart: Similar to a scope chart, except that a line sweeps across the display to separate new data from old data.

Waveform chart: Displays one point at a time on one or more plots.

Waveform graph: Plots single-valued functions or array(s) of values with points evenly distributed along the x-axis.

XY graph: A general-purpose, Cartesian graph used to plot multivalued functions, with points evenly sampled or not.

EXERCISES

E7.1 Open Multiple Graphs Demo.vi found in the Exercises&Problems folder in Chapter 7 of the Learning directory. Right-click on each graph and select Visible Items≫Cursor Legend. Right-click on the **Cursor Legend** and create a single-plot cursor. In the **Attributes** pull-down menu change the line style, point style, and color of the cursor. Also, in the **Attributes** menu, select **Allow Drag**. Again, right-click on the **Cursor Legend** and select Snap To≫Plot0. Add a second cursor for **Plot1** and repeat the editing process described above. When ready, practice moving the cursors to different values on the graph.

E7.2 Create a new VI where temperature data created with the **Simulated Temperature (deg F).vi** in the **Activity** directory is displayed on a graph. Clear the graph at the beginning of each execution of the VI.

E7.3 Create a VI that plots an ellipse

$$r^2 = \frac{A^2 B^2}{A^2 \sin^2 \phi + B^2 \cos^2 \phi}$$

where r, A, and B are input parameters and $0 \leq \phi \leq 2\pi$.

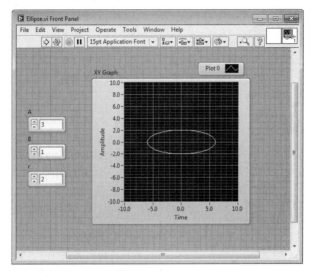

FIGURE E7.3
The ellipse with $r = 2$, $A = 3$, and $B = 1$.

E7.4 Open the NI Example Finder under **Help≫Find Examples…** and open **Waveform Graph.vi** found in the **Fundamentals≫Charts and Graphs** folder. Investigate the block diagram and locate the Build Array and Bundle functions. Run the VI. Note how the Bundle function is used to specify x parameters other than the defaults of $x_0 = 0$ and $\Delta x = 1$. The Build Array function is used to create graphs with multiple plots. Close the VI without saving it when you are finished with your observations.

E7.5 Open the NI Example Finder and go to the **Search** tab. Type in "charts." Open **Charts.vi**. Run the VI and examine the different types of update modes, specifically the differences between the strip chart, scope chart, and sweep chart. When you are done examining the VI, stop and close the VI. Do not save any changes.

E7.6   Open the NI Example Finder and go to the **Browse** tab. Select browsing accord-ing to **Task**. Open Building User Interfaces≫Displaying Data≫Graphs and Charts≫3D Comet Plot.vi. Run the VI. Use your mouse to click on the 3D graph and rotate the position of the graph. Experiment with the projection using the **Projection Palette**. Stop the VI. Switch to the block diagram. The 3D Comet Plot is a math plot for analysing mathematical results. This graph and other types of 3D graphs are available in the **Modern≫Graph≫3D Graph** palette.

PROBLEMS

P7.1   Create a new VI to plot a circle using an XY Graph.

P7.2   Create a new VI where temperature data, created with the Simulated Temper-ature (deg F).vi in the Activity directory, is displayed on a strip chart. Compute and display the running average of the temperature data.

P7.3   Open a blank VI and place a Simulate Signal Express VI on the block dia-gram. Configure the Express VI to generate a 50-Hz sine wave. Click **OK** to exit the Express VI configuration page. Right-click on the **Sine** terminal and select Create≫Graph Indicator. Run the VI.

P7.4   Create a VI that graphs the function $\sin x$ where $x = 0 \ldots n\pi$ and the integral $y = \int_0^{n\pi} \sin x\, dx$. The value of n should be an input on the front panel.

P7.5   Open and run Chart Mistake.vi which can be found in the Exercises& Problems folder in Chapter 7 of the Learning directory. The VI generates two signals: a sawtooth waveform and a triangle waveform. Both signals have

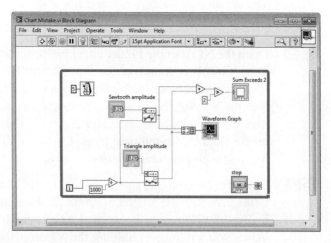

FIGURE P7.5
The Chart Mistake VI block diagram.

their amplitude controlled by front panel inputs. Whenever their sum exceeds 2, an LED lights up.

Take a look at the block diagram (shown in Figure P7.5) after you have run the VI. The designer of this VI intended the Waveform Graph to display the sawtooth and triangle waveforms being generated, with the graph continuously updating on each loop iteration. Find the mistake in the implementation of this goal. Why do the time values on the graph never increment? How can you display two separate plots? Make the necessary adjustments so that this VI does what its designer intended.

P7.6 Open the NI Example Finder and navigate to the Temperature Analysis VI found in the Fundamentals≫Graphs and Charts folder. Run the VI and note the maximum value displayed on the Max indicator on the right. Create an annotation on the Temperature Graph indicator marking the maximum point and specifying the value.

Once the graph is annotated, change the color of the annotation from the default yellow to blue. This color will show up better when you export the image to a file, where the graph has a white, rather than black, background. As a final step, export the image of the graph to a Bitmap file (.bmp) file. When viewed with an external viewer, the .eps file will appear similar to the graphic in Figure P7.6. Close the Temperature Analysis VI without saving any changes.

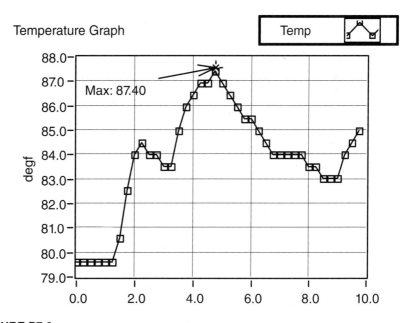

FIGURE P7.6
An Bitmap (.bmp) image of the Temperature Graph annotated with the maximum temperature.

P7.7 Create a VI that plots $\sin\theta$ versus $\cos\theta$ for $0 \leq \theta \leq 2\pi$. Display the output on an XY graph. Label the x- and y-axes "sin(theta)" and "cos(theta)", respectively. Add a cursor and a cursor legend to the plot. Using the cursor, verify that when $\sin\theta = 0$ (along the x-axis) that $\cos\theta = 1$ (along the y-axis). What geometrical figure results from running the VI?

DESIGN PROBLEMS

D7.1 In this problem, you will design a VI which normalizes a waveform of 50 randomly generated points. The completed VI should have a waveform chart and a waveform graph on the front panel. On the waveform chart, display the random points as they are being generated.

To improve the visualization of the waveform chart, use a delay in the loop of 50 milliseconds. Place controls on the front panel so that the offset and scale of the waveform can be adjusted each time the VI is run. When all 50 points have been generated, perform the necessary scaling and shifting on this waveform so that it is centered on the x-axis and has a peak-to-peak amplitude of 1.

Plot both the original and the normalized waveforms on a single waveform graph. Make the **Cursor Legend** and **Graph Palette** visible on the graph. Create two labeled cursors and lock one to the original waveform and the other to the normalized waveform.

Run your VI and use cursors and the zooming tools to verify that your VI is producing the expected output. As a test case, set the offset to 0 and the scale to 2. Then confirm that the original signal varies between 0 and 2, while the normalized signal varies between -1 and 1.

D7.2 Employ the Sine Wave PtByPt.vi to generate a sine wave continuously until a user presses a stop button on the front panel. Allow the user to control the frequency, phase, and amplitude of the sine wave with numeric controls on the front panel. The sine wave should be displayed on a waveform chart and updated with each new data point generated. Then use the Collector Express VI to display the last 1000 points generated on a waveform graph when the user presses the stop button.

   *The Collector Express VI can be found on the **Functions≫Express≫Signal Manipulation** palette.*

D7.3 Create a VI that produces 20 random integers between 0 and 100 and then plots them on a waveform graph. Display the **Cursor Legend** and use cursors to determine the coordinates of the minimum and maximum generated values. Using the Array Max & Min function, display the coordinates of the maximum and minimum on the front panel. Run the VI and compare the values output by the Array

Max & Min function to the values you find with the cursors. Verify that they are equivalent.

D7.4 Create a 3D Stem plot that displays $\sin\theta$ versus $\cos\theta$ for $0 \le \theta \le 2\pi$. Plot $\sin\theta$ along the x-axis, $\cos\theta$ along the y-axis, and θ along the z-axis.

D7.5 Create a VI that generates simulated temperature data in degrees Fahrenheit and converts the temperature to degrees Celsius. The temperature data should be displayed on a waveform chart and on a digital thermometer on the front panel. Use a While Loop with a Conditional Terminal to stop the simulation. When the simulation ends, compute the average temperature and display the result using a digital indicator. Using a Property Node, provide the capability to configure the

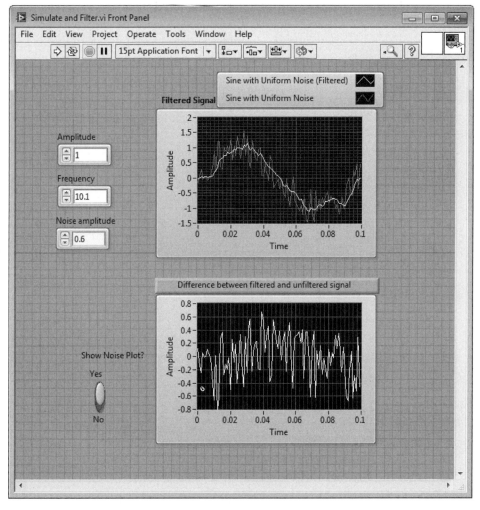

FIGURE D7.6
Filtering a signal and programmatically controlling the visibility of graphs.

plot line color as blue or red as determined by an input on the front panel. The line color should be configurable programmatically while the VI is running.

*You can use the Simulated Temperature (deg F) VI found in the **Activity** folder in the **Learning** directory.*

D7.6   Create a VI that uses the Simulate Signal Express VI to generate a sine wave signal and configure the express VI to add a uniform white noise to the output signal. Add three controls to the front panel to allow the user to specify the noise amplitude, frequency, and amplitude of the sine wave signal. Then use the Filter Express VI to filter the noisy sine wave signal. Display both the filtered and unfiltered signal on the same graph. Create a waveform graph to display the difference between the filtered and unfiltered signal and programmatically make the graph visible or not using a Boolean switch on the front panel. Test the VI with noise amplitude of 0.6 and frequency and amplitude of the sine wave signal of 10.1 Hz and 2, respectively. Use the front panel in Figure D7.6 as a guide.

CHAPTER 8

Data Acquisition

Data Acquisition (DAQ) is the measurement or generation of electrical or physical phenomena. This chapter focuses on the basic notions associated with computer-based analog and digital input and output (I/O) using the LabVIEW DAQ Assistant. Some common terms and concepts associated with data acquisition are also discussed, including the components of a DAQ system, signal conditioning, and the types of signals encountered.

GOALS

1. Review some basic notions of signals and signal acquisition.
2. Introduce the organization of the DAQ VIs.
3. Understand the basics of analog and digital input and output using the DAQ Assistant.

8.1   COMPONENTS OF A DAQ SYSTEM

Data-acquisition systems consist of similar components, whether the application is an electrocardiograph (ECG) heart monitor or a wind-tunnel acquisition system used to tune bicycle performance. A computer-based data acquisition consists of sensors and transducers, signal conditioning, the hardware DAQ device, and a computer-based software suite for acquiring, manipulating, analyzing, storing, and displaying the raw acquired data. Every user of a DAQ system needs to understand the role of each of these components. A DAQ system is illustrated in Figure 8.1. Many different system configurations are available, depending on the unique measurements and the environments in which they will be taken.

In Figure 8.1, the plug-in DAQ device resides in the computer. This computer can be a tower, a desktop model, or a laptop. The DAQ device may be internal, using the PCI, PCIe, PXI, or PXIe computer bus, or may use USB, Firewire, or Ethernet external to the computer. The former approach provides faster acquisition speeds at higher channel counts and control applications, while the latter is typically more practical for remote data acquisition where you want to bring the DAQ system into the field.

Before a computer-based system can measure a physical signal, a sensor (or transducer) must convert the physical signal into an electrical signal (such as voltage or current). Generally, you cannot connect signals directly to a plug-in DAQ device in a computer, unless the output of the sensor is matched to the input of the DAQ device. In many cases, the measured physical signal is very low-voltage and susceptible to noise. The measured signal may then need to be

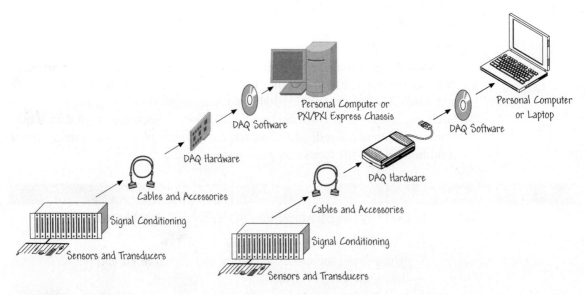

FIGURE 8.1
Typical DAQ system.

amplified and filtered before conversion to a digital format for use in the computer. A signal-conditioning module or accessory conditions measured signals before the DAQ device converts them to digital information. More will be said on signal conditioning later in the chapter. Software controls the DAQ system—acquiring the raw data, analyzing the data, and presenting the results.

The subject of data acquisition cannot be adequately covered in one chapter, although the fundamental ideas can be introduced and discussed in enough detail to generate enthusiasm for pursuing other sources of information.[1] The most effective way to learn about data acquisition is by doing it. Reading about it is not enough. At its most basic level, the task of a DAQ system is the measurement or generation of physical signals.

8.2   TYPES OF SIGNALS

The concepts of signals and systems arise in a wide variety of fields, including science, engineering, and economics. In an effort to develop an analytic framework for studying certain natural phenomena, the notion of an input–output representation has arisen, and is illustrated in Figure 8.2. In the input–output representation, the input signals are operated on by the DAQ system to produce the output signals. Signals are physical quantities that are functions of an independent variable (such as time) and contain information about a natural phenomenon. For example, the input signal from an electrocardiogram (ECG) sensor applied to the skin is very small and contains noise from other nerve signals being sent within the body. The DAQ system includes the hardware and software to acquire, process, and enhance the ECG signals (e.g., detecting the rate of the heart beat or an irregular heart rhythm), and the output signals would be the waveform of the heart beat and a single number indicating the instantaneous heart rate. In general, physical signals are converted to electrical signals (such as

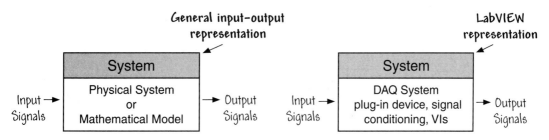

FIGURE 8.2
Modeling physical phenomena with input–output representations.

---

1. A good source of information is the **Measurement Fundamentals Series** (which includes excellent webcasts) available from National Instruments, Inc. on the NI Developer Zone website.

voltage or current) before being amplified and filtered by the signal conditioning and acquired by the DAQ hardware. This conversion is accomplished by a transducer. The list of common transducers includes microphone (sound), camera (images), thermocouple (temperature), strain gauge (bending strain), thermistor (temperature), accelerometer (acceleration), antenna (radio waves), and encoder (rotation or distance). Once the physical signals are measured and converted to electrical signals by the sensor, the information contained in them may be extracted and analyzed.

Five common classes of information that can be extracted from signals are illustrated in Figure 8.3. The signals evolve either continuously or only at discrete points in time and are known as continuous- or discrete-time signals, respectively. Room temperature is an example of a continuous-time signal (although we may discretize the temperature signal by recording the temperature only at specific times or sampling points). The end-of-day closing Dow Jones stock index is an example of a discrete signal, taking on a new value once per working day.

For purposes of discussing data acquisition, we will use the following signal classifications:

- For digital signals, we have two types:
  - On–Off
  - Pulse Train
- For analog signals, we have three types:
  - DC–Static or Slow Changing Signals
  - AC–Fast Changing Signals
  - Frequency-Domain–Frequency Analysis of AC and DC Signals

The five signal types just listed are classified as analog or digital according to how they convey information. A digital signal has only two possible discrete

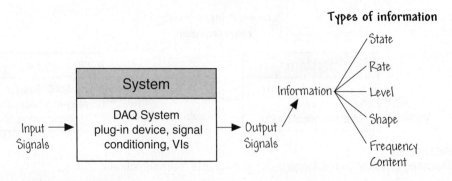

FIGURE 8.3
Measuring and analyzing signals provides information: state, rate, level, shape, and frequency content.

levels: high level (on) or low level (off). An analog signal, on the other hand, contains information that varies continuously with respect to time.

A schematic with the various signal types is shown in Figure 8.4. Each signal type is unique in the information conveyed, and the five signal types closely parallel the five basic types of signal information: state, rate, level, shape, and frequency content.

8.2.1   Digital Signals

The two types of digital signals that we will consider here are the on–off switch and the pulse train signals. The on–off signal illustrated in Figure 8.5 conveys information concerning the immediate digital *state* of the signal. A simple

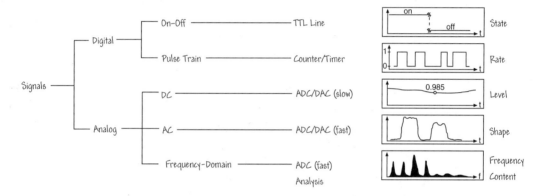

FIGURE 8.4
Signal types (ADC → analog-to-digital converter, DAC → digital-to-analog converter, TTL → transistor-transistor logic).

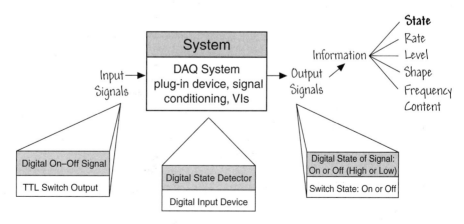

FIGURE 8.5
The on–off signal.

digital state detector is used to measure this signal type. An example of a digital on–off signal is the output of a transistor–transistor logic (TTL) switch.

The second type of digital signal is the pulse train signal illustrated in Figure 8.6. This signal consists of a series of state transitions, and information is contained in the number of state transitions, the *rate* at which the transitions occur, and the time between one or more state transitions. The output signal of an optical encoder mounted on the shaft of a motor is an example of a digital pulse train signal.

8.2.2 Analog DC Signals

Analog DC signals are static (or slowly varying) analog signals that convey information in the *level* (or amplitude) of the signal at a given instant. The analog DC signal is depicted in Figure 8.7 in the context of the input–output representation. Since the analog DC signal is static or varies slowly, the accuracy of the measured level is of more concern than the time or rate at which you take the measurement. Common examples of DC signals include temperature, battery voltage, flow rate, pressure, strain-gauge output, and fluid level, as illustrated in Figure 8.8. The information contained in the signals is often displayed on meters, gauges, strip charts, and numerical readouts for the easiest interpretation by the user.

The DAQ system will return a single value indicating the magnitude of the signal at the requested time. For the measurement to be an accurate representation of the signal, the DAQ system must possess adequate accuracy and resolution capability and adequate sampling rates (usually slow).

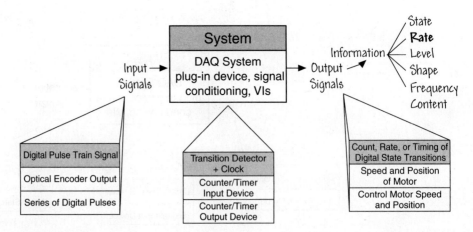

FIGURE 8.6
The pulse train signal.

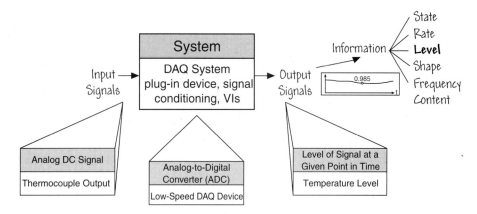

FIGURE 8.7
Analog DC signals are static or slowly varying analog signals.

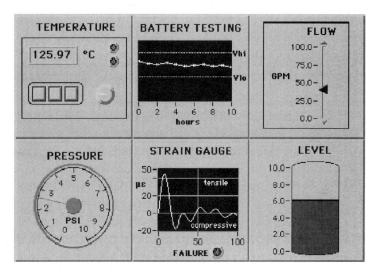

FIGURE 8.8
Common examples of DC signals include temperature, battery voltage, flow rate, pressure, strain-gauge output, and fluid level.

8.2.3   Analog AC Signals

Analog AC signals differ from other signals in that they convey useful information in the signal level (or amplitude) *and* the way this level varies with time. The information associated with an AC signal (also referred to as a waveform) includes such information as time to peak, peak magnitude, time to settle, slope, and shapes of peaks. An analog AC signal is illustrated in Figure 8.9.

You must take a precisely timed sequence of individual amplitude measurements to measure the shape of a time-domain signal. DAQ systems that are used

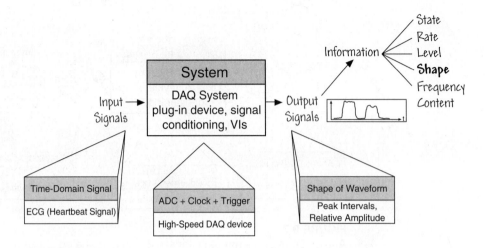

FIGURE 8.9
Analog AC signals convey information in the signal level and in the way this level varies with time.

to measure time-domain signals usually have an **analog-to-digital conversion** (ADC) component, a sample clock, and a trigger.

The physical signal must be sampled and measured at a rate that adequately represents the shape of the signal. Therefore, when acquiring analog time-domain signals, DAQ systems need a high-bandwidth (fast) ADC to sample the signal at high rates. The sampled value must be measured accurately without significant loss of precision.

Generally the signal sampling needs to start at a specfic time to guarantee that an interesting segment of the signal is acquired. Once started, the sample clock accurately times the occurrence of each sample, defined as a single analog-to-digital (A/D) conversion. In many situations, triggering is necessary to initiate the sampling process at a precise time. The trigger starts the acquisition at the proper time based on either the state of the signal or some external condition specified by the user. For example, the acquisition of a sine wave or cosine wave looks the same with a 90-degree phase shift (1/4-period shift in time). A sine wave is triggered when the signal passes through zero with positive slope and the cosine is triggered when the value reaches the first peak. There are an unlimited number of different time-varying signals, some of which are shown in Figure 8.10.

8.2.4 Analog Frequency-Domain Signals

Analog frequency-domain signals are similar to AC time-domain signals in that they also convey information in the way the signals vary with time. However, the information extracted from a frequency-domain signal is based on the signal

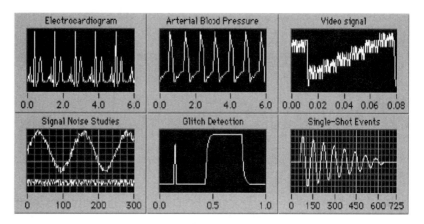

FIGURE 8.10
Six different time-varying signals.

frequency content, as opposed to the shape of the signal. An analog frequency-domain signal is shown in Figure 8.11.

Frequency-domain signals are converted by performing mathematical operations on their time-domain counterparts. As with DAQ systems that measure time-domain signals, a system used to measure a frequency-domain signal must also include the same fundamental hardware components, including an ADC, a sample clock, and a trigger to accurately capture the time-domain waveform. Additionally, the DAQ system must include the necessary analysis capabilities to extract frequency information from the time-domain signal. You can perform

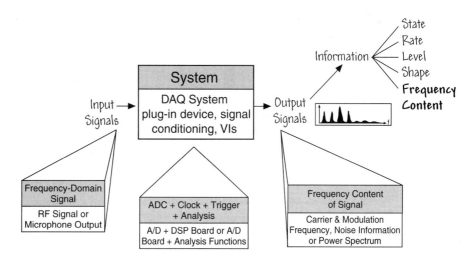

FIGURE 8.11
The information extracted from a frequency-domain signal is based on the signal frequency content.

this type of digital signal processing (DSP) using application software or special DSP hardware designed to analyze the signal quickly and efficiently.

In short, a DAQ system that acquires analog time-domain signals and can compute their frequency-domain representation, should also possess a high-bandwidth ADC capability to sample signals at high rates and an accurate clock to take samples at precise time intervals. Also, it is common to use triggers to initiate the measurement process precisely at prespecified times. Finally, a complete DAQ system should provide a library of analysis functions, including a function to convert time-domain information to frequency-domain information. Chapter 11 discusses some of the LabVIEW analysis functions, but time and space limitations make it impossible to cover all analysis functions available to VI developers.

Figure 8.12 shows some examples of frequency-domain signals. Each example in the figure includes a graph of the originally measured signal as it varies with respect to time, as well as a graph of the signal frequency spectrum. While you can analyze any signal in the frequency domain, certain signals and application areas lend themselves especially to this type of analysis. Among these areas are speech and acoustics analysis, geophysical signals, vibration, and studies of system transfer functions.

8.2.5 One Signal—Five Measurement Perspectives

The five classifications of signals presented in the previous discussions are not mutually exclusive. A particular signal may convey more than one type of

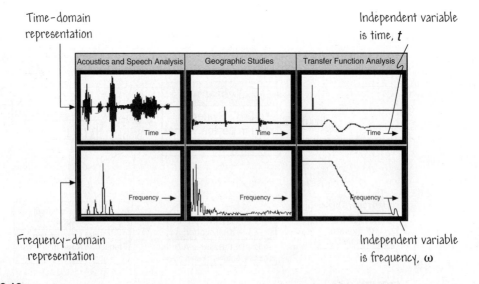

FIGURE 8.12
Three examples of frequency-domain signals: speech and acoustics analysis, geophysical signals, and transfer function frequency response.

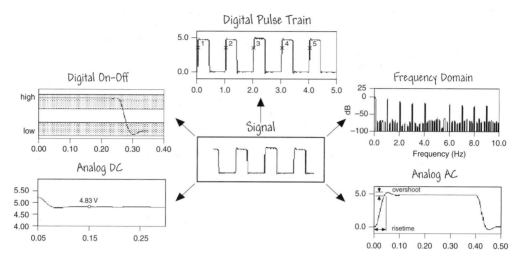

FIGURE 8.13
A series of voltage pulses can provide information for all five signal classes.

information, and you can classify a signal as more than one type and measure it in more than one way. In fact, you can use simpler measurement techniques with the digital on–off, pulse train, and DC signals, because they are just simpler cases of the analog time-domain signals.

The measurement technique you choose depends on the information you want to extract from the signal. In many cases you can measure the same signal with different types of systems, ranging from a simple digital input board to a sophisticated frequency analysis system. Figure 8.13 demonstrates how a series of voltage pulses can provide information for all five signal classes.

8.3   COMMON TRANSDUCERS AND SIGNAL CONDITIONING

When measuring a physical phenomenon, a transducer must convert this phenomenon (such as temperature or force) into a measurable electrical signal (such as voltage or current). Table 8.1 lists some common transducers used to convert physical phenomena into measurable quantities.

All transducers output an electrical signal that is often not well suited for a direct measurement by the ADC (Analog-to-Digital Converter). For example, the output voltage of most thermocouples is very small and susceptible to noise, and often needs to be amplified and filtered before measuring. Figure 8.14 shows some common types of transducer and signal pairs and the required signal conditioning. A highly expandable signal conditioning system—dubbed Signal Conditioning eXtensions for Instrumentation (**SCXI**) and made by National Instruments, Inc.—conditions low-level signals in a noisy environment within an external chassis located near the sensor. The close proximity improves the signal-to-noise ratio (measure of signal quality) of the signals reaching the DAQ

TABLE 8.1   Phenomena and Transducers

Phenomenon	Transducer
Temperature	Thermocouples
	Resistance temperature detectors (RTDs)
	Thermistors
	Integrated circuit sensor
Light	Vacuum tube photosensors
	Photoconductive cells
Sound	Microphone
Force and pressure	Strain gauges
	Piezoelectric transducers
	Load cells
Position (displacement)	Potentiometers
	Linear voltage differential transformer (LVDT)
	Optical encoder
Fluid flow	Head meters
	Rotational flowmeters
	Ultrasonic flowmeters
pH	pH electrodes

device. LabVIEW works with SCXI and many other types of hardware signal conditioning systems.

Some types of **signal conditioning** (such as linearization and scaling) can be performed in the software. LabVIEW provides several VIs for such purposes. The remainder of this section is devoted to short descriptions of some of the basic ideas involved in signal conditioning. Further information on signal conditioning and SCXI hardware (i.e., setup procedures for SCXI hardware, hardware operating modes, and programming considerations for SCXI) can be found in other LabVIEW documentation (check the NI website).

Some common types of signal conditioning follow:

- **Transducer excitation**: Certain transducers (such as strain gauges or microphones) require external voltages or currents to excite their own circuitry in a process known as transducer excitation. The process is similar to a television needing power to receive and decode video and audio signals. The necessary excitation in a DAQ system can be provided by the plug-in DAQ

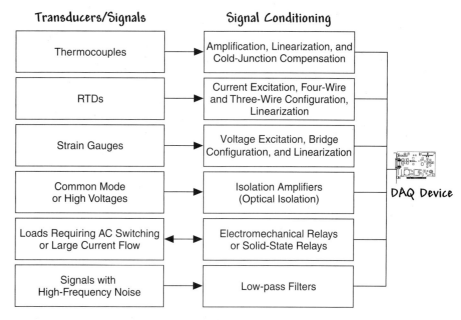

FIGURE 8.14
Common types of transducers/signals and the required signal conditioning.

devices and the signal-conditioning peripherals, and sometimes by external instruments.

- **Linearization**: Common transducers (such as strain gauges, thermistors, RTDs, and thermocouples) generate voltages that are nonlinear with respect to the phenomena they represent. The NI-DAQ driver and LabVIEW software can perform **software linearization** to scale a transducer voltage to the correct units of strain or temperature.

- **Isolation**: Another common use for signal conditioning is to isolate the transducer signals from the computer and other transducers. For example, when the signal being monitored contains large voltage spikes that could damage a computer or harm a person, you should not connect the signal directly to a DAQ device without some type of isolation. Figure 8.15 shows two common methods for isolating signals.

- **Filtering**: Another form of signal conditioning is the filtering of unwanted signals from the desired signal. Notch filters are commonly used to reduce 60 Hz AC powerline noise present in many signals. Other well-known types of filters include low-pass, high-pass, and band-pass filters. Some DAQ devices and signal-conditioning devices have built-in filters.

- **Amplification**: This is the most common type of signal conditioning. Amplification maximizes the use of the available voltage range to increase the accuracy of the digitized signal and to increase the signal-to-noise ratio

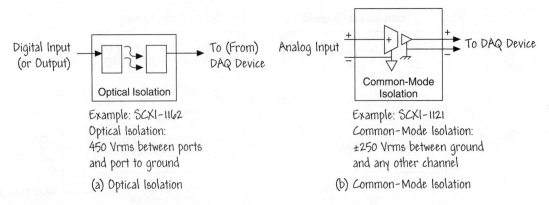

FIGURE 8.15
Two common methods for isolating signals.

(SNR). Low-level signals should be amplified at the DAQ device or at an external signal-conditioning peripheral positioned near the source of the signal, as shown in Figure 8.16. One reason to amplify low-level signals close to the signal source, instead of at the DAQ device, is to increase the signal-to-noise ratio. Consider the case where you amplify the signal only at the DAQ device. Then the DAQ device will also measure and digitize any noise that enters the lead wires along the path as the signal travels from the source to the DAQ device. On the other hand, the ratio of signal voltage to noise voltage that enters the lead wires is larger if you amplify the signal close to the signal source. Table 8.2 shows how the SNR changes with the location of amplification.

 You can minimize the effects of external noise on the measured signal by using shielded or twisted-pair cables and by minimizing the cable length. Keeping cables away from AC power cables and computer monitors will also help minimize 50/60 Hz noise.

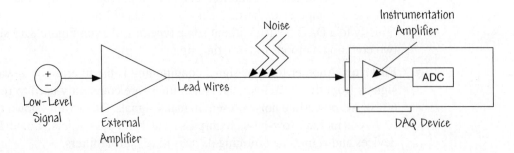

FIGURE 8.16
Low-level signals should be amplified at the DAQ device or at an external signal-conditioning peripheral positioned near the source of the signal.

TABLE 8.2   Effects of Amplification on Signal-to-Noise Ratio (S.C. → Signal-Conditioning Peripheral)

	Signal Voltage	S.C. Amplification	Noise in Lead Wires	DAQ Device Amplification	Digitized Voltage	SNR*
Amplify only at DAQ device	0.01 V	None	0.001 V	×100	1.1 V	10
Amplify at S.C. and DAQ device	0.01 V	×10	0.001 V	×10	1.01 V	100
Amplify only at S.C.	0.01V	×100	0.001 V	None	1.001 V	1000

$$*\text{SNR (dB)} = 20 \log_{10} \frac{\text{Amplitude}_{\text{signal}}}{\text{Amplitude}_{\text{noise}}}$$

8.4   SIGNAL GROUNDING AND MEASUREMENTS

Up to this point, we have discussed three components of the DAQ system: transducers, signals, and signal conditioning. Now you might be tempted to think that all that remains is to wire the signal source to the DAQ device and begin acquiring data. However, a few important items must be considered:

- The nature of the signal source (grounded or floating)
- The grounding configuration of the amplifier on the signal-conditioning hardware or DAQ device
- The cabling scheme to connect all the components together

A DAQ system is depicted in Figure 8.17 highlighting the signals and the cabling.

8.4.1   Signal Source Reference Configuration

Signal sources come in two forms: referenced and nonreferenced. Referenced sources are usually called **grounded signals**, and nonreferenced sources are called **floating signals**. A schematic of a grounded signal source is shown in Figure 8.18(a).

Grounded signal sources have voltage signals that are referenced to a system ground, such as earth or a building ground. Devices that plug into a building ground through wall outlets, such as signal generators and power supplies, are the most common examples of grounded signal sources. Grounded signal sources share a common ground with the DAQ device and the computer.

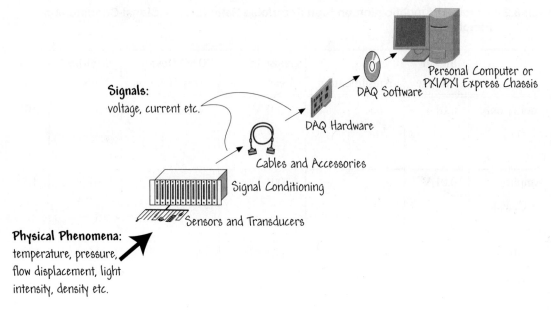

Signals:
voltage, current etc.

DAQ Software

Personal Computer or
PXI/PXI Express Chassis

DAQ Hardware

Cables and Accessories

Signal Conditioning

Sensors and Transducers

Physical Phenomena:
temperature, pressure,
flow displacement, light
intensity, density etc.

FIGURE 8.17
A DAQ system highlighting the signals and the cabling.

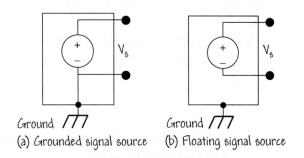

V_s

V_s

Ground

Ground

(a) Grounded signal source   (b) Floating signal source

FIGURE 8.18
Grounded signal sources have voltage signals that are referenced to a system ground; floating signal sources contain a signal that is not connected to an absolute reference.

Floating signal sources contain a signal (e.g., a voltage) that is not connected to an absolute reference, such as earth or a building ground. Some common examples of floating signals are batteries, battery-powered sources, thermocouples, transformers, isolation amplifiers, and any instrument that explicitly floats its output signal. As illustrated in Figure 8.18(b), neither terminal of the floating source is connected to the electrical outlet ground.

8.4.2   Measurement System

A schematic of a measurement system is depicted in Figure 8.19. A measurement system can be placed in one of three categories:

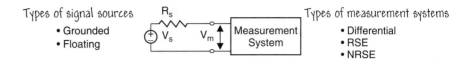

FIGURE 8.19
Types of signal sources and measurement systems.

- Differential
- Referenced single-ended (RSE)
- Nonreferenced single-ended (NRSE)

In a **differential** measurement system, neither the positive or negative terminal of the source or transducer is connected to a ground. National Instruments DAQ devices have advanced instrumentation amplifiers that can be configured as differential measurement systems. Figure 8.20 depicts the 8-channel differential measurement system used in an NI Multifunction DAQ device. A **channel** is a pin or wire lead where analog or digital signals enter or leave a DAQ device. For this configuration, the analog input ground pin labeled AIGND is the measurement system ground. The analog multiplexers (labeled MUX in the figure) increase the number of available measurement channels while still using a single instrumentation amplifier.

An ideal differential measurement system, shown in Figure 8.21, reads only the potential difference between its two terminals—the (+) and (−) inputs. It completely rejects any other voltages present in the system. In other words, an ideal differential measurement system completely rejects the common-mode voltage (voltage between the measurement and system ground).

A **referenced single-ended** (RSE) measurement system measures a signal with respect to the system ground and is often called a ground referenced measurement. Figure 8.22 depicts a 16-channel version of an RSE measurement system.

DAQ devices often use a **nonreferenced single-ended** (NRSE) measurement system, which is a variation of the RSE measurement system. In an NRSE measurement system, all measurements are made with respect to a common reference, because often many transducers share a common power source and ground. Figure 8.23 depicts an NRSE measurement system. AISENSE is the common reference for taking measurements and all signals in the system share this common reference. Note that the signals are not referenced to the system ground, AIGND.

Differential measurements offer the best immunity to noise when measuring grounded and floating signal sources. NRSE offers the next best noise immunity, where all signals share the same ground reference point. An RSE measurement system can be used with a grounded signal source if the signal levels are high and the cabling has a low impedance. RSE measurements are subject

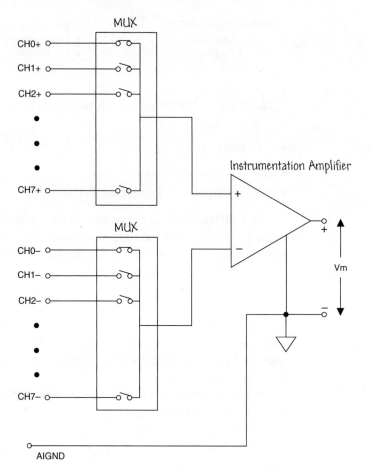

FIGURE 8.20
An 8-channel differential measurement system.

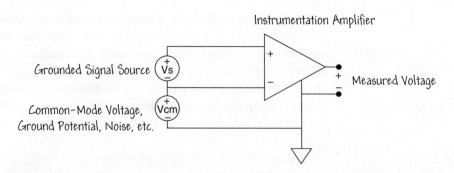

FIGURE 8.21
An ideal differential measurement system completely rejects common-mode voltage.

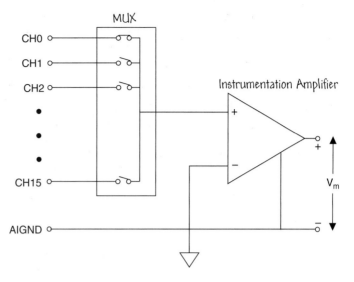

FIGURE 8.22
A 16-channel RSE measurement system.

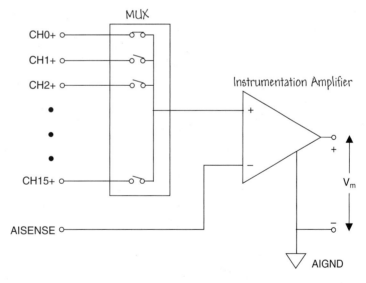

FIGURE 8.23
An NRSE measurement system.

to noise and ground loops (ground is not actually zero volts for all transducers). In general, there is a trade-off between noise immunity and the number of channels. Differential measurements require more connections than NRSE measurements because differential measurements require a single pin for the (−) terminal of each signal whereas (N)RSE measurements share the (−) terminal for each signal. When higher channel counts are needed, you can use single-ended measurement systems when all input signals meet the following criteria:

1. High-level signals (normally, greater than 1 V).
2. Short or properly shielded cabling traveling through a noise-free environment (normally less than 15 ft).
3. All signals can share a common reference signal at the source.

A summary of analog input connections is given in Figure 8.24.

AI Ground-Reference Settings	Signal-Source Type	
	Floating Signal Sources (Not Connected to Building Ground)	Ground-Referenced Signal Sources
	Examples • Ungrounded thermocouples • Signal conditioning with isolated outputs • Battery devices	Example • Plug-in instruments with nonisolated outputs
Differential	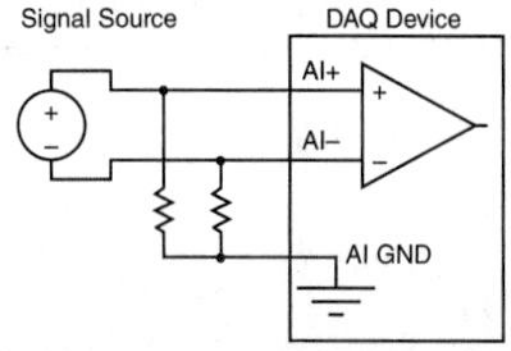	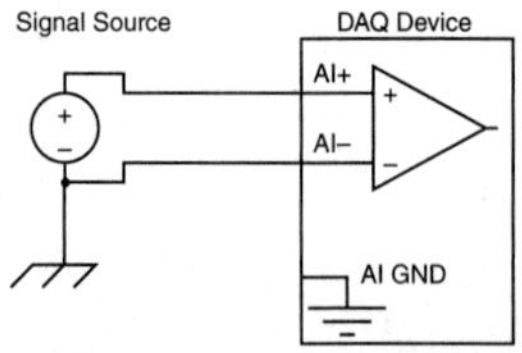
Nonreferenced Single-Ended (NRSE)	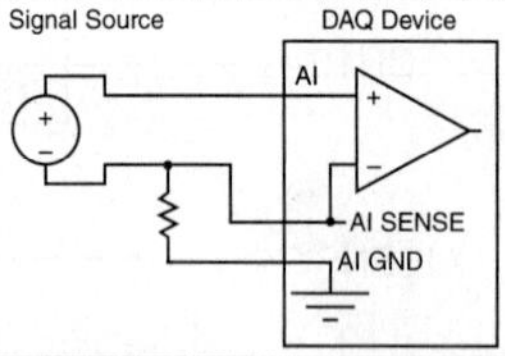	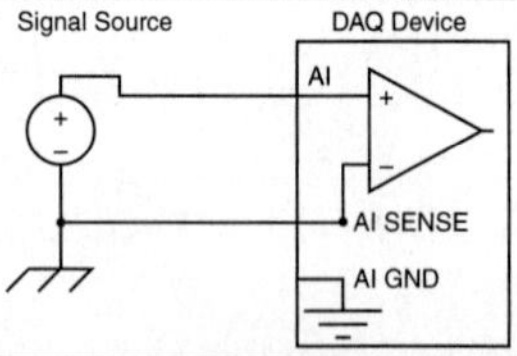
Referenced Single-Ended (RSE)	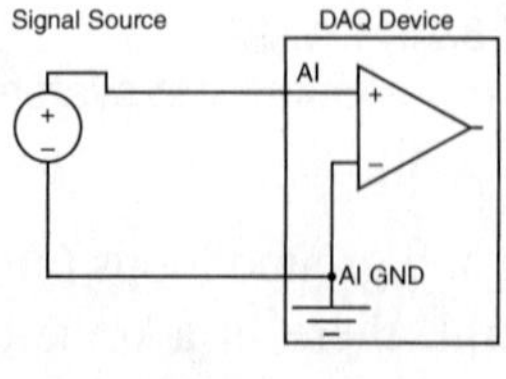	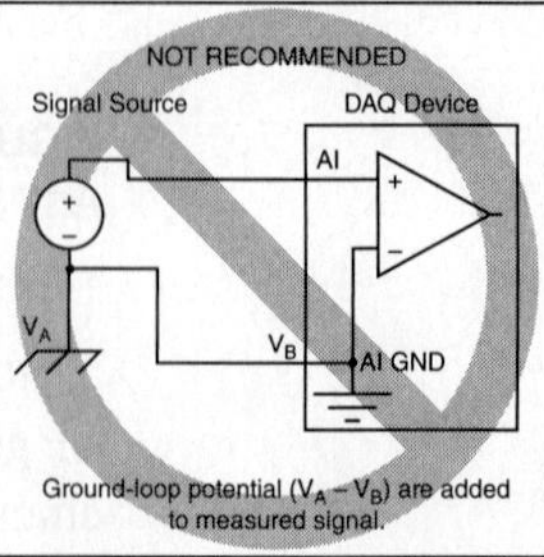

FIGURE 8.24
Summary of analog input connections.

8.5   ANALOG-TO-DIGITAL CONVERSION CONSIDERATIONS

When preparing to configure a DAQ device you need to consider the quality of the analog-to-digital conversion. There are many questions that arise when making analog signal measurements with your DAQ system. For example, what are the signal magnitude limits? How fast does the signal vary with time? The latter question is important because the sample rate determines how often the A/D conversions take place. The four parameters of concern are

- Resolution
- Device range
- Signal input range
- Sampling rate

Depending on the type of DAQ device you have, these four parameters can be set using software by configuring the NI DAQmx channel within LabVIEW or in the **Measurement and Automation Explorer** as discussed later in this chapter.

The number of bits used to represent an analog signal determines the **ADC resolution**. You can compare the resolution on a DAQ device to the number of divisions on a ruler. For a fixed ruler length, the more divisions you have on the ruler, the more precise the measurements can be made. A ruler marked off in millimeters can be read more accurately than a ruler marked off in centimeters. Similarly, the higher the ADC resolution, the higher the number of divisions of the ADC range and, therefore, the more accurately the analog signal can be represented. For example, a 3-bit ADC divides the signal range into 2^3 divisions ($2^3 = 8$), with each division represented by a binary or digital code between 000 and 111 (0 and 7, respectively). The ADC then translates each measurement of the analog signal to one of the digital divisions (rounding to the nearest digital division). Figure 8.25 shows a sine wave digital image obtained by a 3-bit

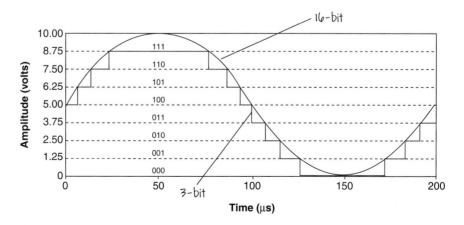

FIGURE 8.25
16-bit versus 3-bit resolution (5 kHz sine wave).

ADC. Clearly, the digital signal does not represent the original signal adequately because there are too few divisions to represent the varying voltages of the analog signal. By increasing the resolution to 16 bits, however, the ADC's number of divisions increases from 2^3 to 2^{16} (8 to 65,536). The ADC can now provide an adequate representation of the analog signal.

The **device range** refers to the minimum and maximum analog signal levels that the ADC can handle. You should attempt to match the range to that of the analog input signal to take best advantage of the available resolution. Fortunately, many DAQ devices feature selectable ranges. Consider the example shown in Figure 8.26(a), where the 3-bit ADC has eight digital divisions in the range from 0 to 10 volts. If you select a range of -10.00 to $+10.00$ volts, as shown in Figure 8.26(b), the same ADC now separates a 20-volt range into eight divisions. The smallest detectable voltage increases from 1.25 to 2.50 volts, and you now have a much less accurate representation of the signal.

The **signal input range** is the maximum and minimum value of the signal you are measuring. The closer the signal input range is to the incoming analog signal maximum and minimum, the more digital divisions will be available to the ADC to represent the signal. Using a 3-bit ADC and a range setting of 0.00 to 10.00 volts, we see in Figure 8.27 the effects of a signal input range between 0 and 5 volts and 0 and 10 volts. With a signal input range of 0 to 10 volts, the ADC uses only four of the eight divisions in the conversion; with a signal input range of 0 to 5 volts, the ADC now has access to all eight digital divisions. This makes the digital representation of the signal more accurate.

The resolution and range of a DAQ device and the signal input range determine the smallest detectable change in the input voltage. This change in voltage represents 1 **least significant bit** (LSB) of the digital value and is often

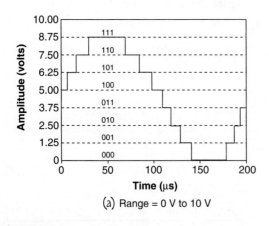

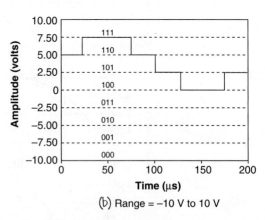

(a) Range = 0 V to 10 V

(b) Range = −10 V to 10 V

FIGURE 8.26
The 3-bit ADC in (a) has eight digital divisions in the range from 0 to 10 volts—if you select a range of −10.00 to 10.00 volts as in (b), the same ADC now separates a 20-volt range into eight divisions.

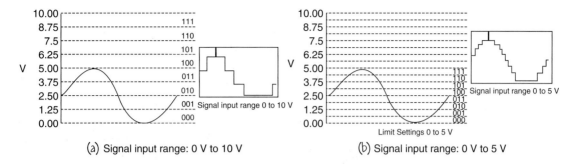

(a) Signal input range: 0 V to 10 V   (b) Signal input range: 0 V to 5 V

FIGURE 8.27
Precise signal input range selection allows the ADC to use more digital divisions to represent the signal.

called the **code width**. The smallest code width is calculated with the following
formula:

$$V_{cw} = \frac{\text{range}}{2^{\text{resolution}}},$$

where the resolution is given in bits. For example, a 12-bit DAQ device with a
0 to 10 V range detects a 2.4-mV change. This is calculated as follows:

$$V_{cw} = \frac{\text{range}}{2^{\text{resolution}}} = \frac{10}{2^{12}} = 2.4 \text{ mV},$$

while the same device with a -10 to 10 V range detects only a change of 4.8 mV:

$$V_{cw} = \frac{\text{range}}{2^{\text{resolution}}} = \frac{20}{2^{12}} = 4.8 \text{ mV}.$$

A high-resolution ADC provides a smaller code width for a given range. For
example, consider the two preceding examples, except that the resolution is now
16-bit. A 16-bit DAQ device with a 0 to 10 V range detects a 0.15-mV change:

$$V_{cw} = \frac{\text{range}}{2^{\text{resolution}}} = \frac{10}{2^{16}} = 0.15 \text{ mV},$$

while the same device with a -10 to 10-V range detects a change of 0.3 mV:

$$V_{cw} = \frac{\text{range}}{2^{\text{resolution}}} = \frac{20}{2^{16}} = 0.3 \text{ mV}.$$

You also need to determine whether your signal is unipolar or bipolar, as this
affects the code width. **Unipolar** signals range from 0 V to a positive value (e.g.,
0 V to 10 V). **Bipolar** signals range from a negative value to a positive value
(e.g., -5 V to 5 V). A smaller code width is obtained by specifying the range to
be unipolar, if indeed the signal is unipolar; and conversely, specifying the range
as bipolar if the signal is bipolar. If the maximum and minimum variation of the

analog signal is smaller than the value of the range, you will need to set the signal input range to more accurately reflect the analog signal variation. If the smallest range on the DAQ devices does not provide enough resolution, signal conditioning or a higher accuracy DAQ device should be considered.

Some confusion may exist about selecting the signal input range rather than selecting the gain. *When you set the signal input range, you are effectively selecting the gain for that signal. Signal input range settings automatically magnify the magnitude of the signal to create more precise analog-to-digital conversions.*

The **sampling rate** is the rate at which the DAQ device samples an incoming analog signal. Figure 8.28 shows an adequately sampled signal as well as the effects of undersampling. The sampling rate determines how often an analog-to-digital (A/D) conversion takes place. Computing the proper sampling rate requires knowledge of the maximum frequency of the incoming signal and the accuracy required of the digital representation of the analog signal. It also requires some knowledge of the noise affecting the incoming signal and the capabilities of your hardware. A fast sampling rate acquires more points in a given time, allowing, in general, a better representation of the original signal than a slow sampling rate would allow. In fact, sampling too slowly may result in a misrepresentation of the incoming analog signal.

The effect of undersampling is that the signal appears as if it has a different shape in the time domain, thus appearing to have a different frequency than it truly does. This misrepresentation is called an *alias*. To prevent undersampling, you must sample greater than twice the rate of the maximum frequency component of the incoming signal. One way to deal with aliasing is to use low-pass filters that attenuate any frequency components in the incoming signal above the

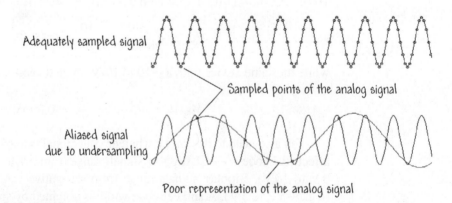

FIGURE 8.28
Sampling too slowly may result in a poor representation of the analog signal.

Nyquist frequency (defined to be one-half the sampling frequency). These filters are known as *antialiasing filters* and these filters must be applied in hardware before the ADC.

Signal Input Range Selection

The kind of calculations required to select the signal input range for a DAQ application are illustrated in this example. Keep in mind that the objective is to minimize the code width while making sure that the entire signal fits within the allowable device range.

Assume your transducer output is a sine wave with an amplitude of ± 30 mV and an offset of 10 mV. Your device has signal input ranges of 0 to $+10$ V, ± 10 V, and ± 5 V. What signal input range would you select for maximum precision if you use a DAQ device with 12-bit resolution?

Table 8.3 shows how the code width of a hypothetical 12-bit DAQ device varies with device range and signal input range. The values in Table 8.3 depend on the hardware, and you should consult your DAQ hardware documentation to determine the available device voltage ranges for your device.

Step 1: Select a device range. Since the signal input is bipolar (the sine wave contains both positive and negative voltages), we must choose one of the bipolar ADC ranges. For our device, there are two, so to start we choose ± 5 V, for a range of 10 V.

Step 2: Determine the code width. Using the formula for determining the code width, we compute

$$V_{\text{cw}} = \frac{\text{range}}{2^{\text{resolution}}} = \frac{10}{2^{12}} = 2.4 \text{ mV}.$$

Step 3: Choose a signal input range. Since our signal has a maximum magnitude of $+30$ mV, we must choose signal input ranges that allow us to read ± 30 mV. Referring to Table 8.3, we find that (for a range of -5 to 5 V) a signal input range of -50 mV to 50 mV will cover the signal variation of ± 30 mV. Choosing this range yields a code width (or precision) of 24.4 μV.

Step 4: Repeat steps 1 through 3 for a range of ± 10 V. We must also try the other range of ± 10 V to see if it yields a smaller code width. Repeating steps 1 through 3 gives us a signal input range of -0.1 V to 0.1 V and a code width of 48.8 μV. This does not improve our accuracy. Thus, we choose the settings as follows:

  □ Device Range $= -5$ to 5 V

  □ Signal Input Range $= -50$ mV to 50 mV

TABLE 8.3   Code Width for Various Device Ranges and Signal Input Ranges

Device Range	Signal Input Range	Code Width*
0 to 10 V	0 to 10 V	2.44 mV
	0 to 5 V	1.22 mV
	0 to 2.5 V	610 μV
	0 to 1.25 V	305 μV
	0 to 1 V	244 μV
	0 to 0.1 V	24.4 μV
	0 to 20 mV	4.88 μV
−5 to 5 V	−5 to 5 V	2.44 mV
	−2.5 to 2.5 V	1.22 mV
	−1.25 to 1.25 V	610 μV
	−0.625 to 0.625 V	305 μV
	−0.5 to 0.5 V	244 μV
	−50 to 50 mV	24.4 μV
	−10 to 10 mV	4.88 μV
−10 to 10 V	−10 to 10 V	4.88 mV
	−5 to 5 V	2.44 mV
	−2.5 to 2.5 V	1.22 mV
	−1.25 to 1.25 V	610 μV
	−1 to 1 V	488 μV
	−0.1 to 0.1 V	48.8 μV
	−20 to 20 mV	9.76 μV

* Code width is the smallest measurable difference in voltage.

8.6   DAQ VI ORGANIZATION

The LabVIEW DAQ VIs are organized in the NI-DAQmx palette, as illustrated in Figure 8.29. These VIs are device drivers for the National Instruments DAQ hardware line. NI-DAQmx, originally released in NI-DAQmx 7.0, is a next-generation DAQ driver architecture that incorporates significant improvements over previous NI-DAQ drivers in both performance and ease of use. For students learning about data acquisition for the first time, the special advantage

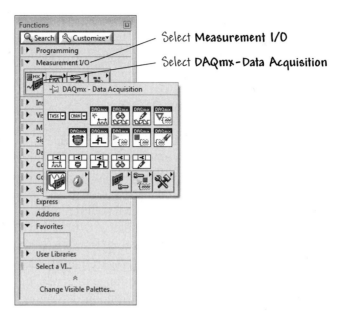

FIGURE 8.29
The **DAQmx-Data Acquisition** palette.

of NI-DAQmx over previous versions of NI-DAQ is the inclusion of the DAQ Assistant for configuring channels and measurement tasks. The task of writing VIs to acquire data is significantly simplified with the DAQ Assistant, as we shall learn later in this chapter. For these reasons, the remainder of this chapter will cover NI-DAQmx.

The NI-DAQmx VIs are a special type of VI called Polymorphic VIs. The result is a core set of VIs that can adapt to different DAQ functionality, such as analog input, analog output, digital I/O, etc. You access the palette by opening the **Functions** palette and clicking on **Measurement I/O≫DAQmx-data Acquisition**, shown in Figure 8.29. In this book, we will focus on the functionality of these three DAQ operations: **Analog Input**, **Analog Output**, and **Digital I/O**.

8.7 CHOOSING YOUR DATA ACQUISITION DEVICE

In this chapter we introduce the NI-DAQmx driver to communicate with the National Instruments USB X Series data acquisition device in LabVIEW. The USB-6341 Series is but one of many high-speed, general-purpose data acquisition options available to users of LabVIEW. National Instruments offers plug-in data acquisition devices that can measure signals in frequency ranges from DC to Gigahertz with digital accuracies of 8 to 24 bits. Many common interfaces for PC connectivity are available including PCI, PCIe, PXI, USB, Firewire,

Ethernet and wireless Ethernet. Engineers at National Instruments designed the NI-DAQmx driver so that the same VIs can be used to communicate with nearly all NI data acquisition devices and even allow many of those devices to be simulated.

8.7.1   X Series Data Acquisition Devices

The X Series multifunction data acquisition devices from National Instruments are optimized for higher accuracy at fast sampling rates for a broad range of applications. These devices are ideal for laboratory environments where a versatile, high-speed, high-quality data acquisition device is needed. X Series devices have 8, 16, or 32 analog inputs with a resolution of 16 or 18 bits and acquisition rates up to 2.0 MS/s. They also include up to 48 digital I/O lines, up to four analog outputs, four 32-bit, 100 MHz counter/timers, programmable input ranges, and digital and analog triggering (analog triggering not available on all models). X Series devices are ideal for a wide range of applications including (i) acquiring, visualizing, and logging data, (ii) control systems, (iii) signal generation, and (iv) test automation.

8.7.2   Low Cost Data Acquisition for Students

As a leader in data acquisition, National Instruments continues to expand its portfolio of low-cost options for students. Although student devices cannot match the performance of an X Series device, they offer a low-cost solution to data acquisition that is well suited for home and laboratory use. Most of the exercises presented in this chapter can be easily completed using USB data acquisition. With plug-and-play USB connectivity, these devices are simple enough for quick measurements but versatile enough for more complex measurement applications, and they serve the same purpose as many traditional desktop instruments. Visit http://www.ni.com/academic/students/ to learn about the resources for students.

You must install the NI-DAQmx device driver before completing the rest of this chapter.

8.7.3   Simulated Data Acquisition

You can create NI-DAQmx simulated devices in NI-DAQmx 7.5 and later. Using NI-DAQmx simulated devices, you can craft VIs for use with data acquisition devices without actually having the hardware. No time is lost when you later plug-in or move to a computer with the hardware you simulated, because you can import the NI-DAQmx simulated device configuration to the physical device using the MAX Portable Configuration Wizard. You can prototype and test your data acquisition applications before you get to the laboratory, knowing you are ready to acquire real-world signals. If you do not have access to a data acquisition device, follow the steps outlined below to complete the exercises and homework problems presented in the rest of the chapter using a simulated data acquisition device.

 Simulated data acquisition is currently only available on the Windows platform.

Creating NI-DAQmx Simulated Devices

Since this book uses the National Instruments USB-6341 data acquisition device for the exercises, we will add that device as an NI-DAQmx simulated device. To accomplish this task, follow these steps:

1. Open the **Measurement and Automation Explorer (MAX)** by clicking on the **Measurement & Automation** icon on your desktop (see Section 8.8.1 for more information on MAX).
2. Right-click on **Devices and Interfaces** and select **Create New**
3. A dialog box prompts you to select a device to add. Select **NI-DAQmx Simulated Device** and click **Finish**.
4. In the **Choose Device** dialog box, select the family of devices for the device you want to simulate. In this case, we will choose **X Series DAQ**.
5. Select **NI USB-6341** and click **OK**.

8.7.4 Macintosh, Linux, and Mobile Devices

NI-DAQmx Base driver software offers a variety of other OS users a programming interface similar to the polymorphic VIs available in the NI-DAQmx driver software as well as increased ease of use in connecting with high-performance NI data acquisition hardware. NI-DAQmx Base offers a clean, concise programming interface, programmatic channel and task creation, and other features previously available only with NI-DAQmx for Windows. The new driver software also includes ready-to-use National Instruments LabVIEW VIs and C function features similar to those included in the full-featured version of NI-DAQmx.

National Instruments built NI-DAQmx Base driver software for these OSs using the NI Measurement Hardware Driver Development Kit and developed the software almost entirely in the multiplatform NI LabVIEW graphical development environment.

For more information on updates to DAQ for the Mac, refer to www.ni. com/mac. For more information on updates to DAQ for Linux, refer to www.ni. com/linux. For more information on updates to DAQ for mobile devices, refer to www.ni.com/labview/mobile.

8.8   DAQ HARDWARE CONFIGURATION

LabVIEW provides utilities designed to help you define which signals are connected to which channels on your data acquisition device and to easily group those into combined tasks. In previous years, significant amounts of time were spent defining the signal types and connections and converting voltage to engineering units—and all this before beginning development and programming of the actual DAQ system. For example, if you are using thermocouples, you must perform cold-junction compensation (CJC) calculations and apply appropriate scaling factors to convert raw measured voltages into actual temperature readings. This process is now one of entering the necessary information in dialog boxes, or through wizards, to define an input signal, the type of transducer being used, any scaling factors required, CJC values, and the conversion factors, as well as the timing and triggering information you desire. You can then reference the channel name or task (that you assign) for the input signal(s) and the conversion from voltage to physical units is performed automatically (and transparently). You can save different configuration files for different settings or systems. Not only can you assign **channel** and **task names**, sensor types, engineering units, and scaling information, as well as timing and triggering information, to each channel using the LabVIEW utility, but you can also define the physical quantities you are measuring on each DAQ hardware channel. Once the software has been properly configured, the hardware will be configured correctly to make the measurement for each channel in terms of the physical quantity.

8.8.1   Windows

LabVIEW for Windows installs a configuration utility for establishing all device and channel configuration parameters. This utility is known as the **Measurement & Automation Explorer**, or MAX for short. MAX is also helpful in troubleshooting and self-calibration. The MAX utility reads the information the Device Manager records in the Windows registry and assigns a logical device name to each National Instruments DAQ device in your system. You use the device name to refer to the device in LabVIEW. Figure 8.30 shows the relationship between the DAQ device and MAX. The Windows Configuration Manager keeps track of all the hardware installed in your system, including National Instruments DAQ devices.

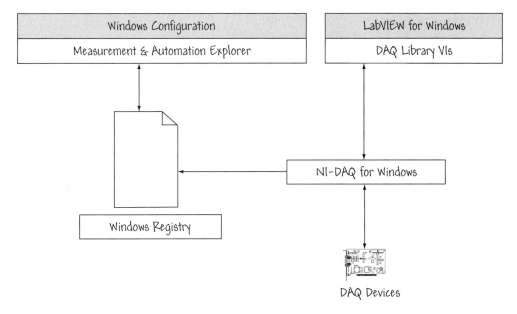

FIGURE 8.30
Windows configuration management.

If your National Instruments DAQ device does not show up in MAX it may not be configured correctly or it may not be supported in the version of NI-DAQ that is installed. The installation guide for your DAQ device can be found on www.ni.com/manuals.

You can check the Windows Configuration by accessing the Device Manager on your computer. You will find **Data Acquisition Devices**, which lists all DAQ devices installed in your computer, as shown in Figure 8.31. Highlight a DAQ device and select **Properties** or double-click on the device, and you see a dialog window with tabbed pages. **General** displays overall information regarding the device. You use **Resources** to specify the system resources to the device such as interrupt levels, DMA, and base address for software configurable devices. **Driver** specifies the driver version and location for the DAQ device.

 Using MAX (Windows)

The objective of this exercise is to use MAX to examine the configuration for the DAQ device in your computer and to configure one NI-DAQmx task.

Start MAX by double-clicking on the desktop icon or by selecting **Measurement & Automation Explorer** from the **Tools** menu in LabVIEW, as illustrated in Figure 8.32.

Expand the **Devices and Interfaces** section, as seen in Figure 8.33, and highlight NI USB-6341. The USB-6341 is an NI-DAQmx device, so open the

Right-click on the
DAQ device

Lists all DAQ devices
installed on your computer

Displays general
information

Specifies the driver
version and location
for the DAQ device

Obtain more information
on DAQ devices

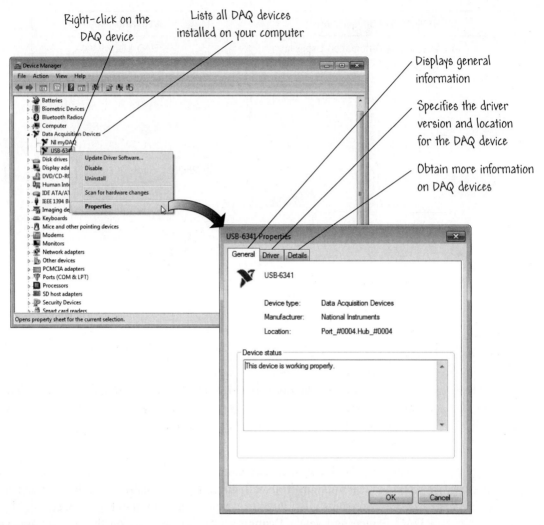

FIGURE 8.31
Checking the Windows Configuration by accessing the Device Manager.

NI-DAQmx Devices section as well. MAX shows the National Instruments devices and software in your system. Note the device name indicated in the quotes after the DAQ device. The LabVIEW DAQ VIs use the device name to determine which device performs DAQ operations. In Figure 8.34, we see that the DAQ device USB-6341 is "Dev1." You may have a different device installed and some of the options shown may be different.

You can get more information about your DAQ device configuration by examining the properties using the **Attributes**, **Device Routes**, and **Calibration** tabs. With the DAQ device highlighted, MAX displays the attributes of the

FIGURE 8.32
Accessing the Measurement & Automation Explorer (MAX).

device, such as the serial number, as shown in Figure 8.33. Clicking the **Device Routes** tab provides detailed information about the internal signals that can be routed to other destinations on the device, as shown in Figure 8.34. This is a powerful resource that gives you a visual representation of the signals that are available to provide timing and synchronization with components that are on the device and shared with other external devices. The **Calibration** tab provides information about the last time the device was calibrated both internally and externally. If you right-click the NI-DAQmx device in the configuration tree and select **Self Calibrate**, the DAQ device will be calibrated using a precision voltage reference source. Once the device has been calibrated, the **Self Calibration** information updates in the **Calibration** tab.

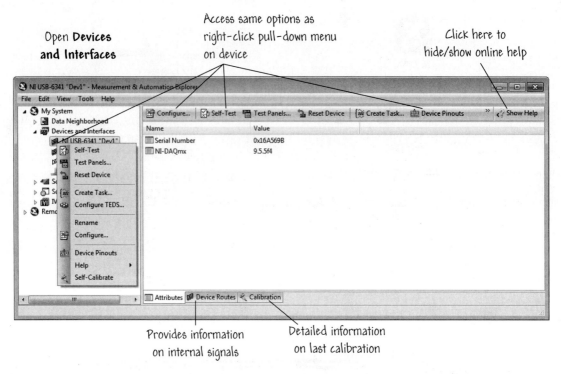

FIGURE 8.33
Checking the properties of the DAQ devices.

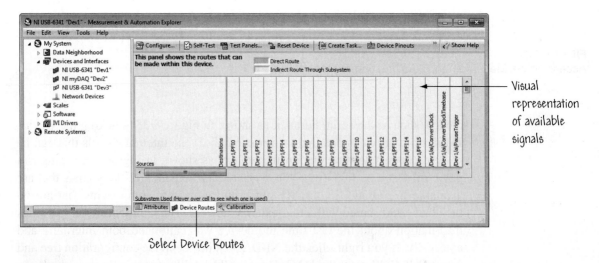

FIGURE 8.34
Accessing information about the DAQ device configuration.

*Press the **Show/Hide** button in the top right corner of the MAX window to hide the online help.*

Next, we want to create and configure an NI-DAQmx task. An NI-DAQmx task is a shortcut to configuring a data acquisition task that allows you to save a configuration complete with channel, timing, and triggering information. Right-click on the **Data Neighborhood** icon and choose **Create New**, as shown in Figure 8.35. The window shown in Figure 8.36 will appear. In the window, select **NI-DAQmx Task** and press the **Next** button, as depicted in Figure 8.36. You will now configure a channel to take a reading from a temperature sensor (Analog Input).

After pressing the **Next** button, a window for configuring the input will appear, and you can select the input as shown in Figure 8.36. We will select the Acquire Signals≫Analog Input≫Temperature≫Thermocouple as shown in Figure 8.37. Once we do so, a new dialog box opens up to permit us to assign a new local channel to the task. In this case, we select **ai0** associated with Dev1— our USB-6341 DAQ device. Once you have completed the job of creating the local channel, select **Next** to proceed.

The next dialog box that appears allows you to name your task, as shown in Figure 8.38. By default, the name **MyTemperatureTask** is provided—we will accept this name this time around. You can change the name to something

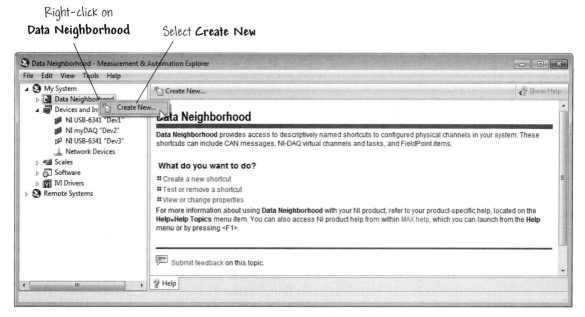

FIGURE 8.35
Configuring a new virtual channel.

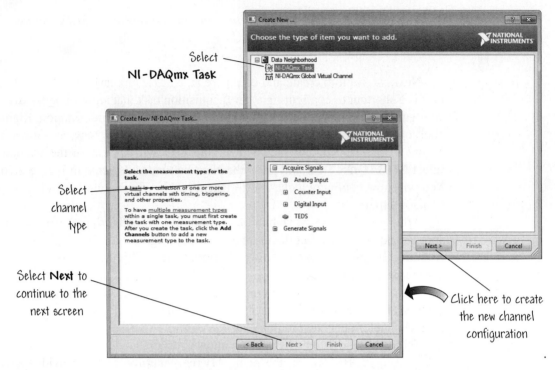

FIGURE 8.36
Creating a new virtual channel configuration.

more descriptive depending on your application. Once you have named the NI-DAQmx task, select **Finish**.

*As you proceed through the configuration process, you can always go back to a previous page using the **Back** button.*

When you are finished defining the channel, MAX entries should appear as shown in Figure 8.39. In future sessions, you can edit the configurations through MAX by highlighting the desired task and right-clicking on **Temperature**. MAX provides important configuration information. For example, as seen in Figure 8.39, the temperature minimum and maximum values are shown to be 0°C and 100°C, respectively. These values can be readily changed by editing the screen directly. Also, if you click on the double arrow ≫ under **Channel Settings**, the screen will expand to show the device type and physical channel. You can right-click on **Temperature** to open a dialog box in which the channel assignment can be changed should the need arise in future sessions. When you are finished configuring the NI-DAQmx task, close the MAX window to exit. ◆

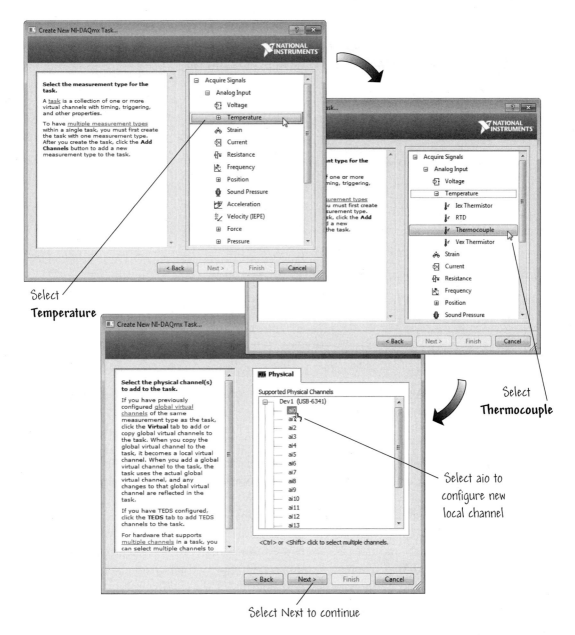

FIGURE 8.37
Selecting the measurement type and creating a new local channel.

8.8.2   Channels and Tasks

In Section 8.8.1, we created an NI-DAQmx task to acquire temperature readings from a thermocouple using an NI USB-6341 DAQ device. In the sections following this one, we will use the DAQ Assistant to create other

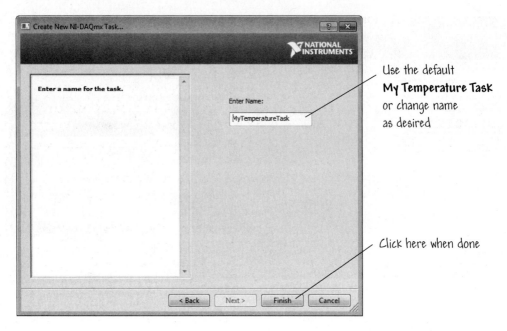

FIGURE 8.38
Naming the task and finishing the task configuration.

NI-DAQmx tasks. But what exactly is an NI-DAQmx task? And what are NI-DAQmx channels?

With LabVIEW 6i (or earlier versions), when developing DAQ applications, a set of *virtual channels* was configured using MAX. This configuration was a collection of property settings that included a physical channel, the type of measurement specified in the channel name, and scaling information. In the traditional NI-DAQ (prior to LabVIEW 7 Express), virtual channels were a simple method to remember which channels were used for different measurements. NI-DAQmx *channels* are similar to the virtual channels of traditional NI-DAQ channels and must be configured accordingly. In NI-DAQmx (introduced with LabVIEW 7 Express), you can configure virtual channels either in MAX or using the DAQ Assistant—in both cases the configuration process is almost identical.

An NI-DAQmx *task* is a collection of one or more channels, timing, triggering, and other properties that apply to the task itself. For example, a task might represent a measurement you want to perform. In the previous section, we created a task to measure temperature from one channel on a DAQ device.

 To complete the exercises in the next section using the DAQ Assistant, you must have installed NI-DAQmx and have installed an NI-DAQmx-supported device, or created an NI-DAQmx simulated device in MAX.

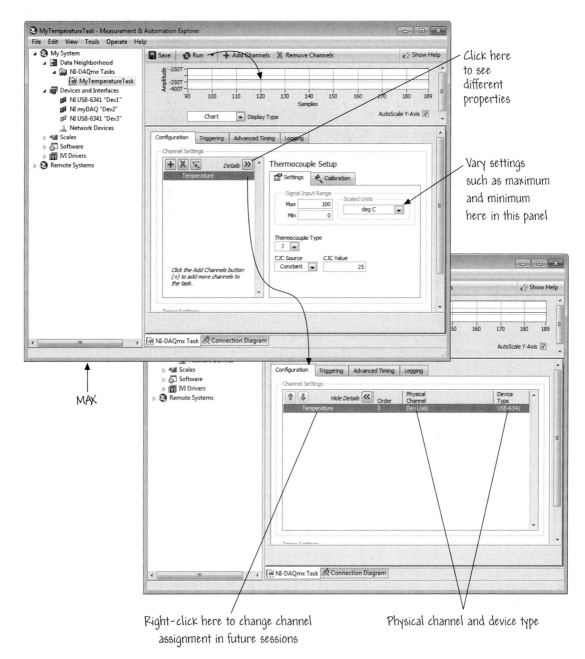

FIGURE 8.39
The MAX final configuration.

8.9 USING THE DAQ ASSISTANT

In this section, we introduce the DAQ Assistant. The DAQ Assistant is a graphical interface that we can use to configure measurement tasks and channels (described in Section 8.8).

The DAQ Assistant is located on the **Functions≫Measurement I/O≫ DAQmx-Data Acquisition** palette. To launch the DAQ Assistant, place it on the block diagram, as illustrated in Figure 8.40. When the DAQ Assistant is placed on the block diagram, a DAQ Assistant dialog box will automatically appear. Once the DAQ Assistant is open, the steps required to configure the NI-DAQmx task are basically the same as described in Section 8.8 using MAX.

The general process of constructing a data acquisition VI using the DAQ Assistant is as follows:

- Open a new VI.

- Place the DAQ Assistant on the block diagram.

- A DAQ Assistant dialog box appears to help you configure the measurement task.

- Configure, name, and test the NI-DAQmx task.

- Click the **OK** button to return to the block diagram.

- Edit the front panel and block diagram to complete the VI.

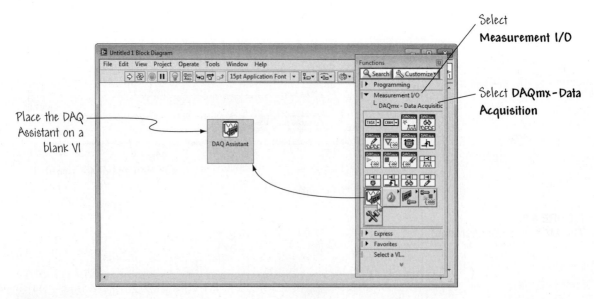

FIGURE 8.40
Locating and placing the DAQ Assistant on the block diagram.

   ▪ If desired, generate an NI-DAQmx Task Name control so that the task can
      be used in other applications.

**Practice
with the
DAQ Assistant**

In this example, you will construct an NI-DAQmx task to acquire voltage
readings from a measurement device attached to a USB-6341 DAQ device.

The first step is to open a new VI and place the DAQ Assistant on the block
diagram. The DAQ Assistant will automatically launch to begin the process
of creating and configuring the NI-DAQmx task. In the first dialog box, click
the **Acquire Signals** button to display all the input options. Click the **Analog
Input** button to display the analog input options as illustrated in Figure 8.41.
Select **Voltage** to create a new voltage analog input task. The dialog box then
displays a list of channels on each DAQ device installed. The number of channels listed depends on the number of channels you have on the DAQ device.
In our case, we find a list **ai0**, **ai1**, ... Select **ai0** and click the **Finish** button.

At this point, the DAQ Assistant opens a new window, shown in Figure 8.42,
which displays the options for configuring the selected channels. Notice that you
can change the **Max** and **Min** values in the **Signal Input Range** and modify the
number of **Samples to Read**.

At this point, the NI-DAQmx task is ready to run. You can run the task to
verify that you correctly configured the channel. Click the **Run** button, as shown
in Figure 8.42, to confirm that you are collecting data in the top half of the DAQ
Assistant. Once you have verified the proper functioning and configuration, click
the **OK** button to return to the block diagram. The DAQ Assistant appears on
the block diagram ready for integration into the VI application.

We are now ready to add a graph to the VI to plot the voltage data acquired
from the DAQ device. On the block diagram, right-click the **data** output on
the DAQ Assistant and select Create≫Graph Indicator, as illustrated in
Figure 8.43. The waveform graph will be added and automatically wired. Switch
the view to the front panel. The waveform graph legend displays the channel
name **Plot 0** until you run the program. The DAQmx task is now ready to acquire
voltage data and to display the output on the waveform graph.

It is possible to rename the channels configured using the DAQ Assistant.
You might, for instance, decide later that a more descriptive name would be
helpful in organizing your configured channels. To accomplish this, return to the
block diagram, right-click the DAQ Assistant, and select **Properties** to rename
the channel, as shown in Figure 8.44. Right-click **Voltage** in the **Channel Settings** box and select **Rename** to display the **Rename a channel or channels**
dialog box. In the **New Name** text box, enter **Extreme Voltage**, then click the
OK button. You are done—the channel has been renamed. To verify the name
change, click the **OK** button and return to the block diagram. Then, switch to the
front panel and run the VI. Notice that **Extreme Voltage** appears in the waveform graph plot legend.

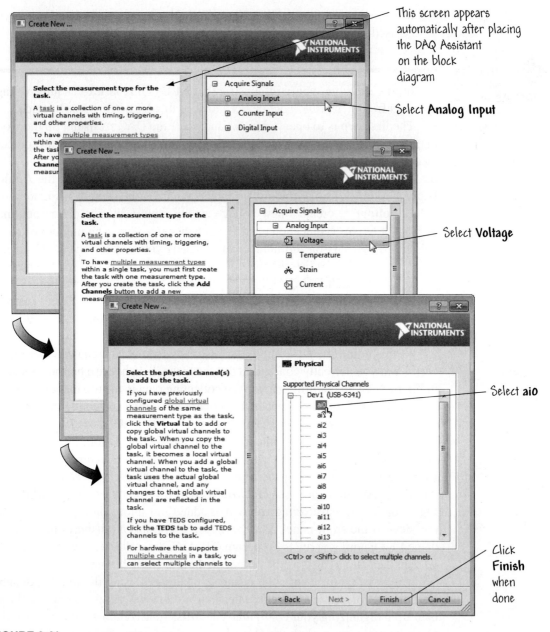

FIGURE 8.41
Using the DAQ Assistant to configure an analog input channel.

*You also can select the name of the channel and press the <F2> key to access the **Rename a channel or channels** dialog box.*

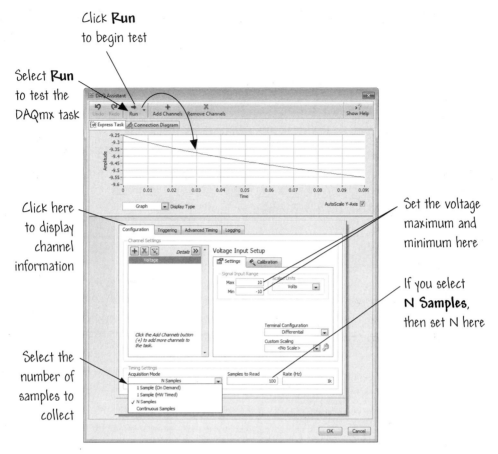

Click **Run**
to begin test

Select **Run**
to test the
DAQmx task

Click here
to display
channel
information

Select the
number of
samples to
collect

Set the voltage
maximum and
minimum here

If you select
N Samples,
then set N here

FIGURE 8.42
Configuring the channel settings and testing the DAQmx task.

Save the VI as Read Voltage.vi in the Users Stuff folder in the Learning
directory.   ◆

8.9.1   DAQmx Task Name Constant

If you configure an NI-DAQmx task using the DAQ Assistant, then the task is
a *local task*; hence, it cannot be saved to MAX for use in other applications. If
you want to make the task available to other applications, you can use the DAQ
Assistant to generate an NI-DAQmx Task Name control so that the task can be
saved to MAX and used in other applications. This is readily accomplished by
right clicking on the DAQ Assistant in the block diagram and selecting **Convert
to NI-DAQmx Task** in the pull-down menu, as illustrated in Figure 8.45. The

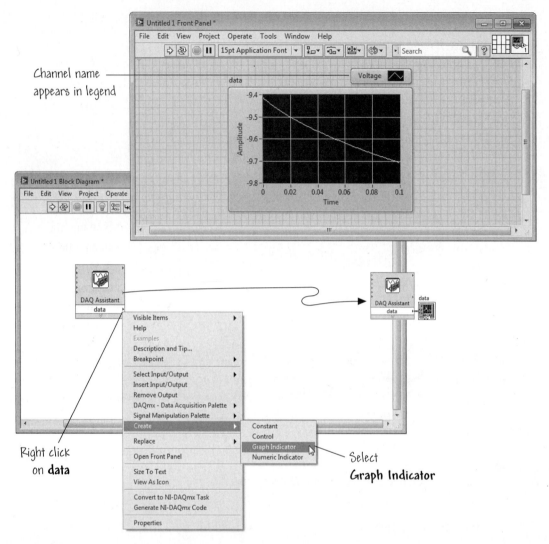

FIGURE 8.43
Adding a waveform graph to the DAQ Assistant.

DAQmx Task Name Constant is a LabVIEW data type used by the DAQ VIs for communicating to the DAQ devices. After converting the DAQ Assistant to a DAQmx Task Name Constant, the DAQ Assistant will appear so that you can re-configure the task, if needed. After exiting the DAQ Assistant, the DAQmx task is available in MAX and hence ready for use in other applications. You will probably want to rename the task.

If you want to use one of the DAQmx tasks from MAX in an application, you first place a DAQmx Task Name Constant on the block diagram. You can find

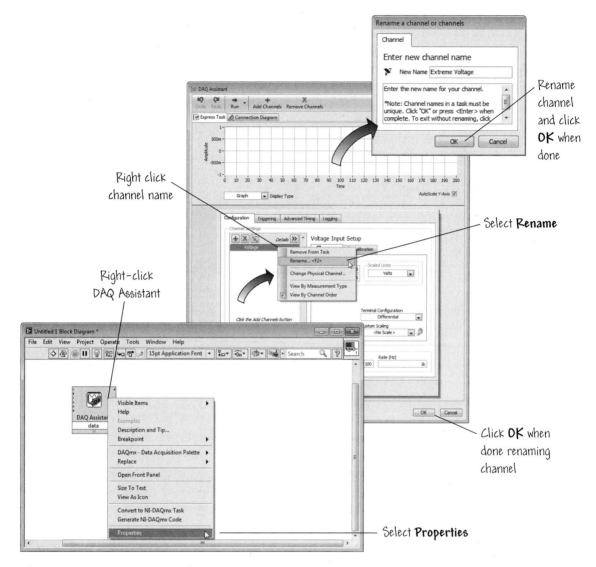

FIGURE 8.44
Renaming a channel.

the DAQmx Task Name Constant on the **Functions≫Measurement I/O≫NI-DAQmx Data Acquisition** palette, as shown in Figure 8.46. The task names can be entered in two different ways. You can use the **Operating** tool to click on the DAQmx Task Name Constant and choose the desired task name found in the MAX utility, as shown in Figure 8.46. Another way to enter the DAQ task name is to use the **Labeling** tool to enter the task channel number. The task name must refer to one of the tasks configured using MAX, as discussed in the previous sections.

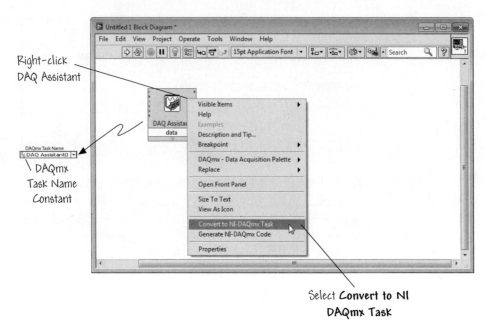

FIGURE 8.45
Create a DAQmx Task Name Constant to make a DAQmx task available to other applications through MAX.

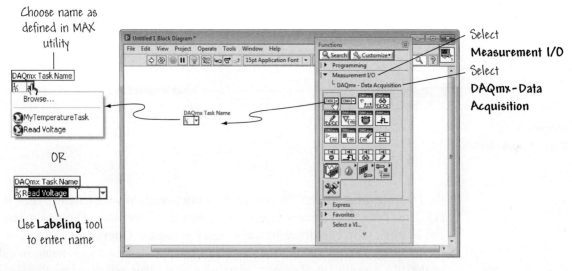

FIGURE 8.46
Entering DAQmx task names.

8.10   ANALOG INPUT

On the first screen of the DAQ Assistant you choose whether to acquire or generate a signal. From these menu options you can choose the measurement type. The available types are analog, counter and digital input and output, and TEDS. In this section we focus on analog input. In subsequent sections we will cover analog output and digital I/O.

The analog input is used to perform analog-to-digital (A/D) conversions. In the DAQ Assistant, clicking on **Analog Input** opens the screen that lists the available analog input measurement types: voltage, temperature, strain, current, resistance, frequency, position, acceleration, custom voltage with excitation, and sound pressure. In the *Practice with the DAQ Assistant* example in Section 8.9, the voltage measurement type was selected. Each measurement type has its own characteristics necessary for turning the acquired voltage to useful data, such as the resistor value for a current measurement or a strain gauge parameter for a strain measurement.

Once you have selected the channels to add to your task, the DAQ Assistant settings and testing screen opens. The DAQ Assistant is shown in Figure 8.42. In the center section of the DAQ Assistant is the **Advanced Timing** box, where you can vary the type of sampling and rate, and the **Triggering** box, where you can configure the start and reference triggers. After setting the signal input range, terminal configuration, and scaling for each channel, the next step is to configure the timing and triggering.

Only one analog input task can run at a time on a single Multifunction DAQ device, although a single task can acquire data on multiple channels.

8.10.1   Task Timing

When performing analog input, the task can be timed to acquire a single sample (on demand), a single sample (hardware timed), *n* samples, or to acquire data continuously. When **1 Sample (On Demand)** is selected (see Figure 8.42), NI-DAQmx acquires one value from an input channel and immediately returns the value to your application. This operation does not require any buffering or hardware timing. Timing is defined by the rate at which the DAQ Assistant runs. For example, if you periodically monitor the fluid level in a tank, you would acquire single data points. You can connect the transducer that produces a voltage representing the fluid level to a single channel on the measurement device and initiate a single-channel, single-point acquisition when you want to know the fluid level.

It is possible to acquire multiple samples for the channels in your task by executing the VI in a repetitive manner. This process is only suggested for measuring static or slow changing signals and point-by-point control applications because the time between samples may vary or it may take too much time to accurately capture your signal. The preferable way to acquire multiple samples is to use hardware timing. This method employs a buffer on the DAQ device and in computer memory to acquire data more efficiently and with very accurate timing. For these types of applications, set the sample mode to **N Samples**. With NI-DAQmx, you can acquire multiple samples on a single channel or multiple samples on multiple channels. For example, you can monitor both the fluid level and the temperature in a tank using two transducers connected to two channels on the device. Each time you execute the DAQ Assistant, *n* samples of data are acquired at the specified rate.

If you want to view, process, or log large amounts of data for more than one iteration of *n* samples, you need to continually acquire samples. For these types of applications, set the sample mode to **Continuous**. In this configuration the DAQ Assistant would typically be placed inside a loop and would transfer small buffers of data back to the computer each time the loop iterate, with no lost data between buffers.

8.10.2   Task Triggering

So far we have discussed the acquisition of data that begins immediately after the DAQ Assistant is run in software. Often it is desirable for the acquisition of data to begin after a stimulus such as detection of a defined voltage level or change in a digital signal. The stimulus used to begin a task is called a **trigger**. The start trigger is used to define when a task starts. The DAQ Assistant provides three start triggers: analog edge, analog window, and digital edge. The reference trigger establishes the reference point (time zero) in a set of input samples. A reference trigger can be set up to acquire data both before and after the trigger is received. Data acquired up to the reference point is known as **pretrigger data**. Data acquired after the reference point is known as **posttrigger data**. Not all data acquisition devices support triggering.

Triggers are only supported when Task Timing is configured to **N Samples**.

Practice with Analog Input

In this example we will construct a VI that acquires an analog signal and displays the output on a meter. For the DAQ device, we are using the DAQ Signal Accessory from National Instruments to produce the temperature measurements. The sensor outputs a voltage proportional to the temperature and is hard-wired to channel 0 of the DAQ device.

If you do not have an NI-DAQ Accessory, you can still follow this example up to the point of actually acquiring temperature data from the device.

To begin the development, open a blank VI and place a Meter on the front panel. The Meter is located on the **Controls≫Silver≫Numeric** palette. Adjust the scales on the meter to read in the range 0.0 to 0.4. Then place a Vertical Toggle Switch on the front panel and name it Power. The Vertical Toggle Switch is located on the **Controls≫Modern≫Boolean** palette. Configure the toggle switch to a default value of False and a mechanical action of **Latch When Pressed**. Create two free labels titled Off and On. The process of building the front panel is illustrated in Figure 8.47. The needed controls and indicators should be on the front panel before continuing to the block diagram. When the front panel is ready, switch to the block diagram to continue.

On the block diagram, place a While Loop and enlarge the loop to accommodate several programming elements. Then place a DAQ Assistant (see Section 8.9 for instructions on using the DAQ Assistant) inside the loop. When the DAQ Assistant appears on the screen, proceed to configure a DAQmx task

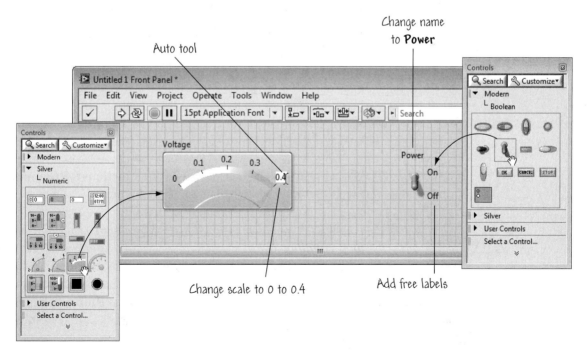

FIGURE 8.47
Configuring the front panel of the temperature acquisition VI.

to read an analog input channel and return the voltage. The steps to accomplish this are illustrated in Figure 8.48. In brief, the steps are as follows:

1. Select Acquire Signals≫Analog Input≫Voltage as the measurement type.
2. Select Dev1≫ai0 as the physical channel.
3. Click the **Finish** button.
4. When the **Analog Input Voltage Task** dialog box appears, configure the **Task Timing** to **1 Sample (On Demand)**.
5. Click the **OK** button to close the **Analog Input Voltage Task Configuration** dialog box.

The settings specified for the task are now saved in the DAQ Assistant. At this point, we need to add several programming elements to the block diagram to complete the VI. First, we want the loop to execute every 100 ms. To accomplish this goal, place a Time Delay function on the block diagram. This function is located on the **Functions≫Programming≫Timing** palette. When the Time Delay function is placed on the block diagram, a dialog box will appear in which the time delay value is set. The default value is a 1 s delay—change the delay to 0.1 s.

It is recommended that you include the capability to stop the program execution in the case where the DAQ Assistant produces an error. The DAQ Assistant outputs include a status indicator that is accessible by the Unbundle by Name function, located on the **Functions≫Programming≫Cluster & Variant** palette. This function can be used to access the **status** from the error cluster when wired to the DAQ Assistant as shown in Figure 8.49. You can also wire the error wire directly to the Or gate—an option that does not require the Unbundle by Name function.

You will also want to be able to halt the program execution using the toggle switch on the front panel. Actually, you will need to be able to halt program execution if an error occurs *or* if the user clicks the power switch (i.e., the toggle switch). To accomplish this task, place the Or function on the block diagram inside the While Loop. The Or function is located on the **Functions≫Programming≫ Boolean** palette. Wire the block diagram using Figure 8.49 as a guide.

*If you located the While Loop on the **Functions≫Express≫Execution Control** palette, the **loop condition** will be automatically wired to a **Stop** button. Delete the **Stop** button and wire the toggle switch instead.*

When you have finished wiring the block diagram and the **Run** button indicates that the program is ready for execution, save the VI as AI Single Point.vi in the Users Stuff folder in the Learning directory.

Return to the front panel and run the VI. The meter will display the voltage and the temperature sensor outputs. Stop the execution of the VI by clicking the power switch. ◆

Place DAQ Assistant on block diagram

Meter

While Loop

Toggle Switch

Select **Analog Input**

Select **Voltage**

Select **Dev1 >> ai0**

Click here when done

Choose **1 Sample (On Demand)**

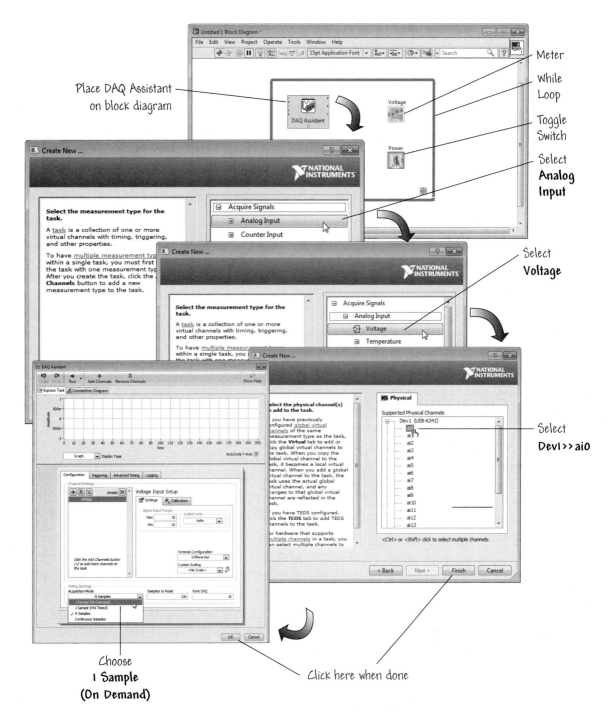

FIGURE 8.48
Configuring the DAQmx task using the DAQ Assistant.

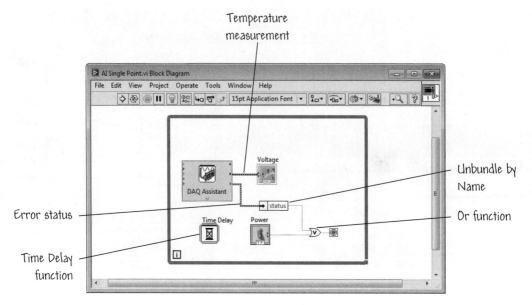

FIGURE 8.49
Acquiring analog input one point at a time.

8.11   ANALOG OUTPUT

Analog output is used to perform digital-to-analog (D/A) conversions. The first screen of the DAQ Assistant is where you choose the analog output measurement type. Refer to Section 8.9 for an introduction to using the DAQ Assistant. You can also configure a DAQmx analog output task using MAX as described in Section 8.8. We will focus on the use of the DAQ Assistant in this section, but actually the process of configuring the DAQmx task is almost the same using the DAQ Assistant as it is with MAX.

Configuring a DAQmx task to perform analog output consists of the same basic steps as for the analog input (see Section 8.10). In the DAQ Assistant, clicking on **Generate Signals** and then selecting **Analog Output** opens a list that shows the available analog output signal generation types: voltage and current.

A compatible device must be installed that can generate a voltage or current task.

Once you have selected the channels to add to your task, the DAQ Assistant settings and testing screen opens, as illustrated in Figure 8.50. In the center section of the DAQ Assistant are the **Advanced Timing** box and the **Triggering**

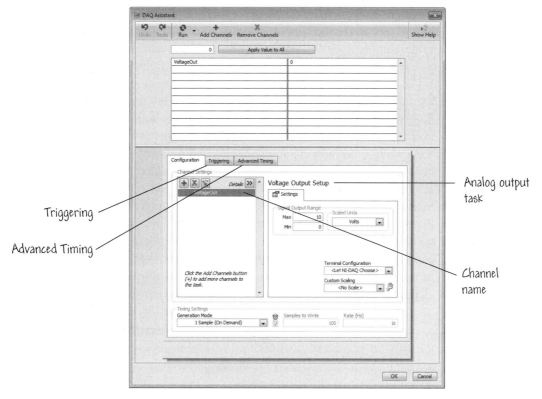

FIGURE 8.50
DAQ Assistant for the analog output DAQmx task.

box. This is where you can vary the type of sampling and rate and configure the start and reference triggers. After setting the Signal Output Range, the next step in configuring the DAQmx task is to configure the timing and triggering.

 Only one analog output task can run at a time on a single Multifunction DAQ device.

8.11.1   Task Timing

The analog output task can be timed to generate a single sample, *n* samples, or continuous samples. How do you decide when to use single-sample timing? If the signal level is more important than the sample generation rate, then choose **1 Sample (On Demand)**. Each time the DAQ Assistant runs, the output value on the analog output channel(s) will update. This method is used when a constant signal, or slowly changing signal is needed. You can implement timing in Lab-VIEW to control the time at which signal value(s) are updated. Generating a single sample does not require any buffering or hardware timing.

You should use **N Samples** if you want to generate a *finite* time-varying signal, such as a 5-second AC sine wave. As discussed in Section 8.10 for analog input, one way to generate multiple samples (representing, for example, a time-varying signal) for one or more channels is to generate single samples in a repetitive manner. Just as before, **1 Sample** is only suggested for generating static or slowly changing signals and point-by-point control applications because the time between samples may vary or take too much time to generate to actually recreate your signal. The best way to generate a time-varying signal is to use hardware timing that uses a buffer in computer and DAQ device memory to generate samples timed more accurately. With hardware timing, a TTL signal, such as a clock on the device, controls the rate of generation. A hardware clock can run much faster and is more accurate than a software loop. As with other functions, you can generate multiple samples on a single channel or multiple channels.

Some devices do not support hardware timing for analog output, such as the NI USB-6008 and USB-6009. Consult the device documentation if you are unsure if the device supports hardware timing.

Continuous sample generation is similar to generating *n* samples, except that multiple buffers or arrays of data can be output without missing any samples between buffers. When performing continuous sample generation, an event must occur to stop the sample generation. If you want to continuously generate signals, such as generating a non-finite signal, set the timing mode to **Continuous**.

8.11.2   Task Triggering

As discussed in Section 8.11.1, two very common output tasks are generating *n* samples and generating continuous samples. Just as with analog input, these tasks can be initiated by triggers. The start trigger starts the generation. The DAQ Assistant provides three start triggers: analog edge, analog window, and digital edge. The reference trigger is not supported for analog output tasks.

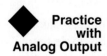

Practice with Analog Output

In this example, we develop a VI that outputs an analog voltage using a DAQ device. The VI will output the voltage from 0 to 9.5 V in 0.5 V increments. For the DAQ device, we are using the DAQ Signal Accessory from National Instruments. The goal is to use the VI developed in this example to output an analog signal to the DAQ device, and to use the analog input VI developed in Section 8.10 to read the data and display the result on a meter.

 *If you have a DAQ Signal Accessory, connect **Analog Out CH0** to **Analog In CH1**. If you do not have a DAQ Signal Accessory, you can still follow this example up to the point of actually generating analog output to the DAQ device.*

In this example, we begin the process of designing a VI to output analog data by starting with a partially completed VI. Open the Voltage Output VI located in the **Chapter 8** folder in the **Learning** directory. You will find the front panel and block diagram shown in Figure 8.51.

On the block diagram there are two interesting functions: the Time Delay function and the Select function. The Time Delay function is configured to cause the For Loop to execute every 500 ms. The Select function checks whether the loop is in its last iteration, and if so, then the DAQ device outputs 0 volts. It is a good idea to reset the output voltage to a known level to prevent damage to connected devices.

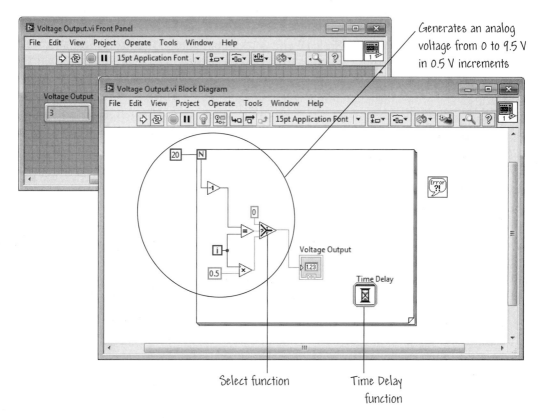

FIGURE 8.51
Generating a voltage output from 0 to 9.5 V in 0.5 V increments.

On the block diagram, place the DAQ Assistant, as illustrated in Figure 8.52. When the DAQ Assistant appears, configure the DAQmx task to generate an analog output voltage. The following steps (shown in Figure 8.53) will produce the desired task:

- Select Generate Signals≫Analog Output≫Voltage on the first DAQ Assistant dialog box.

- Select Dev1≫ao0 as the physical channel and click the **Finish** button.

- In the **Analog Output Voltage Task Configuration** dialog box that appears, configure the **Task Timing** to **1 Sample (On Demand)** and change the output range minimum to 0 and maximum to 10.

- Click the **OK** button to close the **Analog Output Voltage Task Configuration** dialog box. This saves the settings specified for the task in the DAQ Assistant.

Once you are back on the block diagram, wire the DAQ Assistant as shown in Figure 8.54. Save the VI as Voltage Output Done.vi in the Users Stuff folder in the Learning directory.

Open the AI Single Point VI that you completed in Section 8.10. Configure the meter scale minimum to 0.0 and maximum to 10.0. On the block diagram of the AI Single Point VI, double-click the DAQ Assistant to open the **Analog Input Voltage Task Configuration** dialog box. Right-click **Voltage**

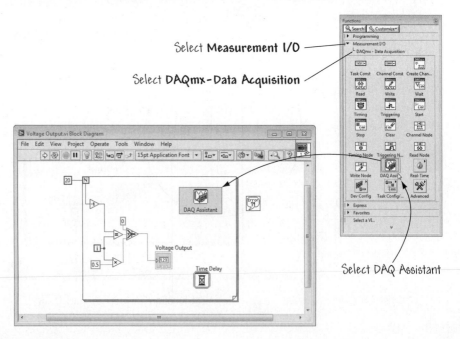

FIGURE 8.52
Placing the DAQ Assistant on the block diagram.

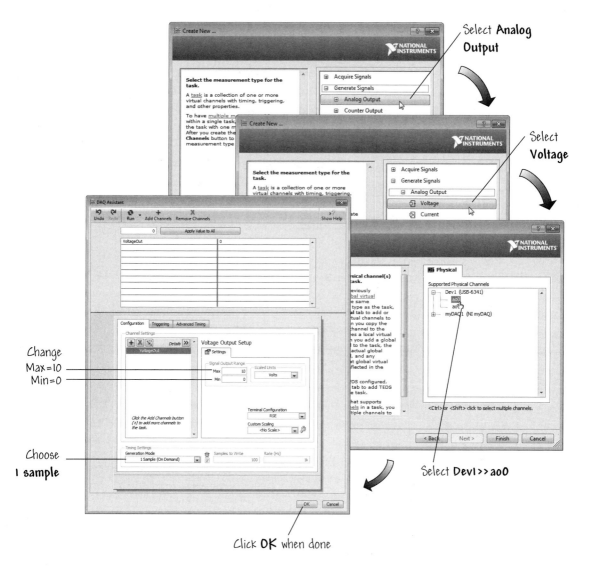

FIGURE 8.53
Configuring the analog output DAQmx task.

in the channel list section and select **Change Physical Channel** as illustrated in Figure 8.55. Select **ai1** for the channel because you wired the DAQ Signal Accessory to output a voltage on **Analog Out CH0** and acquire the voltage from **Analog In CH1**. Change the voltage range to 0 to 10 and then click the **OK** button to close the dialog box.

Run the AI Single Point VI. Recall that this VI will run until there is an error or the toggle switch is used to stop the execution of the program. Once the AI Single Point VI is executing, run the Voltage Output Done VI. Notice that the Voltage Output Done VI outputs the voltage in 0.5 V increments from

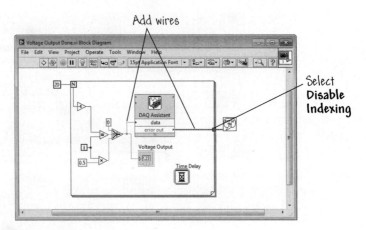

FIGURE 8.54
Wiring the DAQ Assistant.

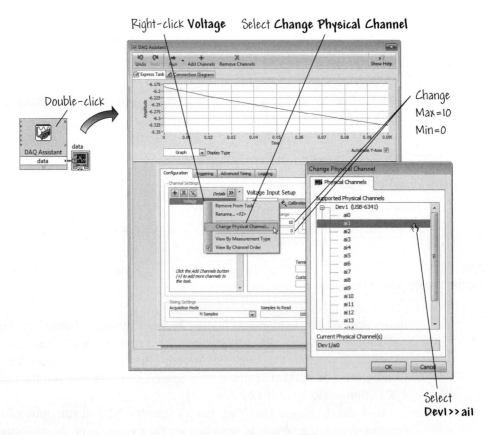

FIGURE 8.55
Reconfiguring the DAQmx task to read from channel ai1.

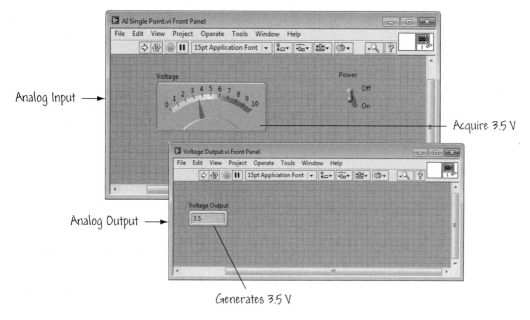

FIGURE 8.56
Generating and acquiring analog signals.

0 to 9.5 V as displayed in the digital indicator in Figure 8.56. At the same time, the meter in the AI Single Point VI displays the voltage readings over the same range. When the For Loop executes its last iteration, the VI outputs 0 V to reset the analog output channel, as desired.

What have we accomplished? The Voltage Output Done VI is generating an analog signal output to the DAQ Accessory. This is a digital-to-analog conversion. The AI Single Point VI is acquiring the signal on the DAQ Accessory and displaying the data on its meter. This is an analog-to-digital conversion. ◆

8.12 DIGITAL INPUT AND OUTPUT

Most multifunction DAQ devices have digital inputs and outputs (I/O). A digital channel or line can generally be configured as an input or output before or while the line is being used. Digital I/O can generally be thought of as analog I/O with only two possible voltage values. Digital I/O hardware generates or accepts binary on/off signals. For example, TTL is one standard of binary signal where on = 5 volts and off = 0 volts. In addition to on/off binary values, you may also find 1/0, high/low, and 5 V/0 V are used to define the binary signal interchangeably. These signals are often used to control processes, generate patterns for testing, and communicate with peripheral equipment.

The binary value of individual digital lines can be written or read programmatically using the DAQ Assistant. Each line corresponds to a channel in the task. Digital lines are often used in groups of four or eight and are referred to as a port. All lines in a port are generally configured for the same function, either input or output. Since a port contains multiple digital lines, by writing to or reading from a port, you can set or retrieve the states of multiple lines simultaneously.

The DAQ Assistant can also be used to read or write port values using an 8-bit integer or an array of Booleans. When using *polled* I/O, software timing is used to decide when to read or write a value to a line or port by executing the DAQ Assistant. Each port corresponds to a channel in the task.

Most M Series devices and other advanced digital I/O devices are capable of hardware-timed digital I/O signal generation and acquisition. These devices can create arbitrary digital waveforms and acquire digital signals at very high speeds.

Practice with Digital I/O

In this example, we plan to control the digital I/O lines on the DAQ device. As in previous examples in Section 8.10 and 8.11, we are using the DAQ Signal Accessory from National Instruments. Our final VI will turn on the LEDs of port 0 on the DAQ Signal Accessory based on the digital value set on the front panel. Each LED is wired to a digital line on the DAQ device. The lines are numbered 0, 1, 2, and 3, starting with the LED on the right.

The LEDs on the DAQ Signal Accessory use negative logic. That is, writing a 1 to the LED digital line turns off the LED. Writing a 0 to the LED digital line turns on the LED.

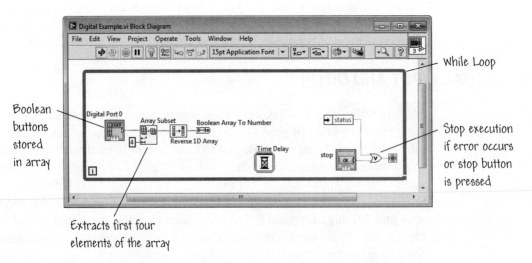

FIGURE 8.57
Digital I/O example block diagram.

Open the Digital Example VI, located in the **Chapter 8** folder in the **Learning** directory. You will find the block diagram shown in Figure 8.57.

Place the DAQ Assistant on the block diagram within the While Loop. When the DAQ Assistant dialog box appears, configure the DAQmx task for Digital I/O. The following steps (shown in Figure 8.58) will produce the desired task:

- Select **Generate Signals≫Digital Output≫Port Output**.

- Select **Dev1≫port0** for the physical channel and click the **Finish** button.

- In the **Digital Port Output Setup** dialog box that appears, select **Invert All Lines In Port** because the LEDs use negative logic.

- Click the **OK** button to close the configuration dialog box.

All of the settings specified for the task are saved internally in the DAQ Assistant. The Boolean buttons on the front panel are stored in an array to

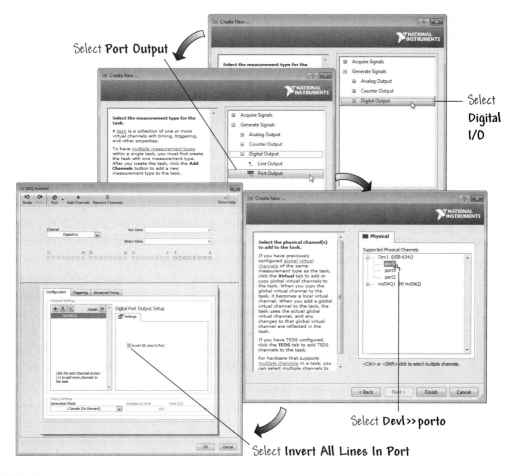

FIGURE 8.58
Configuring the DAQmx task for digital I/O.

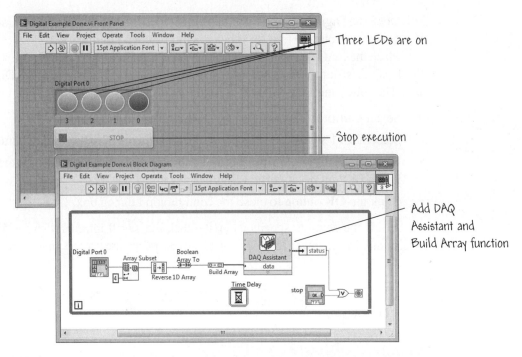

Three LEDs are on

Stop execution

Add DAQ
Assistant and
Build Array function

FIGURE 8.59
The Digital Example VI completed.

simplify the code. The Array Subset function extracts only the first four elements in the array. The output of the array subset needs to be reversed since element 0 of the array is the most significant bit. The array is then converted to a number with the Boolean Array To Number function, which passes the output to the DAQ Assistant to write that value to the port.

Once you are back on the block diagram, wire the DAQ Assistant as shown in Figure 8.59. When the VI is ready, click the **Run** button on the front panel. Turn the Boolean LEDs on and off on the front panel and observe that the LEDs on the DAQ Signal Accessory turn on and off at the same time and in the same order. You are now performing digital I/O!

Save the VI as Digital Example Done.vi in the Users Stuff folder in the Learning directory.   ◆

More Practice with Digital I/O

In this example our goal is to create a VI that reads in a digital input signal then displays it in binary on LEDs. We will generate the input signal using a counter, which is a digital timing device. Counters are typically used for event counting, frequency measurement, period measurement, position measurement, and pulse generation. A counter contains the following four main components:

- **Count Register**—Stores the current count of the counter. You can query the count register with software.

- **Source**—An input signal that can change the current count stored in the count register. The counter looks for rising or falling edges on the source signal. Whether a rising or falling edge changes the count is software selectable. The type of edge selected is referred to as the active edge of the signal. When an active edge is received on the source signal, the count changes. Whether an active edge increments or decrements is also software selectable.

- **Gate**—An input signal that determines if an active edge on the source will change the count. Counting can occur when the gate is high, low, or between various combinations of rising and falling edges. Gate settings are made in software.

- **Output**—An output signal that generates pulses or a series of pulses, otherwise known as a pulse train.

As in the previous example, we will again use the DAQ Signal Accessory. The Signal Accessory has a quadrature encoder which produces pulses as the knob is rotated. Start by connecting the output of Quadrature Encoder A to the source of Counter 0. Now open a blank VI and place the DAQ Assistant on your block diagram.

When the configuration window appears, select Generate Signals≫Counter Input≫Edge Count so that the DAQ Assistant will read the value stored in the counter representing the number of edges on the signal from the quadrature encoder. Select **ctr0** as the channel, since you connected the quadrature encoder to Counter 0 on the Signal Accessory. The default settings for task timing and edge count setup are fine. Press **OK** to save these settings. When LabVIEW prompts you to confirm auto loop creation, select **Yes**. This will automatically create a loop that will continue reading the counter output until the user presses Stop on the front panel.

Create a numeric indicator on the **data** output of the DAQ Assistant and display it on the front panel. Your block diagram should look like the one in Figure 8.60.

Run the VI and turn the knob on the Signal Accessory. You should observe that the counter increments each time you turn the knob. Save the VI as Digital IO 1.vi in the Users Stuff folder in the Learning directory.

Return to the block diagram and place a node to convert the waveform data output to a double. Since there are 4 LEDs on the Signal Accessory, we can only display 2^4, or 16, values using the LEDs. Thus we will divide the output by 16 and display the remainder of this operation on the LEDs. Utilize the Quotient & Remainder function to perform this task. As the counter continues

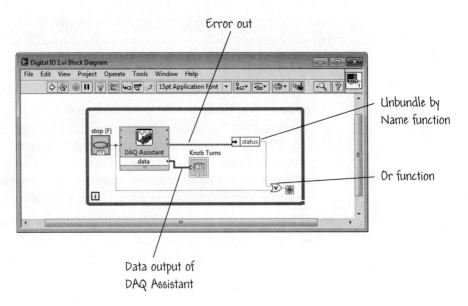

FIGURE 8.60
The Digital 1 I/O VI block diagram.

incrementing by one, the result of this division will increment from 0 up to 15 and then wrap back around to 0. Use the Build Array function to convert the remainder from the division to an array so that it can be a data input to the DAQ Assistant.

Now place a second DAQ Assistant on the block diagram. This time, select **Generate Signals≫Digital Output≫Port Output** as the measurement type. Choose **port0** as the channel, since this is the port the LEDs are on. The default configuration settings are fine and do not need to be updated. Connect the array containing the output value to be displayed on the LEDs to the data input on the second DAQ Assistant. Rewire the error cluster so that it passes through the second DAQ Assistant before it reaches the stop terminal. Your block diagram should now resemble the one seen in Figure 8.61.

Now run your VI and turn the knob on the Signal Accessory. The LEDs on digital port 0 should count from 0–15 in binary. You have now created a VI that performs digital input (from the counter) as well as digital output (to the LEDs).

Save the VI as Digital IO 2.vi in the Users Stuff folder in the Learning directory.

Working versions of Digital IO 1.vi *and* Digital IO 2.vi *can be found in* Chapter 8 *of the* Learning *directory.*   ◆

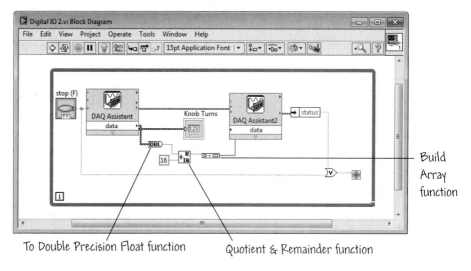

To Double Precision Float function

Quotient & Remainder function

Build Array function

FIGURE 8.61
The Digital I/O 2 VI block diagram.

 BUILDING BLOCK

8.13   BUILDING BLOCKS: PULSE WIDTH MODULATION

In this Building Block, you will develop a VI that outputs a pulse width modulated digital signal to a data acquisition device using the DAQ Assistant. You
will also investigate a more robust approach to generating a pulse width modulated output on a DAQ device using an existing LabVIEW example. To begin,
open the VI you saved as **PWM with Chart.vi** in Chapter 7 and save a new
copy of the VI as **PWM with Digital Output.vi** in the Users Stuff folder in the
Learning directory.

*In case you do not have a working version of the Building Blocks VI from
Chapter 7, a working version of the PWM with Chart VI can be found in the
Building Blocks folder in the Learning directory.*

The Boolean values are currently being written to the waveform chart. You
should also output the Boolean values to a single digital output line. To accomplish this, add the DAQ Assistant inside the While Loop and configure it as
Generate Signals≫Digital Output≫Line Output on **port0/line0**. Next, wire
the Boolean data to the DAQ Assistant. When performing digital output, the
DAQ Assistant expects an array of Boolean values. Therefore, you will need to
add a Build Array function to convert the single Boolean value to an array of

Boolean values. Use the block diagram in Figure 8.62 as a guide to constructing your VI.

*You must have either a DAQ device or a simulated DAQ device with a digital output in order to select the **port0/line0** from the DAQ Assistant.*

8.13.1   Generating Pulse Width Modulated Signals with Hardware Counters

In these past exercises, the sampling rate of the signal has been 100 Hz (10 milliseconds) and is determined by software timing. The maximum possible software timed rate in LabVIEW is 1000 Hz (1 millisecond). This is a limitation of the operating system and not a LabVIEW limitation. The 1000 Hz output rate is not sufficient for many applications.

Most National Instruments data acquisition devices have two built-in counters/timers which are basically digital lines combined with a memory register that can count or generate rising or falling edges from digital clock signals and

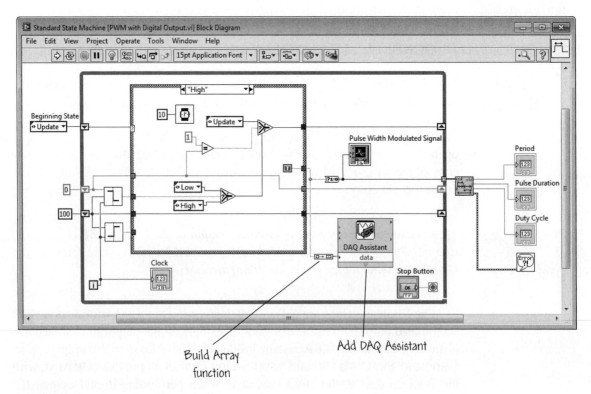

Build Array
function

Add DAQ Assistant

FIGURE 8.62
The PWM with Digital Output VI block diagram.

other digital sources. Because the feature is implemented in hardware, its timing is based on the hardware specifications and not the speed of the computer.

8.13.2   Applications of Pulse Width Modulation

Pulse width modulation is commonly used for motor control and to specify servo position in remote-controlled toy hobby-sized vehicles, such as cars, trucks, and planes. The signal generated by a low current output, such as a data acquisition device or microcontroller, is used to switch a higher current circuit, usually involving a transistor or relay. The higher current is used to actually drive the device. For motor control, the higher the duty cycle, the faster the motor turns, and the lower the duty cycle, the slower the motor turns.

A working version of the PWM with Digital Output VI can be found in the Building Block *folder of the* Learning *directory.*

8.14   RELAXED READING: REDUCING SEISMIC RISK FOR AN ANCIENT ROMAN AMPHITHEATER

In this reading we explore building a distributed monitoring system to identify structural damage and reduce seismic risk of the Arena di Verona—one of the largest open-air opera theaters in the world built in the first century. The challenge is to create a system to communicate with existing measurement sensors in the Arena di Verona that continuously monitors signals from sensor devices, tracks changes in a crack and the onset of structural vibrations related to environmental changes.

The purpose of the structural health monitoring system is to continuously observe static and dynamic measurements from distributed sensors inside the Arena di Verona. The structural health monitoring system is depicted in Figure 8.63(a). The system is composed of ten data acquisition units with small hardware for minimal visual impact in the Roman amphitheater shown in Figure 8.63(b). A monitoring PC was placed at the ticket office entrance. NI hardware and the supervisory Structural-X software—which was developed using LabVIEW—is used to monitor the structural health of the building by acquiring data from uniaxial accelerometers that analyze the vibration characteristics of the structure, potentiometer-type displacement sensors that check the status of cracks and fissures, and environmental sensors that check temperature and humidity at certain points of the structure.

(a)   (b)

FIGURE 8.63
(a) The structural health monitoring system in use; (b) the Arena di Verona.

One programmable sensor network module gateway operates as the collection and management point for all data acquisition units. Three NI EtherCAT RIO chassis with C Series modules collect the dynamic data and communicate using a deterministic Ethernet connection. The dynamic measurement stations are composed of an NI EtherCAT chassis with C Series modules to collect the data. The measurements made by each box varies—the first box measures four accelerometers; the second box measures four accelerometers, two potentiometers, one temperature sensor, and one humidity sensor; and the third box measures eight accelerometers, two potentiometers, one temperature sensor, and one humidity sensor. A PC that monitors data acquisition using the Structural-X supervisory software application is accessible through an Internet connection from remote workstations.

The six wireless sensor network modules collect slow-changing data such as temperature, humidity, and displacement, and communicate wirelessly with the master box. Four modules acquire data from four other potentiometers. The nodes transmit acquired sensor data over an IEEE 802.15.4 wireless network.

Inside each box are other devices such as a 24 V DC power supply, electrical safeguards, an uninterruptible power supply and measurement conditioning for potentiometric sensors. The master box houses the NI 9792 that is used as the master data acquisition controller. The NI 9792 is a programmable WSN gateway that manages the NI WSN nodes on the wireless IEEE 802.15.4 network and the network of NI 9144 EtherCAT chassis: and features centralized data collection and management functions. The master box is also home to a 24 V DC power supply, electrical safeguards, an uninterruptible power supply, and an Ethernet hub. The supervisory PC is a Panel PC placed at the ticket box office for data display, storage, and box configuration.

The Structural-X software application is comprised of two software applications. The first application, known as EMBEDDED, runs on the real-time master and the second application, known as CLIENT, runs on the supervisory panel PC. Structural-X EMBEDDED collects data from all the devices of the system and publishes data on the network. Structural-X CLIENT verifies conversion parameters of the sensors; configures acquisition and storage parameters; and compensates for offset measurements. The software helps the user check the system's status and the graphs of acquired measurements in the time and frequency domains. All data is stored locally with data storage frequency set during configuration and is accessible through an Internet connection from remote workstations.

In summary, by combining the wireless sensor network nodes with the NI CompactRIO EtherCAT system, a structural health monitoring system was created that can acquire, process, and share both static and synchronized dynamic data, and also minimize the installation footprint with its distributed architecture. For further information, visit http://sine.ni.com/cs/app/doc/p/id/cs-14919.

8.15 myDAQ BUILDING BLOCKS

The DAQ Assistant Express VI provides an easy-to-use, dialog-based configuration, but in many cases it is not customizable enough for your particular application. In this myDAQ Building Block, you will explore the more customizable DAQmx code behind the Express VIs you have been using up to this point.

Begin by opening the Read Temp SubVI.vi that you worked on in Chapter 4 and saved in the Support VIs folder in the myDAQ Building Blocks folder within the Learning directory.

A working version of Read Temp subVI.vi can be found in the Chapter 4 folder of the myDAQ Building Blocks≫Solutions folder in the Learning directory.

To guide your exploration of the DAQmx code, follow the steps below.

1. On the block diagram of Read Temp subVI.vi, double-click on the DAQ Assistant to open the configuration dialog, as illustrated in Figure 8.64. Explore the settings but do not make any changes. Select **Cancel** to close the dialog when you are finished.

2. Right-click on the DAQ Assistant and select **Generate NI-DAQmx Code**. Figure 8.65 illustrates the VI after the code generates. Notice that the NChan 1Samp VI corresponds with the DAQ Assistant that was configured to get a single sample from two channels. The DAQmx code is separated into three sections: configuration, read, and close. This structure allows the read,

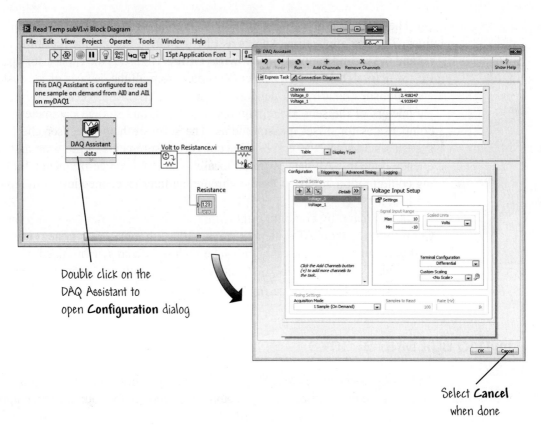

Double click on the
DAQ Assistant to
open **Configuration** dialog

Select **Cancel**
when done

FIGURE 8.64
Configuring the DAQ Assistant by double clicking on the icon.

which will often be performed repeatedly, to be decoupled from the configuration and close, which usually will only need to occur once.

3. Upon generating the DAQmx code, the wiring on the block diagram will be broken. To fix that, wire the output Data of the DAQmx Read.vi to the input myDAQ:AI0, AI1 of the Volt to Resistance.vi as shown in Figure 8.65. Note that the Convert To Dynamic Data function will appear automatically.

4. Most of the DAQmx configuration occurs in the subVI titled Untitled1.vi created during the code generation. Double click this VI and expose the block diagram to examine the channel configuration. Enable the **Context Help** to see the inputs and outputs of each function.

5. Close the VI. Changes will not be used in subsequent myDAQ Building Blocks and may be discarded.

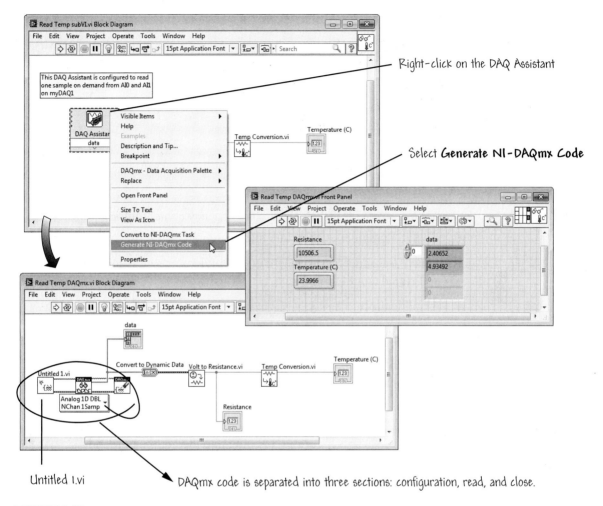

FIGURE 8.65
Configuring the DAQ Assistant by double clicking on the icon.

*A working version of Read Temp DAQmx.vi and the associated Untitled 1.vi can be found in the **Chapter 8** folder of the **myDAQ Building Blocks** folder in the **Learning** directory.*

In this myDAQ Building Block, you used the DAQ Assistant Express VI to provide easy access to a configuration dialog. You also generated NI-DAQmx code from the DAQ Assistant and explored the customization that is possible with the DAQmx functions.

8.16 SUMMARY

DAQ systems consist of the following elements:

- Signals
- Transducers
- Signal-conditioning hardware
- DAQ device or module
- Application software

There are two types of signals:

- Analog—provides level, shape, or frequency content information
- Digital—provides state or rate information

With many transducers it is necessary to provide for signal conditioning. Some types of signal conditioning are

- Amplification
- Transducer excitation
- Linearization
- Isolation
- Filtering

There are two types of signal sources:

- Grounded sources—devices that plug into the building ground
- Floating sources—isolated from the building ground system

There are three types of measurement systems:

- Differential—use this type whenever possible!
- Referenced single-ended—use single-ended types if you require more channels
- Nonreferenced single-ended

Multifunction DAQ devices typically include:

- Analog-to-digital converters (ADCs)
- Digital-to-analog converters (DACs)
- Digital I/O ports
- Counter/timer circuits

When configuring the DAQ device, you should consider how the following parameters will affect the quality of the digitized signal:

- Resolution: Increasing resolution increases the precision of the ADC.
- Device Range: Decreasing range increases precision.
- Signal Input Range: Changing signal input range to reflect the actual signal range increases precision.

The Measurement & Automation Explorer (MAX) is a utility that helps configure the channels on the DAQ device according to the sensors to which they are connected.

The LabVIEW DAQ VIs are organized into palettes corresponding to the type of operation involved—analog input, analog output, counter operations, or digital I/O. Specifically, the DAQ VIs are organized into six palettes (we covered the first three topics in this chapter):

- **Analog Input**
- **Analog Output**
- **Digital I/O**
- **Counter**
- **Calibration and Configuration**
- **Signal Conditioning**

KEY TERMS

ADC resolution: The resolution of the ADC measured in bits. An ADC with 16 bits has a higher resolution (and thus a higher degree of accuracy) than a 12-bit ADC.

Analog-to-digital converter (ADC): An electronic device (often an integrated circuit) that converts an analog voltage to a digital number.

Bipolar: A signal range that includes both positive and negative values (e.g., −5 V to 5 V).

Channel: Pin or wire lead where analog or digital signals enter or leave a data acquisition device.

Channel name: A unique name given to a channel configuration in the Measurement & Automation Explorer or DAQ Assistant.

Code width: The smallest detectable change in an input voltage of a DAQ device.

Data acquisition (DAQ): Process of acquiring data from plug-in devices.

Device Range: The minimum and maximum analog signal levels that the analog-to-digital converter can digitize.

Differential measurement system: A method of configuring your device to read signals in which you do not connect inputs to a fixed reference (such as the earth or a building ground).

Digital-to-analog converter (DAC): An electronic device (often an integrated circuit) that converts a digital number to an analog voltage or current.

Floating signal sources: Signal sources with voltage signals that are not connected to an absolute reference or system ground. *Also* called nonreferenced signal sources.

Grounded signal sources: Signal sources with voltage signals that are referenced to system ground, such as the earth or building ground. *Also* called referenced signal sources.

Handshaked digital I/O: A type of digital I/O where a device accepts or transfers data after a signal pulse has been received. *Also* called latched digital I/O.

Immediate digital I/O: A type of digital I/O where the digital line or port is updated immediately or returns the digital value of an input line. *Also* called nonlatched digital I/O.

Input/output (I/O): The transfer of data to or from a computer system involving communication channels and DAQ interfaces.

Linearization: A type of signal conditioning in which the voltage levels from transducers are linearized, so that the voltages can be scaled to measure physical phenomena.

LSB: Least significant bit.

Measurement & Automation Explorer: Provides access to all National Instruments DAQ and GPIB devices.

Nonreferenced single-ended (NRSE) measurement system: All measurements are made with respect to a common reference, but the voltage at this reference may vary with respect to the measurement system ground.

Referenced single-ended (RSE) measurement system: All measurements are made with respect to a common reference or ground. *Also* called a grounded measurement system.

Sampling rate: The rate at which the DAQ device samples an incoming signal.

SCXI: Signal Conditioning eXtensions for Instrumentation. The National Instruments product line for conditional low-level signals within an external chassis near the sensors so that only high-level signals in a noisy environment are sent to the DAQ device.

Signal conditioning: The manipulation of signals to prepare them for digitizing.

Signal Input Range: The range that you specify as the maximum and minimum voltages on analog input signals.

Unipolar: A signal range that is always positive (e.g., 0 V to 10 V).

Update rate: The rate at which voltage values are generated per second.

EXERCISES

E8.1 Assume you are sampling a transducer that varies between 80 mV and 120 mV. Your device has a voltage range of 0 to +10 V, ±10 V, and ±5 V. What signal input range would you select for maximum precision if you use a DAQ device with 12-bit resolution? Use the signal input range values in Table 8.3 for your calculations.

E8.2 Open the Measurement & Automation Explorer. Locate your DAQ device under Devices and Interfaces≫NI-DAQmx Devices. Right-click on the device and select **Test Panels**. Run a test panel to observe the signals connected to your DAQ device. After running a test panel, go to **Data Neighborhood** and create a DAQmx Global Virtual Channel to measure one of your input signals. Test your DAQmx Global Virtual Channel.

E8.3 Open the NI Example Finder to Search and enter "function." Select Function Generator with FM. To open the **Find Examples**, select Find Examples in the **Help** pull-down menu. This VI is a good example of how virtual instruments emulate actual instruments. Run this VI and vary the inputs to obtain a sawtooth wave signal at 50 Hz and a sampling rate of 500 Hz.

E8.4 Construct a voltmeter VI that reads in a specified number of samples of an analog voltage and displays the input on a waveform graph on the front panel. Connect a sine wave to one of the analog input channels on your DAQ device and configure the DAQ Assistant to acquire n samples.

If you do not have a DAQ device, use the Simulated DAQ as described in Section 8.7.3.

The front panel, illustrated in Figure E8.4, should have two numeric controls allowing the user to programmatically adjust the number of samples acquired and the sampling rate. The block diagram should utilize the DAQ Assistant. You can use the block diagram in Figure E8.4 as a guide to construct the VI. Experiment with different sampling rates and numbers of samples.

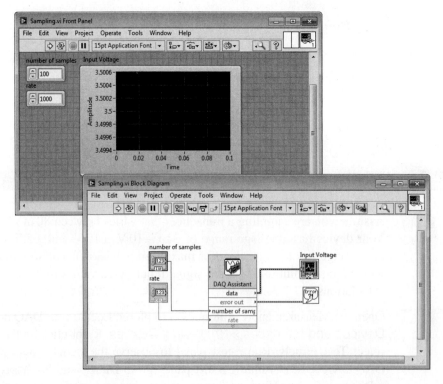

FIGURE E8.4
Voltmeter VI.

PROBLEMS

P8.1 Open a blank VI and place a DAQmx Task Name constant on the block diagram. Right-click the constant and select New NI-DAQmx Task≫Max. Using MAX, configure a voltage analog input named **MyVoltageTask**. After configuring and testing your channel, use the Code Generation feature to generate code for your channel. To do this, right-click on the channel constant and select Generate Code≫Configuration and Example. Run the VI.

P8.2 Suppose that you want to design a VI to acquire analog data measuring the volume of a tank. You only want to take continual measurements when the machinery is on. Turning on the machinery can act as a trigger, specifically a digital trigger, to start measuring data. Construct a VI that acquires analog data only after a digital trigger has occurred.

P8.3 Consider a servo motor that requires a +5 V signal to operate. It is desired that the servo motor be under the control of an operator. Construct a VI such that the operator must set a Boolean toggle switch to True in order to send a +5 V signal to the motor to start the motor. When the operator sets the Boolean switch to False, the motor must stop. In this way, the operator can control the servo motor.

You can use either a digital output line or an analog output line.

P8.4 Construct a VI to monitor a security system that uses a light sensor to determine when a person has entered a restricted zone. When the sensor is activated, it returns a 0 V TTL signal. Write a VI to monitor the state of the light sensor and turn on a Boolean indicator when the sensor is activated.

You can use either a digital output line or an analog output line.

DESIGN PROBLEMS

D8.1 Design a VI which displays an analog voltage input on a numeric indicator with compensation for fluctuations caused by noise spikes. To reduce noise in the measurement, display the average of the 100 most recently acquired data points rather than simply displaying the most recently acquired sample. Your VI should have error handling capabilities.

D8.2 Construct a VI that outputs voltage from 10 V down to 0 V in 0.5 V steps using the DAQ Assistant. Display the output signal you send to the DAQ card on a meter indicator. To verify that the DAQ card is outputting the correct signal, you may wish to connect your DAQ board to an oscilloscope or digital multimeter, or change the settings on the voltmeter VI you created in Section 8.10 so that it can be used to monitor this output voltage.

You can perform both analog output and analog input tasks at the same time, but only one of each.

D8.3 Develop a VI that reads in an analog input voltage and then sets a Boolean output signal if the input voltage exceeds a limit entered by the user on a front panel control. Also display the input voltage on a meter. Use the DAQ Assistant to perform the I/O tasks.

CHAPTER 9

Strings and File I/O

This chapter introduces strings and file input/output (I/O). You have used strings in a limited fashion throughout the book, but here we discuss them more formally. In instrument control applications, numeric data are commonly passed as character strings, and LabVIEW has many built-in string functions that allow you to manipulate the string data. Writing to and reading data from files also utilizes strings. We will discuss how to use File I/O VIs to save data to and retrieve data from a disk file and programmatically find the path to the system directories of the current user. We will also discuss the use of key Express VIs for string manipulation and file I/O.

GOALS

1. Practice creating string controls and indicators.
2. Be able to convert a number to a string, and vice versa.
3. Learn to use the File I/O VIs to write to and read data from a disk file.
4. Understand how to write data to a file in a format compatible with many common spreadsheet applications.

9.1   STRINGS

A **string** is a sequence of characters that can be displayable or nondisplayable. Strings are commonly used in everyday applications, such as **ASCII** text messages. In LabVIEW, strings are also used in instrument control when numeric data is passed as character strings and subsequently converted back to numbers. Another situation requiring the use of strings is in storing numeric data, where numbers are first converted to strings before writing them to a file on disk. In this chapter, we will discuss strings and their various uses in file input/output (I/O).

String controls and indicators are in the **String & Path** subpalettes of the **Controls≫Silver** and **Controls≫Modern** palettes, as illustrated in Figure 9.1. In this textbook, we prefer to use the **Controls≫Silver** palette. They can also be accessed on the **Text Controls** and **Text Indicators** subpalettes of the **Controls≫ Express** palette. As discussed in previous chapters, you enter and change text inside a string control using the **Operating** tool or the **Labeling** tool. If there is not enough room to fit your text in the default size of the string control, you can enlarge the string controls and indicators by dragging a corner with the **Positioning** tool. If front panel space is limited, you can use a scrollbar to minimize the space that a front panel string control or indicator occupies, as illustrated in Figure 9.2. The Visible Items≫Vertical Scrollbar option is located on the string shortcut menu. If there is not enough room to place the scrollbar within the string control or indicator, the scrollbar option will be

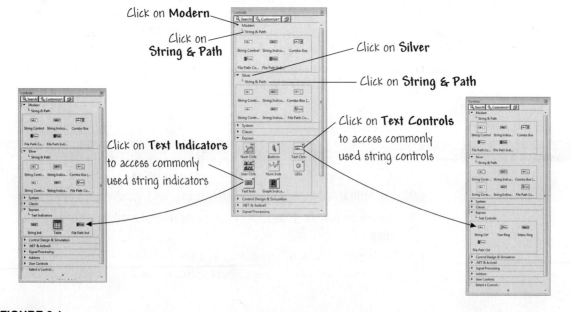

FIGURE 9.1
Accessing string controls and indicators in the **String & Path** subpalettes of the **Controls≫Silver** and **Controls≫Modern** palettes. Common string controls and indicators are also accessible using the **Express** palette.

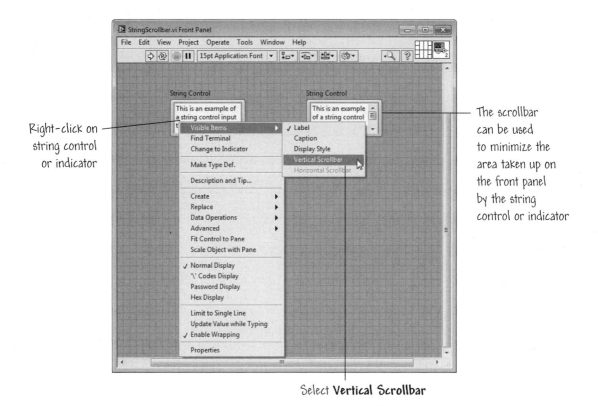

Right-click on string control or indicator

The scrollbar can be used to minimize the area taken up on the front panel by the string control or indicator

Select **Vertical Scrollbar**

FIGURE 9.2
The scrollbar can be used to minimize the size of string controls and indicators.

dimmed. This indicates that you must increase the vertical size if you want a scrollbar. A horizontal scrollbar is also available.

You can configure string controls and indicators for different types of display, such as passwords, \ codes, and hex. String controls and indicators can display and accept characters that are usually **nondisplayable**—backspaces, carriage returns, tabs, and so on. Choose '\' **Codes Display** from the string shortcut menu to display these characters. In the '\' **Codes Display** mode, nondisplayable characters appear as a backslash followed by the appropriate code. A complete list of codes appears in Table 9.1.

To enter a nondisplayable character into a string control, press the appropriate key, like space or <tab>, or type the backslash character \, followed by the code for the character. As shown in Figure 9.3(a), after you type in the string and click the **Enter** button, any nondisplayable characters appear in backslash code format.

The characters in string controls and indicators are represented internally in ASCII format. You can view the ASCII codes in hex by choosing **Hex Display** from the string shortcut menu, as shown in Figure 9.3(b). You can also

TABLE 9.1   A List of Backslash Codes

Code	G Interpretation
\00 – \FF	Hex value of an 8-bit character; must be uppercase.
\b	Backspace (ASCII BS, equivalent to \08)
\f	Form feed (ASCII FF, equivalent to \0C)
\n	Line feed (ASCII LF, equivalent to \0A)
\r	Carriage return (ASCII CR, equivalent to \0D)
\t	Tab (ASCII HT, equivalent to \09)
\s	Space (equivalent to \20)
\\	Backslash (ASCII \, equivalent to \5C)

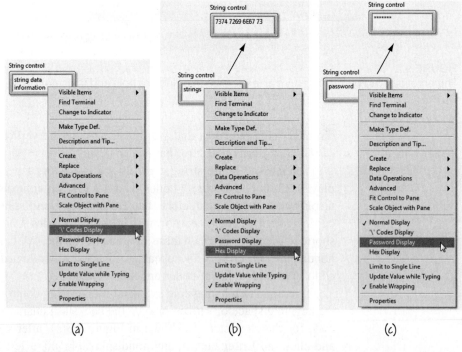

(a)   (b)   (c)

FIGURE 9.3
(a) Displaying characters that are usually nondisplayable; (b) choosing **Hex Display** to view the ASCII codes in hex; (c) choosing the **Password Display** option from the shortcut menu.

choose a password display by enabling the **Password Display** option from the string shortcut menu, as shown in Figure 9.3(c). With this option selected, only asterisks appear in the string front panel display, although on the block diagram the string data reflects the input string. This allows you to set up a security system requiring a password key before proper operation of the VI.

Practice with Manipulating Strings

In this exercise, you will practice with different ways to manipulate strings using three VIs:

- **Format Into String**: Concatenates and formats numbers and strings into a single output string.

- **Scan From String**: Scans a string and converts valid numeric characters (0 to 9, +, −, e, E, and period) to numbers.

- **Match Pattern**: Searches for an expression in a string, beginning at a specified offset, and if it finds a match, splits the string into three substrings—the substring before the matched substring, the matched substring itself, and the substring that follows the matched substring.

The VI that you will create is intended to simulate interaction with a digital multimeter (DMM). Open a new VI and construct a front panel using Figure 9.4 as a guide. Place two string controls, one numerical control, and one string indicator on the front panel. Switch to the block diagram. Select Format Into String in the **String** subpalette, as shown in Figure 9.4. In this exercise, this function converts the number you specify to a string and concatenates the inputs to form a command for the DMM. Using the **Positioning** tool, enlarge Format Into String so that two inputs appear in the lower left corner. Wire the numeric control **Number** to the first input argument (as shown in Figure 9.4). Wire the string control **Units** to the second input argument; the input type for Format Into String will automatically change from **DBL** to **abc** upon wiring. Wire the string control **DMM Command** to the **initial string** input at the top left. Finally, wire the output **resulting string** to the string indicator **Command sent to DMM**.

You can create strings according to a format specified using format strings. With format strings you can specify the format of arguments—the field width, base (hex, octal, and so on), and any text that separates the arguments. The format string can be seen in Figure 9.4 (wired at the top of the Format Into String function). A dialog box can be used to obtain the desired format string. On the block diagram, right-click on the Format Into String function and select **Edit Format String**, as illustrated in Figure 9.5. You can also double-click on the node to access the dialog box. Notice that the **Current format sequence** contains the argument types in the order that you wired them—**Format fractional number** and **Format string**. **Format fractional number** corresponds to the input value from the **Number** control and **Format string** corresponds to the input value from the **Units** control.

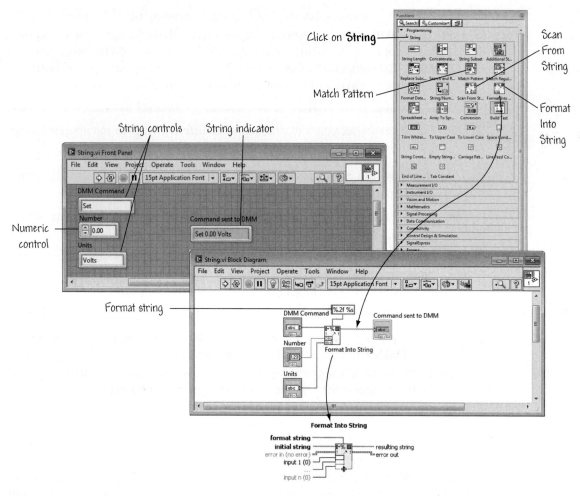

FIGURE 9.4
Constructing a VI using the Format Into String function to build a command to send to the DMM.

The only change we want to make is to set the precision of the numeric to 2. To do this, highlight **Format fractional number** in the **Current format sequence** list box, click the **Use specified precision** check box, and type in the number 2 in the box. When you are finished, press <Enter> (Windows) or <return> (Macintosh) and you should see the **Corresponding format string** (near the bottom of the dialog box). Press the **OK** button to automatically insert the correct format string information and wire the format string to the function, as shown in Figure 9.4.

Return to the front panel and type text inside the two string controls and a number inside the numeric control, and run the VI. You now have a VI that can concatenate strings and numbers to form a single string command that can

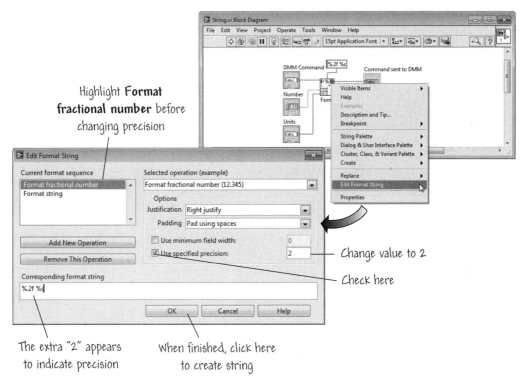

Highlight **Format fractional number** before changing precision

The extra "2" appears to indicate precision

When finished, click here to create string

Change value to 2

Check here

FIGURE 9.5
Setting the desired format string using the dialog box.

be sent to an external instrument, such as the digital multimeter. Save the VI as String.vi in the Users Stuff folder.

To continue the exercise, we want to add the capability to scan a string and convert any valid numeric characters (0 to 9, +, −, e, E, and period) to numbers. Use Figure 9.6 as a guide and add a string control and numeric indicator to the front panel. Then switch to the block diagram and wire the Scan From String function, as shown in Figure 9.6. The function itself is located on the **Programming≫String** palette, as illustrated in Figure 9.4. After the wiring is complete, double-click on the function (or right-click on the function and choose **Edit Format String**) to open the dialog box. The current scan sequence should indicate **Scan number** and the corresponding scan string should show %f. This is fine, so just select **OK** and make sure that the format string input (at the top of the Scan From String function) is properly wired (see Figure 9.6). To test the VI, type 5.56 V in the Present DMM Setting string control and run the VI. You should find that the VI extracts the numeric value 5.56 and displays that value in the numeric indicator DMM Setting - Numeric. Try other voltage values in the input.

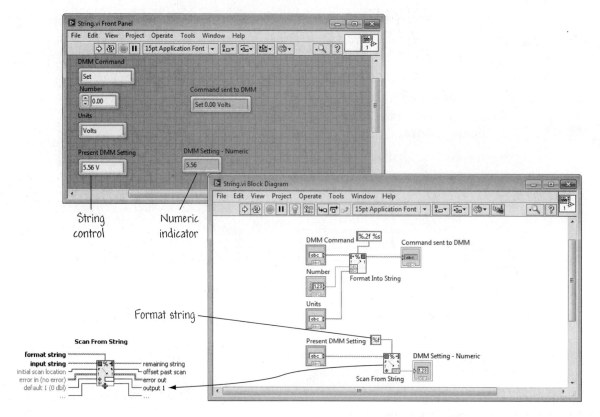

FIGURE 9.6
Constructing a VI using the Scan from String function to extract the present DMM setting from a string.

Now, complete this VI using Figure 9.7 as a guide. The last addition to the VI uses the Match Pattern function in conjunction with the Scan From String function to detect and extract a series of DMM data points. The Match Pattern function detects the matched substring (in this case, a comma) and outputs the substring before the comma to the Scan From String function, which scans the substring and outputs a detected number. The While Loop iterates until the end of the string is detected—in other words, an empty string is found. You can find the empty string constant on the **Strings** palette, as illustrated in Figure 9.4.

When your VI has been wired properly, enter a few numbers in the string control **DMM Data Points**, such as 3.37, 4.56, 6.89, 5.67. Run the VI and verify that the output array **Data Points - Numeric** contains the four numbers.

A working version of **String.vi** *can be found in* **Chapter 9** *of the* **Learning** *directory. Refer to the working version if your VI is not working as you think it should.* ◆

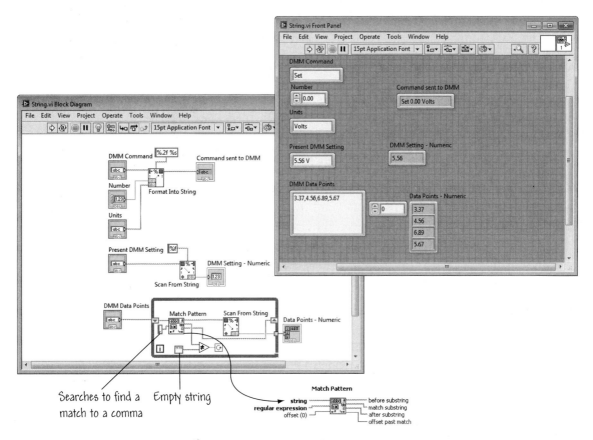

FIGURE 9.7
Constructing a VI using the Match Pattern function to detect and extract a series of DMM data points.

9.1.1   Converting Numeric Values to Strings with Build Text Express VI

The Build Text Express VI can be used to convert numeric values into strings. The Build Text Express VI is located on the **Functions≫Programming≫String** or from the **Functions≫Express≫Output** palette, as shown in Figure 9.8. If the input is not a string, this Express VI converts the input into a string based on the configuration of the Express VI.

When you place the Build Text Express VI on the block diagram, the **Configure Build Text** dialog box appears. The dialog box shown in Figure 9.8 shows the Express VI configured to accept one input, **temperature**, and change it to a fractional. The input temperature concatenates on the end of the string Temperature is. A space has been added to the end of the Temperature is string.

A front panel and block diagram illustrating the use of the Build Text Express VI is shown in Figure 9.9. The Digital Thermometer.vi found in the Activity folder of the Learning directory generates the input temperature data. The

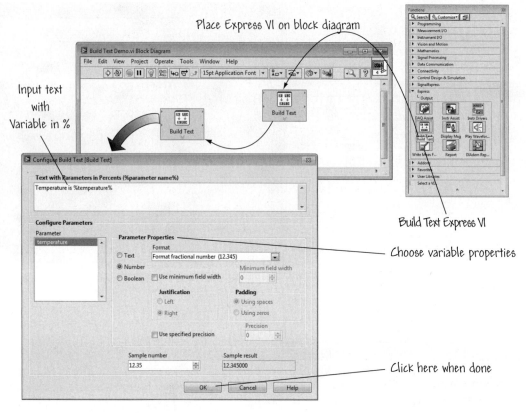

FIGURE 9.8
Using the Build Text Express VI.

output is directed to a string indicator. In the example shown, the displayed temperature is 79.59°.

 The VI shown in Figure 9.9 is called **Build Text Demo.vi** *and is located in* **Chapter 9** *of the* **Learning** *directory.*

9.2   FILE I/O

File I/O operations pass data to and from files. Use the File I/O VIs and functions on the **File I/O** palette to handle all aspects of file I/O, including the following:

- Opening and closing data files.
- Reading data from and writing data to files.

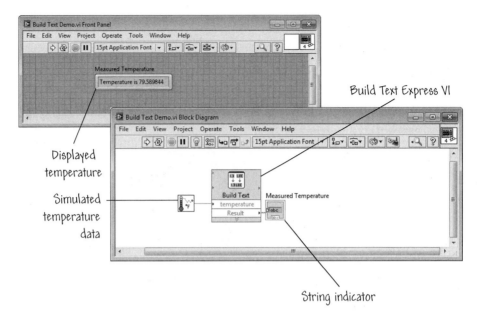

FIGURE 9.9
Converting temperature data to a string.

- Reading from and writing to spreadsheet-formatted files.
- Moving and renaming files and directories.
- Changing file characteristics.

The **File I/O** palette, shown in Figure 9.10, includes VIs and functions designed for common file I/O operations, such as writing to or reading from the following types of data:

- Numeric values to or from spreadsheet text files
- Characters (strings) to or from text files
- Lines from text files
- Data to or from binary files

A typical file I/O operation involves the following process:

1. Create or open a file. Indicate where an existing file resides or where you want to create a new file by specifying a path or responding to a dialog box to direct LabVIEW to the file location. After the file opens, a reference number (refnum) represents the file.
2. Read from or write to the file.
3. Close the file.

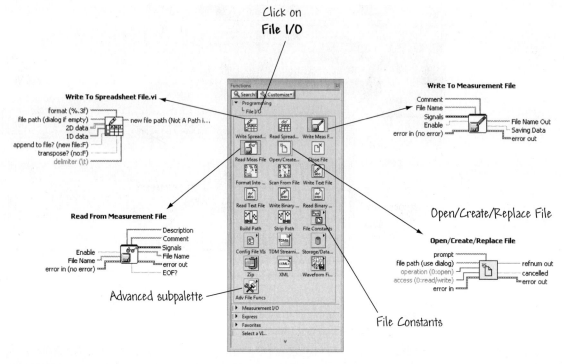

FIGURE 9.10
The **File I/O** palette.

The high-level File I/O VIs are located on the top row of the palette (see Figure 9.10) and can perform all three steps for common file I/O operations. They consist of the following VIs:

- **Write To Spreadsheet File**: Converts a 2D or 1D array of single-precision (SGL) numbers to a text string and writes the string to a new byte stream file or appends the string to an existing file.

- **Read From Spreadsheet File**: Reads a specified number of lines or rows from a numeric text file, beginning at a specified character offset, and converts the data to a 2D, single-precision array of numbers.

- **Write To Measurement File:** Writes data to a text-based measurement file (.lvm) or binary measurement file (.tdm).

- **Read From Measurement File:** Reads data from a text-based measurement file (.lvm) or binary measurement file (.tdm).

The **File I/O** palette also includes functions to control each file I/O operation individually. Use these functions to create or open a file, read data from or write data to a file, and close a file. You also can use them to perform the following tasks:

- Create directories.
- Move, copy, or delete files.
- List directory contents.
- Change file characteristics.
- Manipulate paths.

For example, to create a new file or replace an existing file, you can use the intermediate File I/O function Open/Create/Replace File, as shown in Figure 9.10. To define a path to the file, right-click on the **file path (use dialog)** input on the left side of the function and select **Create Control** to create a File Path control. Once a file is selected and opened, a reference number (refnum) is created, and you can read from or write to the file. Remember to close an open file before your application completes execution. Many of LabVIEW's high-level File I/O VIs automatically open or create new files, read from or write data to those files, and close the files upon completion.

The VIs and functions designed for multiple operations might not be as efficient as the functions configured or designed for individual operations.

You will find a selection of VIs on the **File I/O** palette that work with different data formats. You can read data from or write data to files in three formats—text, binary, and datalog. The format you use depends on the data you acquire or create and the applications that will access that data. Use the following basic guidelines to determine which format to use:

- If you want to make your data available to other applications (such as Microsoft Excel), use text files because they are the most common and the most portable.
- If you need to perform random access file reads or writes or if speed and disk space are crucial, use binary files because they are more efficient than text files in disk space usage and in speed.
- If you want to manipulate complex records of data or different data types in LabVIEW, use datalog files because they are the best way to store data if you intend to access the data only from LabVIEW and you need to store complex data structures.

When dealing with files, you will frequently see the terms **end-of-file**, **refnum**, **not-a-path**, and **not-a-refnum**. The end-of-file (EOF) is the character offset of the end of the file relative to the beginning of the file. Refnum is an identifier that LabVIEW associates with a file when opened. Not-a-path and not-a-refnum are predefined values that indicate that a path is invalid and that a refnum associated with an open file is invalid, respectively.

In the next three subsections we will cover the three most common file I/O operations, including writing to a file continuously, reading from a file, and writing to a spreadsheet file. In the last subsection, we will discuss the use of Express VIs for configuration-based file I/O functionality.

9.2.1   Writing Data to a File

In this section we will discuss writing data to a file using the Write To Text File function. This VI writes a string of characters or an array of strings as lines to a file. The Open/Create/Replace File function is used to create a new file, and the Close File function is used to close the file once all the data is written.

Writing Data to a File

In this exercise, the objective is to create a VI to append temperature data to a file in ASCII format. This VI uses a For Loop to generate temperature values and store them in a file. During each iteration, the VI converts the temperature data from a numeric to a string, adds a comma as a delimiting character, and then appends the string to a file.

Open a new front panel and place the objects as shown in Figure 9.11. The front panel contains a numeric control and a waveform chart. The chart displays the temperature data. Right-click on the waveform chart and select **Visible Items≫Digital Display** and deselect **Visible Items≫Plot Legend**. Also make sure that either the waveform chart has y-axis autoscaling enabled or that the maximum value of the y-axis is set to at least 90 degrees. The **Number of Points** control specifies how many temperature values to acquire and write to file. Right-click on **Number of Points** and choose **Representation≫I32**.

Switch to the block diagram and wire the code as shown in Figure 9.11. Add a For Loop and make it large enough to encompass the various components. Place the **Digital Thermometer.vi** on the block diagram—remember that this VI is located in the **activity** folder of the **Learning** directory, which you can select through **Functions≫Select a VI**, and it returns a simulated temperature measurement from a temperature sensor.

Add the Open/Create/Replace function outside the For Loop. Right-click on the input **operation** and choose **Create≫Constant**. From the list, choose **create**, meaning the VI will prompt the user to create a new file each time the VI is run. This VI then outputs a file refnum, which is passed to the Write To Text File function. Once the For Loop is done executing, the refnum is passed to the Close File function, which closes the file. If you do not close the file, the file stays in memory and is not accessible from other applications or to other users.

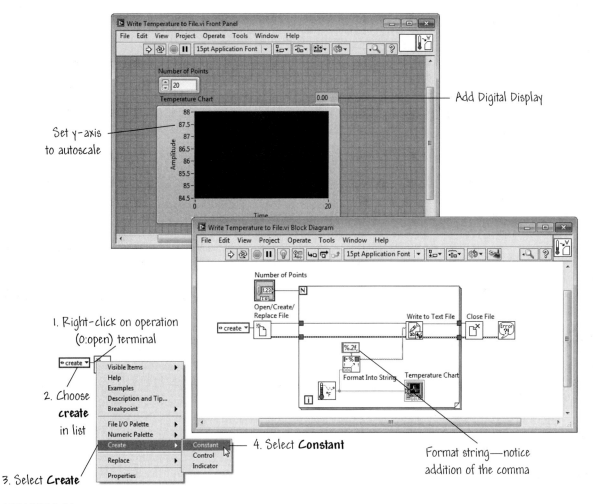

Set y-axis
to autoscale

Add Digital Display

1. Right-click on operation
   (0:open) terminal

2. Choose
   create
   in list

3. Select **Create**

4. Select **Constant**

Format string—notice
addition of the comma

FIGURE 9.11
Using the Write Temperature to File VI to write temperature data to a file.

When you wire the refnum out of the For Loop, auto-indexing is enabled by default. In order to successfully wire the refnum to the Close File function, right-click the tunnel and select **Disable Indexing**.

The Format Into String function is used to convert the temperature measurement (a number) to a string and concatenates the comma that follows it. We need to modify the format string to change the precision to two digits after the decimal point and to add the comma after each data point. The format string can be seen in Figure 9.11 (wired at the top of the Format Into String function). On the block diagram, right-click on the Format Into String function and select **Edit Format String** or double-click on the node to access the dialog

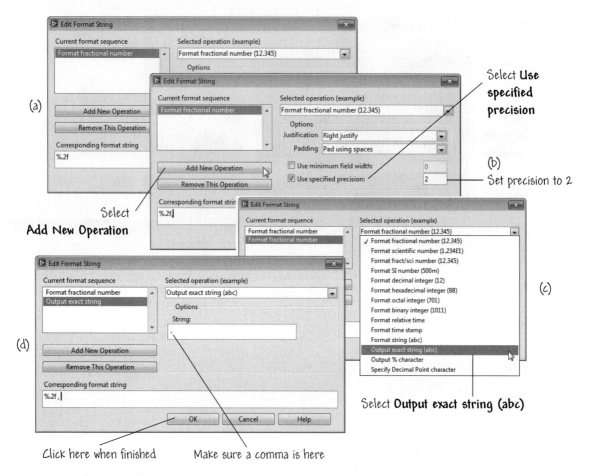

FIGURE 9.12
Setting the desired format string using the dialog box.

box. As shown in Figure 9.12(a), the **Current format sequence** contains **Format fractional number**. The first thing to do is set the precision to 2, as illustrated in Figure 9.12(b). Then select **Add New Operation** and in the **Selected operation** pull-down menu choose **Output exact string (abc)** as shown in Figure 9.12(c). Make sure that the exact string that you input is a comma [as shown in Figure 9.12(d)]. When finished, select **OK** and verify that the string format is "%.2f,"—as desired.

Finish wiring the objects, return to the front panel, and run the VI with the number of points set to 20. A file dialog box prompts you for a file name—select a file name such as Test. When you enter a file name, the VI starts writing the temperature values to that file as each point is generated. Save the VI as Write Temperature to File.vi in the Users Stuff folder. Use any word processing software, such as Notepad for Windows or Teach Text for Macintosh, to open the data file and view the contents. You should get a file containing twenty data values (with a precision of two places after the decimal point) separated by commas.

A working version of Write Temperature to File.vi *can be found in the* Chapter 9 *folder in the* Learning *directory.*   ◆

The VI architecture used in the previous exercise is an example of a disk-streaming operation. Disk streaming is a technique for keeping files open while you perform multiple write operations, for example, within a loop. Typical disk-streaming operations place the Open/Create/Replace File function before a loop, the Read or Write function in the loop, and the Close File function after the loop so continuous writing to a file can occur within the loop without the overhead associated with opening and closing the file in each iteration.

9.2.2   Reading Data from a File

As a natural follow-on to writing data to a file, in this section we discuss reading data from a file using the Read From Text File function. This function reads a specified number of characters or lines from a text byte stream file. By default, this function displays a file dialog box and prompts you to select a file and then closes it afterwards. In the following example, you will get a chance to construct a VI to read the temperature data that you wrote to a file in the previous exercise.

Read Data from a File

The goal here is to construct a VI that reads the temperature data file you wrote in the previous example and displays the data on a waveform graph. When reading from files be careful to read the data in the same data format in which you saved it. In this case, the data was originally saved in ASCII so it must be read in as string data.

Open a new front panel and build the front panel shown in Figure 9.13. You will need to place a string indicator and a waveform graph to display the temperature data that is read.

Build the block diagram as shown in Figure 9.13. The Read From Text File function reads the data from the file and outputs the information in a string. If no path name or refnum is provided, a file dialog prompts you to enter a file name.

The Extract Numbers VI is located in Strings.llb found in general in the examples folder of the LabVIEW directory. This VI takes an ASCII string containing numbers separated by commas, line feeds, or other non-numeric characters and converts them to an array of numbers. It uses the Match Pattern function to convert a spreadsheet string (that is, delimiter-separated columns with end-of-line characters between rows) into an array of numbers (by default) or strings. You can access the Extract Numbers VI using **Functions≫Select a VI**.

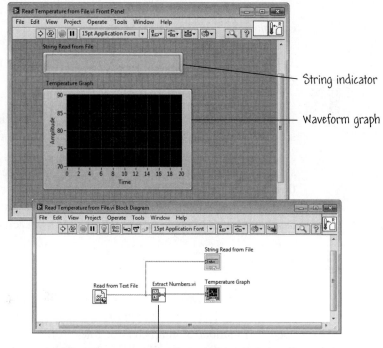

Located in **Strings.llb** found in **general** in the **examples** folder of the **LabVIEW** directory

FIGURE 9.13
Using the Read From Text File function to read temperature data from a file.

Complete the wiring of the block diagram, return to the front panel, and run the VI. When prompted for a file name, select the data file name that contains the temperature data from the previous exercise. You should see the same temperature data values displayed in the graph as you saw in the previous exercise. Save the VI as **Read Temperature from File.vi** in the **Users Stuff** folder and close the VI.

A working version of **Read Temperature from File.vi** *can be found in the* **Chapter 9** *folder in the* **Learning** *directory.*   ◆

9.2.3 Manipulating Spreadsheet Files

In many instances it is useful to be able to open saved data in a spreadsheet. In most spreadsheets, tabs separate columns and EOL (end-of-line) characters separate rows, as shown in Figure 9.14(a). Opening the file using a spreadsheet program yields the table shown in Figure 9.14(b).

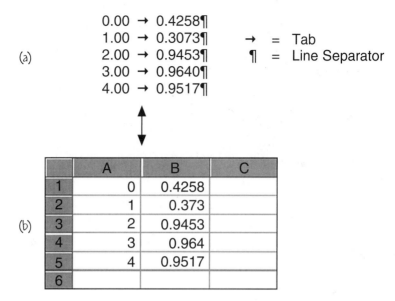

FIGURE 9.14
A common spreadsheet data format.

We will concentrate in this section on writing data to a spreadsheet format, with the idea that you will be accessing that data using a spreadsheet application outside of LabVIEW. However, if you want to read spreadsheet data from within LabVIEW, it is possible to read data in text format from a spreadsheet using the Read From Spreadsheet File VI. This VI reads a specified number of lines or rows from a numeric text file beginning at a specified character offset and converts the data to a 2D, single-precision array of numbers. This is a high-level VI; hence it opens the file beforehand and closes it afterwards.

Write to a Spreadsheet File

The objective of this exercise is to construct a VI that will generate and save data to a new file in ASCII format. This data can then be accessed by a spreadsheet application.

Open a new VI and construct a front panel, as shown in Figure 9.15. This VI generates two data arrays, plots them on a graph, and writes them to a file where each column contains a data array. The front panel contains only one waveform graph.

Open the block diagram and construct the VI by adding the block diagram functions shown in Figure 9.15. The Write To Spreadsheet File VI (for location of this VI see Figure 9.10) converts the 2D or 1D array of numbers to a text string and writes the string to a new text file or appends the string to an existing file. If you have not specified a path name, then a file dialog box appears and

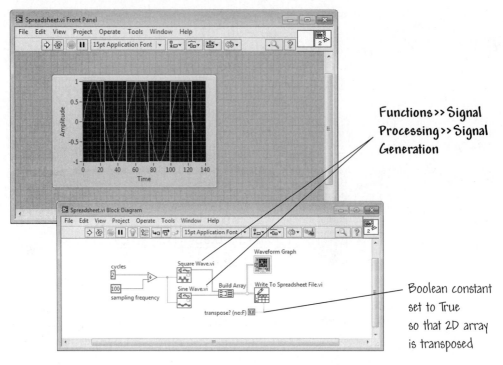

Functions >> Signal
Processing >> Signal
Generation

Boolean constant
set to True
so that 2D array
is transposed

FIGURE 9.15
Using the Write To Spreadsheet File VI to write data in ASCII format to a file.

prompts you for a file name. You can write either a 1D or 2D array to file—in this exercise we have a 2D array of data, so the 1D input is not wired. With this VI, you can use spreadsheet delimiters such as tabs or commas in your data. The format string default is %.3f, which creates a string long enough to contain the number with three digits to the right of the decimal point.

This VI is a high-level VI that opens or creates the file before writing to it and closes it afterwards. You can use the Write To Spreadsheet File VI to create a text file readable by most spreadsheet applications.

The Boolean constant (see Figure 9.15) connected to the Write To Spreadsheet File VI controls whether or not the 2D array is transposed before writing it to file. To change the value of the Boolean constant, click on the constant with the **Operating** tool. Generally in LabVIEW, each row of a 2D array contains a data array (in this example, the first row of the 2D array is the square wave and the second row is the sine wave); thus in this case you want the data transposed because you want the two data arrays to be organized into columns in the spreadsheet.

After finishing up the wiring on the block diagram, return to the front panel and run the VI. After the data arrays have been generated, a file dialog box prompts you for the file name of the new file you are creating. Type in a file name and click on **OK**.

Do not attempt to write data in LLBs. Doing so may result in overwriting your library and losing your previous work.

Save the VI in Users Stuff and name it Spreadsheet.vi. You now can use a spreadsheet application or a text editor to open and view the file you just created.

In this example, the data was not converted or written to file until the fully data arrays had been collected. If you are acquiring large buffers of data or would like to write the data values to disk as they are being generated, then you must use a different File I/O function.

A working version of Spreadsheet.vi can be found in the Chapter 9 folder in the Learning directory. ◆

9.2.4   File I/O Express VIs

The two Express VIs that can be utilized for file I/O are the Read From Measurement File and the Write to Measurement File Express VIs. These two Express VIs can be found on the **Functions≫Programming≫File I/O** palette as illustrated in Figure 9.16(a). They can also be readily found on the **Functions≫Express** palettes. The Read From Measurement File Express VI is located on the **Input** palette and the Write To Measurement File Express VIs Express VI is on the **Output** palette, as shown in Figure 9.16(b). These Express VIs read and write .lvm files—LabVIEW measurement data files. The measurement data file (.lvm) is a tab-delimited text file that can be opened with a spreadsheet application (such as Microsoft Excel) or with a text-editing application (such as Notepad). In addition to the data an Express VI generates, the .lvm file includes information about the data, such as the date and time the data was generated.

You may often find it necessary to permanently store measurement data acquired from your DAQ device. When planning to store data to a file, remember that not all data-logging applications use LabVIEW to process and analyze the stored data. Consider which applications might need to access the data, and write the data in an appropriate format—the data storage format defines which applications can read the file. Since LabVIEW contains standard file operation functions that exist in other languages, you have complete control over the data-logging process. The LabVIEW measurement file, or .lvm file, is an ASCII text file that is easy to create in LabVIEW and easy to read in other applications.

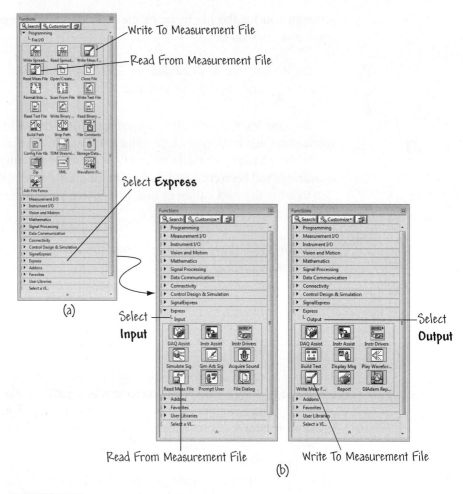

FIGURE 9.16
Locating the LabVIEW data-measurement Express VIs: (a) The **File I/O** palette;
(b) the **Express** palettes.

◆ **Practice with Writing & Reading Measurement Data Files**

In this example, we practice writing and reading LabVIEW measurement files. To begin, open the Write Measurement Data Demo.vi found in Chapter 9 of the Learning directory. The block diagram is shown in Figure 9.17. Central to the process of writing measurement data files is the Write To Measurement File Express VI, which includes open, write, close, and error-handling functions. It also handles formatting the string with either a tab or comma delimiter.

The Merge Signals function found in the Write Measurement Data Demo.vi combines the iteration count and the random number into the dynamic

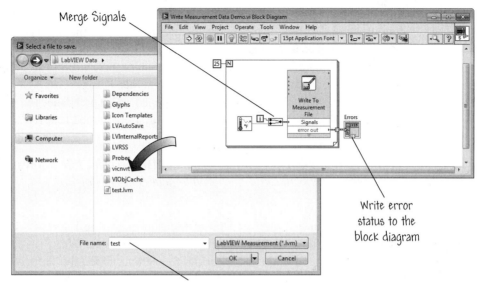

Merge Signals

Write error
status to the
block diagram

Type in file name to write

FIGURE 9.17
Writing measurement data to a file specified by the user.

data type for use by the Express VI (see Chapter 2 for a review of the dynamic data type). The Merge Signals function is located on the **Functions≫Express≫ Signal Manipulation** palette, as shown in Figure 9.18.

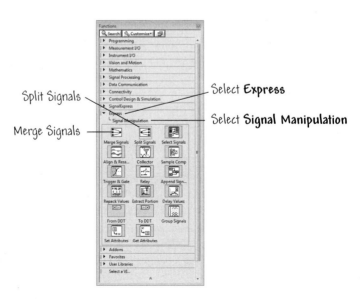

Split Signals

Merge Signals

Select **Express**

Select **Signal Manipulation**

FIGURE 9.18
The Merge Signals and Split Signals functions.

When the Write To Measurement File Express VI is first placed on the block diagram, a dialog box opens automatically to configure the VI. The dialog box is shown in Figure 9.19. The configuration as shown requires the user to choose the output file name. When the VI is run, a dialog box will open, as illustrated in Figure 9.17, for specifying the output file name. This process will lead to the creation of a .lvm file that can be later opened in a spreadsheet or text editor application. Run the VI and save the results in a file named test.lvm.

You can double-click on the Write To Measurement File Express VI in the **Write Measurement Data Demo.vi** *to access the dialog box.*

Now we can use the Read From Measurement File Express VI as a central component of a VI that reads the .lvm file from the **Write Measurement Data Demo.vi** and outputs the results to a graph. Open the **Read Measurement Data Demo.vi** found in **Chapter 9** of the **Learning** directory. The block diagram is shown in Figure 9.20.

Double-click on the Read From Measurement File Express VI to access the configuration dialog box, shown in Figure 9.21. Notice that the following options have been set:

- In the **Action** section, a check mark is placed in the **Ask user to choose file** check box.

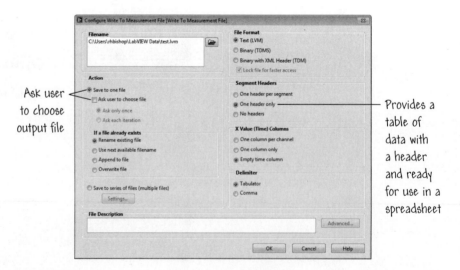

FIGURE 9.19
The Write To Measurement File Express VI dialog box.

- Set the **Segment Size** to **Retrieve segments of original size** so that all the data stored in the file is retrieved.

- Set **Time Stamps** to **Relative to start of measurement**. Because the dynamic data type stores information about the signal timing, this setting aligns the data with the time of the measurement.

- In the **Generic Text File** section, remove the check mark from the **Read generic text files** check box because the data is stored in a LabVIEW measurement file.

Click the **OK** button to close the dialog box when you are finished. Display the front panel and run the VI. In the file name prompt that appears, select the test.lvm file that you created with the Write To Measurement File Express VI. The temperature data that was stored in the LabVIEW measurement file appears in the waveform chart. ◆

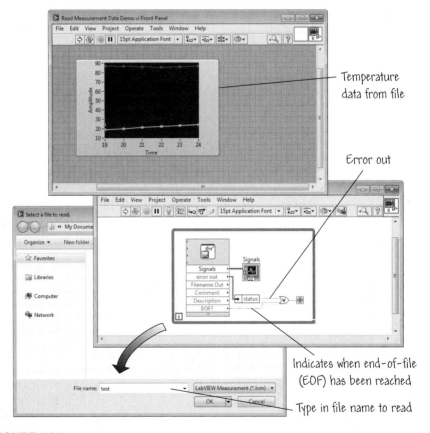

FIGURE 9.20
Reading measurement data from a file specified by the user.

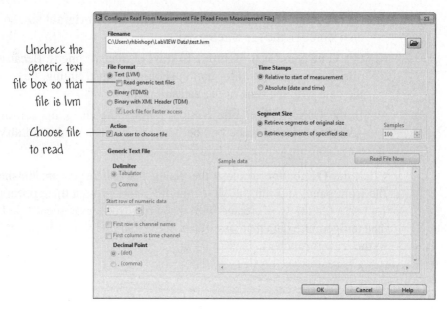

FIGURE 9.21
The Read From Measurement File Express VI dialog box.

9.2.5 Obtaining the Path to the System Directories

You can find the path to the system directories using Get System Directory VI on the File Constants palette, as shown in Figure 9.22. This VI returns the path to the type of system directory specified in the **system directory type** input. For example, as shown in Figure 9.22, if you specify **User Home** in the **system directory type** input, the VI returns the path to the directory that contains the personal files of the current user of the system (in Figure 9.22 this is **C:\Users\rhbishop**). The Get System Directory VI is located on the **Programming≫File I/O≫File Constants** palette.

If the specified directory does not exist, then the Get System Directory VI returns the path where the system expects to find it. To create a directory, you can set the **create directory if not found?** *input to the Get System Directory VI to* **TRUE**.

As shown in Figure 9.22, the directory types include the **User Home** directory, where the personal files of the current user are located, the **User Desktop** directory, which contains files located on the desktop, and the **Application Files** directory, which contains applications installed on the system (including LabVIEW).

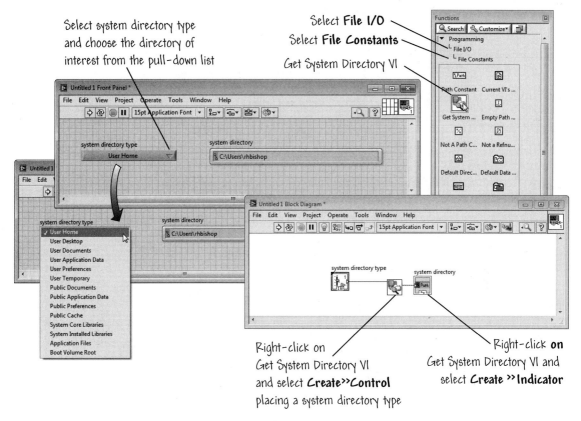

FIGURE 9.22
Using Get System Directory VI to obtain the path to the various system directories.

BUILDING BLOCK

9.3   BUILDING BLOCKS: PULSE WIDTH MODULATION

In this Building Block, you will add a feature to the PWM with Chart VI (developed in Chapter 7) to write the output data of the PWM wave to a file. Outputting a file can save information for further examination or for use with other programs or LabVIEW VIs. To begin, navigate to the Users Stuff folder in the Learning directory and open PWM with Chart.vi. Once you have opened the PWM with Chart VI, save a new copy of the VI as PWM with File Output.vi.

A working version of PWM with Chart.vi can be found in the Building Blocks folder of the Learning directory. This working VI is provided in case you did not save your work in Chapter 7 to the Users Stuff folder.

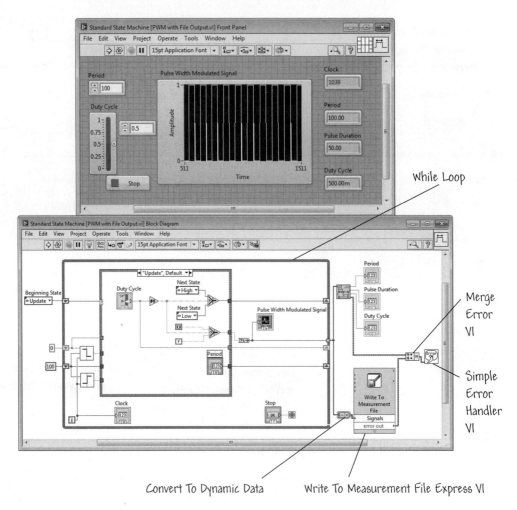

FIGURE 9.23
The PWM with File Output VI.

Using Figure 9.23 as a guide, develop the code to create a LabVIEW measurement data file containing the array output by the While Loop. Configure the Express VI so that LabVIEW overwrites the same data file rather than creating a new one each time the VI runs. You may also wish to change the default file path and file name. Note that when you wire the array to the **Signals** input of the Write To Measurement File Express VI, LabVIEW automatically places a node to convert from numeric data type to **dynamic data type**, a data type used by Express VIs that carries additional information about a signal beyond its defining values, such as name and time of acquisition.

To accomplish the objective of writing data to a file, use the Write To Measurement File Express VI found on the **Functions≫Express≫Output** palette.

You might want to review the material in Section 9.2.4 on using the Express File I/O VIs.

When you place the Write To Measurement File Express VI on the block diagram, the **Configure Write To Measurement File** dialog box will open automatically. When that occurs, make sure to select **Ask user to choose file**. Once you have configured the Write To Measurement File Express VI, use **Quick Drop** to find, place, and wire the Merge Error VI and Simple Error Handler VI. Follow the wiring shown in Figure 9.23.

After wiring the VI correctly, test it out. On the front panel, set the Duty Cycle to 0.5 and the Period to 100. Run the VI, and after it has completed executing, view the file in Notepad (Windows) or TextEdit (Macintosh).

Close the VI when you are done experimenting and save your changes.

A working version of PWM with File Output.vi *can be found in the* Building Blocks *folder of the* Learning *directory.*

9.4   RELAXED READING: ON THE SCIENCE OF CYCLING SPEED

In this reading we explore the development of an accurate way to measure and test the effects of aerodynamic drag on cyclists. Using LabVIEW, NI hardware, and off-the-shelf components, a unique wind tunnel was created to conduct continuous loops of development and testing for cyclists and their bikes by evaluating and optimizing their riding positions.

In a sport in which seconds often separate winning from losing, it is critical to understand and optimize each factor that affects a cyclist's body and bike while riding. Drag caused by wind presents one of the biggest challenges for cyclists. As they pick up speed while riding, their bodies, bikes, and equipment force a separation of air, resulting in a resistance called pressure drag. Since drag increases with speed, riders feel more resistance at higher speeds and must use more power to overcome the forces working against them. Cyclists use between 70% and 90% of the power they generate to overcome aerodynamic drag. Thus, minimizing drag through effective body positioning and aerodynamically refined equipment helps cyclists achieve maximum efficiency and speed. To minimize drag by creating better cycling products and understanding how riders can better position themselves on their bikes, Specialized Bicycle Components needed an accurate way to measure and test the effects of drag on cyclists riding in a real-world environment.

Wind tunnel testing is crucial for bike manufacturers to optimize their equipment. Wind tunnels not specifically designed for cycling tests are time consuming and inconvenient for bike manufacturers to use. Specialized Bicycle Components built a wind tunnel, shown in Figure 9.24(a), to bring all aspects of the aerodynamics testing process to its Morgan Hill facility in California where they now conduct continuous loops of development and testing for all of their bikes and equipment and supports their team of professional cyclists by evaluating and optimizing their riding positions. The tunnel is about 100 feet long with a 30 foot by 10 foot test chamber. The six 75 horsepower fans positioned at the back of the tunnel to move air through the horn-shaped inlet are shown in Figure 9.24(b). The bike and rider are almost never steady in real-world conditions and constantly interact with each other. Testing bikes in this realistic, dynamic way, as opposed to the traditional tests with the bike rigidly bolted to the tunnel will likely lead to new information that would not have surfaced otherwise.

The integrated system design employs a combination of LabVIEW, NI hardware, and commercial off-the-shelf components to form a custom measurement and control system. With LabVIEW, engineers can monitor the data from sensors placed on a bike and cyclist remotely on an iPad using the Data Dashboard for LabVIEW mobile app. Additionally, the cameras capture real-time visual data that easily integrates into the system. The flexibility of the NI PXI chassis enabled additional tests using new sensors and controllers with relatively quick turnaround times. This is especially important since test needs change regularly—for example, performing R&D testing on equipment versus performance testing with professional athletes. This is one of the biggest advantages of the system design approach.

In summary, the entire measurement and control system was developed in just a few months while seamlessly integrating each element of the system that

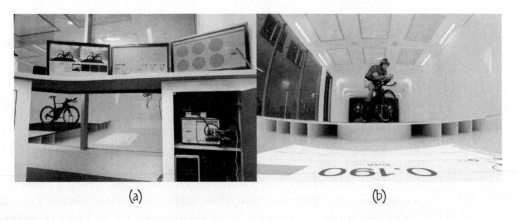

(a)   (b)

FIGURE 9.24
(a) The world's first cycling-specific wind tunnel at Specialized Bicycle Components in Morgan Hill, California; (b) the six 75 horsepower fans positioned at the back of the tunnel to move air through the horn-shaped inlet.

optimized time to market. LabVIEW provides a single software framework to meet the unique requirements of performing control, measurement, and vision acquisition. The consolidation to a single software solution, tightly integrated with reconfigurable hardware, also simplifies the maintenance and supportability of the system. For further information, visit http://sine.ni.com/cs/app/doc/p/id/cs-15562.

9.5   myDAQ BUILDING BLOCKS

In this myDAQ Building Block, you will add functionality to the temperature monitoring system to log data to a text file when the temperature deviates from the set range. Begin by opening Control System Logging unedited.vi found in the Chapter 9 folder in the myDAQ Building Blocks folder within the Learning directory.

Switch to the block diagram, as shown in Figure 9.25. Notice that the VI is very similar to Control System Multiple.vi from Chapter 6 except that a series of file functions has been added outside of the While Loop and inside the False case of the Case structure. The purpose of these file functions is to enable data logging to a file. Use the **Context Help** to investigate these functions. The Open/Create/Replace File function has been configured to replace or create a file named TempLog.txt located in the Application Directory. Recall that the Application Directory is the location where the current VI is saved. You can change this default name, modify the code to choose the name and location with front panel controls, or delete the input entirely to use a dialog box to choose the file. The Write to Text File function creates the header for the log file.

The two constants, Tab Constant *and* End of Line Constant, *can be found on the **Functions≫Programming≫String** palette.*

The Close File function and the Simple Error Handler.vi close the file and display any errors that occurred during the execution of the file functions. Notice that the input to the While Loop conditional terminal has been modified to stop the program if the file functions give an error. The Write To Text File function inside the While Loop is the function that will be logging data while the program is running. It is in the False case of the Case Structure, so the program will only write to file when the temperature is out of range.

Use the steps below as a guide to construct a string within the Case structure that matches the header created outside of the While Loop.

1. Use the Get Date/Time String function to get a time string. Be sure to enable the option to include seconds. Use the Number To Fractional String function to convert the numeric Average Temperature value into a string.

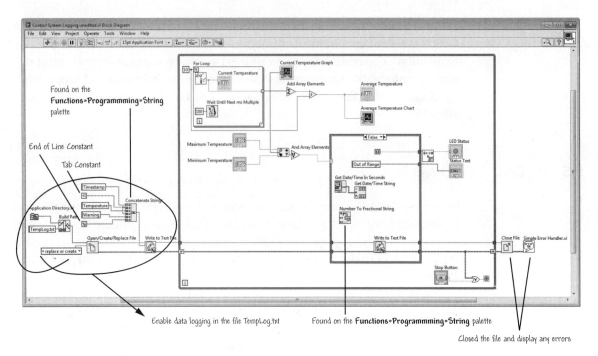

Found on the
Functions»Programmming»String
palette

End of Line Constant

Tab Constant

Enable data logging in the file TempLog.txt

Found on the **Functions»Programmming»String** palette

Closed the file and display any errors

FIGURE 9.25
The block diagram of Control System Logging unedited.vi.

2. Place a Concatenate Strings function, Tab Constant, and End of Line Constant on the block diagram inside the For Loop. Use those constants along with the outputs of the Get Date/Time String and Number to Fractional String functions to build the string that will be written to the file. Use the header structure outside of the While Loop as a guide so that the column order of the data matches the header. Connect the output of the Concatenate Strings function to the **Text** input of the Write To Text File function.

3. Test the VI. Set the **Maximum Temperature** and **Minimum Temperature** so that the measured temperature is outside the range. Stop the VI and open the log file with a text editor to check to see that the out of range temperature was logged as expected. Save the VI as **Control System Logging.vi** in the **Chapter 9** folder in the **myDAQ Building Blocks** folder within the **Learning** directory.

4. What other information might you want to include in a log entry? How would you change the code to add that information?

A working version of **Control System Logging.vi** *can be found in the* **Chapter 9** *folder of the* **myDAQ Building Blocks»Solutions** *folder in the* **Learning** *directory.*

In this myDAQ Building Block, you added functionality to the temperature monitoring system to log data to a text file when the temperature deviates from the set range. Logging data is important for analysis purposes after an experiment or other data acquisition function is complete.

9.6   SUMMARY

File input and output operations store information in and retrieve information from a disk file. LabVIEW supplies you with simple functions that take care of all aspects of file I/O. You can store data in or retrieve data from ASCII byte stream, binary byte stream, or datalog files. This allows you to interact with word-processing programs, spreadsheet programs, or with other VIs in LabVIEW.

KEY TERMS

ASCII: American Standard Code for Information Interchange.

Binary byte stream files: Files that store data as a sequence of bytes.

Datalog files: Files that store data as a sequence of records of a single, arbitrary data type that you specify when you create the file.

EOF: End-of-file. Character offset of the end of a file relative to the beginning of the file (that is, the EOF is the size of the file).

Hex: Hexadecimal. A base-16 number system.

Nondisplayable characters: ASCII characters that cannot be displayed, such as null space, backspace, and tab.

Not-a-path: A predefined value for the path control that means that a path is invalid.

Not-a-refnum: A predefined value that means the refnum associated with an open file is invalid.

Refnum: An identifier that LabVIEW associates with a file when you open it. You use the file refnum to indicate that you want a function or VI to perform an operation on the open file.

String: A sequence of displayable or nondisplayable ASCII characters.

Text (ASCII) byte stream files: Files that store data as a sequence of ASCII characters.

E9.1   Develop a VI that writes data generated by the Generate Waveform VI to a spreadsheet. Use the block diagram in Figure E9.1 as a guide. Use the **Write to Spreadsheet File.vi**. Wire the array of 10 data points from the For Loop to the **1D data** input of the Write To Spreadsheet File VI.

Run the VI, and when prompted, enter the file name to save the data. Upon completion of the VI execution, open the file in Notepad (Windows) or TextEdit (Macintosh) to view the data. Verify that 10 data points are in the file.

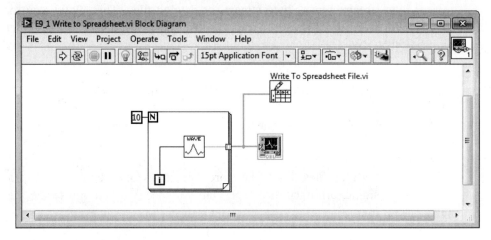

FIGURE E9.1
Use this block diagram as a guide and add the Write To Spreadsheet File VI to output the waveform data.

   You can find the Generate Waveform VI in the activity folder in the Learning directory.

E9.2   Develop a VI that writes data generated by the Generate Waveform VI to a measurement file. Use the block diagram in Figure E9.1 as a guide. Use the Write to Measurement File Express VI. Wire the array of 10 data points from the For Loop to the **Signals** input of the Write To Measurement File VI.

Run the VI, and when prompted, enter the file name to save the data. Upon completion of the VI execution, open the file in Notepad (Windows) or TextEdit (Macintosh) to view the data. Verify that 10 data points are in the file.

E9.3   Create a new VI that acquires analog data and writes the data to a LabVIEW measurement file. Use the DAQ Assistant to configure the analog input operation.

Right-click on the DAQ Assistant and select **Generate NI-DAQmx Code** to automatically create the code for this acquisition. Take 50 measurements and write the measurements using the Write To Measurement File VI. Refer back to Chapter 8 for more information on using the DAQ Assistant.

E9.4 Create a VI that reads the LabVIEW measurement file created in E9.3 and displays that information on the front panel.

E9.5 Modify the VI you created in E9.4 so that data is saved to a new file every time the VI runs. Also, make it so that LabVIEW appends the next sequential number to the file name.

E9.6 Open the VI you created in the Building Blocks exercise for this chapter, which you saved as **PWM with File Output.vi** in the **Users Stuff** folder in the **Learning** directory. On the front panel of the PWM with File Output VI, set the **Duty Cycle** to 0.5 and the **Period** to 100. Then run the VI and check to see that an output file was created at the expected location.

*In case you did not complete the Building Blocks exercise, you can find a working version of **PWM with File Output.vi** in the **Building Blocks** folder in the **Learning** directory.*

Construct a VI that prompts the user for the file name containing the measurement data created by the PWM with File Output VI and then displays the data on a waveform graph. Turn off autoscaling on the horizontal axis and make the scrollbar visible so that you can view large data sets more clearly. Run the VI and verify that it displays the same waveform obtained when running the PWM with File Output VI.

E9.7 Open **Build Text Demo.vi** found in **Chapter 9** of the **Learning** directory. This is the same example discussed in Section 9.1. Modify the VI so that a temperature reading can only be displayed if the user first enters a correct password.

Use the block diagram in Figure E9.7 as a guide. Place a string control on the front panel and right-click to change the display mode from **Normal** to **Password**. Place an LED next to the password control that lights when the correct password is entered. Using a Case structure, provide an empty output display and a message to the user with notification that an incorrect password was entered in the case that an invalid password is given. If the correct password is entered, display the temperature reading on the output indicator.

When you are finished, save the VI as **Password.vi** in the **Users Stuff** folder in the **Learning** directory.

E9.8 Create a VI that takes a string input and outputs the string in reverse. For example, if the input string is **December**, then the output string is **rebmeceD**.

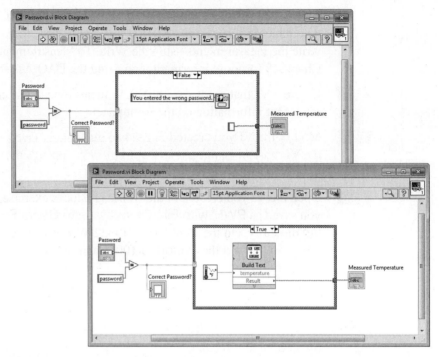

FIGURE E9.7
The Password VI block diagram.

PROBLEMS

P9.1   Although we did not cover the Binary File I/O functions in this chapter, you may find them useful for applications in which you wish to quickly store data in as little space as possible. In the NI Example Finder, find the Write Binary File VI by searching for the keywords "write binary." Once you have located the Write Binary File VI, open it and read the description of the VI and investigate the block diagram. Then run the VI and when prompted enter the file name BinaryFile for the saved data.

Return to the NI Example Finder and hunt for the Read Binary File VI by searching for the keywords "read binary." Once again, read the VI description and explore the block diagram. When you have done this, run the VI and when prompted, enter the same file name BinaryFile that you used to save the binary data created using the Write Binary File VI. You should recognize the data on the waveform graph as the same data that was graphed and saved with the Write Binary File VI.

Now return to the Write Binary File VI and modify this VI so that it creates a LabVIEW measurement file containing the same data. Use the Write To Measurement File Express VI. Use the Get File Size function found on the **Programming≫File I/O≫Advanced File Functions** palette to display the size of both the LabVIEW measurement file and the binary file on numeric indicators on the front panel.

Save the modified file as Write Two Files.vi in the Users Stuff folder in the Learning directory. Run the VI again and observe the size of each file created. Are the files the same size? Why or why not? List the different advantages of the two file formats.

P9.2 Open Basic Spectral Measurements.vi (located in the express subfolder of the examples folder of the LabVIEW directory. Using the Report Express VI found on the **Express≫Output** palette, modify this VI to create a report that displays the graphs of both signals and includes the report title, author, company name, and comments about the report. Give the user the option to enter additional comments. Save the report as an HTML file. Run the VI and view the HTML document that you created in a web browser or word-processing application.

P9.3 Create a VI that concatenates a message string, numeric, and unit string using the Build Text Express VI and writes the data to file. Create the input variables **Number** and **Units** and configure them in the Build Text Express VI. Wire a string control to the **Beginning Text** input of the Build Text Express VI. Write the resulting concatenated string to a datalog file. To find out more about datalog files, refer to the online help and examples.

P9.4 Construct a VI that converts a 2D array into a tab-delimited spreadsheet string. Then convert the tab-delimited spreadsheet string into a comma-delimited spreadsheet string and display both output strings in indicators on the front panel.

P9.5 Create a VI that takes a numeric input (under 3999) and converts it to its Roman numeral equivalent. For example, the year 2010 is MMX. Recall that $I = 1$, $V = 5, X = 10, L = 50, C = 100, D = 500$, and $M = 1000$.

DESIGN PROBLEMS

D9.1 Construct a VI that performs number base conversions and then writes the resulting conversions to a text file. The front panel should have the following components:

- A numeric control into which the user will enter the decimal value to be converted. Change the representation of this control to **Unsigned Integer** (U64) so that the user will not enter any negative or non-integer values.

- A menu ring to select from **Binary**, **Octal**, or **Hexadecimal** conversion.

- A string indicator to display the result of the conversion.

- A file path control to enter the name of the text file to which the results are to be saved.

- A Boolean button which causes the VI to write the results of the most recent conversion to the designated file when true. You will need to change the mechanical action of this button so that it does not remain depressed after a write request is read by the VI.

- A stop button which ends VI execution. The program should perform as many new conversions and file writes as the user requests, until the stop button is pressed.

Each time a new file write is requested, concatenate a sentence to the end of the text file conveying the most recent results. For example, the sentence "Decimal 7 is 111 in binary" would be added to the file that converted the number 7 to binary. Be sure that your VI opens the text file once at the beginning of the VI and then closes it once just before stopping. You should not open and close the file with each loop iteration.

To convert a decimal number to a different base, use a While Loop that performs successive division by 2, 8, or 16 for binary, octal, and hexadecimal, respectively and outputs the remainders from each division. Concatenate each new digit to a result string as it is computed. You will need to use several string functions to convert the division results to characters and then manipulate the characters so that you can create a string that accurately represents the converted number.

Run the VI several times and verify that the correct results are being written to the output file.

D9.2 Create a VI which appends an array containing a time value and a numeric input to the end of a spreadsheet file, whose path is specified by the user. The time value should be recorded in seconds, relative to the time the VI began running. The VI should run continuously, saving the most recent input value to the spreadsheet file each time the user presses a **Save** button on the front panel. End execution when a stop button is pressed.

D9.3 Develop a VI to encode or decode words using a secret code. To encode a word in this code, first separate the word into two substrings – one with the first half of the original word, and one containing the second half. If there is an odd number of letters in the word, then split it so that the first substring has one more letter than the second one.

Create the encoded word by concatenating the first half of the word onto the end of the second half of the word, inserting the second letter of the original word in between these two halves. For example, the word "gravity" would be encoded as "ityrgrav" and the word "master" would be encoded as "teramas."

Place a string control on the front panel and a Boolean switch allowing the user to select whether the input is to be decoded or encoded. Display the output on a string indicator.

D9.4 Suppose we want to represent numbers in coded format. Consider the following code:

$$1 = \text{lv}, 2 = \text{ma}, 3 = \text{po}, 4 = \text{ni}, 5 = \text{be}, 6 = \text{ut}, 7 = \text{ke},$$
$$8 = \text{su}, 9 = \text{lu, and } 0 = \text{ha}$$

For example, the number 987 is written in coded form as lusuke. Create a VI that takes a string as an input and outputs the equivalent number on a numeric indicator. Display the number −1 if the input code does not translate.

D9.5 Consider the code in D9.4. Create a VI that performs mathematical operations of addition and multiplication on three coded number inputs. For example, suppose the inputs are mapo, utlv, and ni. Then,

$$\text{mapo} + \text{utlv} + \text{ni} = 88$$
$$\text{mapo} * \text{utlv} * \text{ni} = 5612.$$

Display −1 if one or more inputs are invalid.

D9.6 Create a VI that compares the current temperature to the max and min temperature allowed. Display the following messages according to the appropriate range: Heatstroke Warning, No Warning, or Freeze Warning. If there is a warning, have a Boolean light up. For example, if the current temperature is 30°F, the temperature above which we have heatstroke warning is 99°F, and the temperature below which we have a freeze warning is 32°F, then the VI should display a Freeze Warning and light the LED, as shown in Figure D9.6. The current temperature is an input to the VI.

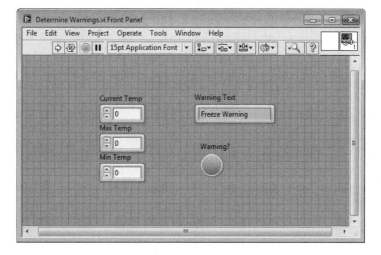

FIGURE D9.6
Checking temperature ranges and displaying warning messages.

CHAPTER 10

NI LabVIEW MathScript RT Module

NI LabVIEW MathScript RT Module adds math-oriented, textual programming to LabVIEW, both at the command-prompt level and within virtual instruments, by providing a high-level, computing language with an easily accessible syntax and ample functionality to address computing tasks related to signal processing, analysis, and mathematics. In this chapter, we discuss the MathScript Interactive Window and the MathScript Node. The essentials of creating user-defined functions and scripts, of saving and loading data files, and of using the MathScript Node are presented. It is anticipated that students will use the MathScript Interactive Window primarily to design and test code for subsequent use in MathScript Nodes as part of the overall VI development strategy.

GOALS

1. Learn to use the MathScript Interactive Window.

2. Learn to define functions and create scripts.

3. Learn to use MathScript Nodes to integrate scripts with virtual instruments.

10.1   WHAT IS MATHSCRIPT RT MODULE?

The virtual instrument (VI) introduced in Chapter 2 is the central programming element of LabVIEW. The VI is the computational workhorse. With VIs and subVIs you can build applications to perform a variety of tasks, including simulation and design of complex systems, acquiring data and prototyping systems with real hardware, and deploying and controlling complete systems. But what if you wanted to test a one-time mathematical operation for possible introduction into a VI? What if you had a sequence of computations that you wanted to perform to test an idea or to assist with your VI development? One can think of many instances when it would be very helpful to have easy, integrated access to a math-oriented, text-based computing environment.

MathScript RT Module provides access to a text-based math-oriented language with a command-prompt from within the LabVIEW development environment. Math-oriented, textual programming, both at the command prompt level and within VIs and subVIs, is enabled through the use of MathScript Nodes.

Working within the LabVIEW environment you can choose a textual approach (with command-line prompt and scripts), a graphical approach (with VIs), or a combination of the two (with MathScript Nodes within the VI). You can choose between the graphical environment and the command-line prompt in the most effective way for you to address your VI design requirements. MathScript RT Module does not require additional third-party software to compile and execute.

LabVIEW MathScript RT Module supports only Windows as part of the Student Edition. The MathScript RT Module must be downloaded as a separate download from LabVIEW.

MathScript RT Module includes hundreds of built-in functions. There are linear algebra functions, curve-fitting functions, digital filters, functions for solving differential equations, and probability and statistics functions. And since MathScript RT Module employs a commonly used syntax, it follows that you can work with many of your previously developed mathematical computation scripts, or any of those openly available in engineering textbooks or on the Internet.

The fundamental math-oriented data types are matrices with built-in operators for generating data and accessing elements. In addition, you can extend the application with custom user-defined functions. An overview of the features of MathScript RT Module is presented in Table 10.1.

You can find more information on MathScript RT Module at ***www.ni.com/ labview/mathscript/,*** *including lists of built-in MathScript functions and links to online examples.*

TABLE 10.1   Overview of the Features of MathScript RT Module

MathScript Feature	Description
Powerful textual math	MathScript RT Module includes more than 800 functions for math, signal processing, and analysis; functions cover areas such as linear algebra, curve fitting, digital filters, differential equations, probability/statistics, and much more
Math-oriented data types	MathScript RT Module uses matrices as fundamental data types, with built-in operators for generating data, accessing elements, and other operations
Data type highlighting	MathScript RT Module analyzes m-files during editing for enhanced script debugging and readability to determine the data types of inputs and constants
Compatible	MathScript RT Module can process certain files utilizing other text-based syntaxes. However, it does not support all m-file script syntaxes; hence not all existing text-based scripts are compatible
Extensible	You can extend MathScript RT Module by defining your own custom functions
Part of LabVIEW	MathScript RT Module does not require additional third-party software to compile and execute

10.2   ACCESSING THE MATHSCRIPT INTERACTIVE WINDOW

The **MathScript Interactive Window** provides a user interface in which you can enter and execute commands and see the result after the commands complete. You can enter commands one-at-a-time through a command line or as a group through a simple text editor. You can access the interactive window from the Getting Started screen or any VI by selecting Tools≫MathScript Window, as illustrated in Figure 10.1. The MathScript Interactive Window is a user interface composed of a **Command Window** (the user command inputs), an **Output Window** echoing the inputs and showing the resulting outputs, a **Script Editor** window (for loading, saving, compiling, and running scripts), a **Variables** window (showing variables, dimensions, and type), and a **Command History** window providing a historical account of commands. A MathScript Interactive Window is shown in Figure 10.2 with the various components highlighted.

The programmatic interface of MathScript RT Module is the MathScript Node, a structure that provides the capability to run scripts from within a VI. The MathScript Node is discussed in Section 10.7.

Select **MathScript Window**

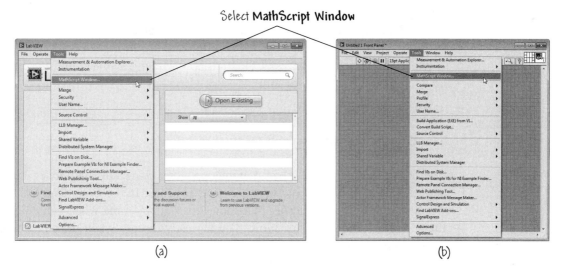

(a)   (b)

FIGURE 10.1
Accessing the MathScript Interactive Window from (a) the Getting Started screen, or (b) the **Tools** pull-down menu on the front panel or block diagram.

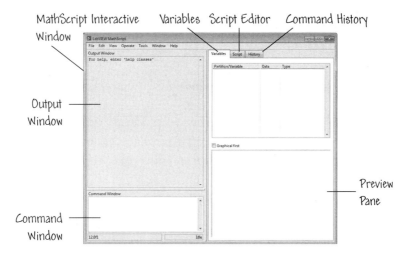

MathScript Interactive   Variables   Script Editor   Command History
Window

Output
Window

Preview
Pane

Command
Window

FIGURE 10.2
The basic components of the MathScript Interactive Window.

As you work, the Output Window updates to show your inputs and the subsequent results. The Command History window tracks your commands. The History view is very useful because there you can access and reuse your previously executed commands by double-clicking a previous command to execute it again. You can also navigate through the history of previous commands (which will appear in the Command Window) by using the <↑> and <↓> keys. In the Script Editor window, you can enter and execute groups of commands and then save the commands in a file (called a script) for use in a later session. Taken together, the various elements of the MathScript Interactive Window constitute an interactive command-line tool.

10.2.1   The Command History and Output Windows

The commands you entered in previous sessions using the MathScript Interactive Window will reappear in subsequent sessions. In the Command History window you will find a header that shows the day and time when you entered the commands. This feature allows you to easily discern when the commands were entered. You can clear the Output Window by typing clc in the Command Window, then pressing <Enter>, as illustrated in Figure 10.3.

You can also copy data from the Output Window and paste it in the Script Editor window or a text editor by highlighting the desired text in the Output Window and selecting Edit≫Copy or pressing the <Ctrl-C> keys to copy the selected text to the clipboard.

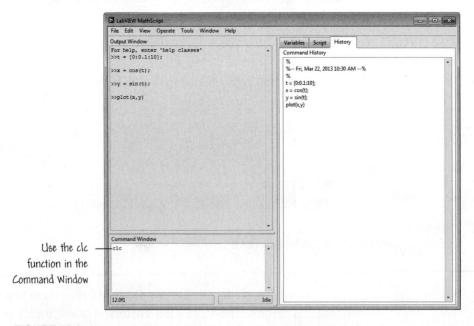

Use the clc function in the Command Window

FIGURE 10.3
Clearing the Output Window.

10.2.2   Viewing Data in a Variety of Formats

An interesting feature of the MathScript Interactive Window is the ability to view your variables in a variety of formats, as shown in Figure 10.4. Table 10.2 shows the available data types in MathScript. Depending on the variable type of your data, the available viewing formats include: numeric, string, graph, XY graph, sound, surface, and picture. You can also edit a variable in the Preview Pane when the display type is **Numeric** or **String**. Selecting **Sound** plays the data as a sound, but works for one-dimensional variables only. The remaining display types show the data as graphs of one sort or another: **Graph** displays the data on a waveform graph, **XY Graph** displays the data on an XY graph, **Surface** displays the data on a 3D surface graph, and **Picture** displays the data on an intensity graph. Chapter 7 provides additional information on plotting in LabVIEW.

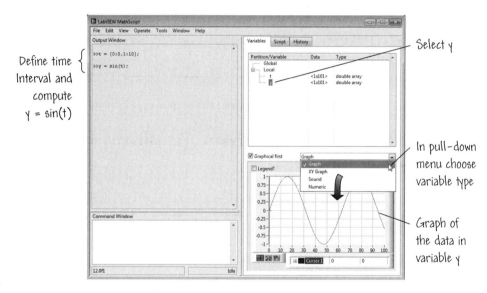

FIGURE 10.4
Showing the data types in various formats.

*The **Graphical first** button specifies whether to display first the numerical or graphical representation of variables in the Preview Pane. You can view only the numerical representation of scalar variables.*

In this example you will use the MathScript Interactive Window to construct a plot of the cosine function. You will also learn how to view the variables in a variety of formats.

TABLE 10.2 Overview of MathScript Data Types and Relationship to Syntax

Data Type	MathScript Syntax			
	Scalar	**1D-Array**	**2D-Array**	**Matrix**
Unsigned integer numeric				
8-bit	Scalar≫U8	1D-Array≫U8 1D	2D-Array≫U8 2D	
16-bit	Scalar≫U16	1D-Array≫U16 1D	2D-Array≫U16 2D	
32-bit	Scalar≫U32	1D-Array≫U32 1D	2D-Array≫U32 2D	
64-bit	Scalar≫U64	1D-Array≫U64 1D	2D-Array≫U64 2D	
Signed integer numeric				
8-bit	Scalar≫I8	1D-Array≫I8 1D	2D-Array≫I8 2D	
16-bit	Scalar≫I16	1D-Array≫I16 1D	2D-Array≫I16 2D	
32-bit	Scalar≫I32	1D-Array≫I32 1D	2D-Array≫I32 2D	
64-bit	Scalar≫I64	1D-Array≫I64 1D	2D-Array≫I64 2D	
Single-precision, floating-point numeric	Scalar≫SGL	1D-Array≫SGL 1D	2D-Array≫SGL 2D	
Double-precision, floating-point numeric	Scalar≫DBL	1D-Array≫DBL 1D	2D-Array≫DBL 2D	
Extended-precision, floating-point numeric	Scalar≫EXT	1D-Array≫EXT 1D	2D-Array≫EXT 2D	
Complex single-precision, floating-point numeric	Scalar≫CSG	1D-Array≫CSG 1D	2D-Array≫CSG 2D	
Complex double-precision, floating-point numeric	Scalar≫CDB	1D-Array≫CDB 1D	2D-Array≫CDB 2D	
Complex extended-precision, floating-point numeric	Scalar≫CXT	1D-Array≫CXT 1D	2D-Array≫CXT 2D	
Boolean	Scalar≫Boolean	1D-Array≫Boolean 1D	2D-Array≫Boolean 2D	
String	Scalar≫String	2D-Array≫Boolean 2D		
Matrix				
Real				Matrix≫Real Matrix
Complex				Matrix≫Complex Matrix

Open a new MathScript Interactive Window from the Getting Started screen by selecting Tools≫MathScript Window, as illustrated in Figure 10.1. In the Command Window enter the time from $t = 0$ seconds to $t = 10$ seconds in increments of 0.1 seconds, as follows:

```
t=[0:0.1:10];
```
.

Then, compute the $y = \cos(t)$ as follows:

```
y=cos(t);
```
.

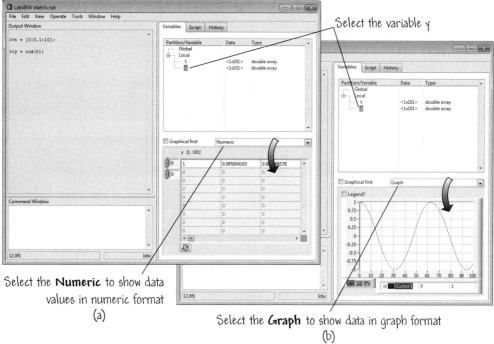

Select the variable y

Select the **Numeric** to show data
values in numeric format

(a)

Select the **Graph** to show data in graph format

(b)

FIGURE 10.5
(a) Entering the time, computing $y = \cos(t)$, and viewing the variable y in numerical form; (b) viewing the variable y in graphical form.

Notice that in the Variables window the two variables t and y appear, as illustrated in Figure 10.5(a). Select the variable y and note that in the Preview Pane the variable appears in the numeric format. Now, in the pull-down menu above the Preview Pane, select **Graph**. The data will now be shown in graphical form, as illustrated in Figure 10.5(b).

 *The graph can be undocked from the Preview Pane for resizing and customization. To undock the graph, right-click on the graph and select **Undock Window**. The window can now be resized and the plot can be customized interactively and printed.*

As an alternative to using the Preview Pane, you can also obtain a plot of y versus t programmatically using the plot command:

plot(t,y) .

The process is illustrated in Figure 10.6. A new window appears that presents the graph of y versus t. Following the same procedure, see if you can obtain a plot of $y = \cos(\omega t)$ where $\omega = 4$ rad/sec. ◆

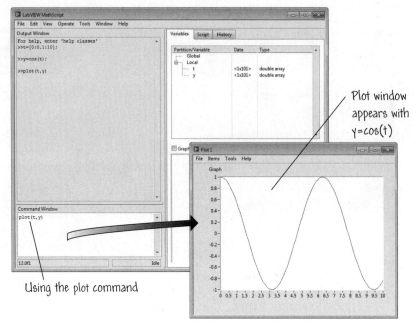

FIGURE 10.6
Obtaining a plot of the cosine function using the plot command.

10.3   MATHSCRIPT HELP

You can display several types of help content for MathScript by calling different help commands from the Command Window. Table 10.3 lists the help commands.

TABLE 10.3   Help Commands for MathScript

Command	Description of Help Provided
help	Provides an overview of the MathScript window
help classes	Provides a list of all classes of MathScript functions and topics as well as a short description of each class
help *class*	Provides a list of the names and short descriptions of all functions in a particular MathScript class Example: help basic
help *function*	Provides reference help for a particular MathScript function or topic, including its name, syntax, description, inputs and outputs, examples to type in the Command Window, and related functions or topics Example: help abs

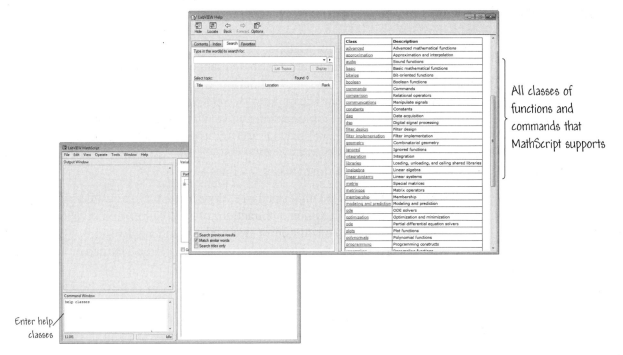

All classes of functions and commands that MathScript supports

Enter help classes

FIGURE 10.7
Accessing the help for MathScript classes, members, and functions.

As illustrated in Figure 10.7, entering **help classes** in the Command Window results in an output showing all classes of functions and commands that MathScript supports. Examples of the classes of functions are **basic** and **matrixops**. Entering **help basic** in the Command Window results in a list of the members of the basic class, including **abs** (i.e., the absolute value), **conj** (i.e., the complex conjugate function), and **exp** (i.e., the exponential function). Then, entering **help abs** in the Command Window will result in an output that contains a description of the **abs** function, including examples of its usage and related topics.

*Students usually use a combination of the help commands shown in Table 10.3 to find functions. For example, if you are looking for a function to perform a filter, the common approach would be to first enter **help classes** to view the available classes of functions. From that list of functions, you would then enter **help dsp** to get a list of DSP functions, including many filter functions. From there, you can choose the functions you want to learn more about and enter **help filter**, for example, to access the help for that function.*

10.4 SYNTAX

The syntax associated with MathScript is straightforward. Most students with some experience programming a text-based language will be comfortable with the programming constructs in MathScript. If you need help getting started with MathScript, you can access help by selecting Help≫Search the LabVIEW Help from the MathScript Interactive Window (or alternatively on the block diagram or front panel of your VI) and typing mathscript in the search window.

Eleven basic MathScript syntax guidelines are:

1. **Scalar operations:** MathScript is ideally suited for quick mathematical operations, such as addition, subtraction, multiplication, and division. For example, consider the addition of two scalar numbers, 16 and 3. This is a simple operation that you might perform on a calculator. This can be accomplished using the MathScript command:

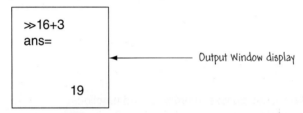

In MathScript, if you perform any calculation or function without assigning the result to a variable, the default variable ans is used. If you want to assign the value of the addition of two scalars to the variable x, enter the following command:

```
≫x=16+3
x=

      19
```

In the same manner, you can add two scalar variables y and z by entering the following commands:

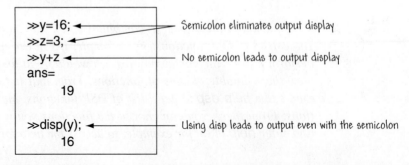

Notice that in the previous example a semicolon was used for the first two lines, and no output was displayed. In MathScript, if you end a command line with a semicolon, the MathScript Interactive Window does not display the output for that command.

Some functions display output even if you end the command line with a semicolon. For example, the **disp** *function displays an output even if followed by a semicolon.*

You use the symbol '-' for subtraction, the symbol '/' for division, and the symbol '*' for multiplication, as illustrated below:

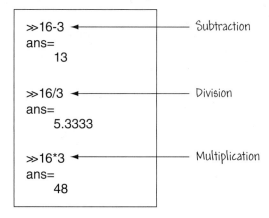

2. **Creating matrices and vectors:** To create row or column vectors and matrices, use white space or commas to separate elements, and use semicolons to separate rows. Consider for example, the matrix **A** (a column vector),

$$A = \begin{bmatrix} 1 \\ 2 \\ 3 \end{bmatrix}.$$

In MathScript syntax, you would form the column vector as

A=[1;2;3] .

Similarly, consider the matrix **B** (a row vector)

$$B = \begin{bmatrix} 1 & -2 & 7 \end{bmatrix}.$$

In MathScript syntax, you would form the row vector as

B=[1,-2,7] or B=[1 -2 7] .

As a final example, consider the matrix **C** (a 3 × 3 matrix):

$$C = \begin{bmatrix} -1 & 2 & 0 \\ 4 & 10 & -2 \\ 1 & 0 & 6 \end{bmatrix}.$$

In MathScript syntax, you would form the matrix as

C=[-1 2 0; 4 10 -2; 1 0 6] or C=[-1, 2, 0; 4, 10, -2; 1, 0, 6] .

3. **Creating vectors using the colon operator:** There are several ways to create a one-dimensional array of equally distributed elements. For example,
you will often need to create a vector of elements representing time. To create a one-dimensional array equally distributed and incremented by 1, use
the MathScript syntax,

```
>>t=1:10
t=

        1   2   3   4   5   6   7   8   9   10
```
.

To create a one-dimensional array equally distributed and incremented
by 0.5, use the MathScript syntax,

```
>>t=1:0.5:10
t=

        1   1.5   2   2.5   3   3.5   4   4.5   5   5.5
   6   6.5   7   7.5   8   8.5   9   9.5   10
```
.

4. **Accessing individual elements of a vector or matrix:** You will often want
to access specific elements or subsets of a vector or matrix. Consider the
3 × 3 matrix **C**:

$$C = \begin{bmatrix} -1 & 2 & 0 \\ 4 & 10 & -2 \\ 1 & 0 & 6 \end{bmatrix}.$$

In MathScript syntax, you can access the element in the second row and third column of the matrix **C**, as follows:

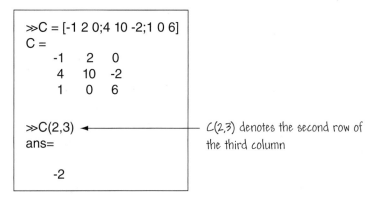

```
>>C = [-1 2 0;4 10 -2;1 0 6]
C =
       -1    2    0
        4   10   -2
        1    0    6

>>C(2,3)
ans=

       -2
```

C(2,3) denotes the second row of the third column

You can assign this value to a new variable by entering the following command:

```
>>F=C(2,3)
F=

     -2
```
.

You can also access an entire row or an entire column of a matrix using the colon operator. In MathScript syntax, if you want to access the entire second row of matrix **C**, enter the following command:

```
>>C(2,:)
ans=

     4   10   -2
```
.

The second row of C

In the same way, if you wish to access the entire third column of matrix **C**, enter the following command:

```
>>C(:,3)
ans=

     0
    -2
     6
```
.

The third column of C

Suppose you want to extract the 2×2 submatrix from **C** consisting of rows 2 and 3 and columns 1 and 2. You use brackets to specify groups of rows and columns to access a subset of data as follows:

```
≫C([2 3], [1 2])         A submatrix of C
ans=

        4    10
        1    0
```

5. **Calling functions in MathScript:** You can call MathScript functions from the Command Window. Consider the creation of a vector of a certain number of elements that are equally distributed in a given interval. To accomplish this in MathScript syntax, you can use the built-in function linspace.

   Using the command help linspace you find that this function uses the syntax

$$\text{linramp}(a, b, n)$$

where a specifies the start of the interval, b specifies the end of the interval, and n identifies the number of elements. Thus, to create a vector of $n = 13$ numbers equally distributed between $a = 1$ and $b = 10$, use the following command:

```
≫G = linramp (1, 10, 13)
G=

      1    1.75    2.5    3.25    4    4.75    5.5    6.25    7
   7.75    8.5    9.25    10
```

If you do not specify a value for n, the linramp command will automatically return a vector of 100 elements.

To select a subset of **G** that consists of all elements after a specified index location, you can use the syntax described in guideline 4 and the end function to specify the end of the vector. For example, the following command will return all elements of **G** from the fifth element to the final element:

```
≫H=G(5:end)
H =

      4    4.75    5.5    6.25    7    7.75    8.5    9.25    10
```

The function linramp is an example of a built-in MathScript function. Calling user-defined functions is covered in Section 10.5.2.

6. **Assigning data types to variables:** MathScript variables adapt to data types. For example, if

$$a = \sin(3^*pi/2)$$,

then a is a double-precision floating-point number. If

$$a = \text{'temperature'}$$,

then a is a string.

7. **Using complex numbers:** You can use either i or j to represent the imaginary unit equal to the square root of -1.

If you assign values to either i or j in your scripts, then those variable names are no longer complex numbers. For example, if you let $y = 4 + j$, then y is a complex number with real part equal to 4 and imaginary part equal to $+j$. If however, you first assign $j = 3$, and then compute $y = 4 + j$, the result is $y = 7$, a real number.

8. **Matrix operations:** Many of the same mathematical functions used on scalars can also be applied to matrices and vectors. Consider adding two matrices **K** and **L**, where

$$\mathbf{K} = \begin{bmatrix} -1 & 2 & 0 \\ 4 & 10 & -2 \\ 1 & 0 & 6 \end{bmatrix} \text{ and } \mathbf{L} = \begin{bmatrix} 1 & 0 & 0 \\ 0 & 1 & 0 \\ 0 & 0 & 1 \end{bmatrix}.$$

To add the two matrices **K** and **L**, element by element, enter the following MathScript command:

```
>>K+L
ans =

    0    2    0
    4   11   -2
    1    0    7
```

In a similar fashion, you can also multiply two matrices **K** and **L**, as follows:

```
≫K*L
ans =

      -1    2    0  .
       4   10   -2
       1    0    6
```

Consider the 3×1 matrix **M** (column vector) and the 1×3 matrix **N** (row vector)

$$\mathbf{M} = \begin{bmatrix} 1 \\ 2 \\ 3 \end{bmatrix} \text{ and } \mathbf{N} = [0 \quad 1 \quad 2].$$

Then, the product $\mathbf{M} * \mathbf{N}$ is the 3×3 matrix

```
≫M*N
ans =

       0    1    2
       0    2    4
       0    3    6
```

and the product $\mathbf{N} * \mathbf{M}$ is a scalar

```
≫N*M
ans =

       8
```

To multiply two matrices, they must be of compatible dimensions. For example, suppose a matrix **M** is of dimension $m \times n$, and a second matrix **N** is of dimension $n \times p$. Then you can multiply $\mathbf{M} \times \mathbf{N}$ resulting in an $m \times p$ matrix. In the example above, the 3×1 matrix **M** (column vector) was multiplied with the 1×3 matrix **N** (row vector) resulting in a 3×3 matrix. You cannot multiply $\mathbf{N} \times \mathbf{M}$ unless $m = p$. In the example above, the 1×3 matrix **N** (row vector) was multiplied with the 3×1 matrix **M** (column vector) resulting in a 1×1 matrix (a scalar), so in this case, $m = p = 1$.

When working with vectors and matrices, it is often useful to perform mathematical operations *element-wise*. For example, consider the two vectors

$$\mathbf{M} = \begin{bmatrix} -1 \\ 4 \\ 0 \end{bmatrix} \text{ and } \mathbf{N} = \begin{bmatrix} 2 \\ -2 \\ 1 \end{bmatrix}.$$

In light of our previous discussion, it is not possible to compute $\mathbf{M} * \mathbf{N}$ since the dimensions are not compatible. However, you can multiply the vectors element-wise using the syntax '.*' for the multiplication operator, as follows:

$$\mathbf{M}.*\mathbf{N} = \begin{bmatrix} -1*2 \\ 4*(-2) \\ 0*1 \end{bmatrix} = \begin{bmatrix} -2 \\ -8 \\ 0 \end{bmatrix}.$$

By definition, matrix addition and subtraction occurs element-wise. However, it is also possible to perform division element-wise. For $\mathbf{M}$ and $\mathbf{N}$ as above, we find that element-wise division yields

$$\mathbf{M}./\mathbf{N} = \begin{bmatrix} -1/2 \\ 4/(-2) \\ 0/1 \end{bmatrix} = \begin{bmatrix} -0.5 \\ -2 \\ 0 \end{bmatrix}.$$

Element-wise operations are useful in plotting functions. For example, suppose that you wanted to plot $y = t \sin(t)$ for $t = [0:0.1:10]$. This would be achieved via the commands

```
t=[0:0.1:10];
y=t .* sin(t);        ←——— Element-wise multiplication using .*
plot(t,y)
```

9. **Logical expressions:** MathScript can evaluate logical expressions such as EQUAL, NOT EQUAL, AND, and OR. To perform an equality comparison, use the statement

```
a = = b
```
.

If *a* is equal to *b*, MathScript will return a 1 (indicating True); if *a* and *b* are not equal, MathScript will return a 0 (indicating False). To perform an inequality comparison, use the statement

$$a \sim = b$$.

If *a* is not equal to *b*, MathScript will return a 1 (indicating True); if *a* and *b* are equal, MathScript will return a 0 (indicating False).

In other scenarios, you may want to use MathScript to evaluate compound logical expressions, such as when at least one expression of many is True (OR), or when all of your expressions are True (AND). The compound logical expression AND is executed using the '**&**' command. The compound logical expression OR is executed using the '**|**' command.

10. **Control flow constructs:** Table 10.4 provides the MathScript syntax for commonly used programming constructs.

The MathScript syntax is similar to the widely used m-file script syntax used by The MathWorks, Inc. MATLAB® software[1], COMSOL Script™, and others. In MathScript, you generally can execute scripts written in the m-file script syntax; however, some scripts or functions may not be compatible with the MathScript engine.

11. **Adding comments:** To add comments to your scripts, precede each line of documentation with a % character. For example, consider a script that has two inputs, *x* and *y*, and computes the addition of *x* and *y* as the output variable *z*,

```
% In this script, the inputs are x and y
% and the output is z.
z=x+y; % z is the addition of x and y
```
⎬ Comments start with %

The script shown above has three comments, all preceded by the % character. In the next section, we will discuss more details on how to use comments to provide help documentation.

Some considerations that have a bearing on your usage of LabVIEW MathScript RT Module follow:

1. You cannot define variables that begin with an underscore, white space, or digit. For example, you can name a variable time, but you cannot name a variable 4time or _time.

---

1. MATLAB® is a registered trademark of The MathWorks, Inc.

TABLE 10.4 MathScript Syntax for Commonly Used Constructs

Construct	Grammar	Example

Case-Switch Statement	switch expression
	case expression
	statement-list
	[**case** expression
	statement-list]
	...
	[otherwise
	statement-list]
	end
		switch mode
		case 'start'
		a = 0;
		case 'end'
		a = −1;
		otherwise
		a = a+1;
		end
		When a case in a case-switch statement executes, LabVIEW does not select the next case automatically. Therefore, you do not need to use break statements as in C.

For Loop	for expression statement-list end	for k = 1:10 a(k) = sin(2*pi*k/10) end
If-Else Statement	if expression statement-list [elseif expression statement-list] ... [else statement-list] end	if b = = 1 c = 3 else c = 4 end
While Loop	while expression statement-list end	while k < 10 a(k) = cos(2*pi*k/10) k = k + 1; end

2. Variables are case sensitive. The variables X and x are not the same variables.

3. LabVIEW MathScript does not support n-dimensional arrays where $n > 2$, cell arrays, sparse matrices, or structures.

10.4.1 Key MathScript Functions

MathScript offers more than eight hundred textual functions for math, signal processing, and analysis. These are in addition to the more than six hundred graphical functions for signal processing, analysis, and math that are available as VIs within LabVIEW. Table 10.5 lists many of the key areas with supporting MathScript functions. For a comprehensive function list, visit the National Instruments website at **http://www.ni.com/mathscript** or see the online help.

TABLE 10.5   MathScript Function Summary

Function Classes	Brief Description
Plots (2D and 3D)	Standard *x*-*y* plot; mesh plot; 3D plot; surface plot; subplots; stairstep plot; logarithmic plot; stem plot and more
Digital Signal Processing (DSP)	Signal synthesis; Butterworth, Chebyshev, Parks-McClellan, windowed FIR, elliptic (Cauer), lattice and other filter designs; FFT (1D/2D); inverse FFT (1D/2D); Hilbert transform; Hamming, Hanning, Kaiser-Bessel and other windows; pole/zero plotting and others
Approximation (Curve Fitting & Interpolation)	Cubic spline, cubic Hermite and linear interpolation; exponential, linear and power fit; rational approximation and others.
Ordinary Differential Equation (ODE) Solvers	Adams-Moulton, Runge-Kutta, Rosenbrock and other continuous ordinary differential equation (ODE) solvers
Polynomial Operations	Convolution; deconvolution; polynomial fit; piecewise polynomial; partial fraction expansion and others
Linear Algebra	LU, QR, QZ, Cholesky, Schur decomposition; SVD; determinant; inverse; transpose; orthogonalization; solutions to special matrices; Taylor series; real and complex eigenvalues and eigenvectors; polynomial eigenvalue and more
Matrix Operations	Hankel, Hilbert, Rosser, Vandermonde special matrices; inverse; multiplication; division; unary operations and others
Vector Operations	Cross product; curl and angular velocity; gradient; Kronecker tensor product and more
Probability and Statistics	Mean; median; Poisson, Rayleigh, chi-squared, Weibull, T, gamma distributions; covariance; variance; standard deviation; cross correlation; histogram; numerous types of white noise distributions and other functions
Optimization	Quasi-Newton, quadratic, Simplex methods and more
Advanced Functions	Bessel, spherical Bessel, Psi, Airy, Legendre, Jacobi functions; trapezoidal, elliptic exponential integral functions and more
Basic	Absolute value; Cartesian to polar and spherical and other coordinate conversions; least common multiple; modulo; exponentials; logarithmic functions; complex conjugates and more
Trigonometric	Standard cosine, sine and tangent; inverse hyperbolic cosine, cotangent, cosecant, secant, sine and tangent; hyperbolic cosine cotangent, cosecant, secant, sine and tangent; exponential; natural logarithm and more
Boolean and Bit Operations	AND, OR, NOT and other logic operations; bitwise shift, bitwise OR and other bitwise operations
Data Acquisition/ Generation	Perform analog and digital I/O using National Instruments devices
Other	Programming primitives such as If, For and While Loops; unsigned and signed datatype conversions; file I/O; benchmarking and other timing functions; various set and string operations and more

10.5   DEFINING FUNCTIONS AND CREATING SCRIPTS

You can define functions and create scripts to use in the MathScript Interactive Window or in the MathScript Node. In this section, we will focus on the use of the MathScript Interactive Window.

Functions and scripts can be created in the Script Editor Window on the MathScript Interactive Window (see Figure 10.2). You can also use your favorite text editor to create functions and scripts. Once your function or script is complete, you should save it for use later. The filename for a function must be the same as the name of the function and must have a lowercase .m extension. For example, the filename for a user-defined starlight[2] function must be starlight.m. Remember that MathScript is case sensitive.

Use unique names for all functions and scripts and save them in a directory that you specified in the Path section of the File≫LabVIEW MathScript Properties dialog box. Entering the Path command in the Command Window returns the current search path list. The command path(a) changes the current path to a and path(path, b) adds b to the end of the search path list.

10.5.1 User-Defined Functions

As noted in Section 10.4, MathScript offers more than eight hundred textual functions for math, signal processing, and analysis. But what if you have a special-purpose function that you want to add to your personal library? This function may be particular to your area of study or research, and is one that you need to call as part of a larger program. With MathScript it is simple to create a function once you understand the basic syntax.

A MathScript function definition must use the following syntax:

> function *outputs* = *function_name(inputs)*
> *% documentation*
> *script*

An example of a user-defined function definition utilizing the proper syntax is

> function ave = compute_average(x,y)
> % compute_average determines the average of the two inputs x and y.
> ave = (x + y)/2;

Begin each function definition with the term function. The *outputs* term lists the output variables of the function. If the function has more than one output variable, enclose the variables in square brackets and separate the variables with white space or commas. The *function_name* is the name of the function you want to define and is the name that you use when calling the function. The *inputs* term lists the input variables to the function. Use commas to separate the input variables. The *documentation* is the set of comments that you want MathScript

---

2. The name starlight does not represent a real function. It is used here for illustrative purposes only.

to return for the function when you execute the help command. Comments are preceded with a % character. You can place comments anywhere in the function; however, LabVIEW returns only the first comment block in the Output Window to provide help to the user. All other comment blocks are for internal documentation. The *script* defines the executable body of the function.

Checking the help on the function compute_average.m and then executing the function with $x = 2$ and $y = 4$ as inputs yields

```
≫help compute_average
compute_average determines the average of the two inputs x and y.

≫x=2;y=4;compute_average(x,y)
ans=

        3
```

Note that there is a MathScript function named mean that can also be used to compute the average of two inputs, as follows:

```
≫mean([2 4])
ans =

      3
```

Custom functions can be edited in the Script Editor window and saved for later use. In Figure 10.8, the buttons **Load, Save As, Run Script**, and **New Script** are shown. Selecting **Load** will open a window to browse for the desired function (or script) to load into MathScript. Similarly, selecting **Save As** will open a browser to navigate to the desired folder to save the function.

In Figure 10.8, the function compute_average is used to compute the average of two arrays. Notice that the function computes the average element-wise. If compute_average had inadvertently been named mean, then LabVIEW would execute the user-defined function instead of the built-in function. Generally it is not a good idea to redefine LabVIEW functions, and students should avoid doing so.

If you define a function with the same name as a built-in MathScript function, LabVIEW executes the function you defined instead of the original MathScript function. When you execute the help command, LabVIEW returns help content for the function you defined and not the help content for the original MathScript function.

Select **Open Script** to browse for a
function to load in MathScript

Select **New Script** to
clear script editor window

Select **Save Script** to browse for a
directory to save the function

Select **Run Script** to
execute m-file

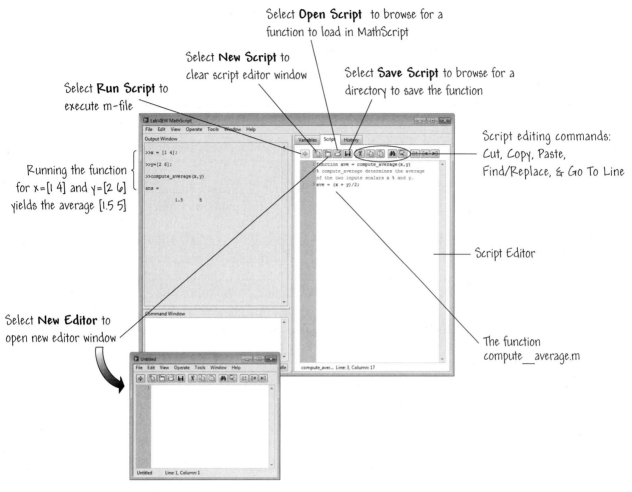

Script editing commands:
Cut, Copy, Paste,
Find/Replace, & Go To Line

Running the function
for x=[1 4] and y=[2 6]
yields the average [1.5 5]

Script Editor

Select **New Editor** to
open new editor window

The function
compute__average.m

FIGURE 10.8
Loading and saving functions.

Other examples of valid function syntax for the **starlight** function include:

function starlight	% No inputs and no outputs
function a = starlight	% No inputs and one output
function [a b] = starlight	% No inputs and two outputs
function starlight(g)	% One input and no outputs
function a = starlight(g)	% One input and one output
function [a b] = starlight(g)	% One input and two outputs
function starlight(g, h)	% Two inputs and no outputs
function a = starlight(g, h)	% Two inputs and one output
function [a b] = starlight(g, h)	% Two inputs and two outputs

There are several restrictions on the use of functions. First, if you define multiple functions in one MathScript file, all functions following the first are **subfunctions** and are accessible only to the main function. A function can call only those functions that you define below it. Second, you cannot call functions recursively. For example, the function starlight cannot call starlight. And third, LabVIEW does not allow circular recursive function calls. For example, the function starlight cannot call the function bar if bar calls starlight.

10.5.2 Scripts

A script is a sequence of MathScript commands that you want to perform to accomplish a task. For convenience and reusability, once you have created a script, you can save it and load it into another session of LabVIEW at a later time. Also, often you can use a script designed for a different task as a starting point for the development of a new script. Since the scripts themselves are saved as common ascii text and editable with any text editor (including the one found in the MathScript Interactive Window), it is easy to do this. The MathScript functions as well as the user-defined functions can be employed in scripts.

Continuing the example above, suppose that we used a script to compute the average of two numbers. The compute_average function could be used within the script. Once saved, the script can subsequently be loaded into MathScript for use in another session. A script using the compute_average function is shown in Figure 10.9.

FIGURE 10.9
Editing, saving, and running a script to compute the average of two arrays element-wise.

Saving and Loading Scripts

Being able to save scripts is an important feature giving you the capability to develop a library of scripts that you can readily access in future MathScript sessions. To save a script that you have created in the Script Editor window, select File≫Save, as illustrated in Figure 10.10(a). You can also save your script by clicking the **Save As** button on the Script Editor window of the MathScript Interactive Window, as illustrated in Figure 10.10(b). In both cases, a file dialog box will appear for you to navigate to the directory in which you want to save the script. Enter a name for the script in the **File name** field. The name must have a lowercase .m extension if you want LabVIEW to run the script (in this example, we use the name average_example.m). Click the **OK** button to save the script.

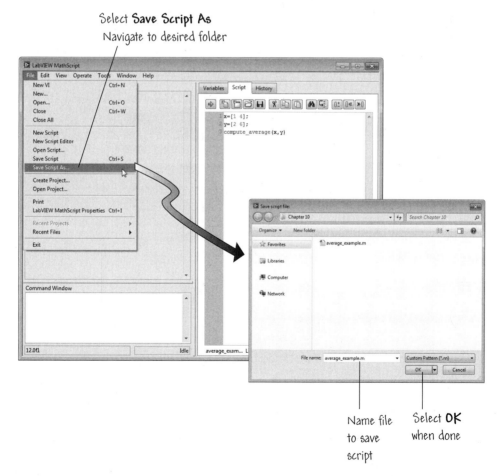

FIGURE 10.10
Saving a script using the File≫Save Script As pull-down menu.

You can load existing scripts into the MathScript Interactive Window. This will be useful upon returning to a MathScript session or if you want to use a script in the current session that was developed in a previous session. To load an existing script, select **Operate≫Load Script** or click the **Load** button on the Script Editor of the MathScript Interactive Window.

Figure 10.11 illustrates the process of loading scripts. In the example, the script compute_average.m is loaded into a MathScript session, and then using the **Run Script** button, the script is executed.

*You can find a working version of the script **average_example.m** and the function **compute_average.m** in the **Chapter 10** folder of the **Learning** directory.*

Practice with Scripts

In this example you will load an existing MathScript script and use it to smooth a noisy signal. To begin, open the MathScript Interactive Window (see

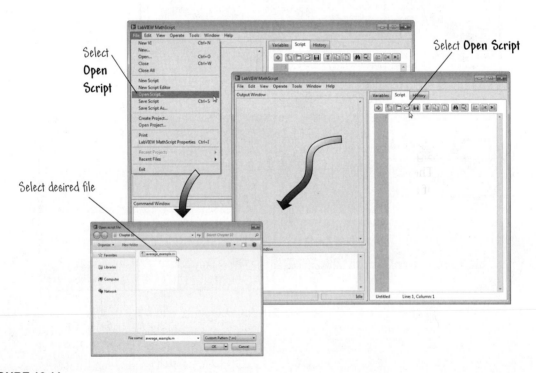

FIGURE 10.11
(a) Loading a script using the Operate≫Load Script pull-down menu; (b) loading a script using the **Load** button on the MathScript Interactive Window.

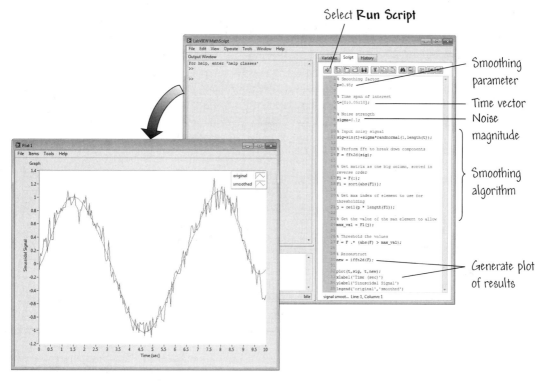

FIGURE 10.12
Smoothing a noisy signal using a script.

Section 10.2 for more information on accessing MathScript). Load in the script **signal smoothing.m** found in the **Chapter 10** folder of the **Learning** directory. The script is shown in Figure 10.12. Selecting **Run Script** will produce a plot of a noisy signal and a smoothed version of the same signal.

The script has three inputs: smoothing factor, p, time span of interest, t, and noise strength, *sigma*. By default, the values are set at $p = 0.95$; $t=[0:0.05:10]$; and *sigma* $= 0.1$.

It is easy to investigate the effect of varying the smoothing parameter. In the Script Editor window, change the value of $p = 0.95$ to $p = 0.90$. Then click the **Run Script** button (see Figure 10.12). You should find that the smoothed signal is smoother for $p = 0.95$ than for $p = 0.90$, as illustrated in Figures 10.13(a) and 10.13(b). In fact, if you set the value of $p = 0.001$, the smoothed signal will be almost identical to the original noisy signal. Changing the value of *sigma* will produce signals with more or less noise, depending on the size of *sigma*. Raising the value of *sigma* to *sigma* $= 0.5$ results in a very noisy signal, as shown in Figure 10.13(c). ◆

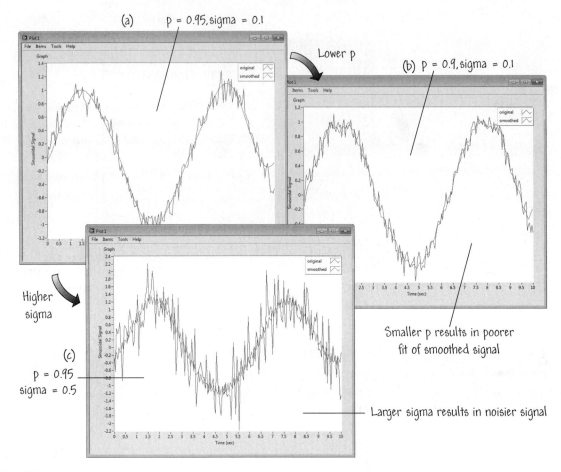

FIGURE 10.13
(a) Smoothing with $p = 0.95$ and *sigma* $= 0.1$; (b) smoothing with $p = 0.90$ and *sigma* $= 0.1$; (c) smoothing with $p = 0.95$ and *sigma* $= 0.5$.

10.6   SAVING, LOADING, AND EXPORTING DATA FILES

10.6.1   Saving and Loading Data Files

In MathScript you can save and load data files in the MathScript Interactive Window. A data file contains numerical values for variables. Being able to save and load data gives you the flexibility to save important data output from a MathScript session for use in external programs. There are two ways to save data files. The first method saves the data for *all* the variables in the workspace, and the second method allows you to select the variables to save to a file.

To save all the variables in the workspace, select Operate≫Save Data in the MathScript Interactive Window. In the file dialog box, navigate to the directory in which you want to save the data file. Enter a name for the data file in the **File name** field and click the **OK** button to save the data file.

The second method allows you to select the variables to save. In this case, in the Command Window, enter the command save filename var1, var2, ..., varn, where filename is the name of the file to store the data in and var1, var2, ..., varn are the variables that you want to save. In this case, the data will be saved in filename in the LabVIEW Data folder in the path specified in File≫LabVIEW MathScript Properties. In Figure 10.14 the process of saving data is illustrated. In Figure 10.14(a), all the variables are saved in the file save_all.mlv after navigating to the folder LabVIEW Data. In Figure 10.14(b), the variable x is saved in the file save_x.mlv.

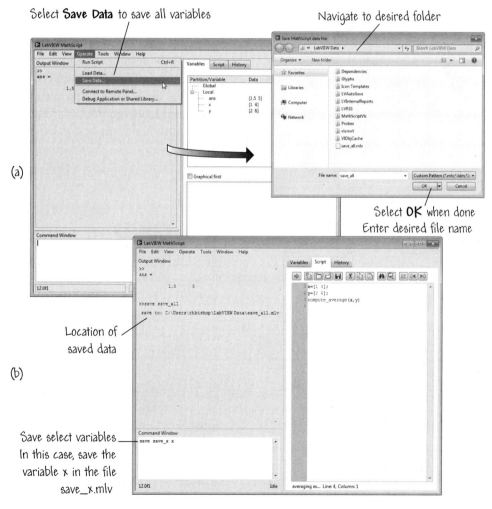

FIGURE 10.14
Saving data files: (a) Saving all the variables in the workspace; (b) saving select variables.

You also can load existing data files into your MathScript session. In the MathScript Interactive Window, select Operate≫Load Data as illustrated in Figure 10.15.

 You must save data files before you can load them into the MathScript Interactive Window.

MathScript can also load .mat and .mlv files created by other technical computing tools that support the m-file script syntax, as long as they only include data types that MathScript supports. Refer to Table 10.2 for MathScript data types.

10.6.2 Exporting Data

In some instances, it may be useful to export your data to an external application. In MathScript this is readily accomplished. On the MathScript Interactive Window, navigate to the Variables window. Select one of the variables that you desire to export. Exporting is accomplished from the Preview Pane and can be done if the display type is **Numeric**, **Graph**, **XY Graph**, or **Picture**. To do so,

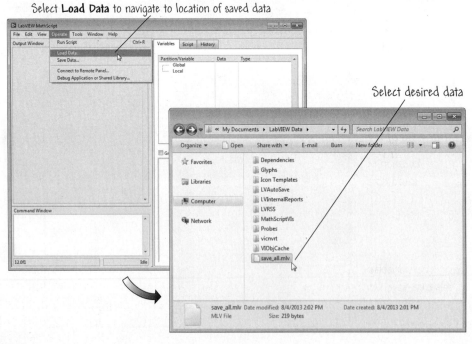

FIGURE 10.15
Loading data from a previous MathScript session.

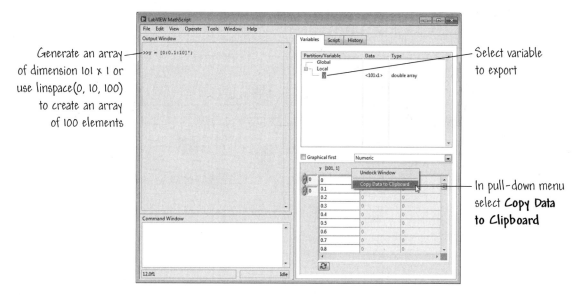

Generate an array of dimension 101 × 1 or use linspace(0, 10, 100) to create an array of 100 elements

Select variable to export

In pull-down menu select **Copy Data to Clipboard**

FIGURE 10.16
Saving data for external applications.

right-click the Preview Pane and select **Copy Data to Clipboard** from the short-cut menu to copy the data for the desired variable, as illustrated in Figure 10.16. You then can paste the data in your external application. In Figure 10.16, the time vector t is being exported. The next step would be to highlight the variable y and export it. Then, both t and y would exist in the spreadsheet for further analysis.

 Since data in spreadsheets is generally in column format, make sure that the variables that you export are in array format of size $n \times 1$, where n is the length of the array. If the array is a $1 \times n$ array, then when exported to a spreadsheet, it will paste in a row rather than a column.

10.7   MATHSCRIPT NODES

Working in the MathScript environment, you can choose a textual approach (with command-line prompt and scripts) as described in the previous sections, or you can choose a graphical approach by employing scripts within VIs in a special node called a **MathScript Node**. The MathScript Node provides a means of combining graphical and textual code within LabVIEW by entering script text directly or importing it from your .m or text files. The MathScript Inter-active Window is used primarily to design and test code for subsequent use in MathScript Nodes. An example of a MathScript Node is shown in Figure 10.17.

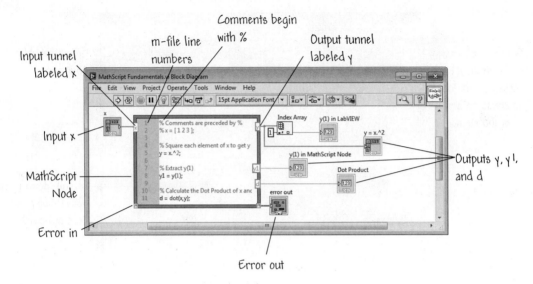

FIGURE 10.17
The MathScript Node.

As seen in Figure 10.17, you define inputs and outputs on the border of the MathScript Node to specify the data that is to be transferred in and out of the node (between the graphical data flow programming and the textual Math-Script code). Your scripts within the MathScript Node can access features from traditional LabVIEW graphical data flow programming. In Figure 10.17, the MathScript Node has a variable input on the left side of the node, named **x**. The right side of the MathScript Node contains the output variables **y**, **y1**, and **d**. By default, the MathScript Node also includes an **error in** input and an **error out** output on the bottom left and right of the node frame, respectively.

The MathScript Node shows line numbers by default. The line numbers are illustrated in Figure 10.17. If you want to show or hide the line numbers, right-click inside the MathScript Node and select **Visible Items≫Line Numbers** from the shortcut menu.

You can find the MathScript Fundamentals VI in Figure 10.17 in **Chapter 10** *of the* **Learning** *directory.*

In many cases, your text-based algorithms that have been verified and used successfully in other situations may be utilized directly rather than re-coded using graphical programming constructs. The idea of the MathScript Node is to provide a mechanism for introducing text-based programming into the graphical programming arena. In the discussion of Structures in Chapter 5, we discussed the use of nodes, with emphasis on the Formula Node, which is another implementation of the same idea of using text-based sequential programming from within the VI.

10.7.1 Accessing the MathScript Node

The MathScript Node is accessed on the **Programming≫Structures** palette or **Mathematics≫Scripts & Formulas** palette, as illustrated in Figure 10.18. First select the MathScript Node from the palette and place it on the block diagram. Keeping the left mouse button depressed you can extend the node frame as illustrated in Figure 10.19. You can also use the **Positioning** tool to extend the MathScript Node to the desired size once it has been placed on the block diagram.

10.7.2 Entering Scripts into the MathScript Node

There are three ways to place a script in the MathScript Node. You can use the cursor tool to type the script directly in the MathScript Node. For short and simple scripts this is a convenient way to utilize the MathScript Node. The syntax is the same as described in Section 10.4.

You can also import an existing script from a file into the MathScript Node. To import a script, right-click the MathScript Node border and select **Import** from the shortcut menu to display the **Select a Script** dialog box. When the dialog box appears, select the file you want to import and click **Open**. The MathScript text will then appear in the script node, as shown in Figure 10.20. You

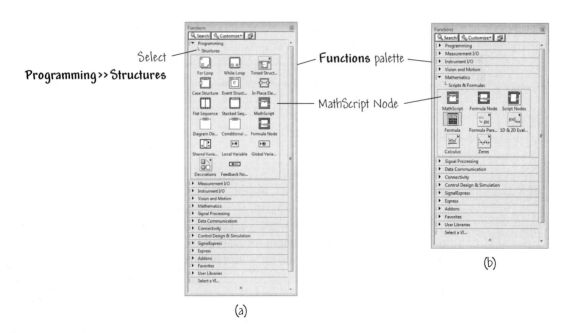

FIGURE 10.18
The MathScript Node is on the (a) **Programming≫Structures** palette and (b) **Mathematics≫Scripts & Formulas** palette.

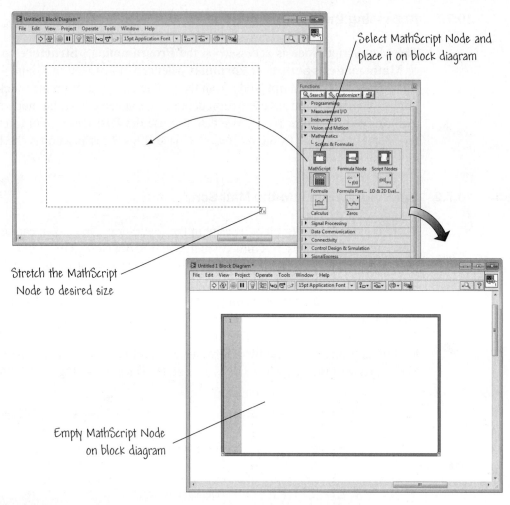

FIGURE 10.19
Placing the MathScript Node on the block diagram.

can also cut and paste a script into a MathScript Node using any text editor, including the one in the MathScript Interactive Window.

It is suggested that you write your script and run it within the MathScript Interactive Window environment for testing and debugging purposes before you import it into LabVIEW. Once the script is in the node, you will need to wire up the required inputs and outputs.

10.7.3 Input and Output Variables

The script within the MathScript Node most likely will require inputs to carry out the desired computations and the results will need to be output.

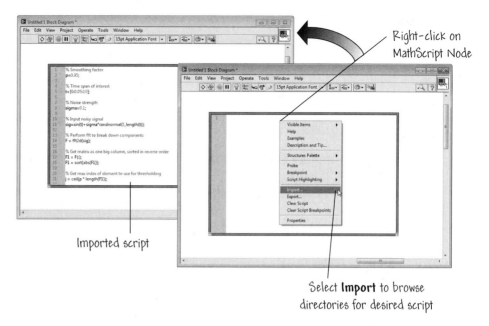

FIGURE 10.20
Importing a script into the MathScript Node.

To add an input variable, right-click the MathScript Node border and select **Add Input** from the shortcut menu, as shown in Figure 10.21. When the input variable appears on the node, you should add the name of the variable. For example, in Figure 10.21, the input p was added to provide a mechanism for changing the smoothing parameter (see the example Practice with Scripts in Section 10.5.2 for a description of the smoothing parameter). You can also use the **Labeling** tool to edit the variable names at any time by double-clicking on the variable name inside of the tunnel and editing, as desired.

The data types of the input and output terminals on the MathScript Node are determined at edit time. Only the data types shown in Table 10.2 can be executed within the node. Selection is handled differently for the input and output terminal data types.

For the output terminals, by default, the data type is automatically selected when you click outside a MathScript Node. To manually change the data type, right-click the terminal of the output and select **Choose Data Type≫All Types** from the shortcut menu. A list of the available data types appears, reflecting those shown in Table 10.2. From this list you can select the data type you want to use. Once you have performed the configuration manually, the automatic selection feature for that terminal is disabled. To re-enable it, right-click the output terminal and select **Choose Data Type≫Auto Select Type** from the shortcut menu.

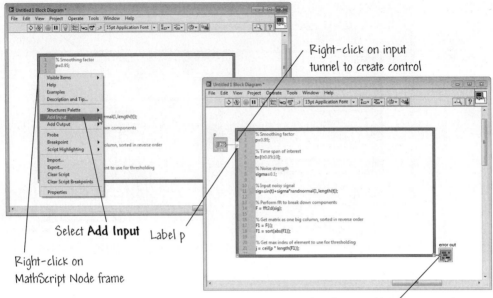

FIGURE 10.21
Adding inputs to the MathScript Node.

On input terminals, you cannot change the data type. Instead, if you wire an unsupported data type to an input terminal on a MathScript Node, LabVIEW displays a broken wire and you must fix the error before the VI will run.

By default the MathScript Node includes one input and one output for the error in and error out parameters. To take advantage of the error-checking parameters for debugging information, it is suggested that you create an indicator for the error out terminal on the MathScript Node before you run the VI. This allows you to view the error information generated at runtime.

To add an output variable, right-click the MathScript Node border and select **Add Output** from the shortcut menu, as illustrated in Figure 10.22. You can create controls and indicators for each input and output on the MathScript Node. For example, to create an indicator on an output terminal, right-click the output terminal and select Create≫Indicator from the shortcut menu. LabVIEW will create an indicator on the front panel and wire terminals to the output on the block diagram, as shown in Figure 10.21 and Figure 10.22.

The script in Figure 10.22 is the signal smoothing.m script discussed in the previous section and is too long to conveniently display in a MathScript Node. In this case, you can display a scrollbar within the script node by right-clicking inside the MathScript Node and selecting Visible Items≫Scrollbar.

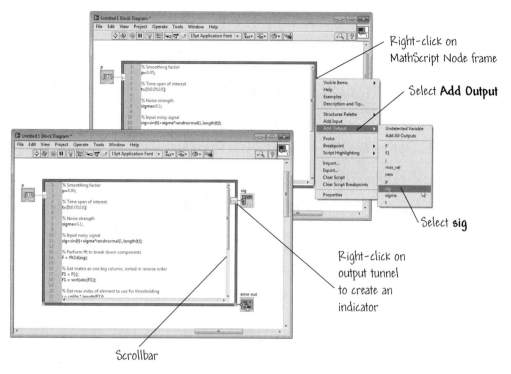

FIGURE 10.22
Adding outputs to the MathScript Node.

Practice with MathScript Nodes

In this example you will load an existing VI that employs the signal smoothing script used in Section 10.5 to smooth a noisy signal. To begin go to the Getting Started screen and select **Open**. When the dialog box appears, navigate to **Chapter 10** of the **Learning** directory and open **smoothing.vi**. The front panel and block diagram are shown in Figure 10.23.

On the front panel, you can input the smoothing parameter, **p**, the time interval of interest, **t**, and the noise level, **sigma**. The time interval is defined by the initial time (usually $t_0 = 0$), the final time, t_f, and the time increment, **dt**. As in Section 10.5, the nominal values are:

$$p = 0.95$$
$$t = [t_0 : dt : t_f] = [0 : 0.1 : 10]$$
$$sigma = 0.1$$

Execute the VI by selecting the **Run** button on the VI toolbar, as shown in Figure 10.23. You should notice that the results are the same as obtained when using the script file in the MathScript Interactive Window.

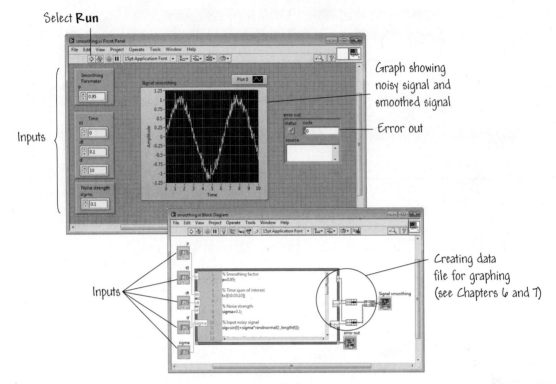

FIGURE 10.23
Using the signal smoothing.m script within the MathScript Node.

It is a straightforward task to vary the smoothing parameter, p, and observe the effects on the signal smoothing. The same can be said about varying the time interval and the noise level. By clicking the **Run Continuously** button, you can change any of the input variables and see the results simultaneously because the VI is being repeated over and over.   ◆

10.7.4   Script Highlighting

Script highlighting uses colors and font characteristics to differentiate elements of a script in a MathScript Node. This improves readability and helps with debugging of the script. There are two types of script highlighting. With **syntax highlighting** the various script elements are displayed in different colors. For example, keywords are displayed in blue and comments in green. **Data type highlighting** shows the data types of variables (see Table 10.2 for more information on available data types) in different colors and font attributes. For example, floating-point data types are shown in orange and integer data types in blue.

You can use **Context Help** to identify the data type of a variable in your script. To accomplish this, click outside the MathScript Node and then move the cursor over the variable of interest within the node. The data type will be displayed in the Context Help window. You also gather information on built-in MathScript and user-defined functions by moving the cursor over the function of interest to display helpful information in the Context Help window. The user-defined function will display the first group of commented lines in the script.

*For more details on the selected colors and font attributes that MathScript is using for the various script elements and data types, navigate to the **Script Highlighting** section on the **MathScript** page of the **Tools≫Options** dialog box. Here you can inspect the default colors and font attributes, and you can even select new colors and font attributes to suit your own needs and desires.*

To enable syntax highlighting or script highlighting, right-click inside the MathScript Node and select **Script Highlighting≫Syntax** or **Script Highlighting≫Data Types** from the shortcut menu, as illustrated in Figure 10.24. Syntax highlighting is enabled by default. Any color and style settings you apply to scripts with the **Text Settings** pull-down menu on the toolbar are discarded if you enable syntax or data type highlighting.

Data type highlighting is not applied if a node contains an error, represented by a red glyph in the gray region on the left side of the node, as shown in Figure 10.24.

If you want to disable script highlighting on a particular MathScript Node, right-click inside the MathScript Node and select **Script Highlighting≫None**. If you want to disable (or re-enable) script highlighting for all MathScript Nodes, use the **Script Highlighting** section on the **MathScript** page of the **Tools≫Options** dialog box. Now if you use the **MathScript** page to disable script highlighting for all MathScript Nodes, the **Script Highlighting** item is dimmed in the shortcut menus of MathScript Nodes, and you cannot enable it without going back to the **Tools≫Options** dialog box. Similarly, if you disable script highlighting for a particular MathScript Node, you cannot use the **MathScript** page to enable script highlighting for that node.

10.7.5   Debugging Scripts

Many key features of the MathScript Node are available to aid in debugging scripts. These include line numbers (see Figure 10.17), syntax and data type

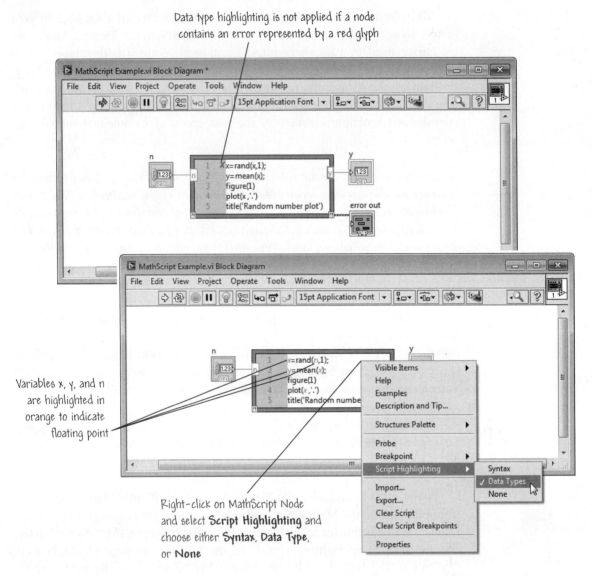

FIGURE 10.24
Script highlighting.

highlighting (see Section 10.7.4), probes, execution highlighting, breakpoints, and error checking before run time. It is anticipated, however, that many students will use the MathScript Interactive Window primarily to design and test code for subsequent use in MathScript Nodes as part of the overall VI development strategy. In this case, much of the debugging and testing of the actual script may have occurred beforehand. However, issues may remain related to integrating the script into the VI, and these will likely require further debugging and testing.

 *It is highly recommended that you take advantage of the error-checking features of the MathScript Node. Create an indicator for the **error out** terminal on the MathScript Node before you run the VI, so that you can view any generated error information at run time.*

When you first copy-and-paste or import a script into the MathScript Node, you will immediately see what is and is not supported because of the error checking that occurs before run time. A red glyph in the gray region on the left side of the MathScript Node indicates that the associated line of script contains an error or unsupported syntax. Of course, such errors must be fixed before the VI can run. If you have enabled script highlighting (as discussed in Section 10.7.4), then the syntax highlighting will let you see whether a user-defined function or a variable has overridden a built-in MathScript function, and the data type highlighting will let you to see whether the script has redefined a variable from one data type to another. Once the VI is able to run, then the process of debugging and testing of the script gets underway. The objective is to make sure that the script is performing the tasks as desired and producing the expected outputs.

The debugging features discussed in Chapter 3 for VIs generally apply to scripts in the MathScript Nodes. For example, you can use probes to view the data in the script as the VI runs and use execution highlighting to see which line of script is executing currently. You can single-step through the script to view each each of its actions and set breakpoints to pause execution.

The **LabVIEW MathScript Probe** (referred to here simply as the probe) provides a way to view the values of the various variables in the script in a MathScript Node as the VI runs. The MathScript probe is not the same as the Probe Watch Window described in Chapter 3, but instead is oriented toward supporting the MathScript environment. Figure 10.25 illustrates the LabVIEW MathScript Probe. The probe displays in the Variables window a list of variables that you define and use in your script. The probe also displays the output generated by the script in the Output window. Notice that the Variables window in the probe includes the same components as the Variables window in the MathScript Interactive Window (see Section 10.2). You can select a variable from among the list of variables and graph the data in the Variables window just as you did with data in the MathScript Interactive Window. As data becomes available, the probe immediately updates and displays it with execution highlighting, single-stepping, and breakpoints to help you determine whether and where data is incorrect in the script.

To display the probe, right-click the MathScript Node and select **Probe** from the shortcut menu, as illustrated in Figure 10.25. A floating LabVIEW

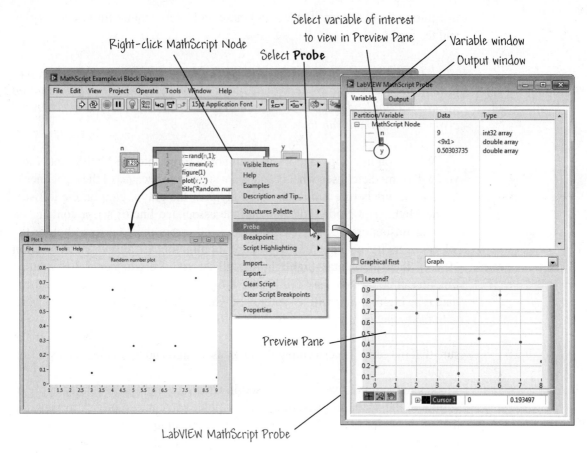

FIGURE 10.25
LabVIEW MathScript Probe.

MathScript Probe window appears. You can also use the **Probe** tool and click the MathScript Node to display the probe. Once you run the VI, the probe window will display data for the variables in the script. Select a variable in the Variable List to view it in the Preview Pane. Depending on the data type, you can view the data in numeric, graph, XY graph, and sound format.

If your VI has more than one MathScript Node, you can add a probe to each node. As you add each additional probe, the associated MathScript Node is added to the Variable List on the Variables page on the probe window. A new Output page for each node is also displayed. It is suggested that you name each MathScript Node in the VI in a unique way to distinguish between them, as the probe window will then show the nodes by their unique names. Otherwise, each MathScript Node in the probe window is sequentially numbered, making them more difficult to associate with their corresponding scripts. To remove the LabVIEW MathScript Probe, close the probe window.

*If your desktop gets cluttered and you cannot readily find the probe, you can right-click the MathScript Node and select **Find Probe** from the shortcut menu.*

Execution highlighting is another important tool in the available arsenal to debug and test VIs employing MathScript Nodes. As discussed in earlier chapters, execution highlighting provides an animation of the VI execution on the block diagram. It is very helpful in visualizing the data flow as part of the debugging and testing process. Execution highlighting can be used to show the progression from one line of the script to the next in MathScript Nodes. The execution animation displays a blue arrow that blinks next to the line that is currently executing. Remember that execution highlighting greatly reduces the speed at which the VI runs.

To enable execution highlighting, click the **Highlight Execution** button on the block diagram toolbar. When you run the VI, you can watch the animation inside the MathScript Node on the block diagram. Click the **Highlight Execution** button again at any time to disable execution highlighting.

The final debugging feature we will discuss in this section is the breakpoint. As explained in Chapter 3, breakpoints are very useful tools in the debugging of VIs. For example, when you set a breakpoint using the **Breakpoint** tool on a wire, execution pauses after data passes through the wire, and the **Pause** button appears red to indicate it is now the **Continue** button. Fortunately, you also can set a breakpoint on a line of script in a MathScript Node.

Use the **Breakpoint** tool to click on a line of script in the MathScript Node to set the breakpoint, as illustrated in Figure 10.26. A red dot will appear to the right of the line number (or immediately to the left of the line of code in the gray box if line numbers are disabled). As shown in Figure 10.26, you also can right-click the line of code and select **Breakpoint≫Set Breakpoint** from the shortcut menu. When the script pauses at the breakpoint, a marquee appears around the MathScript Node and a flashing blue arrow appears to the left of the red dot, as shown in Figure 10.27. When you reach a breakpoint during execution, you can choose among several possible actions. You can single-step through execution using the single-stepping buttons, check intermediate values on probes, change values of front panel controls, or click the **Pause/Continue** button on the block diagram toolbar to continue execution or the **Stop** button to abort execution of the VI. Remember that the **Pause** and the **Continue** buttons have the same symbol but are black and red, respectively.

You cannot place breakpoints on certain lines of script in MathScript Nodes. If you try to do so, the breakpoint is automatically moved to the next line that supports breakpoints.

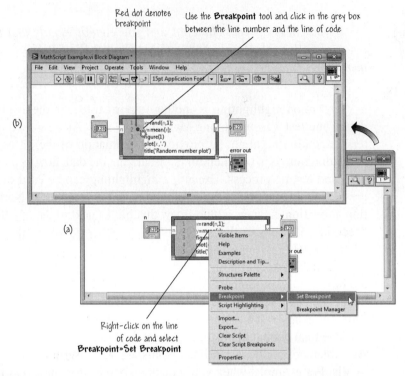

FIGURE 10.26
Setting breakpoints in the script using (a) the shortcut menu; (b) the **Breakpoint** tool.

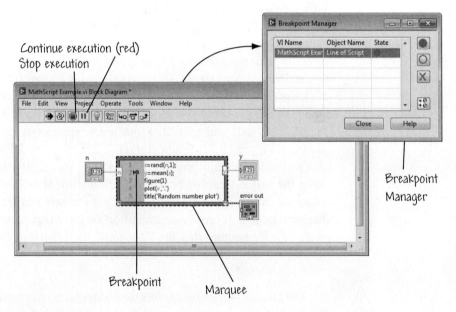

FIGURE 10.27
Pausing the execution of a script using a breakpoint.

To locate and manage all breakpoints in the VI hierarchy (including in the MathScript Node), select **View≫Breakpoint Manager**. Double-click any item in the listbox to highlight the breakpoint in the script.

To disable a breakpoint in the script, right-click the line of script with a breakpoint and select **Breakpoint≫Disable Breakpoint** from the shortcut menu, as shown in Figure 10.28. The red dot breakpoint indicator will transition to a red circle (with no color fill). To enable a breakpoint that you previously disabled, right-click the associated line of script and select **Breakpoint≫Enable Breakpoint** from the shortcut menu. You can disable or enable breakpoints individually or all at once using the Breakpoint Manager window.

There are several ways to remove breakpoints from the MathScript Node. To remove all breakpoints in a script, right-click the MathScript Node and select **Clear Script Breakpoints** from the shortcut menu. To remove a particular breakpoint, click the red dot breakpoint indicator with the **Breakpoint** tool or right-click the line of the script and select **Breakpoint≫Clear Breakpoint** from the shortcut menu. To remove all breakpoints in the VI hierarchy, select

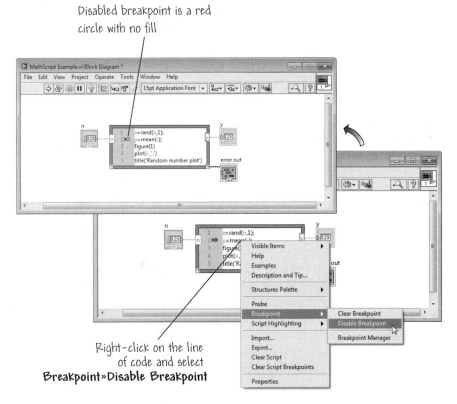

FIGURE 10.28
Disabling breakpoints in the script.

Edit≫Remove Breakpoints from Hierarchy. You also can remove all breakpoints in the VI hierarchy using the Breakpoint Manager window.

10.7.6   Saving Scripts from within the MathScript Node

You may want to save your script contained in a MathScript Node to a text file. In this way, you can later open this text file in a MathScript Interactive Window for further analysis and use. To save the text-based script contained in a MathScript Node, right-click the MathScript Node border and select **Export** from the shortcut menu to display the **Name the Script** dialog box, as illustrated in Figure 10.29. Enter the desired new file name or select the file you want to overwrite and click **Save**. MathScript script files are text files, and although text files usually have a .txt extension, MathScript files should have a .m extension.

Analyzing Random Numbers

In this exercise you will construct a VI that utilizes a MathScript Node. The purpose of the VI is to generate and plot a given number of random numbers. The script will also compute the average of the random numbers for output. The VI is shown in Figure 10.30.

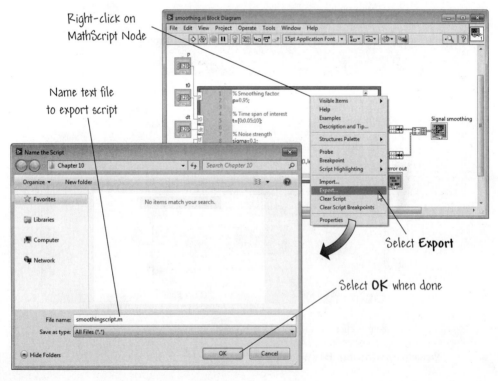

FIGURE 10.29
Exporting MathScript Node scripts to an external text file.

Open a new VI and begin by placing a MathScript Node on the block diagram. Size it large enough to allow for the required lines of code (refer to the block diagram in Figure 10.30). Add an input, n, representing the number of random numbers to generate and plot. Add an output, y, to the script node representing the average of the n random numbers. By default, the input and output variables are double-precision floating-point numerics. Input variables will automatically adapt to the data type provided them. To change the data type of an output variable, right-click it and select Scalar≫DBL, as illustrated in Figure 10.31. Also, use an error out indicator to show any error out messages generated.

Run the VI when ready. Vary the number of random numbers in the set and see how the average varies correspondingly. The MathScript Node will generate a plot of the random numbers using LabVIEW graphics. When you are finished experimenting, save the VI as MathScript Example.vi in the Users Stuff folder.

A working version of MathScript Example.vi can be found in the **Chapter 10** *folder in the* **Learning** *directory.*   ◆

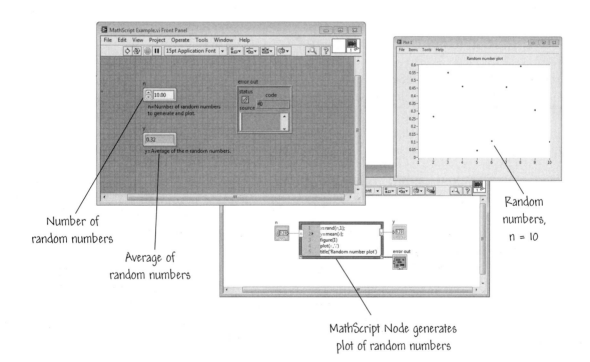

FIGURE 10.30

Using the MathScript Node to generate, plot, and analyze a sequence of random numbers.

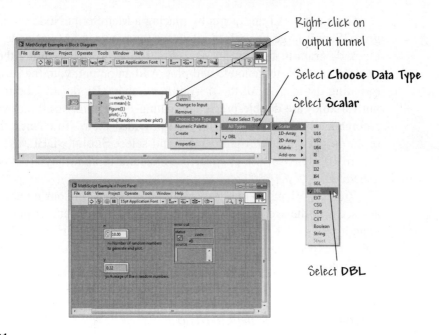

FIGURE 10.31
Choosing the data type for input and output variables on MathScript Nodes.

10.8 APPLICATIONS OF MATHSCRIPT RT MODULE

The MathScript Interactive Window and the MathScript Node provide a means to utilize text-based programming in both a command-line setting as well as interacting with graphical programming constructs in VIs. There are several likely applications of MathScript that are worth noting here.

10.8.1 Instrument Your Algorithms

One use of the MathScript Node is to "instrument your algorithms." Taking advantage of the LabVIEW front panel and powerful graphical user interface (GUI) capability, users can explore and interface with the values of several variables to better understand the detailed workings of their algorithms. The process involves importing your scripts into a MathScript Node and then associating the variables in the algorithm with the controls and indicators available in LabVIEW (e.g., knobs, sliders, and buttons). Figure 10.32 shows the instrumentation of the sine function within a MathScript Node. Of course, the sine function is a simple algorithm (relatively speaking), but the same process applies to more complicated algorithms.

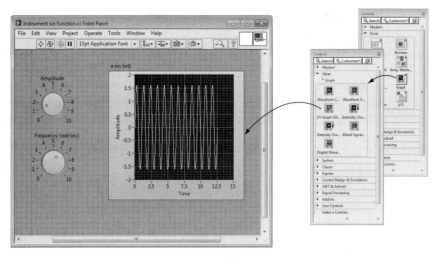

FIGURE 10.32
Instrumenting scripts using LabVIEW controls and indicators.

The Instrument sine function VI in Figure 10.32 can be found in Chapter 10 *of the* Learning *directory.*

10.8.2   Graphical Signal Processing, Analysis, and Mathematics

As discussed in Section 10.4.1, MathScript provides access to an extensive library of algorithms that can be used in scripts and interactively in the Math-Script Interactive Window. The other side of the coin is that the MathScript Node provides connectivity of the text-based algorithms within the MathScript Node to an extensive library of graphical tools for signal processing, analysis, and mathematics. The LabVIEW library of greater than eight hundred graphical VIs provides functionality covering a diverse set of areas including signal generation, signal conditioning, monitoring, digital filtering, windowing, spectral analysis, transforms, curve fitting, interpolation and extrapolation, probability and statistics, optimization, ordinary differential equations, geometry, polynomial functions, 1D/2D evaluation, calculus, and more.

Using the power of the algorithms associated with MathScript coupled with the power of the graphical programming VIs available in LabVIEW will lead to advanced applications in signal processing, analysis, and mathematics. Additional detail on the analysis functionality of LabVIEW can be found in Chapter 11.

10.8.3   Integrating Measurement Hardware

Experienced users of LabVIEW know that the graphical environment can be readily used to manage data acquisition operations. Integrating text-based programming with graphical programming naturally leads to simplified data acquisition, signal generation, and signal analysis. Scripts that execute from within MathScript Nodes can take advantage of the hardware control capabilities of the LabVIEW development environment.

An important feature of LabVIEW is that variables can be shared between the MathScript Interactive Window and the MathScript Node environments. The capability to share variables allows real-world signals to be manipulated in the MathScript Interactive Window for algorithm development before deploying the algorithm in the MathScript Node.

The capability to share variables is accomplished using **global variables**. Global variables are declared using the command global with the following syntax:

> global stars;

In this usage the variable stars has been declared a global variable.

As an example, consider the VI shown in Figure 10.33. The block diagram uses a Simulate Signal Express VI, a Convert from Dynamic Data Express VI, a waveform graph, and a MathScript Node. The Simulate Signal data type must be converted from a variant to a 1D array of DBL using the Convert from Dynamic Data Express VI (use **Quick Drop** to find unfamiliar VIs) with the default configuration.

The MathScript Global.vi pictured in Figure 10.33 is located in Chapter 10 of the Learning directory.

The global command should appear in the MathScript Node, as shown in Figure 10.33, to declare the desired variables as being global and therefore accessible in the MathScript Interactive Window. To illustrate the global variable feature, execute the MathScript Global VI containing the MathScript Node. When the VI execution ends, navigate to the Command Window in the MathScript Interactive Window environment, type global stars and press **Enter**. This will allow you to see the data passed to the variable stars on the MathScript Node.

In the Variables window of the MathScript Interactive Window, you will see the variable stars, as illustrated in Figure 10.33. Select stars, and then (following the discussion in Section 10.2.3) choose to view the stars data in graphical form. The XY Graph of the data, as shown in the Preview Pane of Figure 10.33, resembles the plot of the data obtained programmatically in the

Warning indicates MathScript Node
runs with slower performance due to
global variable

Add "global stars" in the MathScript Node   **stars** is now a global variable

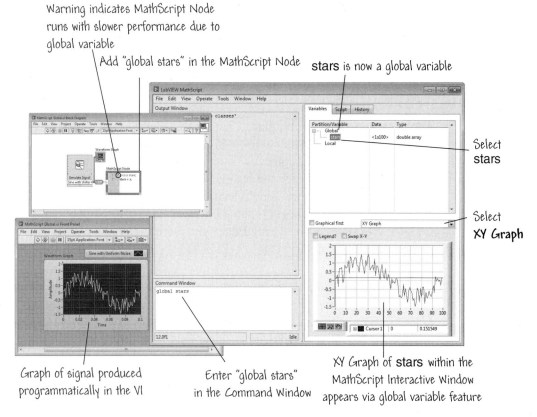

Select
stars

Select
XY Graph

Graph of signal produced
programmatically in the VI

Enter "global stars"
in the Command Window

XY Graph of **stars** within the
MathScript Interactive Window
appears via global variable feature

FIGURE 10.33
Sharing a global variable between the MathScript Interactive Window and the MathScript Node.

MathScript Global VI. The data is identical, but the plot parameters (such as color) may differ between the Preview Pane and the Waveform Graph.

Having ready access to the same variables in the graphical programming VI and in the MathScript Interactive Window environment provides a superb environment in which to design, test, and verify algorithms.

BUILDING BLOCK

10.9 BUILDING BLOCKS: PULSE WIDTH MODULATION

In this Building Block, you will modify both the Rising Edge and Falling Edge VIs that you constructed in Chapter 2. Then you will utilize a MathScript Node to compute the rising and falling edge. This is an alternative way of creating the Rising and Falling Edge VIs using MathScript.

The basic function needed by the MathScript Node to achieve the desired task is the modulus function. The mod(A,B) function computes the remainder of A/B, when the quotient is rounded toward $-\infty$. The modulus function syntax in MathScript is

$$X = mod(A,B)$$.

The mod function will be placed within the MathScript Node to replace the graphical programming functions (e.g., Round to Nearest) employed in the Building Blocks section in Chapter 2.

The block diagram for the Rising Edge VI using the MathScript Node is shown in Figure 10.34 and should be used as a guide in the development of this Building Block VI. To begin, open the Rising Edge VI that you constructed in Chapter 2. Save the VI as Rising Edge–MathScript VI in the **Users Stuff** folder in the **Learning** directory.

 If you successfully completed Building Blocks in Chapter 2, you should be able to find working versions of **Rising Edge.vi** *and* **Falling Edge.vi** *in the* **Users Stuff** *folder of the* **Learning** *directory. If not, you can find working versions in the* **Building Blocks** *folder of the* **Learning** *directory.*

Switch to the block diagram of Rising Edge–MathScript VI. At this point, it should be exactly the same as the block diagram for Rising Edge VI from the Chapter 2 Building Block. Delete all the functions and wires on the block diagram, but leave the inputs and outputs unchanged. The functions deleted include a Multiply, Subtract, Divide, Equal?, and Round to Nearest.

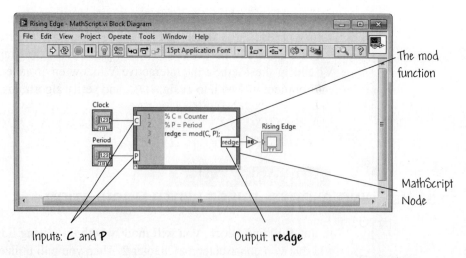

FIGURE 10.34
The Rising Edge VI using the mod function within a MathScript Node.

Using Figure 10.19 as a guide, place a MathScript Node on the block diagram and enlarge it sufficiently to allow the placement of three lines of code, as illustrated in Figure 10.34. As discussed in Section 10.7.2, add the script within the MathScript Node. The inputs to the script are Clock and Period, and the output is redge.

To add input and output variables to the MathScript Node, right-click on the border and select **Add Input**, as described in Section 10.7.3. Add two input variables denoted C and P, for Clock and Period, respectively. Follow the same procedure to add the output variable redge.

The block diagram for the Rising Edge VI is illustrated in Figure 10.34. Once the VI is wired, set the inputs to P = 100 and C = 100 and select **Run**. Verify that the VI is operating as expected by comparing the output with the Rising Edge VI when considering the same inputs. When done, save the VI as Rising Edge–MathScript.vi in the Users Stuff folder in the Learning directory.

Following the same procedure, edit the Falling Edge VI that you created in Chapter 2 Building Blocks. Verify that the VI is operating as expected by comparing the output of the Falling Edge–MathScript VI with the Falling Edge VI when considering the same inputs. Remember to save the VI as Falling Edge–MathScript.vi in the Users Stuff folder in the Learning directory.

Working versions of Rising Edge–MathScript.vi *and* Falling Edge–Math Script.vi *are in the* Building Blocks *folder of the* Learning *directory.*

10.10   RELAXED READING: ACQUIRING AND ANALYZING THE BIOACOUSTIC COMMUNICATION OF KILLER WHALES

In this reading, we highlight a data acquisition and analysis system to study how killer whales use echolocation in their natural habitat in the cold, rough waters of the North Sea. The software was modified to monitor the birth and echolocation development of the first harbor porpoise calf ever to survive in captivity.

Killer whales use biological sonar, also known as echolocation, for communication, identification, and navigation. The whale emits short ultrasonic "clicks" that travel long distances because of the acoustical properties of water. The echoes of the clicks are returned from objects in the environment to provide the whale with information regarding its surroundings. To understand how whales use echolocation in their natural habitat, we need to understand the properties of the clicks. For example, the echolocation clicks of a sperm whale have very high source levels and the potential to operate in long-range biosonar systems. On the other hand, low-amplitude sounds such as those produced by small-toothed whales work only with short-range biosonar systems.

Understanding the ultrasonic emissions of the killer whales requires that the signals be collected in their natural habitat—the very cold, rough waters of the North Sea above the Arctic Circle. Historically, it has been challenging to find reliable equipment for use in these types of environments. Sophisticated software and hardware are required that can stand up to the harsh environmental elements while being easy to use in the field. The signal-collection system must also overcome the challenge of interference from other noise sources, including boats and other animals.

A system based on LabVIEW software and NI data-acquisition hardware was developed to meet the challenge. The recording gear consists of a linear hydrophone array, a multichannel preamplifier, and a bandpass filter connected to a digital recorder. High-speed digital recorders are used with signal-processing software based on LabVIEW. The hardware includes the NI USB-6251 high-speed M Series multifunction DAQ module to provide the required resolution and channel count. The USB interface is employed to manage up to 30 to 50 GB of data each day, and it has a dedicated software solution for logging, alarming, playback, and analysis. Additionally, to avoid noise interference from other electrical units onboard the ship, all the equipment is battery powered.

The recorded wave files are loaded and prefiltered to automatically find and measure click locations using peak detection. Integration with LabVIEW provides more functionality for logging, monitoring, and analysis. Functions available include amplitude demodulation and envelope detection as well as filtering and squelch. The results can be displayed and saved in the same application with the LabVIEW MathScript routine.

FIGURE 10.35
High-speed digital recorders are used with signal-processing software based on LabVIEW to monitor the birth and echolocation development of the first harbor porpoise calf ever to survive in captivity. (Courtesy of Fjord & Bælt.)

The software was modified to monitor the birth and echolocation development of the first harbor porpoise calf ever to survive in captivity, as illustrated in Figure 10.35. Simultaneously recording 10 signals sampled at 500 kHz each on one laptop for several hours every day was accomplished without any software or hardware problems. Researchers logged hundreds of gigabytes of unique data on whale echolocation. For more information, see http://sine.ni.com/cs/app/doc/p/id/cs-11543.

10.11   myDAQ BUILDING BLOCKS

Some calculations may be more easily accomplished with text-based code than graphical code. Begin by opening **Read Temp with MathScript unedited.vi** found in the **Chapter 10** folder in the **myDAQ Building Blocks** folder within the Learning directory. In this myDAQ Building Block, you will replace the **Temp Conversion.vi** with a MathScript node.

The **Temp Conversion.vi** converts the resistance of the thermistor (as determined with a voltage divider) into temperature using the Steinhart-Hart equation

$$\frac{1}{T} = k_a + k_b \, (\ln R) + k_c (\ln R)^3$$

where T is temperature (Kelvin), R is resistance (Ω), and the k_a, k_b, and k_c terms are the Steinhart-Hart coefficients for the thermistor. The Steinhart-Hart coefficients can be experimentally determined, but you can find standard coefficients for the EPCOS B57164K103J thermistor on the front panel of **Temp Conversion.vi**. Double-click on the subVI **Temp Conversion.vi** in Figure 10.36 to find that $k_a = 0.00108403$, $k_b = 0.000246475$, and $k_c = -1.24014E-9$. Use the steps below as a guide to construct a conversion formula using the MathScript Node.

1. Right-click the left border of the MathScript Node and select **Add Input**. Name this input r (to indicated resistance).

2. Using the Steinhart-Hart equation and the given constants, write a script to calculate temperature from resistance. Remember to include the conversion from Kelvin to Celsius ($T_{Celsius} = T_{Kelvin} - 273.15$).

The MathScript Interactive Window, accessed by selecting **Tools**≫ **MathScript Window,** *can be used to test and debug your script, then copy and paste it into the MathScript Node.*

3. Right-click the right border of the MathScript Node and select **Add Output**. Choose the variable that you assigned to the temperature value (in degrees Celsius).

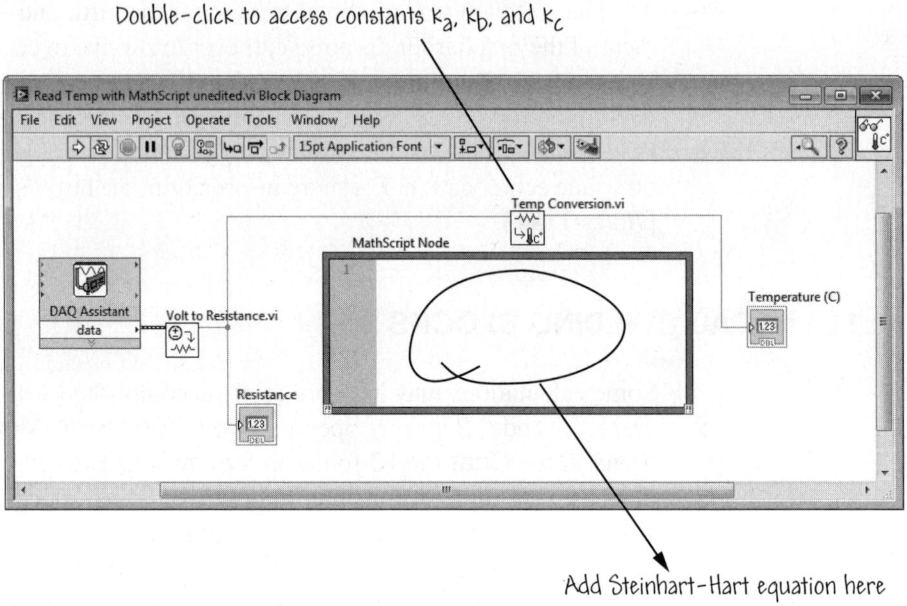

Double-click to access constants k_a, k_b, and k_c

FIGURE 10.36
The block diagram of the Read Temp with MathScript unedited.vi.

4. Delete the **Temp Conversion.vi**, connect the resistance output of **Volt to Resistance.vi** to the MathScript Node input, and connect the temperature output to the **Temperature (C)** indicator.

5. Run the VI several times to make sure that you get expected results.

6. Save the VI as **Read Temp with MathScript.vi** in the Chapter 10 folder in the **myDAQ Building Blocks** folder within the **Learning** directory.

 A working version of **Read Temp with MathScript.vi** *can be found in the* **Chapter 10** *folder of the* **myDAQ Building Blocks»Solutions** *folder in the Learning directory.*

In this myDAQ Building Block, you created a VI that converted the resistance of the thermistor into temperature using the Steinhart-Hart equation within a MathScript Node. The MathScript Interactive Window was found to be a useful tool to create and test scripts before copying and pasting into the MathScript Node.

10.12   SUMMARY

A math-oriented, textual programming language called MathScript was introduced. We discussed the MathScript Interactive Window and the MathScript Node. With the MathScript Interactive Window students can interact with Lab-VIEW through a command prompt, and with MathScript Nodes students can integrate scripts with virtual instruments. The essentials of creating user-defined functions and scripts, of saving and loading data files, and of using the Math-Script Node were presented. The debugging and testing tools associated with VIs were applied to MathScript Nodes, including line numbers, syntax and data type highlighting, probes, execution highlighting, breakpoints, and error checking before run time. It is anticipated that students will use the MathScript Interactive Window primarily to design and test code for subsequent use in MathScript Nodes as part of the overall VI development strategy.

KEY TERMS

Breakpoint: A device used to programmatically pause the execution of a VI, including scripts in a MathScript Node, for debugging and testing purposes.

Class: A grouping of MathScript functions that perform mathematical or signal processing calculations and analysis using a text-based language.

Command History Window: Displays a history of the commands you executed.

Command Line: The field (or line) in the Command Window where commands are entered.

Command Window: Specifies the MathScript command you want LabVIEW to execute.

Data type highlighting: Data type highlighting shows the data types of variables in different colors and font attributes. For example, floating-point data types are shown in orange and integer data types in blue.

Execution highlighting: A feature used to show the progression from one line of the script to the next in MathScript Nodes. The execution animation displays a blue arrow that blinks next to the line currently executing.

Function: MathScript includes more than five hundred built-in functions, and users can develop new functions by creating **m**-file script text files that contain text that specifies the desired functionality.

Global variables: A variable that can be accessed and passed among several VIs and between the MathScript Interactive Window and MathScript Nodes.

LabVIEW MathScript Probe: A window to view the values of variables in the script in a MathScript Node as the VI runs.

MathScript: A high-level, text-based programming language with an easily accessible syntax and ample functionality to address programming tasks related to signal processing, analysis, and mathematics.

MathScript Interactive Window: An interactive interface used to edit and execute mathematical commands, create mathematical scripts, and view numerical and graphical representations of variables.

MathScript Node: A LabVIEW structure that is used to programmatically execute scripts within a VI.

Output Window: Displays the commands you entered in the Command Window and the output that MathScript generates from those commands.

Preview Pane: Displays in numercial, string, graphical, or sound format the variables you select in the variable list (in the Variables window).

Script: A sequential program containing groups of commands (generally) saved in a file called an **m**-file available for use in LabVIEW sessions.

Script highlighting: A feature of MathScript Nodes that employs colors and font characteristics to differentiate elements of a script and includes syntax highlighting and data type highlighting.

Script Editor window: Displays the scripts that you create.

Subfunction: All functions following the first function in a MathScript file are subfunctions and are accessible only to the main function.

> **Syntax highlighting:** With syntax highlighting the various script elements are displayed in different colors. For example, keywords are displayed in blue and comments in green.
>
> **Variables window:** Displays a list of all variables you define and previews variables that you select.

EXERCISES

E10.1 Open the MathScript Interactive Window. In the Command Window, use the help command to learn about the different usages of the plot function. In your own words, how will m and n be plotted with the syntax plot(m,n)?

E10.2 Write a script to generate a 3 × 2 matrix **M** of random numbers using the rand function. Use the help command for syntax help on the rand function. Verify that each time you run the script the matrix **M** changes. Save the script in the Users Stuff folder in the Learning directory.

E10.3 In the MathScript Interactive Window, create a script that generates a time vector over the interval 0 to 10 with a step size of 0.5 and creates a second vector, y, according to the equation

$$y = e^{-2t}\left(0.3\sin(t) + 0.7\cos(0.1t)\right)$$

Then, add the plot function to generate a graph of y versus t.

Type the script in the Script Editor window and, when done, use the **Save As** button in the Script Editor to save the script in the Users Stuff folder in the Learning directory. Clear the script from the Script Editor window, and then **Load** the script back into the Script Editor and select **Run Script**.

E10.4 Using a MathScript Node, generate a stem plot of a cosine function, $y = \cos(t)$, where t varies from 0 to π, with an increment of $\pi/20$.

E10.5 Open the MathScript Interactive Window. In the Command Window, create the matrices **A** and **B**:

$$\mathbf{A} = \begin{bmatrix} 1 & 5 \\ 2 & 9 \\ 5 & 1 \end{bmatrix} \text{ and } \mathbf{B} = \begin{bmatrix} 7 & 4 & 3 \\ 4 & 3 & 1 \end{bmatrix}$$

Is it possible to perform the following math operations on the matrices? If so, what is the result?

(a) **A*B**

(b) **B*A**

(c) **A+B**

(d) **A+B′** (where **B′** is the transpose of **B**)

(e) **A./B′**

PROBLEMS

P10.1   Using MathScript, generate a plot of a sine wave of frequency $\omega = 10$ rad/sec. Use the linramp function to generate the time vector starting at $t = 0$ and ending at $t = 10$. Label the x-axis as **Time (sec)**. Label the y-axis as **sin(w*t)**. Add the following title to the plot: **Sine wave with frequency w = 10 rad/sec**.

P10.2   The rand function generates uniformly distributed random numbers between 0 and 1. This means that the average of all of the random values generated by the rand function should approach 0.5 as the number of random numbers increases. Using the rand function, generate random vectors of length 5, 100, 500, and 1000. Confirm that as the number of elements increases, the average of the random numbers approaches 0.5. Generate a plot of the average of the random numbers as a function of the number of random numbers. Use both rand and mean functions in your script.

P10.3   Using the MathScript Interactive Window, generate a random 2D matrix using the syntax z = peaks(100). View the variable z as a surface using the Variables Window in the MathScript Interactive Window. Undock the 3D plot and rotate the plot in 3D space with the mouse so that you can estimate the maximum value of the plot on the z-axis. Use the max function to calculate the value you just estimated.

   (**Hint**: You may want to use the max function twice.)

P10.4   Open a new VI and place a MathScript Node on the block diagram. Create a numeric input on the MathScript Node frame and name it w. Generate the sine wave using Problem 10.1 as a reference and using MathScript syntax within the node. Replace the frequency with the variable w as an input controlled from a front panel knob. Create two outputs on the MathScript Node frame named t and y, respectively, where t is the time history and y is the sine function associated with t. Plot the sine wave using an XY Graph. Explain the behavior of the sine wave as you increase the frequency.

DESIGN PROBLEMS

D10.1 The MathScript script in Figure D10.1 simulates the trajectory of a projectile fired at a specified angle. Using Figure D10.1 as a guide, develop a VI using a MathScript Node to create an interactive simulator for calculating projectile trajectory. The two inputs to the simulator are the initial angle of the trajectory and the initial velocity. Utilize knob controls on the front panel and wire them to input variables **deg** and **v0**. The output vectors **x** and **y** should be combined and graphed on an XY Graph to show the trajectory.

What angle would allow you to shoot over a 3 meter wall standing 1 meter from the wall while hitting a target on the ground about 6 meters away in 1.5 seconds? What is the initial velocity of the projectile to accomplish this targeting goal?

```
% D10.1 -- Projectile Trajectory Simulator
t_max = 1.5;          % final simulation time
dt = 0.1;             % length of simulation vector
t = 0:dt:t_max;       % create the time vector for the simulation
deg = 79;             % initial angle of the trajectory
g = 9.81;             % acceleration due to gravity (m/s^2)
v0 = 15.0;            % initial velocity
theta =deg * pi/180;  % convert angle of trajectory to degrees
v0x = v0*cos(theta);  % calculate initial velocity in the x plane
v0y = v0*sin(theta);  % calculate initial velocity in the y plane
x = v0x*t;            % calculate x position with respect to time
y = v0y*t - g*t.^2;   % calculate y position with respect to time
plot(x,y);            % plot the result
grid on;              % turn on the grid so that you can see the zero
                        crossing
```

FIGURE D10.1
MathScript to simulate the motion of a projectile.

D10.2 The script shown in Figure D10.2 simulates an AM modulated signal. Verify that the script functions properly using the MathScript Interactive Window. You will need to type the script into the Script Editor. Once you have verified that the script runs properly, open a new VI and using a MathScript Node, develop a VI simulation of the AM modulated signal.

Create front panel knob controls for the **signal_freq** and **carrier_freq** variables labeled **Signal Frequency** and **Carrier Frequency**, respectively. Use separate Waveform Graphs for both the **mod** and modified **real_freq** result. When the VI is ready, set **signal_freq** = 20 and **carrier_freq** = 100 and run the

```
% D10.2 -- Amplitude Modulation (AM) Simulator
t_max = 1;                          % final simulation time
dt = 0.001;                         % length of simulation vector
t = 0:dt:t_max;                     % create the time vector for the
                                      simulation
signal_freq = 20;                   % signal frequency
carrier_freq = 100;                 % carrier frequency
signal = sin(signal_freq*2*pi*t);   % create signal wave (typically music
                                      or speech)
carrier = sin(carrier_freq*2*pi*t); % create carrier wave
mod = carrier.*(1+signal);          % AM modulate the signal with the
                                      carrier
subplot(2,1,1)                      % define a single window with two
                                      plots
plot(t,mod);                        % plot the time domain modulated
                                      signal
real_freq = abs(fft(mod));          % calculate the frequency domain
subplot(2,1,2);                     % define a single window with two
                                      plots
plot(real_freq);                    % plot the frequency domain modulated
                                      signal
```

FIGURE D10.2
AM modulated signal simulation.

VI. What happens to the signals on either side of the carrier signal (the center frequency) as you move the **signal_freq** control on the plot of **real_freq**? What changes when you move the **carrier_freq** control?

CHAPTER 11

Analysis

LabVIEW is an excellent environment for mathematical analysis and signal processing. It is ideally suited to developing programs for solving linear algebraic systems of equations, curve fitting, integrating ordinary differential equations, computing function zeroes, computing derivatives of functions, integrating functions, generating and analyzing signals, computing discrete Fourier transforms, and filtering signals. The Express VIs for curve fits, filters, signal generation, and spectral analysis provide the tools necessary to quickly prototype and solve problems. This chapter presents an overview of these topics.

GOALS

1. Introduce some of the LabVIEW capabilities for mathematical analysis.

2. Study some of the VIs available for analyzing signals and systems.

11.1   LINEAR ALGEBRA

LabVIEW provides a number of important VIs dedicated to solving systems of linear algebraic equations. These types of systems arise frequently in engineering and scientific applications. An entire branch of mathematics is dedicated to the study of *linear algebra*. We will only be able to touch on a few important subjects within the linear algebra arena in the space available to us here.

11.1.1   Review of Matrices

The basic element used in the computations is the *matrix*. A matrix is represented by an $m \times n$ array of numbers:

$$\mathbf{A} = \begin{bmatrix} a_{0,0} & a_{0,1} & \cdots & a_{0,n-1} \\ a_{1,0} & a_{1,1} & \cdots & a_{1,n-1} \\ \cdots & \cdots & \cdots & \cdots \\ a_{m-1,0} & a_{m-1,1} & \cdots & a_{m-1,n-1} \end{bmatrix},$$

where n is the number of columns, and m is the number of rows. When $m \neq n$, the matrix is called a *rectangular* matrix; conversely when $m = n$, the matrix is called a *square* matrix. An $m \times 1$ matrix is called a *column vector*, and a $1 \times n$ matrix is called a *row vector*. Other special forms of the matrix are the *diagonal* matrix, the *zero* matrix, and the *identity* matrix. Examples of these three types of matrices are

$$\mathbf{A} = \begin{bmatrix} 2 & 0 & 0 \\ 0 & 4 & 0 \\ 0 & 0 & -6 \end{bmatrix}, \quad \mathbf{0} = \begin{bmatrix} 0 & 0 & 0 \\ 0 & 0 & 0 \\ 0 & 0 & 0 \end{bmatrix}, \quad \mathbf{I} = \begin{bmatrix} 1 & 0 & 0 \\ 0 & 1 & 0 \\ 0 & 0 & 1 \end{bmatrix},$$

respectively. In LabVIEW, matrices can be created using the matrix control seen in Figure 11.1. The Real Matrix control can be found on the **Controls≫Modern ≫Array, Matrix & Cluster** palette.

The elements of a matrix, **A**, are denoted by $a_{i,j}$, where i indicates row number and j indicates column. Matrices can contain real numbers or complex numbers. If the elements of the matrix are complex numbers, the Complex Matrix control, found on the same palette as the Real Matrix control, should be used.

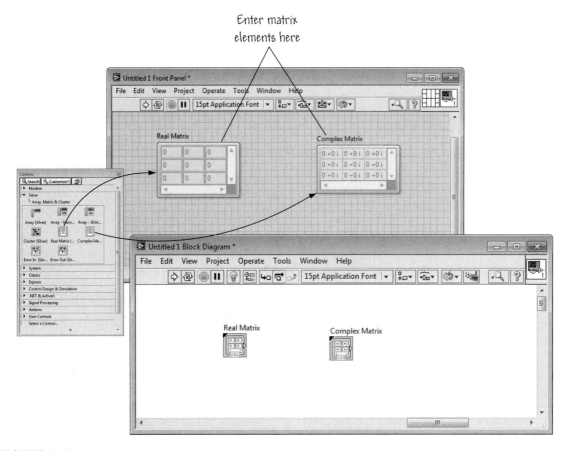

FIGURE 11.1
The Real Matrix control as seen from the front panel and the block diagram.

Addition of two matrices is performed element by element. For example,

$$
\begin{bmatrix} 2 & 3 & 1 \\ -7 & 4 & 5 \\ 2 & 0 & -6 \end{bmatrix} + \begin{bmatrix} 2 & -7 & 2 \\ 2 & 4 & 1 \\ 1 & 5 & -1 \end{bmatrix} = \begin{bmatrix} 4 & -4 & 3 \\ -5 & 8 & 6 \\ 3 & 5 & -7 \end{bmatrix}.
$$

Matrix addition can be computed in LabVIEW using the primitive Add function as shown in Figure 11.2. Addition, subtraction, multiplication, and division are all done using the primitive arithmetic functions.

You can easily check that $\mathbf{A} + \mathbf{B} = \mathbf{B} + \mathbf{A}$ and that $(\mathbf{A} + \mathbf{B}) + \mathbf{C} = \mathbf{A} + (\mathbf{B} + \mathbf{C})$. Therefore, matrix addition is commutative and associative.

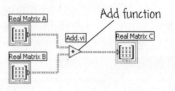

FIGURE 11.2
Addition of two real matrices using the Add function.

If we multiply a matrix by a scalar, α, the result is obtained by multiplying each element of the matrix by the scalar, yielding

$$\alpha\mathbf{A} = \begin{bmatrix} \alpha a_{0,0} & \alpha a_{0,1} & \cdots & \alpha a_{0,n-1} \\ \alpha a_{1,0} & \alpha a_{1,1} & \cdots & \alpha a_{1,n-1} \\ \cdots & \cdots & \cdots & \cdots \\ \alpha a_{m-1,0} & \alpha a_{m-1,1} & \cdots & \alpha a_{m-1,n-1} \end{bmatrix}.$$

Multiplication of matrices ($\mathbf{AB}$) requires the two matrices to be of compatible dimensions—the number of columns of $\mathbf{A}$ must be equal to the number of rows of $\mathbf{B}$. Thus if $\mathbf{A}$ is an $m \times n$ matrix, and $\mathbf{B}$ is an $n \times p$ matrix, the product $\mathbf{C} = \mathbf{AB}$ is an $m \times p$ matrix. The element $c_{i,j}$ of the matrix $\mathbf{C}$ is given by

$$c_{i,j} = a_{i,1}b_{1,j} + a_{i,2}b_{2,j} + \cdots + a_{i,m}b_{m,j}.$$

In general, matrix multiplication is not commutative; that is,

$$\mathbf{AB} \neq \mathbf{BA}.$$

The *transpose* of a real matrix (that is, a matrix comprised of only real numbers) is formed by interchanging the rows and columns. For example, if

$$\mathbf{A} = \begin{bmatrix} 2 & 3 & 1 \\ -7 & 4 & 5 \end{bmatrix}, \quad \text{then} \quad \mathbf{A}^T = \begin{bmatrix} 2 & -7 \\ 3 & 4 \\ 1 & 5 \end{bmatrix},$$

where the matrix $\mathbf{A}^T$ is the transpose of $\mathbf{A}$.

There are over 30 matrix operations available in LabVIEW. Important matrix operations, such as determinant and trace, can be found on the **Functions≫Mathematics≫Linear Algebra** palette. One example is the Transpose Matrix function shown in Figure 11.3. You can use **Context Help** for more information on the specific matrix operation of interest. Figure 11.3 shows the **Context Help** for the Transpose Matrix function.

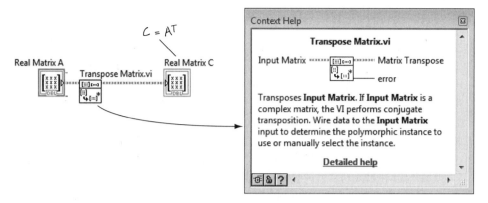

FIGURE 11.3
Performing a matrix transpose operation.

A real matrix is called a *symmetric* matrix if $\mathbf{A}^T = \mathbf{A}$. If the elements of the matrix $\mathbf{C}$ are complex numbers, then we extend the notion of a transpose to a *complex conjugate transpose*. This means that we transpose the matrix and then replace each element with its own complex conjugate. We denote the complex conjugate transpose as $\mathbf{C}^H$. A complex matrix is called a *Hermitian* matrix if $\mathbf{C}^H = \mathbf{C}$.

For square matrices, we can define the operations **trace**, **determinant**, and **inversion**. The trace of an $n \times n$ square matrix $\mathbf{A}$ is the sum of the diagonal elements,

$$\text{tr } \mathbf{A} = a_{1,1} + a_{2,2} + \cdots + a_{n,n}.$$

The determinant of a 2×2 matrix is given by

$$\det \mathbf{A} = \left| \begin{bmatrix} a_{1,1} & a_{1,2} \\ a_{2,1} & a_{2,2} \end{bmatrix} \right| = a_{1,1}a_{2,2} - a_{1,2}a_{2,1}.$$

If the determinant is identically equal to zero, then we say that the matrix is *singular*. In general, the determinant can be computed as a function of the *minors* and *cofactors* of the matrix. The minor of an element $a_{i,j}$ is the determinant of an $n - 1 \times n - 1$ matrix formed by removing the ith row and the jth column of the original matrix $\mathbf{A}$. For example, if

$$\det \mathbf{A} = \left| \begin{bmatrix} a_{1,1} & a_{1,2} & a_{1,3} \\ a_{2,1} & a_{2,2} & a_{2,3} \\ a_{3,1} & a_{3,2} & a_{3,3} \end{bmatrix} \right| ,$$

then the minor of the $a_{2,3}$ element is

$$M_{2,3} = \left\| \begin{bmatrix} a_{1,1} & a_{1,2} \\ a_{3,1} & a_{3,2} \end{bmatrix} \right\| = a_{1,1}a_{3,2} - a_{1,2}a_{3,1}.$$

The cofactor of $a_{i,j}$ is defined as

$$\gamma_{i,j} = \text{cofactor } a_{i,j} = (-1)^{i+j}M_{i,j}.$$

In general, we can compute the determinant of an $n \times n$ square matrix $\mathbf{A}$ as

$$\det \mathbf{A} = \sum_{j=1}^{n} a_{i,j}\gamma_{i,j}$$

for any row i. Similarly, we can compute the determinant as

$$\det \mathbf{A} = \sum_{i=1}^{n} a_{i,j}\gamma_{i,j}$$

for any column j.

The *adjoint matrix* of an $n \times n$ square matrix $\mathbf{A}$ is formed by transposing the matrix and replacing each element $a_{i,j}$ with the cofactor $\gamma_{i,j}$. Therefore, we have

$$\text{adjoint } \mathbf{A} = \begin{bmatrix} \gamma_{1,1} & \gamma_{1,2} & \gamma_{1,3} \\ \gamma_{2,1} & \gamma_{2,2} & \gamma_{2,3} \\ \gamma_{3,1} & \gamma_{3,2} & \gamma_{3,3} \end{bmatrix}^{T} = \begin{bmatrix} \gamma_{1,1} & \gamma_{2,1} & \gamma_{3,1} \\ \gamma_{1,2} & \gamma_{2,2} & \gamma_{3,2} \\ \gamma_{1,3} & \gamma_{2,3} & \gamma_{3,3} \end{bmatrix}.$$

The matrix inverse is denoted by $\mathbf{A}^{-1}$ and can be computed as

$$\mathbf{A}^{-1} = \frac{\text{adjoint } \mathbf{A}}{\det \mathbf{A}}.$$

The matrix inverse must satisfy the relationship

$$\mathbf{A}^{-1}\mathbf{A} = \mathbf{A}\mathbf{A}^{-1} = \mathbf{I}.$$

The matrix inverse does not exist (that is, it is singular) when $\det \mathbf{A} = 0$. If the matrix $\mathbf{A}$ is singular, then there exists a nonzero vector $\mathbf{v}$ such that $\mathbf{A}\mathbf{v} = 0$.

11.1.2   Systems of Algebraic Equations

Suppose that we want to solve the following system of algebraic equations:

$$4x_1 + 6x_2 + x_3 = 4$$
$$x_1 + 2x_2 + 3x_3 = 0$$
$$5x_2 - x_3 = 1$$

The unknowns are the variables x_1, x_2, and x_3. We can identify the two column vectors $\mathbf{x}$ and $\mathbf{b}$ as

$$\mathbf{x} = \begin{bmatrix} x_1 \\ x_2 \\ x_3 \end{bmatrix} \quad \text{and} \quad \mathbf{b} = \begin{bmatrix} 4 \\ 0 \\ 1 \end{bmatrix}.$$

Then we can write the system of algebraic equations as

$$\mathbf{Ax} = \mathbf{b},$$

where

$$\mathbf{A} = \begin{bmatrix} 4 & 6 & 1 \\ 1 & 2 & 3 \\ 0 & 5 & -1 \end{bmatrix}.$$

Thus, we have rewritten the problem in a compact matrix notation. Can we solve for the vector $\mathbf{x}$? In this case, the matrix $\mathbf{A}$ is invertible (i.e., the inverse exists), so the solution is readily obtained as

$$\mathbf{x} = \mathbf{A}^{-1}\mathbf{b} = \begin{bmatrix} 0.9123 \\ 0.1228 \\ -0.3860 \end{bmatrix}.$$

If the matrix $\mathbf{A}$ is singular, then the number of solutions depends on the vector $\mathbf{b}$. If a solution does exist in the singular case, it is not unique!

For any given $n \times n$ square matrix $\mathbf{A}$, we would like to know if there exists a scalar λ and a corresponding vector $\mathbf{v} \neq 0$ such that

$$\lambda \mathbf{v} = \mathbf{Av}.$$

This scalar λ is called the *eigenvalue*, and the corresponding vector $\mathbf{v}$ is called the *eigenvector*. Rearranging yields

$$\lambda \mathbf{v} - \mathbf{A}\mathbf{v} = (\lambda \mathbf{I} - \mathbf{A})\,\mathbf{v} = 0.$$

Therefore, a solution exists (for $\mathbf{v} \neq 0$) if and only if

$$\det\,(\lambda \mathbf{I} - \mathbf{A}) = 0.$$

If $\mathbf{A}$ is an $n \times n$ matrix, then $\det\,(\lambda \mathbf{I} - \mathbf{A}) = 0$ is an nth-order polynomial (known as the *characteristic equation*) whose solutions are called the *characteristic roots* or eigenvalues. The eigenvalues of a square matrix are not necessarily unique and may be complex numbers (even for a real-valued matrix!). Given the eigenvalues of a matrix, we can compute the trace and determinant as

$$\operatorname{tr} \mathbf{A} = \sum_{i=1}^{n} \lambda_i$$

$$\det \mathbf{A} = \prod_{i=1}^{n} \lambda_i$$

We see that if any eigenvalue is zero, then the determinant is zero, and thus it follows that the matrix is singular.

   An interesting fact regarding eigenvalues is that if a matrix is real and symmetric, then its eigenvalues will be real. If we compute the "square" of a real matrix $\mathbf{A}$ according to $\mathbf{B} = \mathbf{A}\mathbf{A}^T$, then $\mathbf{B}$ is real and symmetric. In fact, the matrix $\mathbf{B}$ is nonnegative, in the sense that, for any nonzero column vector $\mathbf{v}$ it follows that the scalar value $\mathbf{v}^T \mathbf{B} \mathbf{v} \geq 0$. Now if we compute the eigenvalues of $\mathbf{B}$, we find that they are all nonnegative and real. Taking the square root of each eigenvalue yields quantities known as *singular values*, denoted here by β_i for $n = 1, 2, \ldots, n$. An $n \times n$ matrix has n nonnegative singular values. Two important singular values are the maximum and minimum values, $\beta_{\max}$ and $\beta_{\min}$, respectively. Singular values are important in computational linear algebra because they tell us something about how close a matrix is to being singular. We know that if the determinant of a matrix is zero, then it is a singular matrix. But what about if you compute the determinant numerically with the computer and determine that the determinant is 10^{-10}? Is this close enough to zero to say that the matrix is singular? The answer to this is provided by the **condition number** of a matrix. It can be defined in different ways, but it is commonly defined as

$$\operatorname{cond} \mathbf{A} = \frac{\beta_{\max}}{\beta_{\min}}.$$

The condition number can vary from 0 to ∞. A matrix with a condition number near 1 is closer to being nonsingular than a matrix with a very large condition number. The condition number is useful in assessing the accuracy of solutions to systems of linear algebraic equations.

As a final practical note, it is not generally a good idea to explicitly compute the matrix inverse when solving systems of linear algebraic equations, because inaccuracies are associated with the numerical computations—especially when the condition number is high. The preferred solution technique involves using *matrix decompositions*. Popular techniques include

- Singular Value Decomposition (SVD)
- Cholesky decomposition (or QR)

The idea is to decompose the matrix into component matrices that have "nice" numerical properties.

For example, suppose we decompose the matrix

$$\mathbf{A} = \mathbf{QR},$$

where $\mathbf{Q}$ is an *orthogonal* matrix (that is, $\mathbf{Q}^T\mathbf{Q} = \mathbf{I}$), and $\mathbf{R}$ is upper triangular (all the elements below the diagonal are zero). Then,

$$\mathbf{Ax} = \mathbf{QRx} = \mathbf{b}.$$

Multiplying both sides by $\mathbf{Q}^T$ yields

$$\mathbf{Rx} = \mathbf{Q}^T\mathbf{b}.$$

Then you use the fact that $\mathbf{R}$ is upper triangular to *back-substitute* and solve for $\mathbf{x}$ without ever computing a matrix inverse.

11.1.3   Linear System VIs

LabVIEW comes with a complete set of VIs that can be used to perform all the matrix computations discussed previously (and much more!). You can find the VIs on the palette shown in Figure 11.4.

The Linear Algebra Calculator.vi is a useful Example VI that can be used to perform a variety of linear algebra operations on a matrix or to solve systems of linear algebraic equations. This VI can take data as input on the front panel or read the data from a file (that is, from a spreadsheet file). The next example gives you the opportunity to experiment with linear algebra computations.

Linear Algebra Calculator

Open Linear Algebra Calculator.vi using the NI Example Finder. In LabVIEW select **Find Examples** in the **Help** pull-down menu. When the NI Example Finder opens, browse by task to the Analyzing and Processing Signals folder and select Mathematics. Within the Mathematics folder you will

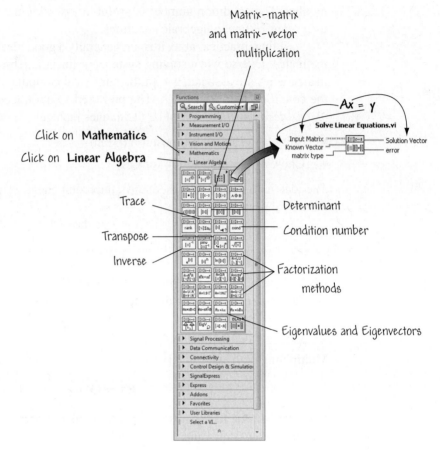

FIGURE 11.4
The linear algebra VIs.

find a variety of interesting examples, including the Linear Algebra Calculator VI. The front panel and block diagram are shown in Figure 11.5.

By default, the VI is set up to solve the system of linear algebraic equations

$$\mathbf{Ax} = \mathbf{b},$$

where

$$\mathbf{A} = \begin{bmatrix} 4 & 2 & -1 \\ 1 & 4 & 1 \\ 0.1 & 1 & 2 \end{bmatrix}, \quad \text{and} \quad \mathbf{b} = \begin{bmatrix} 2 \\ 12 \\ 10 \end{bmatrix}.$$

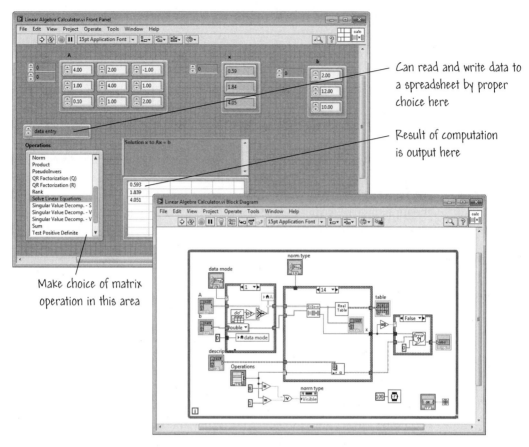

Can read and write data to a spreadsheet by proper choice here

Result of computation is output here

Make choice of matrix operation in this area

FIGURE 11.5
The linear algebra calculator.

Choose **Solve Linear Equations** in the lower left side of the front panel and run the VI. Verify that you obtain the solution

$$
\mathbf{x} = \begin{bmatrix} 0.59 \\ 1.84 \\ 4.05 \end{bmatrix}.
$$

Use the calculator to compute the determinant of the matrix **A**. You should get the result det **A** = 23.6. Compute the condition number, the inverse, and the trace of the matrix **A** (condition number = 4.418 and trace = 10). The VI runs until the stop button is pressed.

Modify the input matrix to be

$$
\mathbf{A} = \begin{bmatrix} 4 & 2 & -1 \\ 1 & 4 & 1 \\ 2 & 8 & 2 \end{bmatrix}.
$$

Compute the condition number. You should find that the condition number is very, very large! Now compute the matrix inverse (by selecting **Inverse** from the menu). What happens? The system of equations cannot be solved because the input matrix is singular. Vary the (3,3) element and observe the effect on the condition number. When the condition number reduces to less than 10, compute the matrix inverse again. Does it work in this case?

Set the (3,3) term of the matrix to 2.01:

$$
\mathbf{A} = \begin{bmatrix} 4 & 2 & -1 \\ 1 & 4 & 1 \\ 2 & 8 & 2.01 \end{bmatrix}.
$$

Verify that the determinant is 0.14. Now compute the condition number—it should be quite high (above 2,500). What does this result lead you to conclude about the advisability of solving the system of linear algebraic equations by inverting the **A** matrix? Basically, when the condition number is high, it is not advisable to solve the system of equations by matrix inversion.   ◆

11.2 STATISTICS AND CURVE FITTING

Statistics and curve fitting are common techniques used in science, engineering, business, medicine, and other fields in the analysis of data. Based on a set of assumptions, statistics attempt to determine trends in sets of related data. Statistics are commonly used every day in applications ranging from computing averages on test grades to predicting the weather. Curve fitting is another way to analyze related data, and involves extracting a set of curve parameters (or coefficients) from the data set to obtain a functional description of the data set. Using curve fitting, digital data can be represented by a continuous model. For example, you may want to fit the data with a straight-line model. The curve-fitting procedure would provide values for the linear curve fit in terms of slope and axis offset. In this section, we will cover curve fitting and discuss statistical methods for analyzing sets of data.

11.2.1   Curve Fits Based on Least Squares Methods

The main algorithm used in the curve-fitting process is known as the least squares method. Define the error as

$$e(\mathbf{a}) = [f(x, \mathbf{a}) - y(x)]^2,$$

where $e(\mathbf{a})$ is a measure of the difference between the actual data and the curve fit, $y(x)$ is the observed data set, $f(x, \mathbf{a})$ is the functional description of the data set (this is the curve-fitting function), and $\mathbf{a}$ is the set of curve coefficients that best describes the curve. For example, let $\mathbf{a} = (a_0, a_1)$. Then the functional description of a line is

$$f(x, \mathbf{a}) = a_0 + a_1 x.$$

The least squares algorithm finds $\mathbf{a}$ by solving the Jacobian system

$$\frac{\partial e(\mathbf{a})}{\partial \mathbf{a}} = 0.$$

LabVIEW's curve-fitting VIs solve the Jacobian system automatically and return the set of coefficients that best describes the input data set. The automatic nature of this process provides the opportunity to concentrate on the results of the curve fitting rather than dealing with the mechanics of obtaining the curve-fit parameters.

When we curve fit data, we generally have available two input sequences, Y and X. The sequence X is usually the independent variable (e.g., time) and the sequence Y is the actual data. A point in the data set is represented by (x_i, y_i), where x_i is the ith element of the sequence X, and y_i is the ith element of the sequence Y. Since we are dealing with samples at discrete points, statistics are appropriate for gathering additional information from the data. The VI calculates the mean square error (MSE), which is a relative measure of the residuals between the expected curve values and the actual observed values, using the formula

$$\text{MSE} = \frac{1}{n} \sum_{i=0}^{n-1} (f_i - y_i)^2,$$

where f_i is the sequence of fitted values, y_i is the sequence of observed values, and n is the number of input data points.

LabVIEW offers a number of curve-fitting algorithms, including:

- Linear Fit:

$$y_i = a_0 + a_1 x_i$$

- Exponential Fit:

$$y_i = a_0 e^{a_1 x_i}$$

- General Polynomial Fit:

$$y_i = a_0 + a_1 x_i + a_2 x_2 + \cdots$$

- General Linear Fit:

$$y_i = a_0 + a_1 f_1(x_i) + a_2 f_2(x_i) + \cdots$$

where y_i is a linear combination of the parameters $a_0, a_1, \ldots$. This type of curve fit provides user-selectable algorithms (including SVD, House-holder, Givens, LU, and Cholesky) to help achieve the desired precision and accuracy.

- Nonlinear Levenberg-Marquardt Fit:

$$y_i = f(x_i, a_0, a_1, a_2, \ldots)$$

where $a_0, a_1, a_2, \ldots$ are the parameters. This method does not require y to have a linear relationship with $a_0, a_1, a_2, \ldots$. Although it can be used for linear curve fitting, the Levenberg-Marquardt algorithm is generally used for nonlinear curve fits.

- B-Spline Fit VI:

The B-spline (basis spline) fit is used to smooth a given data-set input sequence (X, Y), where X is comprised of the data points $X = (x_1, x_2, x_3, \ldots, x_n)$ and Y is comprised of the associated data points $Y = (y_1, y_2, y_3, \ldots, y_n)$ where n is the number of data points. The B-Spline Fit VI calculates the best basis spline fit $(\hat{X}, \hat{Y})$ by minimizing the performance function

$$J = \frac{1}{n} \sum_{i=1}^{n} w_i \left[\left(x_i - \hat{x}_i \right)^2 + \left(y_i - \hat{y}_i \right)^2 \right],$$

where w_i is the ith element of weight (an input to the VI), (x_i, y_i) is the ith pair of the input sequences (X, Y), and $(\hat{x}_i, \hat{y}_i)$ is the ith pair of best B-Spline fit $(\hat{X}, \hat{Y})$. One of the inputs to the B-Spline Fit VI is the degree of the approximating polynomial, set to 3 by default. If you do not set the weights w_i as inputs to the VI, all input data points are equally weighted.

For an example of the use of the B-Spline Fit VI, open the B Spline Fitting Demo VI in the regressn.llb in the analysis subfolder of the examples folder in the LabVIEW directory.

The curve-fitting VIs can be found on the **Fitting** palette, as illustrated in Figure 11.6. The Advanced Curve Fitting VIs found on the **Mathematics≫ Fitting** palette provide additional fit statistics and coefficients associated with the various fitting VIs. The advanced VIs include Exponential Fit Coefficients and Exponential Fit Intervals, Gaussian Peak Fit Coefficients and Gaussian Peak Fit Intervals, General Polynomial Fit Coefficients, Goodness of Fit, Linear Fit Coefficients and Linear Fit Intervals, Logarithm Fit Coefficients and Logarithm Fit Intervals, Nonlinear Curve Fit intervals, Polynomial Fit Intervals, and Remove Outliers. For example, the Exponential Fit Coefficients VI returns the amplitude and damping of the exponential fit (that is, a_0 and a_1 as discussed above), and the Exponential Fit Intervals VI calculates the statistical intervals of the best exponential fit. The Goodness of Fit VI calculates statistical parameters that describe how well a fitted model matches the original data set. The Remove Outliers VI removes data points that fall outside a given range or within a given index array. The Advanced Curve Fitting VIs provide valuable information about how well the fitting VIs are matching the input data and also return key parameters of the fit that can be used for analysis.

For an example of the use of some of the advanced curve-fitting VIs, Open the Parametric Curve Fitting VI in the curvefit.llb *in the* math *subfolder of the* examples *folder in the* LabVIEW *directory.*

◆ **Practicing with Curve Fitting**

Open Regressions Demo.vi in the folder Chapter 11 in the Learning directory. The front panel is shown in Figure 11.7. The VI generates noisy data

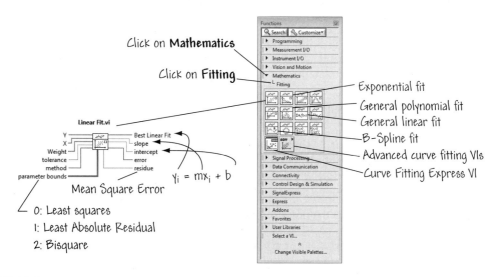

FIGURE 11.6
The curve-fitting VIs.

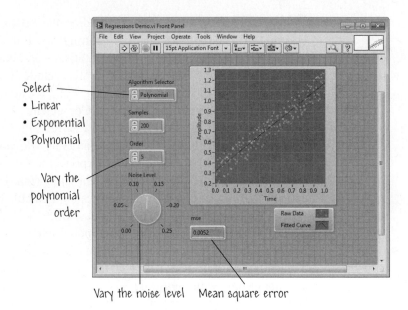

Select
• Linear
• Exponential
• Polynomial

Vary the polynomial order

Vary the noise level   Mean square error

FIGURE 11.7
A demo to investigate curve fitting with linear, exponential, and polynomial fits.

samples that are approximately linear, exponential, or polynomial and then uses the appropriate curve-fitting VIs to determine the best parameters to fit the given data. You can control the noise level with the knob on the front panel. You can also select an algorithm and the number of samples to fit. For the polynomial curve fits, the order of the polynomial can be varied via front panel input.

Select Linear in the Algorithm Selector control and set Noise Level to around 0.05. Run the VI and make a note of the computed error displayed in the mse indicator. Increase the noise level to 0.1 and again make a note of the computed error. Continue this process for noise levels of 0.15, 0.2, and 0.25. Did you detect any trends? You should have seen the MSE increase as the noise level increased.

Select Polynomial in the Algorithm Selector control and set Noise Level to around 0.05. Set the Order control to 2. Run the VI and make a note of the computed error. Follow the same procedure as above. Did you obtain the same trends as the noise increased? Reduce the noise level to 0.15. This time run an experiment with the noise level fixed but increase the polynomial order from 2 to 6. Did you detect any trends as the polynomial order increased? In this case, as you increase the polynomial order, the computed error fluctuates, but remains basically on the same order of magnitude.

With the Algorithm Selector control set to Polynomial, run the VI with the polynomial order set to 0. Then change the polynomial order to 1 and run the VI. With polynomial order equal to 0 and 1, the fitted curve is a horizontal line and a straight line with a (generally) nonzero slope, respectively. Experiment with the Regressions Demo VI and see if you can discover new trends. Consider comparing the linear fit with the exponential fit.   ◆

Open the NI Example Finder to search for examples that can be used to illustrate the curve fitting capability of LabVIEW. When the NI Example Finder opens, browse by task to the Analyzing and Processing Signals folder and select Curve Fitting and Interpolation. Within the Curve Fitting and Interpolation folder you will find a variety of interesting examples, including the Linear, Exp, and Power Fitting VI and the General LS Fitting VI.

11.2.2   Fitting a Curve to Data with Normal Distributions

Real-world data is very often normally (or Gaussian) distributed. The mathematical description of a normal distribution is

$$f(x) = \frac{1}{\sigma\sqrt{2\pi}} \exp\left[-\frac{1}{2}\left(\frac{x-m}{\sigma}\right)^2\right],$$

where m is the mean and σ is the standard deviation. Figure 11.8 shows the normal distribution with $m = 0$ and $\sigma = 0.5$, 1.0, and 2.0. As seen in the figure, the normal distribution is bell shaped and symmetric about the mean, m. The area under the bell-shaped curve is unity. The two parameters that completely describe the normally distributed data are the mean, m, and the standard deviation, σ. The peak of the bell-shaped curve occurs at m. The smaller the value of σ, the higher the peak at the mean and the narrower the curve.

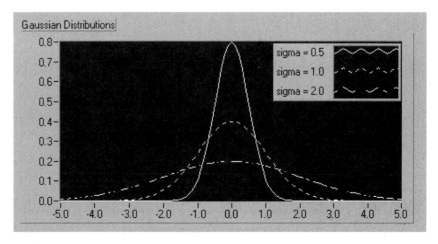

FIGURE 11.8
The normal distribution shape for three values of σ: 0.5, 1.0, and 2.0.

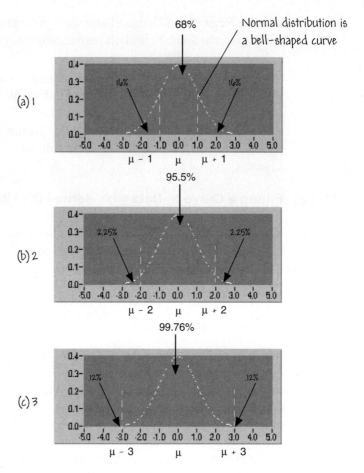

FIGURE 11.9
The normal distribution for 1σ, 2σ, and 3σ.

The normal distribution is illustrated again in Figure 11.9. The standard deviation is an important parameter that defines the "bounds" within which a certain percentage of the data values is expected to occur. For example:

- About two-thirds of the values will lie between $m - \sigma$ and $m + \sigma$.
- About 95% of the values will lie between $m - 2\sigma$ and $m + 2\sigma$.
- About 99% of the values will lie between $m - 3\sigma$ and $m + 3\sigma$.

Therefore, an interpretation of these values is that the probability that a normally distributed random value lies outside $\pm 2\sigma$ is approximately 0.05 (or 5%).

Normal Distributions

In the folder **Chapter 11** in the **Learning** directory you will find a VI called **Normal (Gaussian) Fit.vi**. This VI generates a random data set and then plots the distribution. The front panel is shown in Figure 11.10. If you want to

Found on
**Express >> Signal
Analysis** palette

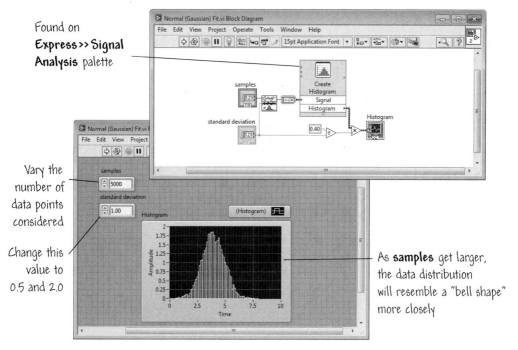

Vary the
number of
data points
considered

Change this
value to
0.5 and 2.0

As **samples** get larger,
the data distribution
will resemble a "bell shape"
more closely

FIGURE 11.10
Normally distributed data.

experiment with normally distributed data, open and run the VI. You can vary the number of samples and the standard deviation (that is, the σ). Set $\sigma = 0.5$, 1.0, and 2.0, and experiment.

Run the VI and vary the number of samples from 10 to 10,000. You should notice that as the number of samples increases, the shape of the data distribution becomes more and more bell shaped. Does this exercise relate in any way to your experience with grade distributions in class? Would you expect to have a bell-shaped grade distribution in a class of 5 students? 100 students? 500 students? (Answers: no, maybe, yes)

Using **Normal (Gaussian) Fit.vi** as a starting point, construct a VI to curve-fit the normal distribution and to compute the mean and sigma (σ) from the curve-fit parameters. ◆

11.2.3 The Curve Fitting Express VI

LabVIEW provides a significant number of Express VIs designed for analysis purposes. The **Signal Analysis** palette is shown in Figure 11.11. We will discuss only a subset of the Express VIs: Curve Fitting, Spectral Measurements, Filter, and Simulate Signal. In this section, we present the Curve Fitting Express VI.

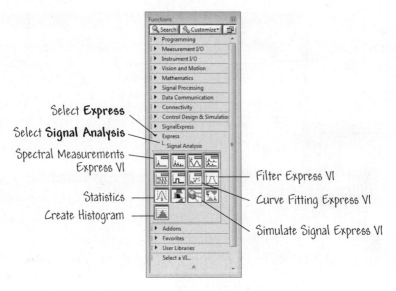

Select **Express**

Select **Signal Analysis**

Spectral Measurements
Express VI

Statistics

Create Histogram

Filter Express VI

Curve Fitting Express VI

Simulate Signal Express VI

FIGURE 11.11
The Signal Analysis Express VIs.

To use the Curve Fitting VI, you first place it on the block diagram. As illustrated in Figure 11.12, once the Curve Fitting Express VI is placed on the block diagram, a dialog box automatically appears to configure the VI. The curve fitting options include a linear fit, a quadratic fit, a spline fit, a polynomial fit (of an order specified by the user), and a general least squares fit. In Figure 11.12, a fifth order polynomial fit is selected. Once the Curve Fitting VI is configured, click **OK** to return to the block diagram.

On the block diagram, the Express VI will expand to display the inputs and outputs. The key input is the signal, and the key output is the best fit signal. You can also access the curve fit residuals and mean squared error. These are useful in quantifying the accuracy of the curve fit itself. The Curve Fitting VI is now ready for inclusion in a VI for signal analysis.

**Practice
with the
Curve Fitting
Express VI**

Consider the VI shown in Figure 11.13. Notice on the block diagram that two Express VIs—the Curve Fitting Express VI and the Simulate Signal Express VI—are utilized. This VI was developed to assist in investigating the impact of selecting various curve-fitting strategies to fit a noisy sine wave signal.

The Curve Fitting Express Demo.vi is located in Chapter 11 of the Learning directory.

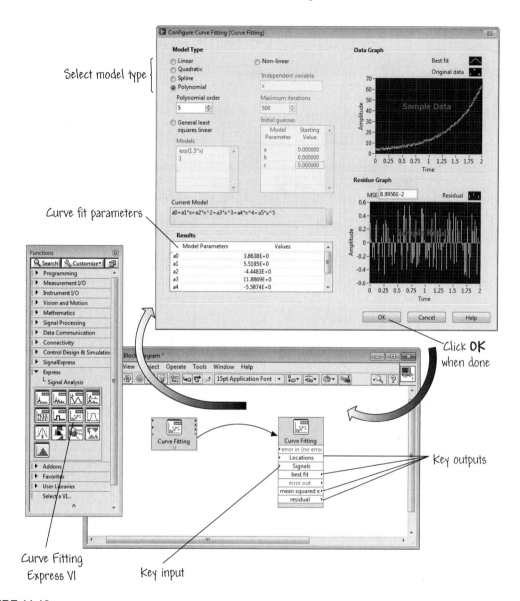

FIGURE 11.12
Configuring the Curve Fitting Express VI.

There are four elements on the block diagram:

- The Curve Fitting Express VI—this can be configured by double-clicking on the icon to access the configuration dialog box.

- The Simulate Signal Express VI—this VI will be described in Section 11.6 in more detail. It has been configured here to provide a sine wave signal at a frequency of 10.1 Hz, and normally distributed white noise is added to the signal.

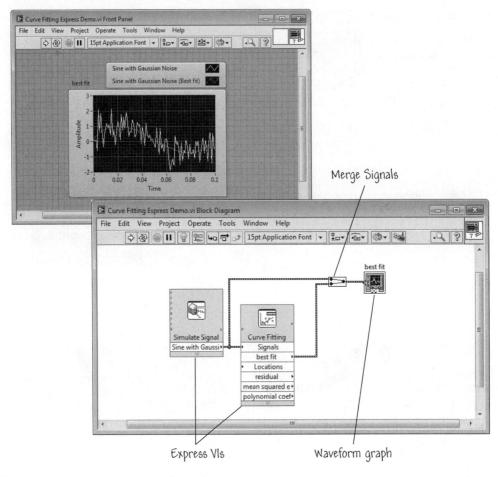

FIGURE 11.13
The Curve Fitting Express Demo VI for investigating noisy sine wave signals.

- The Merge Signals function—this function was described in Section 9.2.4. Its purpose is to merge the raw signal and the best fit signal (after curve fitting) into a format acceptable to the waveform graph.

- The waveform graph—this is used to graphically display the smoothed signal and the raw signal.

Run Curve Fitting Express Demo.vi and observe the results on the waveform graph. Recall that the Curve Fitting Express VI was configured to use a fifth-order polynomial fit.

Double-click the Curve Fitting Express VI to access the configuration dialog box. Reduce the order of the polynomial fit to second-order. You should find that the curve fit is now much poorer than before with the fifth-order polynomial. Try the linear fit and see what happens. ◆

11.3   DIFFERENTIAL EQUATIONS

In LabVIEW, you can solve linear and nonlinear ordinary differential equations (ODEs) using one of seven VIs for solving first- and higher-order differential equations. Figure 11.14 shows the palette with the available ODE VIs.

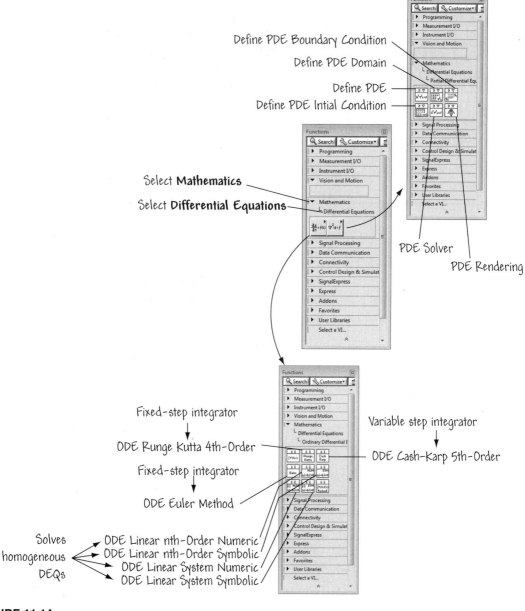

FIGURE 11.14
Solving differential equations using LabVIEW VIs.

The order of a differential equation is the order of the highest derivative in the differential equation.

You can also solve partial differential equations with the assistance of the VIs on the **Mathematics≫Partial Differential Equations** palette as shown in Figure 11.14. The study of partial differential equations is a rich field. The hunt for numerical solutions to partial differential equations is characterized by computational complexities beyond the scope of this book. The strategy here is to focus on using LabVIEW to obtain numerical solutions to ordinary differential equations using common integration methods (such as Runge Kutta methods). Remember, however, that should you need LabVIEW support for partial differential equation solutions, you can search the LabVIEW help to get started.

In the PDE.llb in the math subfolder of the examples folder in the LabVIEW directory there are examples of solving PDEs for flexible elements (two-dimensional Helmholtz equation), string vibrations (one-dimensional wave equation), and thermal distributions (two-dimensional heat equation).

Suppose that we have a set of first-order ordinary differential equations:

$$\dot{\mathbf{x}}(t) = \mathbf{f}(\mathbf{x}(t), \mathbf{u}(t)),$$

where $\mathbf{x}$ is the vector $(x_1, x_2, \cdots x_n)^T$, sometimes known as the *state vector*, and $\mathbf{u}$ is the vector $(u_1, u_2, \cdots u_m)^T$ of inputs to the system. When $\mathbf{u} = 0$, the differential equation is termed a *homogeneous* differential equation, and when

$$\mathbf{f}(\mathbf{x}(t), \mathbf{u}(t)) = \mathbf{A}\mathbf{x}(t) + \mathbf{B}\mathbf{u}(t),$$

the system is a system of *linear* ordinary differential equations.

An example of a *nonhomogeneous* system is

$$\dot{x}_1(t) = x_1^2(t) + \sin x_2(t) + u_1(t)$$
$$\dot{x}_2(t) = x_1(t) + x_2^3(t) + 3u_2(t),$$

where we write

$$\mathbf{x} = \begin{pmatrix} x_1 \\ x_2 \end{pmatrix} \quad \mathbf{u} = \begin{pmatrix} u_1 \\ u_2 \end{pmatrix} \quad \mathbf{f}(\mathbf{x}(t), \mathbf{u}(t)) = \begin{pmatrix} x_1^2(t) + \sin x_2(t) + u_1(t) \\ x_1(t) + x_2^3(t) + 3u_2(t) \end{pmatrix}.$$

To compute a solution we need to specify the initial conditions. For a system with n first-order differential equations, we need to specify n initial conditions, $x_1(0), x_2(0), \ldots, x_n(0)$.

We can also represent physical systems by higher-order differential equations. An example of a second-order mass-spring-damper system is

$$m\frac{d^2y(t)}{dt^2} + b\frac{dy(t)}{dt} + ky(t) = g(t),$$

where the system parameters are m = mass, b = damping coefficient, k = spring constant, and the input function is $g(t)$. When $g(t) = 0$, the system is homogeneous. To compute the solution, we need two initial conditions: $y(0)$ and $\dot{y}(0)$.

You can describe an nth-order differential equation equivalently by n first-order differential equations. Consider the second-order differential equation presented above and define

$$x_1 = y \quad \text{and} \quad x_2 = \dot{y} \quad \text{and} \quad u = g(t).$$

Taking time-derivatives of x_1 and x_2 yields

$$\dot{x}_1 = \dot{y} = x_2$$

$$\dot{x}_2 = \ddot{y} = -\frac{b}{m}\dot{y} - \frac{k}{m}y + \frac{1}{m}g(t) = -\frac{b}{m}x_2 - \frac{k}{m}x_1 + \frac{1}{m}u(t),$$

or

$$\dot{\mathbf{x}}(t) = \mathbf{A}\mathbf{x}(t) + \mathbf{B}u(t) = \begin{bmatrix} 0 & 1 \\ -\frac{k}{m} & -\frac{b}{m} \end{bmatrix} \mathbf{x}(t) + \begin{bmatrix} 0 \\ \frac{1}{m} \end{bmatrix} u(t)$$

LabVIEW has VIs to solve sets of first-order differential equations, and VIs to solve nth-order differential equations. Each VI uses a different numerical method for solving the differential equations. Each method has an associated *step size* that defines the time intervals between solution points. To solve differential equations of the form

$$\dot{\mathbf{x}}(t) = \mathbf{f}(\mathbf{x}(t), \mathbf{u}(t)),$$

use one of the following VIs:

- ODE Cash Karp 5th Order: Variable-step integrator that adjusts the step size internally to reduce numerical errors.

- ODE Euler Method: A very simple fixed-step integrator—that is, it executes fast, but the integration error associated with the Euler method is usually unacceptable in situations where solution precision is important.

- ODE Runge Kutta 4th Order: Fixed-step integrator that provides much more precise solutions than the Euler method.

To solve differential equations of the form

$$\dot{\mathbf{x}}(t) = \mathbf{A}\mathbf{x}(t)$$

where the coefficients of the matrix $\mathbf{A}$ are constant, use one of the following VIs:

- ODE Linear System Numeric: Generates a numerical solution of a homogeneous linear system of differential equations.

- ODE Linear System Symbolic: Generates a symbolic solution of a homogeneous linear system of differential equations.

By a symbolic solution, we mean that the output of the VI is actually a formula rather than an array of numbers. The solution is presented as a formula displayed on the VI front panel.

For solving homogeneous, higher-order differential equations of the form

$$a_0 \frac{d^n y}{dt^n} + a_1 \frac{d^{n-1} y}{dt^{n-1}} + \cdots + a_{n-1} y = 0,$$

LabVIEW has two VIs:

- ODE Linear nth Order Numeric: Generates a numeric solution of a linear system of nth-order differential equations.

- ODE Linear nth Order Symbolic: Generates a symbolic solution of a linear system of nth-order differential equations.

Which VI should you use? A few general guidelines follow:

- Use the ODE Euler Method VI sparingly and only for very simple ODEs.

- For most situations, choose the ODE Runge Kutta 4th Order or the ODE Cash Karp 5th Order VI.

  - If you need the solution at equal intervals, choose the ODE Runge Kutta 4th Order VI.

  - If you are interested in a global solution and fast computation, choose the ODE Cash Karp 5th Order VI.

The Pendulum

The objective of this exercise is to build a VI that solves a second-order differential equation that models the motion of a pendulum. Because the equations must be entered in symbolic form, you will need to incorporate the Substitution Variables VI, which allows you to vary the model parameters more easily. The pendulum model is given by

$$\frac{d^2\theta}{dt^2} + \frac{c}{ml}\frac{d\theta}{dt} + \frac{g}{l}\sin\theta = 0,$$

where m is the mass of the pendulum, l is the length of the rod, $g = 9.8$ is the acceleration due to gravity, and θ is the angle between the rod and a vertical line passing through the point where the rod is fixed (that is, the equilibrium position).

The pendulum model is a homogeneous, nonlinear equation (notice the $\sin\theta$ term!)—you have three choices of VIs:

- ODE Cash Karp 5th Order
- ODE Euler Method
- ODE Runge Kutta 4th Order

The first step is to formulate the pendulum model as two first-order differential equations. All three integration VIs listed above are for first-order systems. This is achieved by making the following substitution:

$$x_1 = \theta \quad \text{and} \quad x_2 = \dot{\theta}.$$

Go ahead and convert the second-order differential equation to two first-order differential equations. You should end with the resulting system:

$$\dot{x}_1 = x_2$$
$$\dot{x}_2 = -\frac{c}{ml}x_2 - \frac{g}{l}\sin x_1$$

Construct a VI to simulate the motion of the pendulum. Use the front panel of the VI shown in Figure 11.15 as a guide.

A block diagram is shown in Figure 11.16 that can be used as a guide in the VI development. To begin the development, use the ODE Euler Method VI. The inputs that you will need follow:

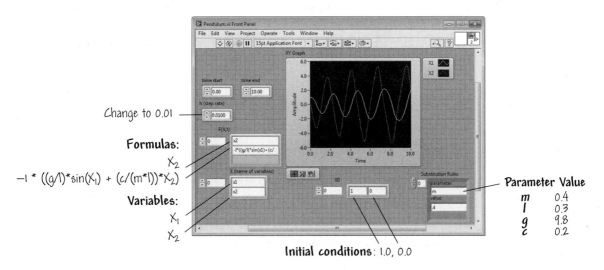

FIGURE 11.15
Simulating the motion of a pendulum—front panel.

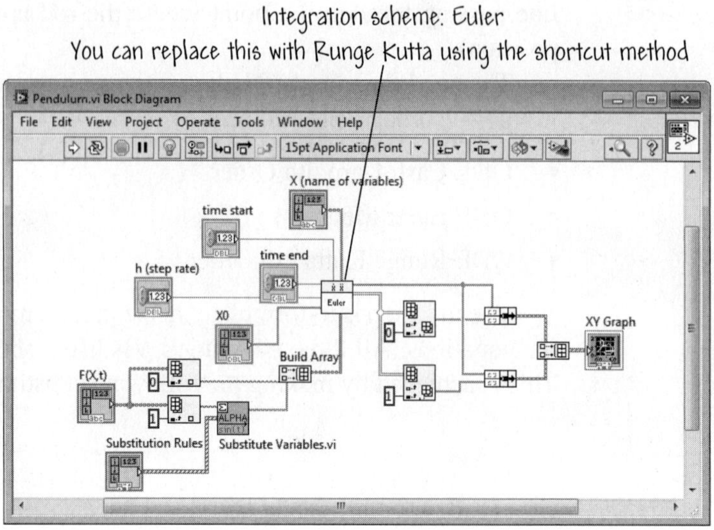

FIGURE 11.16
Simulating the motion of a pendulum—block diagram.

- **X:** An array of strings listing the dependent variables (x_1 and x_2).
- **time start:** The point in time at which to start the calculations ($t_0 = 0$).
- **time end:** The point in time at which to end the calculations ($t_f = 10$).
- **h (step rate):** The time increment for the calculations ($h = 0.01$).
- **X0:** The initial conditions ($x_1(0) = 1$, and $x_2(0) = 0$).

The pendulum differential equations are typed in the **F(x,t)** control. The Substitute Variables VI is used in this exercise as a vehicle for easily varying specific pendulum parameters. This allows you to enter the pendulum model on the front panel and then to substitute numerical values for m, l, g, and c. The Substitute Variables VI can be found on the **Functions≫Mathematics≫Scripts & Formulas≫Formula Parsing** palette.

When the VI is ready to accept inputs, enter the right side of the pendulum model in the **F(X,t)** control. You will enter x_2 and $-(g/l)\sin x_1 - (c/ml)x_2$. For the substitution rules use the following:

- In element 0 of the **Substitution Rules** control, type in
  - parameter: m
  - value: 0.4
- In element 1 of the **Substitution Rules** control, type in
  - parameter: l
  - value: 0.3

- In element 2 of the Substitution Rules control, type in

  □ parameter: g

  □ value: 9.8

- In element 3 of the Substitution Rules control, type in

  □ parameter: c

  □ value: 0.2

Run the VI and observe the waveform on the graph display. You should see a nice, stable response damping out around 6 seconds.

On the front panel, change the h (step rate) from 0.01 to 0.1. Run the VI. Did you detect a problem? The system response is no longer stable! Nothing has changed with the physical model, so this must be due to the Euler integration scheme. Now, switch to the block diagram and right-click the ODE Euler Method VI and **Replace** it with the ODE Runge Kutta 4th Order VI. Both VIs have the same inputs and outputs. Run the VI. What happens? The expected smooth, stable response is obtained. This demonstrates the benefit of the Runge Kutta method over the Euler method—you can run with larger time steps and obtain more accurate solutions.

Back on the front panel, switch the time step back to h (step rate) = 0.01. Run the VI and verify that the results remain essentially the same. Investigate the effect that varying the pendulum mass has on the response.

When you are finished, save the VI in User's Stuff and call it Pendulum.vi.

A working version of Pendulum.vi can be found in the Chapter 11 folder in the Learning directory. If you run this VI, make sure to verify that all the input parameters are correct. ◆

**The
Pendulum
with MathScript**

The pendulum problem considered in the previous example can also be solved using MathScript (see Chapter 10). The ordinary differential equation solver in MathScript based on the Runge Kutta method requires that the system of differential equations be described in a user-defined function in a predefined format. In this example we will employ the well-known ode_rk45 function, which is a one-step solver implementing the Runge Kutta (4,5) formula. For help with ode_rk45 open the MathScript Interactive Window and enter help ode_rk45 in the Command Window.

The ode_rk45 function is used with the following syntax:

$$[t, y] = ode\_rk45('function', times, y0)$$

where the inputs and outputs are defined as follows:

- **Inputs**
  - function—Specifies the name of the function that describes the ODE system. The function must have the following form:

    $$\text{function dy} = \text{fun(times, y)}$$

    where fun is a string.
  - times—Specifies either the endpoints for the time span, such as [0 20], or the time points at which to approximate the y-values, such as 0:20. The input times must be a strictly monotonic, real, double-precision vector.
  - y0—Specifies the y-value at the starting time. The input y0 is a real, double-precision vector.

- **Outputs**
  - t—Returns the time values at which LabVIEW evaluates the y-values in a vector
  - y—Returns the y-values that LabVIEW computes. The output y is a real matrix.

To begin, open the MathScript Interactive Window. Navigate to the Script Editor window where you will define the pendulum system using the same differential equations describing x_1 and x_2 considered in the previous example. Using the script in Figure 11.17 as a guide, enter the script in the Script Editor and when done **Save** the script to pendulum_system.m. See Section 10.5.1 for a review of user-defined functions.

It is suggested that you save pendulum_system.m in the **LabVIEW Data** folder so that MathScript can readily locate it. If needed, look back into Section 10.5 for a review of specifying path names. Since the ode_rk45 solver calls the user-defined function pendulum_system.m as a source for the differential equations, it must be able to locate the file.

Again using Figure 11.17 as a guide, in the Script Editor window construct the calling script. The calling script calls the ode_rk45 integrator which in turn uses the pendulum_system.m script. To visualize the pendulum motion, add a call to the plot function in the script and graph the output vector y as a function of t. The output vector y contains the time-history of x_1 and x_2. Save the calling script as pendulum.m.

Run the pendulum.m script by selecting **Run** at the bottom of the Script Editor window. A plot will be produced as illustrated in Figure 11.17.

*A working version of pendulum_system.m and pendulum.m can be found in the **Chapter 11** folder in the **Learning** directory.* ◆

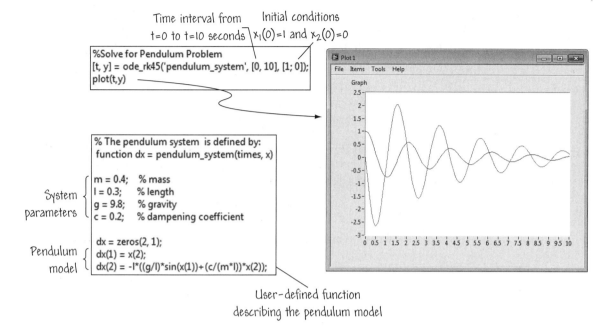

Time interval from t=0 to t=10 seconds

Initial conditions $x_1(0)=1$ and $x_2(0)=0$

```
%Solve for Pendulum Problem
[t, y] = ode_rk45('pendulum_system', [0, 10], [1; 0]);
plot(t,y)
```

```
% The pendulum system is defined by:
function dx = pendulum_system(times, x)
```

System parameters
```
m = 0.4;   % mass
l = 0.3;   % length
g = 9.8;   % gravity
c = 0.2;   % dampening coefficient
```

Pendulum model
```
dx = zeros(2, 1);
dx(1) = x(2);
dx(2) = -l*((g/l)*sin(x(1))+(c/(m*l))*x(2));
```

User-defined function describing the pendulum model

FIGURE 11.17
Using MathScript to simulate the motion of a pendulum.

11.4   FINDING ZEROES OF FUNCTIONS

LabVIEW provides VIs that can be used to compute zeroes of functions. The VIs can be used to determine the zeroes of general functions of the following form:

$$f(x, y) = 0$$
$$g(x, y) = 0$$

For example, you could use the symbolic VIs to find the zeroes of $\sin(x) + \cos(x)$ in the range $-10 \le x \le 10$.

Very often in the course of studying mathematics, engineering, business, and science it is necessary to compute the *zeroes* of a polynomial. An nth-order polynomial has the form

$$f(x) = x^n + a_{n-1}x^{n-1} + a_{n-2}x^{n-2} + \cdots + a_1x + a_0$$

The zeroes of the polynomial are the values of x such that $f(x) = 0$. The zeroes are also known as the *roots* of the polynomial. For example, we discussed in previous sections that the characteristic equation associated with an $n \times n$ matrix **A** is an nth-order polynomial, and the zeroes of the characteristic equation are the eigenvalues of the matrix. Eigenvalues can be real and imaginary. But in this discussion, when we talk about zeroes of a function, we mean the real roots.

LabVIEW provides many VIs for finding zeroes of functions. As shown in Figure 11.18, on the **Mathematics≫Scripts & Formula≫Zeroes** palette you will find the (1) Find All Zeroes of f(x), (2) Newton Raphson Zero Finder, (3) Ridders Zero Finder, (4) nD Nonlinear System Single Solution, and (5) nD Nonlinear System Solver VIs. On the **Mathematics≫Polynomial** palette you will find (6) Polynomial Roots and (7) Polynomial Real Zeroes Counter VIs. These seven VIs perform the following functions:

1. Find All Zeroes of f(x) is a polymorphic VI with the instances Find All Zeroes of f(x) (Formula) and Find All Zeroes of f(x) (VI) that determines all the zeroes of a 1D function in a specified interval.

2. Newton Raphson Zero Finder is a polymorphic VI with the instances Newton Raphson Zero Finder (Formula) and Newton Raphson Zero Finder (VI) that uses derivatives to assist in determining a zero of a 1D function in a specified interval.

3. Ridders Zero Finder is a polymorphic VI with the instances Ridders Zero Finder (Formula) and Ridders Zero Finder (VI) that computes a zero of a function in a given interval, but the function must be continuous and, when evaluated at the edges of the interval, must have different signs.

4. nD Nonlinear System Single Solution is a polymorphic VI with the instances nD Nonlinear System Single Solution (Formula) and nD Nonlinear System

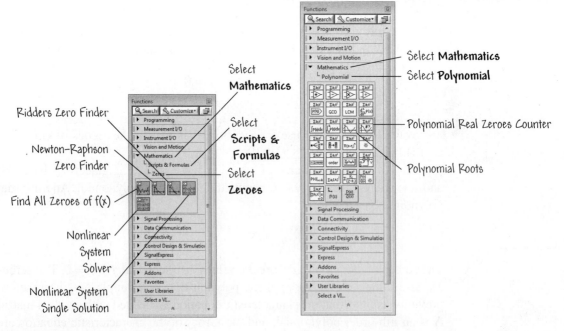

FIGURE 11.18
VIs for finding zeroes of functions.

Single Solution (VI) that computes the zeroes of a nonlinear function where an approximation is provided as input.

5. nD Nonlinear System Solver is a polymorphic VI with the instances nD Nonlinear System Solver (Formula) and nD Nonlinear System Solver (VI) that computes the zeroes of a set of nonlinear functions.

6. Polynomial Roots finds the complex roots of a complex polynomial.

7. Polynomial Real Zeroes Counter determines the number of real zeroes of a polynomial in an interval without actually computing the zeroes.

Finding Roots of a Polynomial Function

The objective of this exercise is to build a VI to compute the roots of a polynomial function. Construct a VI using the front panel and block diagram shown in Figure 11.19 as a guide. Use **Polynomial Roots.vi** to compute the roots. By default, this VI uses the Simple Classification option (if not wired). This means that both real and complex roots are estimated and returned. If your polynomial has repeated roots, you may need to consider wiring the **option** input of the Polynomial Roots VI with a more advanced classification (see the **Context Help** for more information).

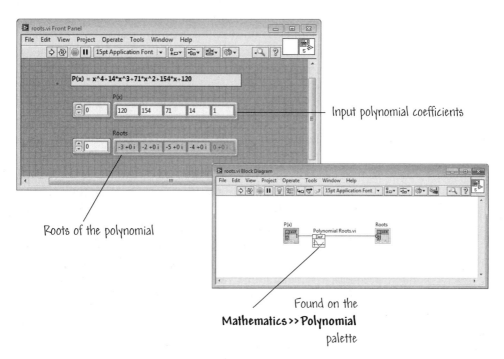

FIGURE 11.19
Computing the roots of a polynomial.

The main input to the Polynomial Roots VI is a 1D array input composed of the coefficients of the polynomial. For example, if the polynomial whose roots you want to compute is given by

$$P(s) = 1 + 2x - 4x^2 + 9x^3 - x^4 = 0,$$

then the corresponding input array is

$$[1 \quad 2 \quad -4 \quad 9 \quad -1].$$

When your VI is ready to accept inputs, enter the following coefficients: [120 154 71 14 1]. Run the VI. Where are the roots? You should determine them to be $x = -5, -4, -3$, and -2.

When you are finished, save the VI as roots.vi in the **Users Stuff** folder in the **Learning directory**.

A working version of roots.vi *can be found in the* **Chapter 11** *folder in the* **Learning** *directory.*   ◆

11.5   INTEGRATION AND DIFFERENTIATION

LabVIEW VIs for integration and differentiation are shown in Figure 11.20. The figure shows that there are many other VIs available for working with 1D functions—unfortunately we cannot cover all the analysis capabilities of LabVIEW in this book.

The Integral x(t) VI uses discrete integration methods. The user can select the method to use to perform the numeric integration from among those provided, including the trapezoidal rule, Simpson's rule (the default), Simpson's 3/8 rule, and Bode rule. The Derivative x(t) VI uses discrete differentiation methods. The user can select the method to perform the numeric differentiation from among those listed, including the second-order central difference (the default), fourth-order central difference, forward difference, and backward difference. The Quadrature VI performs numerical integration using adaptive quadrature approaches. You can select one of six instances of the VI, including 1D Quadrature (VI), 1D Quadrature (Formula), 2D Quadrature (VI), 2D Quadrature (Formula), 3D Quadrature (VI), and 3D Quadrature (Formula). The Integral x(t) VI, Derivative x(t) VI, and Quadrature VI represent a good set of tools for integration and differentiation.

A straightforward implementation of the integration and differentiation VIs is shown in Figure 11.21. You can find the VI shown in Figure 11.21 in the **Chapter 11** folder within the **Learning** directory—it is called **Derivative and Integration.vi**. Open and run the VI. You will find that three plots appear on the

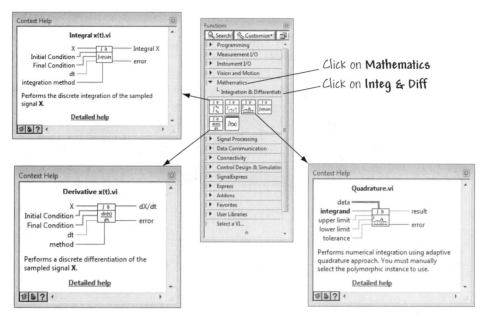

FIGURE 11.20
VIs for integration and differentiation.

graph: the function, the derivative of the function, and a plot of the integral of the function. The plot of the derivative of the sine function should appear as a cosine function. Use the graph palette zoom tool to zoom in on the derivative plot. If you run the VI in **Run Continuously** mode, you can vary the parameter **# of sample** and watch the sine function become smoother as the number increases and, conversely, become less smooth as the number of points decreases.

Open the VI Reference Based Quadrature.vi in the mathxmpl.llb in the analysis subfolder of the examples folder in the LabVIEW directory for examples of integrating using the Quadrature VI.

11.6   SIGNAL GENERATION

You can use LabVIEW to generate signals for testing algorithms and other purposes when real-world signals are not available. These signals can be generated using mathematical equations, arrays of data points, and by using signal generation VIs for common signals. To create signals that represent real-world signals you often need to accurately control the signal characteristics (such as magnitude, frequency and phase of periodic signals, and so on). In this section we discuss some of the possibilities for generating signals using VIs. The discussion

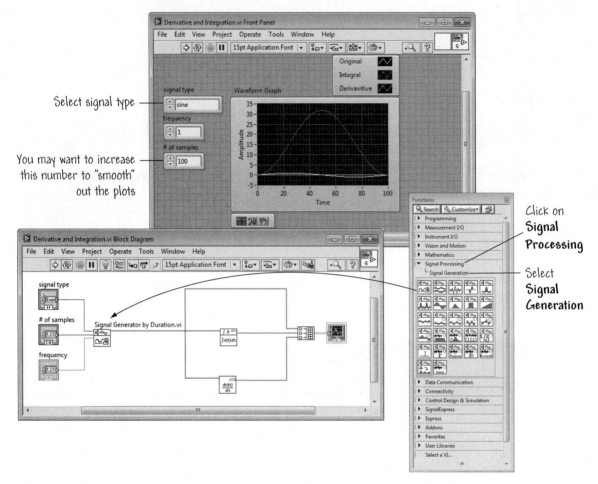

Select signal type

You may want to increase
this number to "smooth"
out the plots

Click on
**Signal
Processing**

Select
**Signal
Generation**

FIGURE 11.21
A VI for integration and differentiation of a 1D function.

begins by introducing normalized frequency, the discrete-time terminology used when generating periodic signals.

11.6.1   Normalized Frequency

In the analog or real world, signals vary continuously with time. When we analyze them in a computer we digitize them, changing them from a continuous signal to a sampled, discrete-time signal. For this reason, in the digital signal world (and with many Signal Generation VIs) we often use the so-called **digital**

frequency or **normalized frequency** (in units of cycles/sample) to describe the frequency (1/period) of a signal. This is computed as

$$f = \text{digital frequency} = \frac{\text{analog frequency}}{\text{sampling frequency}}.$$

The analog frequency is generally measured in units of Hz (or cycles per second) and the sampling frequency in units of samples per second. The normalized frequency is assumed to range from 0.0 to 1.0 corresponding to a frequency range of 0 to the sampling frequency, denoted by f_s. The normalized frequency wraps around 1.0, so that a normalized frequency of 1.2 is equivalent to 0.2. As an example, a signal sampled at the **Nyquist frequency** (that is, at $f_s/2$) is sampled twice per cycle (that is, two samples/cycle). This corresponds to a normalized frequency of $1/2$ cycles/sample = 0.5 cycles/sample. Therefore, we see that the reciprocal of the normalized frequency yields the number of times that the signal is sampled in one cycle (more on sampling in Section 11.7.4).

The following VIs utilize frequencies given in normalized units:

1. Sine Wave

2. Square Wave

3. Sawtooth Wave

4. Triangle Wave

5. Arbitrary Wave

6. Chirp Pattern

When using these VIs, you will need to convert the frequency units given in the problem to the normalized frequency units of cycles/sample. The VI depicted in Figure 11.22 illustrates how to generate two cycles of a sine wave and then convert cycles to cycles/sample.

In the example shown in Figure 11.22, the number of cycles (2) is divided by the number of samples (50), resulting in a normalized frequency of $f = 2/50$ cycles/sample. This implies that it takes 50 samples to generate two cycles of the sine wave. What if the problem specifies the frequency in units of Hz (cycles/sec)? In this case, if you divide the frequency in Hz (cycles/sec) by the sampling rate given in Hz (samples/sec), you obtain units of cycles/sample:

$$\frac{\text{cycles/sec}}{\text{samples/sec}} = \frac{\text{cycles}}{\text{sample}}.$$

The illustration in Figure 11.23 shows a VI used to generate a 60 Hz sine signal and to compute the normalized frequency when the input is in Hz. The

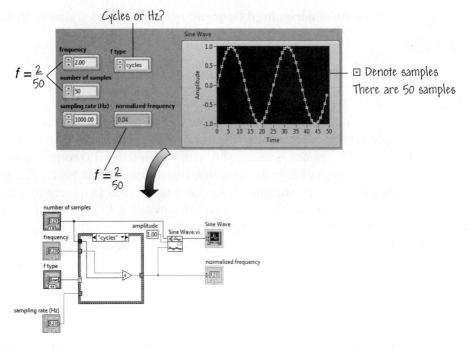

FIGURE 11.22
Generating two cycles of a sine wave and converting cycles to cycles/sample.

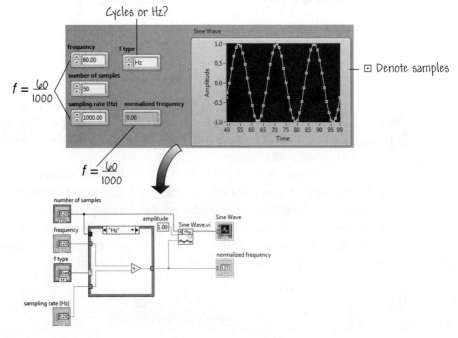

FIGURE 11.23
Generating a 60 Hz sine wave and computing normalized frequency.

normalized frequency is found by dividing the frequency of 60 Hz by the sampling rate of 1,000 Hz to get the normalized frequency of $f = 0.06$ cycles/sample:

$$f = \frac{60}{1,000} = 0.06 \; \frac{\text{cycles}}{\text{sample}}.$$

Therefore, we see that it takes almost 17 samples to generate one cycle of the sine wave. The number 17 comes from computing the reciprocal of $f = 0.06$.

Normalized Frequencies

Open the VI called **Normalized Frequency.vi** located in the **Chapter 11** folder in the **Learning** directory. The front panel and block diagram are shown in Figure 11.24. You can use this VI to experiment with calculating the normalized frequency when the input is in cycles and in Hz. Make sure that the f type is selected as Hz and run the VI. You should find that the normalized frequency is $f = 0.01$ with the default VI input values.

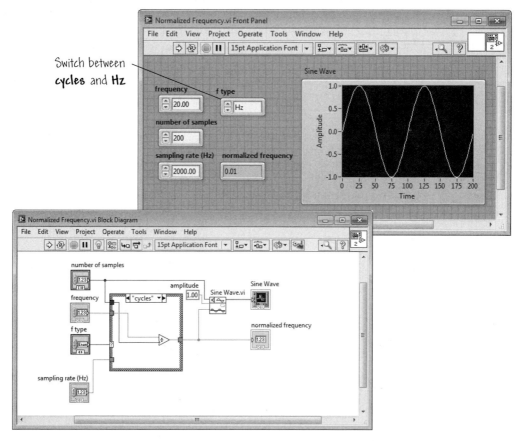

FIGURE 11.24
A VI to compute the normalized frequency.

Manually compute the normalized frequency for **f type** $=$ **Hz**, **frequency** $=$ 10, and **sampling rate (Hz)** $=$ 1,000. Modify the VI input parameters accordingly and verify that you obtain the same answer as by hand. The answer should be $f = 0.01$, computed as

$$f = \frac{10}{1,000} = 0.01 \; \frac{\text{cycles}}{\text{sample}}.$$

You can also work in cycles (rather than Hz) by selecting **f type** to be cycles. Then, in this situation, the normalized frequency is computed as a ratio of **frequency** to **number of samples**.   ◆

11.6.2   Wave, Pattern, and Noise VIs

The basic difference in the operation of the Wave or Pattern VIs is how signal timing is specified and tracked inside the VI. Wave VIs track phase internally generating continuous signals based on normalized frequency and on the number of samples. Pattern VIs generate signals based on cycles and do not track phase. You can distinguish between the two types of VIs by recognizing that the VI names contain either the word *wave* or *pattern*, as illustrated in Figure 11.25.

The Wave VIs operate with normalized frequencies in units of cycles/sample. The only Pattern VI that uses normalized units is the Chirp Pattern VI.

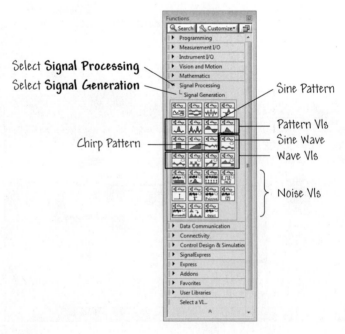

FIGURE 11.25
Signal generation VIs.

Since the Wave VIs can keep track of the phase internally, they allow the user to control the value of the initial phase. The phase in control specifies the initial phase (in degrees) of the first sample of the generated waveform and the phase out indicator specifies the phase of the next sample of the generated waveform. In addition, a reset phase control dictates whether or not the phase of the first sample generated when the wave VI is called is the phase specified at the phase in control, or whether it is the phase available at the phase out control when the VI last executed. A True value of reset phase sets the initial phase to phase in—a False value sets it to the value of phase out when the VI last executed.

The noise VIs generate random noise sequences with different characteristics. The more common types of noise models include the Gaussian white noise and the uniform white noise. The Gaussian White Noise VI generates a Gaussian distributed white noise sequence with zero mean and user-specified standard deviation. The Uniform White Noise VI generates a uniformly distributed random sequence whose values are in the range $[-a{:}a]$, where a is the absolute value of the noise amplitude and is a user input. Other interesting noise types include Bernoulli noise, Gamma noise, and Poisson noise.

All of the noise VIs (Gaussian white noise, uniform white noise, Bernoulli noise, Gamma noise, and Poisson noise) include an initialize? input and seed input. The input initialize? controls the seeding of the underlying random noise generator. If initialize? is TRUE, the internal seed for the random number generator is set according to the value of the input seed. If initialize? is FALSE, then the noise samples continue from the previous noise sequence. The default value for the input initialize? is TRUE. The seed input specifies how to generate the internal seed for the random number generator. When initialize? is TRUE, then the input seed is used to set the internal seed. If the value of seed is greater than 0, it is used as the seed for the random number generator. If the value of seed is less than or equal to 0, then the seed is generated using a random number. If initialize? is FALSE, then the input seed is ignored. The default value for seed is -1.

Practice with Signal Generation

In this exercise you will construct a VI that uses the two types of signal generation VIs. The front panel and block diagram for the VI are shown in Figure 11.26. The two main VIs used on the block diagram are the Sine Pattern.vi and Sine Wave.vi. These VIs are located on the **Signal Generation** palette, as shown in Figure 11.25. The VI also calculates and displays the normalized frequency.

Construct the VI using Figure 11.26 as a guide and run the VI when it is ready. Vary the VI inputs, paying particular attention to the effect that changing the control phase in (degrees) has on the Sine Pattern waveform graph. You should notice that the waveform begins to shift left and right as you vary the

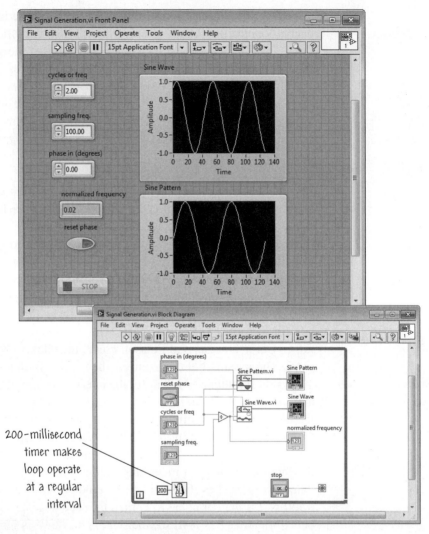

200-millisecond
timer makes
loop operate
at a regular
interval

FIGURE 11.26
Using the Sine Wave and Sine Pattern VIs to construct a signal generation VI.

phase up and down. When you set the reset phase switch to True (On), the initial phase is reset to the value specified by phase in each time the VI is called in the loop; otherwise, the initial phase is set to the previous phase output. While the VI is running, set the reset phase switch to the On position. The sine wave should be rendered stationary, whereas with the reset button set to the Off position, the sine wave varies with the varying phase.

Stop the execution of the VI using the Stop button located at the bottom left of the VI front panel in Figure 11.26. When you are finished experimenting with the VI, save it as Signal Generation.vi in the Users Stuff folder in the Learning directory.

You can find a working version of Signal Generation.vi in the Chapter 11 folder in the Learning directory. ◆

11.6.3   The Simulate Signal Express VI

The Simulate Signal Express VI is located on the **Signal Analysis** palette shown in Figure 11.11. This VI simulates sine waves, square waves, triangular waves, sawtooth waves, and noise signals. In Section 11.2.3, we used the Simulate Signal Express VI to simulate a noisy sine wave.

As with all Express VIs, to use the Simulate Signal Express VI, you first place it on the block diagram. As illustrated in Figure 11.27, once the Simulate

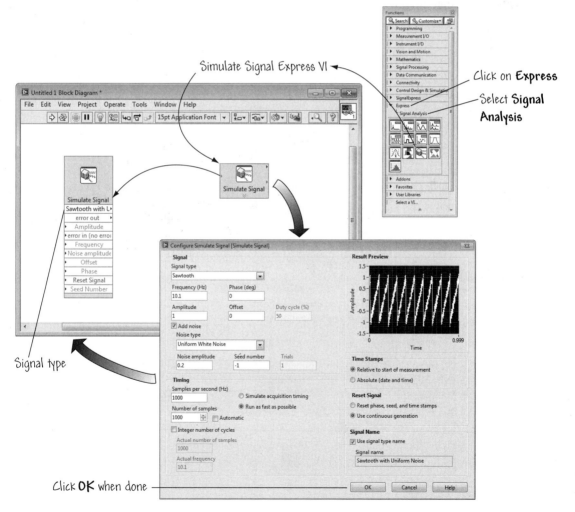

FIGURE 11.27
Configuring the Simulate Signal Express VI.

Signal Express VI is placed on the block diagram, a dialog box automatically appears to configure the VI. In Figure 11.27, a sawtooth wave is selected. The **Frequency (Hz)** is set to 10.1 Hz, although in the forthcoming example we will wire a control to the corresponding frequency input to programmatically vary the frequency from the front panel. We do the same with the **Amplitude**, set to 1 in Figure 11.27, but ultimately wired to a control to the corresponding input to programmatically vary the amplitude. We do the same as well for the **Offset** variable. Once the Simulate Signal Express VI is configured as desired, click **OK** to return to the block diagram.

Practice with the Simulate Signal Express VI

Consider the VI shown in Figure 11.28. Simulate Signal Express Demo.vi was developed to assist in investigating various signals generated by the Simulate Signal Express VI. Notice that the Simulate Signal Express VI is the central element on the block diagram in Figure 11.28. The Express VI is configured to generate a sawtooth signal. The remaining elements on the block diagram are placed there to permit easy access to key parameters of the sawtooth signal: frequency, amplitude, and offset. Noise has also been added to the signal

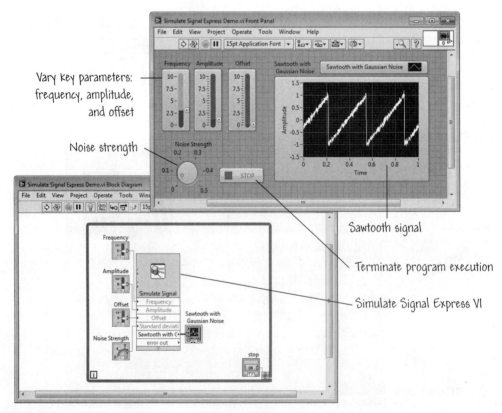

FIGURE 11.28
The Simulate Signal Express Demo VI for investigating signal generation.

and the control **Noise Strength** has been wired to the **Noise Standard Devia-tion** input.

*The Simulate Signal Express Demo.vi is located in **Chapter 11** of the **Learning** directory.*

Run the Simulate Signal Express Demo VI and observe the results on the waveform graph. Vary the frequency, amplitude, and offset and examine the resulting impact on the sawtooth signal.

Double-click the Simulate Signal Express VI to access the configuration dialog box. Select a different signal, such as the DC signal, and repeat the above investigation. What happens? You should find that the Broken Run indicator appears and there are broken wires on the block diagram. Remember that when you change the signal type, the inputs also change. In the case of a DC signal, it makes no sense to input a frequency; hence, the wire to that input generates an error.   ◆

11.7   SIGNAL PROCESSING

In this section we discuss three main topics: the Fourier transform (including the discrete Fourier transform and the fast Fourier transform), smoothing windows, and a brief overview of filtering.

11.7.1   The Fourier Transform

In Chapter 8 we covered the subject of data acquisition, where we discussed the fact that the samples of a measured signal obtained from the DAQ system are a time-domain representation of the signal, giving the amplitudes of the sampled signal at the sampling times. A significant amount of information is coded into the time-domain representation of a signal—maximum amplitude, maximum overshoot, time to settle to steady-state, and so on. The signal contains other useful information that becomes evident when the signal is transformed into the frequency domain. In other words, you may want to know the frequency content of a signal rather than the amplitudes of the individual samples. The representation of a signal in terms of its individual frequency components is known as the **frequency-domain representation** of the signal.

A common practical algorithm for transforming sampled signals from the time domain into the frequency domain is known as the **discrete Fourier trans-form**, or DFT. The relationship between the samples of a signal in the time domain and their representation in the frequency domain is established by the DFT. This process is illustrated in Figure 11.29.

If you apply the DFT to a time-domain signal represented by N samples of the signal, you will obtain a frequency-domain representation of the signal of length N. We denote the individual components of the DFT by $X(i)$. If the signal

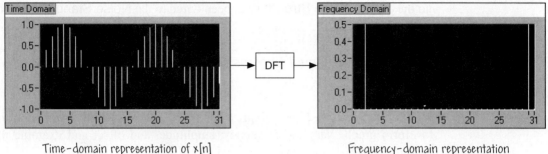

Time-domain representation of x[n]   Frequency-domain representation

FIGURE 11.29
The DFT establishes the relationship between the samples of a signal in the time domain and their representation in the frequency domain.

is sampled at the rate f_s Hz, and if you collect N samples, then you can compute the frequency resolution as $\Delta f = f_s/N$. This implies that the ith sample of the DFT occurs at a frequency of $i\,\Delta f$ Hz. We let the pth element $X(p)$ correspond to the Nyquist frequency. Regardless of whether the input signal is real or complex, the frequency-domain representation is always complex and contains two pieces of information—the amplitude and the phase.

For real-valued time-domain signals (denoted here by $X(i)$), the DFT is symmetric about the index $N/2$ with the following properties:

$$|X(i)| = |X(N-i)| \quad \text{and} \quad \text{phase}(X(i)) = -\text{phase}(X(N-i)).$$

The magnitude of $X(i)$ is **even symmetric**, that is, symmetric about the vertical axis. The phase of $X(i)$ is **odd symmetric**, that is, symmetric about the origin. This symmetry is illustrated in Figure 11.30. Since there is repetition of information contained in the N samples of the DFT (due to the symmetry properties), only half of the samples of the DFT need to be computed, since the other half can be obtained from symmetry.

Figure 11.31(a) depicts a *two-sided transform* for a complex sequence with $N = 8$ and $p = N/2 = 4$. Since $N/2$ is an integer, the DFT contains the Nyquist frequency. When N is odd, $N/2$ is not an integer, and thus there is no component

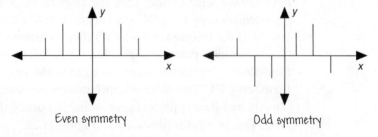

Even symmetry   Odd symmetry

FIGURE 11.30
Even and odd symmetric signals.

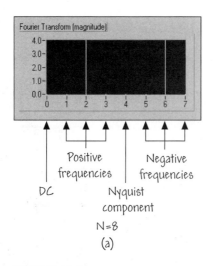
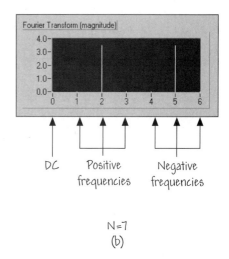

FIGURE 11.31
A two-sided transform representation of a complex sequence.

at the Nyquist frequency. Figure 11.31(b) depicts a two-sided transform when $N = 7$.

The computationally intensive process of computing the DFT of a signal with N samples requires approximately N^2 complex operations. However, when N is a power of 2, that is, when

$$N = 2^m \quad \text{for} \quad m = 1 \text{ or } 2 \text{ or } 3 \text{ or } \cdots,$$

you can implement the so-called **fast Fourier transforms** (FFTs), which require only approximately $N\log_2(N)$ operations. In other words, the FFT is an efficient algorithm for calculating the DFT when the number of samples (N) is a power of 2.

Practice with FFTs

In this example you will practice with FFTs by opening an existing VI and experimenting with the input parameters. Locate and open FFT_2sided.vi. You will find this VI in the **Chapter 11** folder in the **Learning** directory. The front panel is shown in Figure 11.32.

The VI demonstrates how to use the FFT VI to analyze a sine wave of user-specified frequency. The VI block diagram, shown in Figure 11.33, contains three main VIs:

- **FFT.vi**: This VI computes the fast Fourier transform (FFT) or the discrete Fourier transform (DFT) of the input sequence. FFT.vi will execute FFT

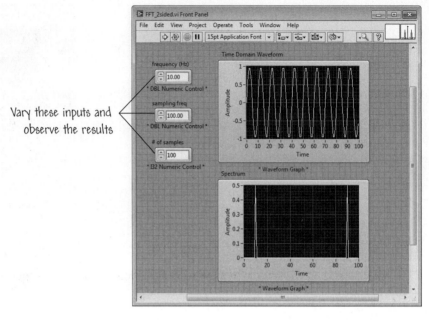

Vary these inputs and
observe the results

FIGURE 11.32
The front panel of FFT_2sided.vi.

routines if the size of the input sequence is a power of 2. If the size of the
input sequence is not a power of 2, then an efficient DFT routine is called.

- **Sine Wave.vi**: This VI generates an array containing a sine wave. The VI
  is located in **Functions≫Signal Processing≫Signal Generation** palette.

- **Complex To Polar.vi**: Separates a complex number into its polar components represented by magnitude and phase. The input can be a scalar number, a cluster of numbers, an array of numbers, or an array of clusters. In the
  case of Figure 11.33, the input is an array of numbers.

Run the VI and experiment with the input parameters. Run several numerical
experiments using the **Run Continuously** mode. What happens to the spectrum
when you vary the signal frequency? For example, set the input signal frequency
to 10 Hz, the sampling frequency to 100 Hz, and the number of samples to 100.
In this case, $\Delta f = 1$ Hz. The spectrum should have two corresponding peaks.
Check this using the VI.

For another experiment, set the signal frequency to 50 Hz, the sampling frequency to 100 Hz, and the number of samples to 100, and run the VI. With
the VI running in **Run Continuously** mode, set the sampling frequency to 101
and observe the effects on the time-domain sequence waveform and the corresponding spectrum. Now, slowly increase the sampling frequency and see what
happens!

When you are finished, close the VI and do not save any changes.

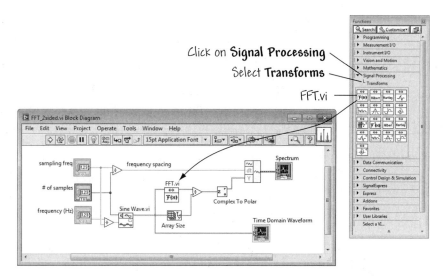

FIGURE 11.33
The block diagram of FFT_2sided.vi.

11.7.2   Smoothing Windows

When using discrete Fourier transform methods to analyze a signal in the frequency domain, it is assumed that the available data of the time-domain signal represents at least a single period of a periodically repeating waveform. Unfortunately, in most realistic situations, the number of samples of a given time-domain signal available for DFT analysis is limited and this can sometimes lead to a phenomenon known as **spectral leakage**. To see this, consider a periodic waveform created from one period of a sampled waveform, as illustrated in Figure 11.34.

The first period shown in Figure 11.34 is the sampled portion of the waveform. The sampled waveform is then repeated to produce the periodic waveform. Sampling a noninteger number of cycles of the waveform results in

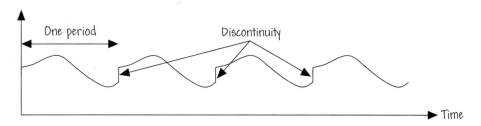

FIGURE 11.34
A periodic waveform created from one period of a sampled waveform.

discontinuities between successive periods! These discontinuities induced by the process of creating a periodic waveform lead to very high frequencies (higher than the Nyquist frequency) in the spectrum of the signal—frequencies that were not present in the original signal. Therefore, the spectrum obtained with the DFT will not be the true spectrum of the original signal. In the frequency domain, it will appear as if the energy at one frequency has "leaked out" into all the other frequencies, leading to what is known as spectral leakage.

Figure 11.35 shows a sine wave and its corresponding Fourier transform. The sampled time-domain waveform is shown in Graph 1 in the upper left corner. In this case, the sampled waveform is an integer number of cycles of the original sine wave. The sampled waveform can be repeated in time, and a periodic version of the original waveform thereby constructed. The constructed periodic version of the waveform is depicted in Graph 2 in the upper middle section of Figure 11.35. The constructed periodic waveform does not have any discontinuities because the sampled waveform is an integer number of cycles of the original waveform. The corresponding spectral representation of the periodic waveform is shown in Graph 3. Because the time record in Graph 2 is periodic, contains no discontinuities, and is an accurate representation of the true waveform, the computed spectrum is correct.

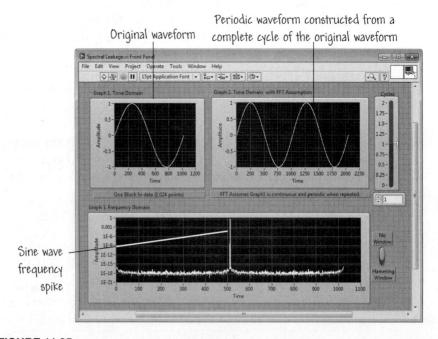

FIGURE 11.35
One complete period of a sine wave is repeated to obtain a periodic signal with no discontinuities; the corresponding Fourier transform shows no leakage.

In Figure 11.36 a spectral representation of another periodic waveform is shown. However, in this case a noninteger number of cycles of the original waveform is used to construct the periodic waveform, resulting in the discontinuities in the waveform shown in Graph 2. The corresponding spectrum is shown in Graph 3. The energy is now "spread" over a wide range of frequencies—compare this result to Graph 3 in Figure 11.35. The smearing of the energy is called spectral leakage, as mentioned earlier.

Leakage results from using only a finite time sample of the input signal. One (unpractical) solution to the leakage problem is to obtain an infinite time record, from $-\infty$ to $+\infty$, yielding an ideal FFT solution. In practice, however, we are limited to working with a finite time record. A practical approach to the problem of spectral leakage is the so-called **windowing** technique. Since the amount of spectral leakage depends on the amplitude of the discontinuity, the larger the discontinuity, the more the leakage. Windowing reduces the amplitude of the discontinuities at the boundaries of each period by multiplying the sampled original waveform by a finite length window whose amplitude varies smoothly and gradually towards zero at the edges.

One such windowing technique uses the *Hamming window*, as illustrated in Figure 11.37. The sinusoidal waveform of the windowed signal gradually tapers to zero at the ends—see the bottom graph in Figure 11.37. When computing the

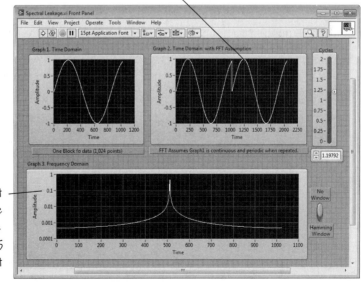

FIGURE 11.36
A portion of a sine wave period is repeated to obtain a periodic signal with discontinuities; the corresponding Fourier transform shows leakage.

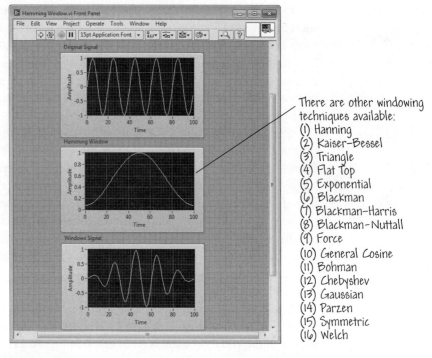

The are other windowing techniques available:
(1) Hanning
(2) Kaiser–Bessel
(3) Triangle
(4) Flat Top
(5) Exponential
(6) Blackman
(7) Blackman–Harris
(8) Blackman–Nuttall
(9) Force
(10) General Cosine
(11) Bohman
(12) Chebyshev
(13) Gaussian
(14) Parzen
(15) Symmetric
(16) Welch

FIGURE 11.37
A sinusoidal signal windowed using a Hamming window.

discrete Fourier transform on data of finite length, you can use the windowing technique applied to the sampled waveform to minimize the discontinuities of the constructed periodic waveform. This approach will minimize the spectral leakage.

The Hamming Window

Open a new front panel and place four waveform graphs and one numeric control, as shown in Figure 11.38.

Construct a block diagram using Figure 11.39 as a guide. In the block diagram, we use three main VIs:

- Hamming Window.vi: This VI applies the Hamming window to the input sequence. It is located in **Functions≫Signal Processing≫Windows**, as shown in Figure 11.39. If we denote the input sequence as **X** (with n elements) and the output sequence of the Hamming window as **Y**, then

$$\mathbf{Y}(i) = \mathbf{X}(i)\,[0.54 - 0.46\cos\omega]$$

where

$$\omega = \frac{2\pi i}{n}.$$

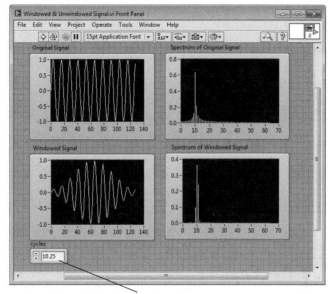

FIGURE 11.38
The front panel for a VI to investigate the use of windows.

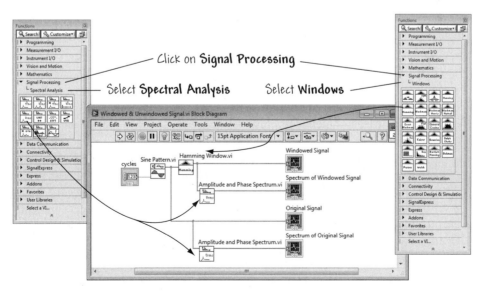

FIGURE 11.39
The block diagram for a VI to investigate the effect of windowing.

- The Amplitude and Phase Spectrum.vi computes the amplitude spectrum of the windowed and nonwindowed input waveforms. You can find this particular VI in the palette **Functions≫Signal Processing≫Spectral Analysis**, as shown in Figure 11.39.

- The Sine Pattern.vi generates a sine wave with the number of cycles specified in the control labeled **cycles**. It is located in **Functions≫Signal Processing≫Signal Generation**.

As interesting numerical experiment, make the following two runs:

- Set the control cycles to 10. Since this is an integer number, when you repeat the waveform to construct a periodic waveform, you will not have any discontinuities. You should observe that the spectrums of the windowed and the nonwindowed waveforms are both centered at 10, and that the spectrum of the original signal displays no spectral leakage.

- Set the control cycles to 10.25. Since this is not an integer, you should observe that the spectrums of the windowed and the nonwindowed waveforms are different. The nonwindowed spectrum should show distinct signs of spectral leakage due to the discontinuities when constructing the periodic waveform. The windowed waveform, while not a perfect spike centered at 10, displays significantly less leakage.

Save your VI as Windowed & Unwindowed Signal.vi in the Users Stuff folder in the Learning directory.

You can find a working version of Windowed & Unwindowed Signal.vi *in* Chapter 11 *of the* Learning *directory.* ◆

11.7.3 The Spectral Measurements Express VI

The **Express≫Signal Analysis** palette shown in Figure 11.11 contains the Spectral Measurements Express VI. This Express VI performs spectral measurements, such as spectral power density, on signals.

To use the Spectral Measurements VI, you first place it on the block diagram. As illustrated in Figure 11.40, once the Spectral Measurements Express VI is placed on the block diagram, a dialog box automatically appears to configure the VI. The spectral measurements options include choosing the measurement type, such as magnitude (peak or RMS), power spectrum, and power spectral density, and selecting the windowing technique from among Hanning, Hamming, Blackman–Harris, Low Sidelobes, and more. In Figure 11.40, the magnitude RMS is selected as the measurement type and the Hanning window is selected for windowing. Once the Spectral Measurements VI is configured, click **OK** to return to the block diagram.

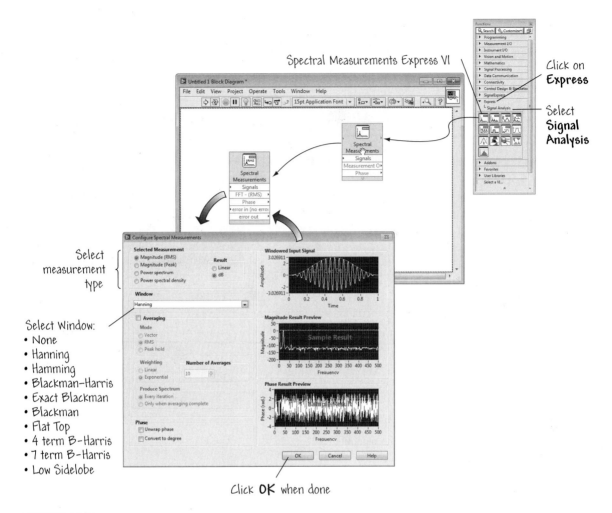

FIGURE 11.40
Configuring the Spectral Measurements Express VI.

On the block diagram, you can expand the Express VI to display the inputs and outputs. The key input is the signal, and the key output is the spectral measurement, such as FFT-(RMS). The Spectral Measurements VI is now ready for inclusion in a VI for signal analysis.

Practice with the Spectral Measurements Express VI

Consider the VI shown in Figure 11.41. Notice on the block diagram that two Express VIs, the Spectral Measurements Express VI and the Simulate Signal Express VI, are utilized. This VI was developed to assist in investigating the power spectral density of a sine wave signal.

Original signal

Power spectral density of the windowed signal

Power spectral density of the sine wave

Generates a sine wave of frequency 10.25 Hz

Configured with a Hamming window

Configured with no windowing

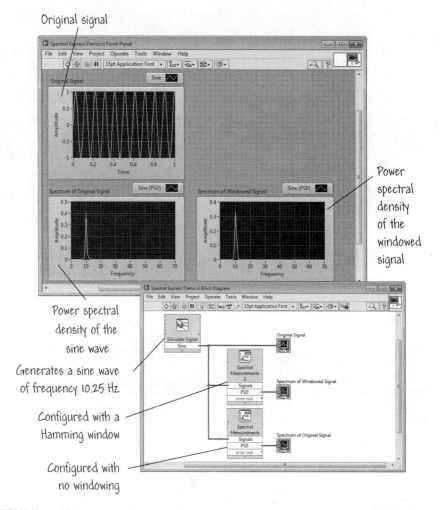

FIGURE 11.41
The Spectral Express Demo VI for investigating the power spectral density of sine wave signals.

Spectral Express Demo.vi is located in Chapter 11 of the Learning directory.

There are four elements on the block diagram:

- Two Spectral Measurements Express VIs—these can be configured by double-clicking on their icons to access the configuration dialog boxes.

- The Simulate Signal Express VI—this VI was described in Section 11.6. It has been configured here to provide a sine wave signal at a frequency of 10.25 Hz.

- The waveform graphs—these are used to graphically display the power spectral density of the sine wave signal, both the windowed and unwindowed versions.

Run the Spectral Express Demo VI and observe the results on the waveform graphs. Double-click the Simulate Signal Express VI to access the configuration dialog box. Change the frequency of the sine wave signal to 50 Hz, run Spectral Express Demo.vi, and observe the effects on the power spectral density. You should observe the peak of the power spectral density shift to 50 Hz. Test the VI using different windowing methods.   ◆

11.7.4   Filtering

There are two main types of filters: analog and digital. Analog filters can be built using analog components like resistors, capacitors, inductors, and amplifiers. Digital filters can be implemented in software, changed programmatically on the fly, are often more stable and predictable because they do not drift with changes in external environmental conditions, and generally have superior performance-to-cost ratios compared to their analog counterparts. In this section we will consider digital filters only.

LabVIEW can be used to implement digital filters and control digital filter parameters (such as filter order, cutoff frequency, stopband and passbands, amount of ripple, and stopband attenuation). You can envision a LabVIEW-based DAQ system wherein data is acquired from external sources (such as an accelerometer sensor) and filtered in the VI software. The results can be easily analyzed and studied using the graphics provided by LabVIEW and written to a spreadsheet. The key is that by using LabVIEW you can utilize digital filters, allowing the VIs to handle the design issues, computations, memory management, and the actual data filtering.

The theory of filters is a rich and interesting subject, and one that cannot be dealt with here in any depth. Please refer to other reference materials for in-depth coverage of filtering theory. A brief discussion of terms is needed, however, to give you a better understanding of the filter parameters and how they relate to the VI inputs.

The **sampling theorem** states that a continuous-time signal can be reconstructed from discrete, equally spaced samples if the sampling frequency is at least twice that of the highest frequency in the time signal. The sampling interval is often denoted by δt. The **sampling frequency** is computed as the inverse of the sampling interval:

$$f_s = \frac{1}{\delta t}.$$

Thus, according to the sampling theorem, the highest frequency that the filter can process—the **Nyquist frequency**—is

$$f_{nyq} = \frac{f_s}{2}.$$

As an example, suppose that you have a system with sampling interval $\delta t = 0.01$ second. Then the sampling frequency is $f_s = \frac{1}{0.01} = 100\,\text{Hz}$. From the sampling theorem we find that the highest frequency that the system can process is $f_{nyq} = \frac{f_s}{2} = 50\,\text{Hz}$. If we expect that the signals that we need to process have components at frequencies higher than 50 Hz, then we must upgrade the system to allow for shorter sampling intervals, say for example, $\delta t = 0.001$ second. What is f_{nyq} in this case?

One main use of filters is to remove unwanted noise from a signal—usually if the noise is at high frequencies. Depending on the frequency range of operation, filters either pass or attenuate input signal components. Filters can be classified into the following types:

1. A **lowpass filter** passes low frequencies and attenuates high frequencies. The ideal lowpass filter passes all frequencies below the *cutoff frequency* f_c.

2. A **highpass filter** passes high frequencies and attenuates low frequencies. The ideal highpass filter passes all frequencies above f_c.

3. A **bandpass filter** passes a specified band of frequencies. The ideal bandpass filter only passes all frequencies between f_{c1} and f_{c2}.

4. A **bandstop filter** attenuates a specified band of frequencies. The ideal bandstop filter attenuates all frequencies between f_{c1} and f_{c2}.

The ideal frequency response of these filters is illustrated in Figure 11.42.

The frequency points f_c, f_{c1}, and f_{c2} are known as the cutoff frequencies and can be viewed as filter design parameters. The range of frequencies that is passed through the filter is known as the **passband** of the filter. The gain of a filter at a specific frequency is typically represented in units of Decibels (dB), where dB is calculated as $10 \, \text{Log}_{10} \, (V_{out}/V_{in})$. The decibel is used to scale very large and very small numbers so that they can all be plotted on the same graph. For instance, when dB $= 0$ then $V_{out} = V_{in}$ indicating no gain. When dB $= 20$ then $V_{out} = V_{in} * 100$ indicating a gain of 100. An ideal filter has a gain of one (0 dB) in the passband—that is, the amplitude of the output signal is the same as the amplitude of the input signal. Similarly, the ideal filter completely attenuates the signals in the stopband—that is, the stopband attenuation is $-\infty$ dB. The lowpass and highpass filters have one passband and one stopband. The range of frequencies that do not pass through the filter is known as the **stopband** of the filter. The stopband frequencies are rejected or attenuated by the filter. The passband(s) and the stopband(s) for the different types of filters are shown in Figure 11.42. The bandpass filter has one passband and two stopbands. Conversely, the bandstop filter has two passbands and one stopband.

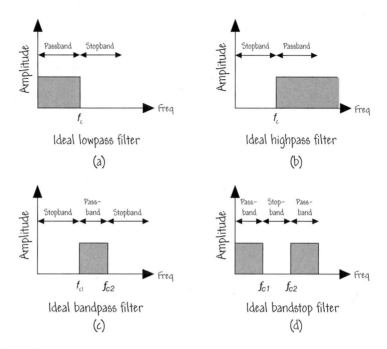

FIGURE 11.42
The ideal frequency response of common ideal filters.

Suppose you have a signal containing component frequencies of 1 Hz, 5 Hz, and 10 Hz. This input signal is passed separately through lowpass, highpass, bandpass, and bandstop filters. The lowpass and highpass filters have a cutoff frequency of 3 Hz, and the bandpass and bandstop filters have cutoff frequencies of 3 Hz and 8 Hz. What frequency content of the signal can be expected? The output of the filter in each case is shown in Figure 11.43. The lowpass filter passes only the signal at 1 Hz because this is the only component of the input signal lower than the 3-Hz cutoff. Conversely, the highpass filter attenuates the 1-Hz component and passes a signal with components at 5 and 10 Hz. The bandpass filter passes only the signal component at 5 Hz, and the bandstop filter filters out the signal component at 5 Hz and passes a signal with frequency content at 1 and 10 Hz.

Ideal filters are not achievable in practice. It is not possible to have a unit gain (0 dB) in the passband and a gain of zero ($-\infty$ dB) in the stopband—there is always a *transition region* between the passband and stopband. A more realistic filter will have the passband, stopband, and transition bands as depicted in Figure 11.44. As you can see, a real filter does not always have a unit gain (0 dB) across the passband and a gain of zero ($-\infty$ dB) across the stopband. The variation in the passband is called the *passband ripple*; see the bandpass filter in Figure 11.44. The *stopband attenuation*, also depicted in Figure 11.44, cannot be infinite (as it would be for an ideal filter).

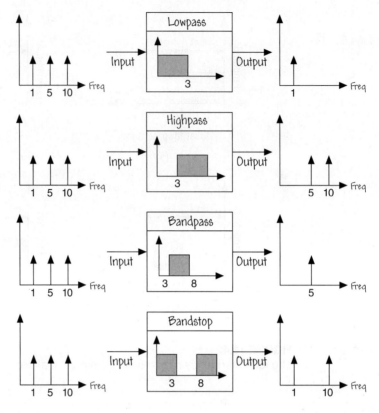

FIGURE 11.43
The output of the four filters in the case where the input signal contains component frequencies of 1 Hz, 5 Hz, and 10 Hz.

If we view the filter as a linear system, then we can consider the response of the system (that is, the filter) to various types of inputs.[1] One interesting input is the *impulse*. If the input to the digital filter is the sequence $x(0), x(1), x(2), \ldots,$ then the impulse is given by $x(0) = 1, x(1) = x(2) = \cdots = 0$. The **impulse response** of a filter (that is, the output of the filter when the input is an impulse) provides another classification system for filters. The Fourier transform of the impulse response is known as the **frequency response**. The frequency response of a system provides a wealth of information about the system, including how it will respond to periodic inputs at different frequencies. Therefore, the frequency response will tell us about the filter characteristics: How does the filter respond in the passbands and stopbands? How accurately does the filter cut off and attenuate high-frequency components?

---

1. A good source of information on systems and system response to various inputs is *Modern Control Systems*, by Richard C. Dorf and Robert H. Bishop, Pearson, Upper Saddle River, NJ, 2010.

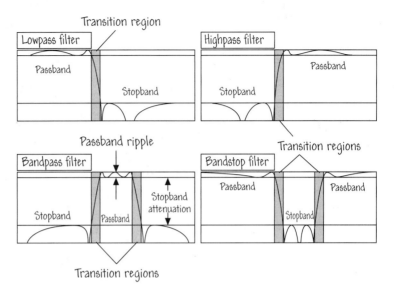

FIGURE 11.44
A realistic frequency response of a filter.

All digital filters can be classified into one of two types based on their impulse response, as either **finite impulse response** (FIR) filters or **infinite impulse response** (IIR) filter. For an IIR filter, the impulse response continues indefinitely (in theory), and the output depends on current and past values of the input signal and on past values of the output. In practical applications, the impulse response for stable IIR filters decays to near zero in a finite time. For an FIR filter the impulse response decays to zero in a finite time, and the output depends only on current and past values of the input signal. Take the example of processing noisy range measurements (that is, distance measurements) to a fixed target. Suppose that you want to determine the distance to the fixed object and you have available a ranging device (e.g., a laser ranging device) that is corrupted by random noise (as is the case for most realistic sensors!). One way to estimate the range is to take a series of range measurements $x(0), x(1), x(2), \ldots, x(k)$ and filter them by computing a running average:

$$x_{\text{ave}}(k) = \frac{1}{k} \sum_{i=1}^{k} x(i).$$

The output of the filter is $x_{\text{ave}}(k)$. This is an FIR filter because the output depends only on previous values of the input $(x(0), x(1), x(2), \ldots, x(k-1))$ and on the current value of the input $(x(k))$. Now, we can rewrite the filter as

$$x_{\text{ave}}(k) = \frac{k-1}{k} x_{\text{ave}}(k-1) + \frac{1}{k} x(k).$$

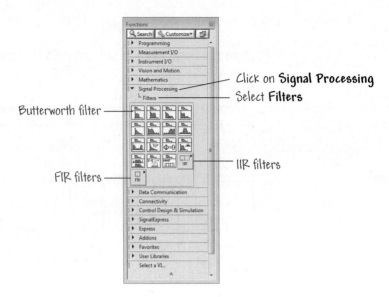

FIGURE 11.45
IIR and FIR filter choices are located in the **Functions** palette.

This is an IIR filter because the output depends on current and previous values of the input as well as on previous values of the output (that is, the $x_{\text{ave}}(k-1)$ term). Mathematically, the two filters provide the same output, but they are implemented differently. The FIR filter is sometimes referred to as a *nonrecursive* filter; the IIR filter is known as a *recursive* filter.

One disadvantage of IIR filters is that the phase response is nonlinear. You should use FIR filters for situations where a linear phase response is needed. A strong advantage of IIR filters is that they are recursive, thus reducing the memory storage requirements.

Some of the most common digital filters, such as Butterworth and Chebyshev, share their name with an analog counterpart because they share many of the same passband and stopband characteristics. Other filters are unique digital filters and are not typically used in an analog implementation. One such well-known filter, the Kalman filter, can be classified as an IIR filter (it is actually a bit more complicated to implement than an IIR filter) and was used successfully to filter navigation data acquired by the Apollo spacecraft to find its way to the moon and to rendezvous around the moon for the long journey home.[2] LabVIEW provides many different types of filters, as shown in Figure 11.45.

---

2. A well-told story on the use of Kalman filters during Apollo can be found in *An Introduction to the Mathematics and Methods of Astrodynamics*, by R. H. Battin, AIAA Education Series, 1999.

The IIR filter VIs available in LabVIEW include the following:

- Butterworth: Provides a smooth response at all frequencies and a monotonic decrease from the specified cutoff frequencies. Butterworth filters are maximally flat—the ideal response of unity in the passband and zero in the stopband—but do not always provide a good approximation of the ideal filter response because of the slow rolloff between the passband and the stopband.

- Chebyshev: Minimizes the peak error in the passband by accounting for the maximum absolute value of the difference between the ideal filter and the filter response you want (the maximum tolerable error in the passband). The frequency response characteristics of Chebyshev filters have an equiripple magnitude response in the passband, a monotonically decreasing magnitude response in the stopband, and a sharper rolloff than for Butterworth filters.

- Inverse Chebyshev: Also known as Chebyshev II filters. They are similar to Chebyshev filters, except that inverse Chebyshev filters distribute the error over the stopband (as opposed to the passband) and are maximally flat in the passband (as opposed to the stopband). Inverse Chebyshev filters minimize peak error in the stopband by accounting for the maximum absolute value of the difference between the ideal filter and the filter response you want. The frequency response characteristics have an equiripple magnitude response in the stopband, a monotonically decreasing magnitude response in the passband, and a rolloff sharper than for Butterworth filters.

- Elliptic: Minimize the peak error by distributing it over the passband and the stopband. Equiripples in the passband and the stopband characterize the magnitude response of elliptic filters. Compared with the same-order Butterworth or Chebyshev filters, the elliptic design provides the sharpest transition between the passband and the stopband. For this reason, elliptic filters are widely used.

- Bessel: Can be used to reduce nonlinear phase distortion inherent in all IIR filters. Bessel filters have maximally flat response in both magnitude and phase. Furthermore, the phase response in the passband of Bessel filters, which is the region of interest, is nearly linear. Like Butterworth filters, Bessel filters require high-order filters to minimize the error.

The FIR filter VIs available in LabVIEW include the following:

- Windowed: The simplest method for designing linear-phase FIR filters is the window design method. You select the type of windowed FIR filter you want—lowpass, highpass, bandpass, or bandstop—via input to the FIR Windowed Filter VI.

- Optimum filters based on the Parks–McClellan algorithm: Offers an optimum FIR filter design technique that attempts to design the best filter possible for a given filter complexity. Such a design reduces the adverse effects at the cutoff frequencies. It also offers more control over the approximation errors in different frequency bands—control that is not possible with the window method. The VIs available include

  □ Equiripple Lowpass

  □ Equiripple Highpass

  □ Equiripple Bandpass

  □ Equiripple Bandstop

Which filter is best suited for your application? Obviously, the choice of filter depends on the problem at hand. Figure 11.46 shows a flowchart that can serve as a guide for selecting the best filter for your needs. Keep in mind that you will probably use the flowchart to determine several candidate filters, and you will have to experiment to make the final choice.

Extracting a Sine Wave

Open a new front panel and construct a front panel similar to the one shown in Figure 11.47. You will need to place one numeric control, two vertical slides, and two waveform graphs. Label them according to the scheme shown in the figure.

Construct a block diagram similar to the one shown in Figure 11.47. In the block diagram, we use three main VIs:

- Butterworth.vi: This VI is used to filter the noise. It is located in **Functions≫Signal Processing≫Filters**, as shown in Figure 11.45.

- Uniform White Noise.vi generates a white noise that is added to the sinusoidal signal. You can find this particular VI in the **Functions≫Signal Processing≫Signal Generation** palette.

- Sine Pattern.vi generates a sine wave of the desired frequency. It is located in the **Functions≫Signal Processing≫Signal Generation** palette.

With this VI you are generating 10 cycles of a sine wave (this value can be varied on the front panel), and there are 1,000 samples. Select a cutoff frequency of 25 Hz and a filter order of five. Note that we did not previously discuss filter order—it is a measure of filter complexity and is related to the number of terms retained in the filter. Run the VI. Vary the cutoff frequency and observe the effects. What happens when the cutoff frequency is set to 50? Does the filtered signal contain noise components? When you are finished exploring, save your VI as Extract the Sine Wave.vi in the Users Stuff folder.

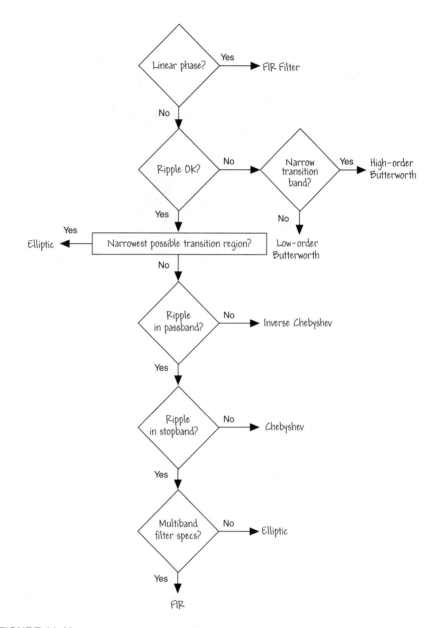

FIGURE 11.46
Flowchart that can serve as a guide for selecting the best filter.

 You can find a working version of the VI in folder Chapter 11 in the Learning directory. It is called Extract the Sine Wave.vi.   ◆

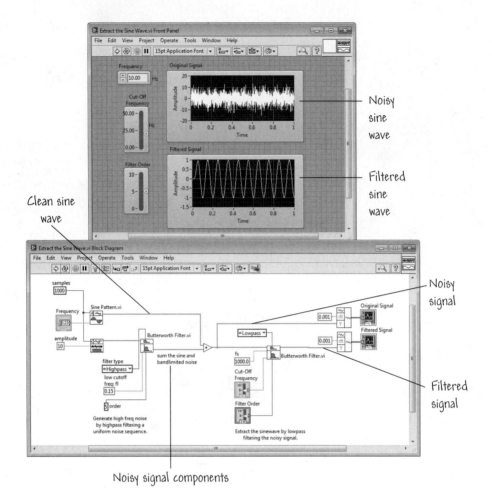

FIGURE 11.47
The front panel and block diagram for a VI to filter a noisy sine wave.

11.7.5   The Filter Express VI

The Filter Express VI is located on the **Express≫Signal Analysis** palette, shown in Figure 11.11. The Filter Express VI processes signals through filters and windows. The filter options include lowpass, highpass, bandpass, bandstop, and smoothing. To use the Filter Express VI, you first place it on the block diagram. As illustrated in Figure 11.48, once the Filter Express VI is placed on the block diagram, the **Configure Filter** dialog box automatically appears to configure the VI. Figure 11.48 shows a lowpass filter selected with a cutoff frequency of 100 Hz. Once the Filter Express VI is configured, click **OK** to return to the block diagram. The Filter Express VI is now ready for inclusion in a VI for signal analysis. On the block diagram, you can expand the Filter Express VI

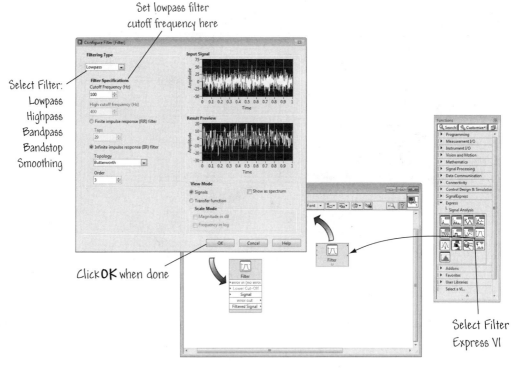

Set lowpass filter
cutoff frequency here

Select Filter:
Lowpass
Highpass
Bandpass
Bandstop
Smoothing

Click **OK** when done

Select Filter
Express VI

FIGURE 11.48
Configuring the Filter Express VI.

to display the inputs and outputs. The key input is the signal and the key output is the filtered signal.

**Practice
with the
Filter Express VI**

Consider the VI shown in Figure 11.49. Notice on the block diagram that three Express VIs are used: the Filter Express VI, the Simulate Signal Express VI, and the Spectral Measurements Express VI. The Filter Express Demo VI was developed to assist in investigating the lowpass filtering of a sinusoidal signal.

Filter Express Demo.vi is located in Chapter 11 *of the* Learning *directory.*

The four Simulate Signal Express VIs have been configured to provide sine wave signals at frequencies of 1, 10, and 100 Hz and a DC signal of magnitude 10. The sinusoidal signal that is to be filtered is a sum of the four individual signals from the Simulate Signal Express VIs. The Filter Express VI has been configured to provide a lowpass filter with default cutoff frequency of 25 Hz. A numeric control is employed to vary the cutoff frequency directly from the front panel. The Spectral Measurements Express VIs appearing in Figure 11.49 have

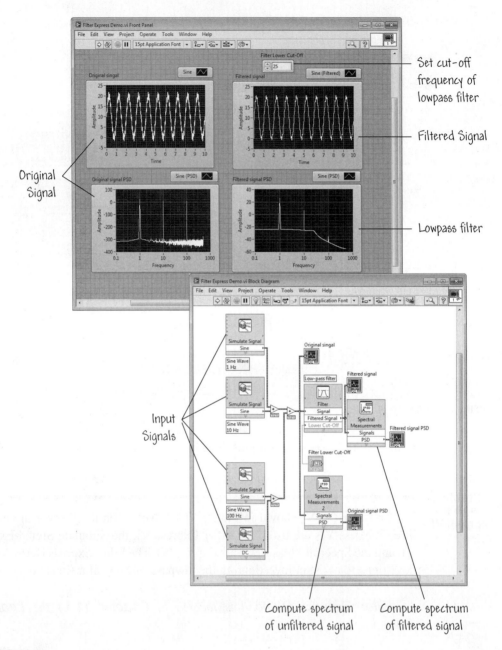

FIGURE 11.49
The Filter Express Demo VI for investigating the filtering of a sinusoidal signal containing three main frequencies at 1, 10, and 100 Hz.

been configured to provide power spectral densities of the incoming signals. The waveform graphs display the power spectral density of the filtered and unfiltered sinusoidal signals.

Run the Filter Express Demo VI and observe the results on the waveform graphs. When the cutoff frequency is set to the default value of 25 Hz, you should observe that the filtered signal has a significantly reduced frequency component at 100 Hz. This is because the lowpass filter has essentially removed the 100 Hz component of the sinusoidal signal.

You can investigate the effect of varying the lowpass filter cutoff frequency on the filtering process. The numeric control located on the front panel will be employed to vary the lowpass filter cutoff frequency. Start the VI by pressing the **Run Continuously** button. Slowly reduce the cutoff frequency and observe the power spectral density graph of the filtered signal (lower right-hand side of the block diagram). Notice that as the cutoff frequency is reduced to a value under 10 Hz, the component of the filtered sinusoidal signal at 10 Hz is correspondingly reduced.

Experiment with increasing the lowpass filter cutoff frequency above 100 Hz. Can you find a value of the cutoff frequency for which the filtered signal power spectral density shows the 100 Hz component?   ◆

BUILDING BLOCK

11.8 BUILDING BLOCKS: PULSE WIDTH MODULATION

LabVIEW's signal generation VIs can be used to simulate real-world signals (as described in Section 11.6). In this Building Block exercise, you will use the Square Wave VI to construct a VI that outputs a pulse width modulated waveform. Your VI will have a similar output to the pulse width modulation VIs in previous Building Blocks, but will be considerably less complicated in its implementation. As you become increasingly familiar with LabVIEW and its capabilities you will find simpler and more efficient ways to reach design objectives. Generally, the objectives of a VI can be met with a variety of solutions. Each solution comes with its own set of advantages and disadvantages.

Open a new VI and save it as **PWM with Square Wave VI.vi** in the **Users Stuff** folder in the **Learning** directory. The front panel and block diagram shown in Figure 11.50 can be used as a guide to developing your VI. The objective of the VI is to output a Square Wave using a vertical slide to adjust the duty cycle. The Square Waveform VI is found on the **Signal Processing≫Waveform Generation** palette. The Reciprocal function is found on the **Programming≫ Numeric** palette. One key difference between the PWM with Square Wave VI and the pulse width modulation VIs in previous Building Blocks is that with each loop iteration, this new VI generates an entire output cycle rather than one individual point.

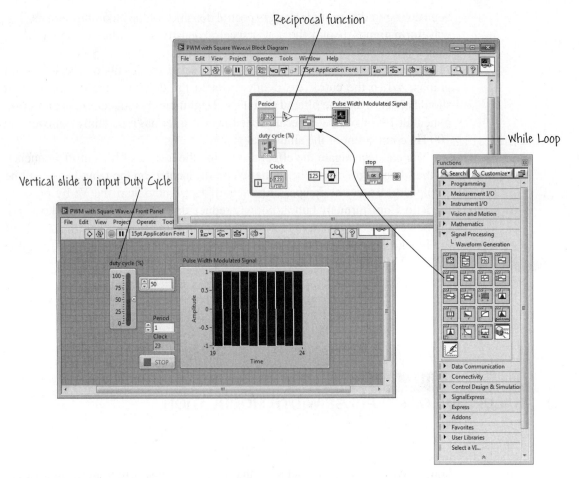

FIGURE 11.50
The PMW with Square Wave VI.

Add the While Loop, Wait (ms) function, and other elements of the VI depicted in Figure 11.50. Once the VI is properly wired and ready to run, set the Duty Cycle = 50, and the Period = 100. Run the VI and observe the outputs displayed in the waveform chart. What happens when you set the Period = 10? When you are finished, push the Stop button.

Using the Spectral Measurements Express VI and a waveform graph can you describe how the values of Period and Duty Cycle affect the frequency content of the signal? (Hint: The frequency is equal to 1 over the period.)

A working version of PWM with Square Wave VI.vi *is found in the Building Blocks folder of the Learning directory.*

11.9   RELAXED READING: HIGH-SPEED CONTROL SYSTEM TO TEST MEMS MICROSHUTTERS

In this reading we explore the challenge of synchronizing the motion of a magnet moving more than 1 meter per second with the opening and closing of tens of thousands of tiny microelectromechanical system microshutters.

The NASA James Webb Space Telescope (JWST) will be stationed at a stable Lagrange point approximately 1 million miles from Earth. This telescope is the next stepping stone toward understanding the universe and studying the Big Bang theory. The near infrared spectrometer developed by the European Space Agency with major NASA contributions, is the primary instrument on the telescope. It observes thousands of distant galaxies to probe the epoch of initial galaxy formations in the universe. To measure numerous faint objects, the instrument must simultaneously observe a large number of objects in previously unknown positions.

To observe objects at these positions, NASA developed the microshutter array, a 171 by 365 matrix of 100 by 200 μm shutters that can open under random access control. Four microshutter arrays in a 2×2 matrix create a programmable transmission mask of about 250,000 shutters so that more than 100 faint objects can be targeted simultaneously. The microshutter is a rectangular door that opens and closes to block light or let it pass through. The shutters pivot on a silicon nitride flexure, actuate magnetically with the help of magnetic coating, and latch electrostatically through electrical connections. The shutters are manufactured in arrays with 365 columns and 171 rows for a total of more than 62,000 shutters per array. The shutters are mounted on a substrate and wired the array in a grid so that we can assert its rows and columns to address each shutter. To open a shutter, a magnet is passed across the front of the array while applying high voltage to the row and column of each shutter. The magnetic field opens the shutter, and the static charge at the intersection of the row and column holds it open.

Each shutter array was fabricated to test an aspect of the overall design to inform further definition of the fabrication processes. Using the NI PCI-7344 four-axis stepper motor controllers and the NI MID-7604 power motor drivers, the software was developed to control the vacuum chamber, shutter control instrumentation, cameras, and other apparatuses to evaluate array performance. Testing with this system revealed that uncontrolled shutter release limits shutter performance. In this uncontrolled approach, one closed a shutter by turning off the power to the row and column of the shutter. With each approach, the shutter impacts its light baffle in a way that significantly limits its lifetime. It was decided that the shutters should be released in synchronization with a passing magnet so that the magnetic field cushions the impact as the shutter closes.

FIGURE 11.51
Fully Functional 1/6th Scale Model of the James Webb Space Telescope Mirror in an Optics Test Bed.

The microshutters must function reliably for up to 100,000 cycles on different shutter designs. Instead of testing for years, the new test chamber must cycle the shutters rapidly. The motor rotates at up to 240 rpm; thus, the sled, connected to the motor with off-center cables, crosses back and forth in front of the shutter array four times per second. The control system needs to latch or release each of the 365 columns of the shutter array exactly as the magnet passes. To control the shutters, communication with the control electronics and custom high-voltage shift registers is required. The new system also needs to rapidly communicate and provide utilities to test and verify many operations of the 584 chips. The system must meet all of these control requirements and be fail-safe. The tests run for days at a time, opening and closing all 62,000 shutters 240 times per minute. If the system loses synchronization, the loss can damage the shutters in just a few minutes. In order to meet these requirements, there were two choices: design and manufacture a custom chip or use the LabVIEW FPGA Module. The solution implemented used the LabVIEW FPGA Module to perform shutter control along with NI hardware.

The entire system contains a host computer that controls the test chamber, a field programmable gate array FPGA host program that runs on the PXI controller, and FPGA software that runs on the PXI-7813R. With the FPGA host interface, engineers can calibrate the system and perform manual control functions, create and download bitmaps to write to the arrays, and run self-test diagnostics on the other functions of the 584 chips. The FPGA software reads the position of the magnet from a quadrature encoder or an absolute encoder. The encoder-decoding algorithm is placed in a single-cycle loop running at 40 MHz to ensure it does not miss any steps. After some filtering to remove jitter, the position value is placed in a first-in-first-out memory (FIFO) buffer. Another loop on the FPGA reads the FIFO and determines what to do with the shutters based on the current location of the magnet. This state machine communicates with the 584 chips using the protocol to turn the appropriate rows and columns on or off. If the FIFO overflows, the state machine controlling the shutters is not going fast enough. The software indicates a synchronization error to the host computer so the system can shut down.

In summary, the testing system has become the foundation for control experimentation on the shutter arrays. As engineers develop new ideas to improve shutter operation, algorithms can be easily added or changed in the state machine block. The LabVIEW FPGA Module and PXI-7813R saved hundreds of man-hours and thousands of dollars over developing a custom chip. In addition to saving costs, the control algorithm is also inexpensively modified to improve testing, explore shutter issues, and further the development of the NASA micro-shutter arrays. For further information, visit http://sine.ni.com/cs/app/doc/p/id/cs-11556.

11.10   myDAQ BUILDING BLOCKS

In this myDAQ Building Block, you will examine the calibration process used to determine the Steinhart-Hart coefficients for your thermistor. Recall in the myDAQ Building Block for Chapter 3, you used a subVI called **Temp Conversion.vi** to convert a resistance measurement into a temperature measurement. A closer examination of **Temp Conversion.vi** shows that it uses the Steinhart-Hart equation

$$\frac{1}{T} = k_a + k_b (\ln R) + k_c (\ln R)^3$$

where T is temperature (Kelvin), R is resistance (Ω), and the k_a, k_b, and k_c terms are the Steinhart-Hart coefficients for the thermistor. The values of the Steinhart-Hart coefficients were experimentally determined using the process you will follow in this building block.

The subVI **Temp Conversion.vi** *can be found in the* **myDAQ Building Blocks≫Support VIs** *folder in the* **Learning** *directory.*

Since you have three unknown coefficients, you need three known data points to solve for the values of the coefficients. In this case, that means that you need to record the resistance of your thermistor at three known temperatures. You can use the myDAQ DMM to measure resistance and a calibrated thermometer to measure temperature to collect these known values. Suppose you have collected the data shown in Table 11.1.

Open Calculate Calibration Coefficients.vi found in the Chapter 11 folder in the myDAQ Building Blocks folder within the Learning directory. On the block diagram, shown in Figure 11.52, the data flows from the controls

TABLE 11.1   **Sample data of resistance versus temperature**

Temperature	Resistance (Ω)	Temperature (°C)
Low	29,400	3.2
Mid	10,080	24.9
High	1,940	65.9

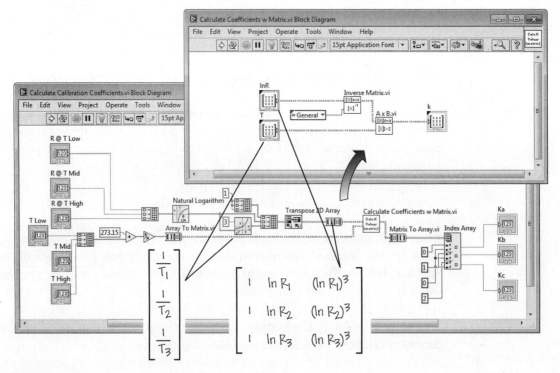

FIGURE 11.52
The LED circuit for the myDAQ Building Blocks.

named R @ T Low, R @ T Mid, and R @ T High to create the 3×3 matrix

$$InR = \begin{bmatrix} 1 & (\ln R_1) & (\ln R_1)^3 \\ 1 & (\ln R_2) & (\ln R_2)^3 \\ 1 & (\ln R_3) & (\ln R_3)^3 \end{bmatrix},$$

where $R_1 = RT_{\text{Low}}$, $R_2 = RT_{\text{Mid}}$, and $R_3 = RT_{\text{High}}$. Similarly, data flows from the controls named T Low, T Mid, and T High to create the 3×1 matrix

$$T = \begin{bmatrix} \frac{1}{T_1} \\ \frac{1}{T_2} \\ \frac{1}{T_3} \end{bmatrix},$$

where $T_1 = T_{\text{Low}}$, $T_2 = T_{\text{Mid}}$, and $T_3 = T_{\text{High}}$. Note that the temperature values are converted to Kelvin. Solving for the Steinhart-Hart coefficients yields

$$\begin{bmatrix} k_a \\ k_b \\ k_c \end{bmatrix} = \begin{bmatrix} 1 & (\ln R_1) & (\ln R_1)^3 \\ 1 & (\ln R_2) & (\ln R_2)^3 \\ 1 & (\ln R_3) & (\ln R_3)^3 \end{bmatrix}^{-1} \begin{bmatrix} \frac{1}{T_1} \\ \frac{1}{T_2} \\ \frac{1}{T_3} \end{bmatrix}.$$

Open the subVI Calculate Coefficients w Matrix.vi to see how it uses the built in matrix functions to solve for the coefficients. The Steinhart-Hart coefficients are passed back up to the top level VI, which converts it back to array form and displays the values on the front panel.

Enter the measured resistance and temperature values from Table 11.1 in the corresponding controls and run Calculate Calibration Coefficients.vi. Your calculated results should return the following Steinhart-Hart coefficients: $k_a = 0.00108403$, $k_b = 0.000246475$, and $k_c = -1.24014E\text{-}9$. For more accurate results associated with your specific thermistor, repeat the calibration process multiple times using your own measured data, discard any outliers and calculate average coefficient values. To use these new values, you can replace the constants on the block diagram of Temp Conversion.vi.

In this myDAQ Building Block, you examined the matrix functions in calibration process used to determine the Steinhart-Hart coefficients for your thermistor. Using measured data provided, the Steinhart-Hart coefficients used in Chapter 10 were confirmed and using your own measured data, coefficients specific to your thermistor can be calculated.

11.11   SUMMARY

LabVIEW provides a great computational environment for analysis of signals and systems. In this chapter, we presented some applications of the many VIs available for analysis of signals, systems, functions, and systems of equations. The material was intended to motivate you to look further into the capabilities of LabVIEW in developing your own VIs for solving linear algebraic systems of equations, curve fitting, integrating ordinary differential equations, computing function zeroes, computing derivatives of functions, integrating functions, generating and analyzing signals, computing discrete Fourier transforms, and filtering signals.

KEY TERMS

Bandpass filter: A system that passes a specified band of frequencies.

Bandstop filter: A system that attenuates a specified band of frequencies.

Condition number: A quantity used in assessing the accuracy of solutions to systems of linear algebraic equations. The condition number can vary from 0 to ∞—a matrix with a condition number near 1 is closer to being nonsingular than a matrix with a very large condition number.

Determinant: A characteristic number associated with an $n \times n$ square matrix that is computed as a function of the minors and cofactors of the matrix. When the determinant is identically equal to zero, we say that the matrix is singular.

Digital frequency: Computed as the analog frequency divided by the sampling frequency. Also known as the **normalized frequency**.

Discrete Fourier transform (DFT): A common practical algorithm for transforming sampled signals from the time domain into the frequency domain.

Even symmetric signal: A signal symmetric about the y-axis.

Fast Fourier transform (FFT): The FFT is a fast algorithm for calculating the DFT when the number of samples (N) is a power of 2.

FIR filter: A finite impulse response filter in which the impulse response decays to zero in a finite time and the output depends only on current and past values of the input signal.

Frequency-domain representation: The representation of a signal in terms of its individual frequency components.

Frequency response: The Fourier transform of the impulse response.

Highpass filter: A system that passes high frequencies and attenuates low frequencies.

Homogeneous differential equations: A differential equation that has no input driving function (that is, $u(t) = 0$).

IIR filter: An infinite impulse response filter in which the impulse response continues indefinitely (in theory) and the output depends on current and past values of the input signal and on past values of the output.

Impulse response: The output of a system (e.g., a filter) when the input is an impulse.

Lowpass filter: A system that passes low frequencies and attenuates high frequencies.

Nyquist frequency: The highest frequency that a filter can process, according to the sampling theorem.

Odd symmetric signal: A signal symmetric about the origin.

Passband: The range of frequencies that is passed through a filter.

Sampling theorem: The statement that a continuous-time signal can be reconstructed from discrete, equally spaced samples if the sampling frequency is at least twice that of the highest frequency in the time signal.

Spectral leakage: A phenomenon that occurs when you sample a noninteger number of cycles, leading to "artificial" discontinuities in the signal that manifest themselves as very high frequencies in the DFT/FFT spectrum, appearing as if the energy at one frequency has "leaked out" into all the other frequencies.

Stopband: The range of frequencies that do not pass through a filter.

Trace: The sum of the diagonal elements of an $n \times n$ square matrix.

Windowing: A method used to reduce the amplitude of sampled signal discontinuities by multiplying the sampled original waveform by a finite-length window whose amplitude varies smoothly and graduates towards zero at the edges.

EXERCISES

E11.1 (a) Open a new VI and place a While Loop on the block diagram by going to **Functions≫Programming≫Structures**.

(b) Within the While Loop, place the Simulate Signal Express VI on the block diagram by navigating to **Functions≫ProgrammingExpress≫Input**. The **Configuration Window** will open in LabVIEW by default, but if it does not, double-click on the Simulate Signal Express VI to view it.

(c) In the **Configuration Window**, alter the **Signal Type** (i.e., Sine, Square, Triangle, Sawtooth, DC) and the additional inputs associated with the signal (i.e., Frequency, Amplitude, Phase, etc.). Notice that the **Result Preview** graph changes according to these controls.

(d) Add noise to the signal by adding a check mark to the box **Add noise** and view how the various types of noise affect the signal.

(e) Feel free to experiment with the other options in the **Configuration Window** to create the simulated signal of your choice. Press the **OK** button to return to the block diagram of your VI.

(f) Right-click on the signal output of your Simulate Signal Express VI and select Create≫Graph Indicator. Wire a Boolean data source (such as a button) to the Conditional Terminal of the While Loop.

(g) Run the VI and confirm that the signal that you configured in the Simulate Signal Express VI matches the signal displayed on the front panel of your VI.

(h) Stop the VI and go back to the **Configuration Window** to make changes to the simulated signal in order to observe other types of signals that can be generated.

E11.2 Open a new VI and go to Help≫Find Examples. .... Click on the **Search** tab, type "express" into the string labeled **Enter keyword(s)**, and press the **Search** button.

(a) Open Express Comparison.vi, which shows how to compare values using the Express VI. Using the Value A and Value B controls, you can modify the signals. If Value A is greater than Value B, the result will be indicated by a 1 on the graph. If Value A is less than Value B, the result will be indicated by a 0.

(b) Open Express Vibration Lab.vi, which simulates a vibratory system of two masses and springs. The parameters of the system, such as the mass and rigidity of the springs, can be changed. Refer to **Context Help** for a more detailed explanation of this example VI, and explore the block diagram.

E11.3 Navigate to the windxmpl.llb in the analysis subfolder in the examples folder in the LabVIEW directory and open the Window Comparison.vi. Observe the effects of the native LabVIEW window algorithms, including Hanning, Hamming, Triangle, and describe in your own words your observations.

E11.4 Navigate to the fltrxmpl.llb in the analysis subfolder in the examples folder in the LabVIEW directory and open the IIR Filter Design.vi. Observe the effects of the native LabVIEW IIR filters, including Elliptic, Bessel, Butterworth, and describe in your own words your observations.

PROBLEMS

P11.1 Create a VI that generates a sine wave, which is displayed on a waveform graph. The user should have the ability to change the frequency and the amplitude of this sine wave programmatically. A power spectrum analysis should be performed on the sine wave and displayed on a waveform graph.

P11.2 Continue your analysis on the sine wave that is being generated from your solution to Problem 11.1. Determine the positive peak value of the signal and the root mean square value of the signal, and display these numeric values on the front panel. Use a Hanning window and a Flat Top window to perform the spectral analysis and display both on the same waveform graph.

Hint: Explore the Express VIs by navigating to **Functions≫Express≫ Signal Analysis**.

P11.3 Continue your analysis on the sine wave that is being generated from your solution to Problem 11.2. Add uniform white noise with noise amplitude of 3 to the sine wave being generated. Filter the noisy sine wave through a lowpass Butterworth filter (Order = 4), and display the filtered sine wave to compare the effect the filter has on the noise.

DESIGN PROBLEMS

D11.1   Consider the following linear, time-invariant system:

$$\dot{\mathbf{x}} = \mathbf{A}\mathbf{x} + \mathbf{B}\mathbf{u}$$

where

$$\mathbf{A} = \begin{bmatrix} k_0 & 1 \\ -1 & -1 \end{bmatrix} \text{ and } \mathbf{B} = \begin{bmatrix} 0 \\ 1 \end{bmatrix}.$$

The vector $\mathbf{x} = [x_1 \quad x_2]^T$ is known as the state vector. The parameter k_0 is a constant. The input to the system is $\mathbf{u}$. Develop a script using MathScript to perform the following computations:

(a) For $0 \leq k_0 \leq 5$, compute the eigenvalues of $\mathbf{A}$. The eigenvalues can be complex (imaginary) numbers. Generate a plot of the real part of the eigenvalues versus the imaginary part.

(b) With the input $\mathbf{u} = 2$ for $t \geq 0$ and $k = 0.5$, find the solution using numerical integration and plot $\mathbf{x}$ versus t for $0 \leq t \leq 20$. Use the initial conditions $\mathbf{x}(0) = [1 \quad 0]^T$.

D11.2   Design a VI to serve as a filter for sinusoidal input signals. The VI should separate the low frequency and high frequency components of the input signal. Make the filter cutoff frequency an input to the VI via a front panel control. Plot the original signal and its components on a waveform graph. Use Simulate Signal Express VIs to generate the input signals for testing.

APPENDIX A

Instrument Control

This appendix introduces the concept of communicating with and controlling external instruments. We focus on the GPIB (General Purpose Interface Bus) and RS-232 (a serial interface bus standard). A brief introduction to the main components of an instrument control system is given. The application of the **Measurement & Automation Explorer (MAX)** in detecting instruments and installing instrument drivers is discussed. The notion of an instrument driver is an important topic woven throughout the appendix. The Instrument I/O Assistant is also introduced.

GOALS

1. Learn about GPIB and serial instrument control.

2. Understand how to interact with your instruments using MAX.

3. Gain some experience with instrument drivers (for the AGILENT 34401 multimeter).

4. Gain experience with the Instrument I/O Assistant.

A.1   COMPONENTS OF AN INSTRUMENT CONTROL SYSTEM

LabVIEW communicates with and controls external instruments (such as oscilloscopes and digital multimeters) using GPIB (General Purpose Interface Bus), RS-232 (a serial interface bus standard), VXI (VME eXtensions for Instrumentation), Ethernet, USB, and other hardware standards. In this appendix we focus on GPIB and serial communication.

A.1.1   What Is GPIB?

Hewlett Packard developed the General Purpose Interface Bus (or **GPIB**) standard in the late 1960s to interconnect and control its line of programmable instruments. The interface bus was originally called HP-IB. In this context, a **bus** is the means by which computers and instruments transfer data. National Instruments made GPIB available to users of non-Hewlett-Packard equipment.

At the time it was developed, GPIB provided a much-needed specification and protocol to govern this communication. Figure A.1 shows a typical GPIB system. While the GPIB is one way to bring data into a computer, it is fundamentally different from data acquisition with boards that plug into the computer. Using a special protocol, GPIB brings data that has been acquired by another computer or instrument into the computer using a "handshake," while data acquisition involves connecting a signal directly to a DAQ device in the computer.

The original purpose of GPIB was to provide computer control of test and measurement instruments. GPIB was soon applied to intercomputer communication and control of scanners, film recorders, and other peripherals because of its 1 Mbyte/sec maximum data transfer rates. The Institute of Electrical and Electronics Engineers (**IEEE**) standardized the GPIB in 1975, and it became

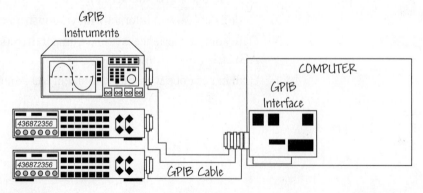

FIGURE A.1
A typical GPIB system.

accepted as IEEE Standard 488-1975. The standard has since evolved into ANSI/ IEEE Standard 488.2-1987. The GPIB functions for LabVIEW follow the IEEE 488.2 specification. The terms GPIB, HP-IB, and IEEE 488 are synonymous.

A.1.2   GPIB Messages

GPIB carries two types of messages:

- **Device-dependent messages** contain device-specific information such as programming instructions, measurement results, machine status, and data files. These are often called **data messages**.

- **Interface messages** manage the bus itself, and perform such tasks as initializing the bus, addressing and unaddressing devices, and setting device modes for remote or local programming. These are often called **command messages**.

Physically, GPIB is a digital, 24-conductor, parallel bus. It comprises 16 signal lines and 8 ground-return lines. The GPIB connector is depicted in Figure A.2. The 16 signal lines are divided into three groups:

- **Eight data lines**: The eight data lines (denoted DIO1 through DIO8) carry both data and command messages. GPIB uses an eight-bit parallel, asynchronous data transfer scheme where whole bytes are sequentially handshaked across the bus at a speed determined by the slowest participant in the transfer. Because the GPIB sends data in bytes (1 byte = 8 bits), the messages transferred are frequently encoded as ASCII character strings. All commands and most data use the 7-bit ASCII or International Standards Organization (ISO) code set, in which case the eighth bit, DIO8, is unused or is used for parity.

- **Three handshake lines**: The three handshake lines asynchronously control the transfer of message bytes among devices. These lines guarantee that message bytes on the data lines are sent and received without transmission error.

- **Five interface management lines**: The five interface management lines manage the flow of information across the interface from device to computer.

 The high-speed GPIB, proposed by National Instruments, is known as ***HS488*** *and was accepted as an addition to the IEEE 488.1 standard in 2003. Speed increases up to 8 Mbyte/sec have been achieved by removing the propagation delays associated with the three handshake lines.*

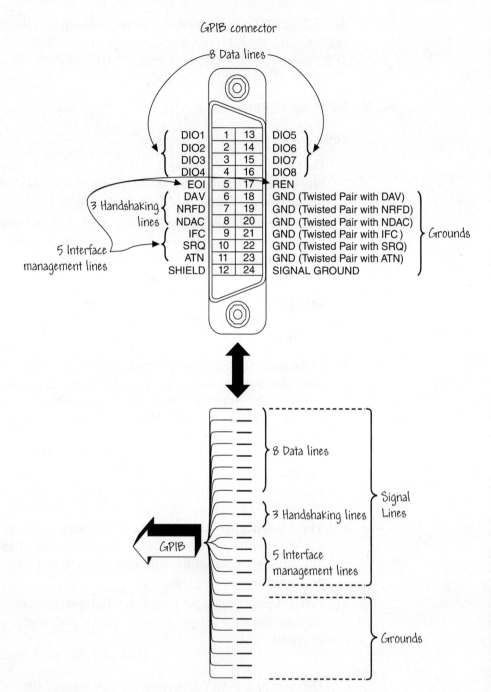

FIGURE A.2
The GPIB connector.

A.1.3   GPIB Devices and Configurations

You can have many GPIB devices (that is, many instruments and computers) connected to the same GPIB. Typical GPIB linear and star configurations are illustrated in Figure A.3. You can even have more than one GPIB device in your computer. A typical multiboard GPIB configuration is illustrated in Figure A.4. GPIB devices are grouped in three categories:

- **Talkers**: send data messages to one or more listeners.
- **Listeners**: receive data messages from the talker.
- **Controllers**: manage the flow of information on the GPIB by sending commands to all devices.

GPIB devices can fall into multiple categories. For example, a digital voltmeter can be both a talker and a listener.

The GPIB has one controller (usually a computer) that controls the bus. The role of the GPIB controller is similar to the role of a central processing unit in a computer. A good analogy is the switching center of a city telephone system. The communications network (the GPIB) is managed by the switching center (the controller). When a party (a GPIB device) wants to make a call (that is, to send a data message), the switching center (the controller) connects the caller

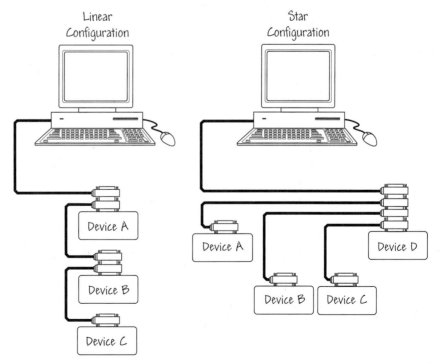

FIGURE A.3
Typical GPIB configurations.

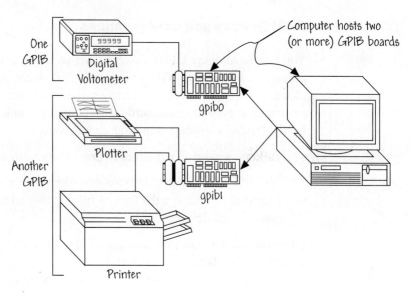

FIGURE A.4
A typical multiboard GPIB configuration.

(the talker) to the receiver (the listener). To transfer instrument commands and data on the bus, the controller addresses one talker and one or more listeners. The controller must address the talker and a listener before the talker can send its message to the listener. The data strings are then handshaked across the bus from the talker to the listener(s). After the talker transmits the message, the controller may unaddress both devices. LabVIEW provides VIs that automatically handle these GPIB functions.

Some bus configurations do not require a controller. For example, one device may always be a talker (called a talk-only device) and there may be one or more listen-only devices. A controller is necessary when you must change the active or addressed talker or listener. A computer usually handles the controller function. With the GPIB board and its software, the personal computer plays all three roles: controller, talker, and listener.

There can be multiple controllers on the GPIB but only one controller at a time is active. The active controller is called the **controller-in-charge** (CIC). Active control can be passed from the current CIC to an idle controller. Only the system controller (usually the GPIB board) can make itself the CIC.

With GPIB, the physical distance separating the GPIB devices matters. To achieve high-rate data transfers, we must accept a number of physical restrictions. Typical numbers are as follows:

- A maximum separation of 4 meters between any two devices and an average separation of 2 meters over the entire bus. For high-speed applications, you should have at least one device per meter of cable.

- A maximum total cable length of 20 meters. A maximum of 15 meters is desirable for high-speed applications.

- A maximum of 15 devices connected to each bus, with at least two-thirds of the devices powered on. For high-speed applications, all devices should be powered on.

Bus extenders and expanders can be used to increase the maximum length of the bus and the number of devices that can be connected to the bus. You can also communicate with GPIB instruments through a TCP/IP network. For more information, refer to the National Instruments website.

A.1.4   Serial Port Communication

Serial communication is another popular means of transmitting data between a computer and another computer, or between a computer and a peripheral device. However, unlike GPIB, a serial port can communicate with only one device, which can be limiting for some applications. Serial port communication is also very slow.

Most computers and many instruments have built-in serial port(s). Since serial communication uses the built-in serial port in your computer, you can send and receive data without buying any special hardware. Serial communication uses a transmitter to send data—one bit at a time—over a single communication line to a receiver. This method works well when sending or receiving data over long distances and when data transfer rates are low. It is slower and less reliable than GPIB, but you do not need a special board in your computer, and your instrument does not need to conform to the GPIB standard. Figure A.5 shows a typical serial communication system.

There are different serial port communication standards. Developed by the Electronic Industries Association (EIA) to specify the serial interface between equipment known as Data Terminal Equipment (DTE), such as modems and

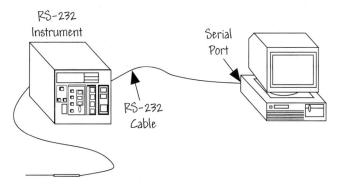

FIGURE A.5
A typical serial communication system.

plotters, and Data Communications Equipment (DCE), such as computers and terminals, the RS-232 standard includes signal and voltage characteristics, connector characteristics, individual signal functions, and recipes for terminal-to-modem connections. The most common revision to this standard is the RS-232C used in connections between computers and printers.

The RS-232 serial port connectors come in two varieties: the 25-pin connector and the 9-pin connector. Both connectors are depicted in Figure A.6. The 9-pin connector has two data lines (denoted TxD and RxD in the figure) and five handshake lines (denoted RTS, CTS, DSR, DCD, and DTR in the figure). This compact connector is occasionally found on smaller RS-232 laboratory equipment and has enough pins for the "core" set used for most RS-232 interfaces. The 25-pin connector is the "standard" RS-232 connector with enough pins to cover all the signals specified in the RS-232 standard. Only the "core" set of pins is labeled in Figure A.6.

Serial communication requires that you specify four parameters: the baud rate of the transmission, the number of data bits encoding a character, the sense of the optional parity bit, and the number of stop bits. Each transmitted character is packaged in a character frame that consists of a single start bit followed by the data bits, the optional parity bit, and the stop bit or bits. The baud rate

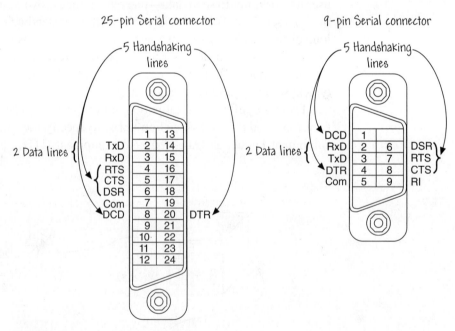

FIGURE A.6
The RS-232 serial port connectors.

is a measure of how fast data is moving between instruments that use serial communication.

A.1.5   Other Bus Technologies

In addition to GPIB and serial-port communication, several other key technologies enable connectivity to instruments, including Ethernet, USB, and IEEE 1394 ports. These buses provide ease of use (USB), connectivity (Ethernet), and high speed (IEEE 1394). IEEE 1394 is also known as firewire.

USB plug-n-play technology was designed primarily to connect peripheral devices to PCs. An increasing number of devices, however, now incorporate it. USB delivers an inexpensive and easy-to-use connection by automatically detecting new devices, querying device identification, and configuring device drivers appropriately. Its ease of use makes it highly attractive for instrument control applications. The use of USB does, however, have drawbacks. USB cables are thin and flexible—nice feature when connecting a keyboard to a PC. In instrument control applications, though, since the cables are not industrial grade, there is a potential for data loss in noisy environments. Moreover, the cables can be pulled fairly easily out of their jacks. Most importantly, there is no industry-standard protocol for instrument control using USB, so that instrument manufacturers must provide individual implementations.

Ethernet is a mature technology, and over one hundred million computers worldwide are Ethernet capable. The advantage of using the extensive existing Ethernet networks makes Ethernet appealing for instrument control applications. The theoretical transfer rate (up to 100 Mb/s) and the determinism of Ethernet networks are also excellent features. These advantages, however, may be overshadowed by the drawbacks. The use of existing Ethernet networks in instrument control may involve network administrators, imposing an additional personnel burden. Also, achievement of the theoretical data transfer rates is hindered by network traffic, overhead, and inefficient data transfer. This uncertainty in the transfer rates reduces the determinism in communicating across Ethernet networks. Finally, additional security measures may be required to ensure data integrity and privacy.

The IEEE Standard 1394-1995 is a high-performance serial bus developed by Apple in the 1980s. Data transfer rates up to 400 Mb/s make this an attractive option for instrument control. Unlike USB, with IEEE 1394 there is an existing protocol defined for controlling instruments over the bus. As with USB and Ethernet, there are drawbacks to the use of IEEE 1394 serial bus for instrument control. Many digital cameras and other consumer electronics products already include IEEE 1394 ports for transferring data, but as yet very few instruments are available with 1394 ports. The main shortcoming is that 1394 ports are not built into Intel PC peripheral chip sets, so that Intel PC users must acquire a separate 1394 controller (typically a PCI board) to communicate. Note that all Macintosh computers have built-in 1394 ports. Finally, like USB cables, the cables for IEEE 1394 are thin and flexible and not of industrial grade, so that data may be lost in some test and measurement applications.

Currently, only a handful of instrument manufacturers include built-in instrument control options for USB, Ethernet, or IEEE 1394. The widespread availability of USB and the existence of large Ethernet networks may motivate designers to begin including a separate communication bus for instrument control, but for the time being the GPIB and serial-port communication options are the most likely to be encountered.

A.2   DETECTING AND CONFIGURING INSTRUMENTS

In this section NI-GPIB configuration for Windows and Mac OS X will be discussed. National Instruments provides a wizard for helping you configure your NI-GPIB device.

A.2.1   Windows

With the **Measurement & Automation Explorer (MAX)** you can automatically detect connected instruments, install the required instrument drivers, and manage the instrument drivers already installed. The software architecture for GPIB instrument control using LabVIEW is similar to the architecture for DAQ (see Chapter 8). Figure A.7 shows the software architecture on the Windows platforms. Instrument drivers are LabVIEW applications written to control a specific instrument. More on instrument drivers in the next section. You can use the instrument drivers or parts of the instrument drivers to develop your own application quickly. You can also set naming aliases for your instruments for easier instrument access.

A.2.2   Macintosh OS X

If you are using a Macintosh, the NI-488.2 Configuration Utility is called the GPIB Explorer. To start the GPIB Explorer from the Finder, double-click on Applications≫National Instruments≫NI-488.2≫Explore GPIB. Macintosh OS automatically recognizes plug-and-play GPIB devices. You can view or modify the default configuration settings using the GBIB Explorer.

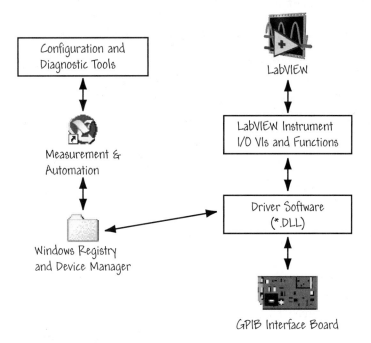

FIGURE A.7
Software architecture for Windows platforms.

**Practice
with the
Instrument
Wizard**

In this example, you will use MAX to configure and test the GPIB interface. MAX interacts with the various diagnostic and configuration tools installed with the driver and also with the Windows Registry and Device Manager (see Figure A.7). The driver-level software is in the form of a dynamically linked library (DLL) and contains all the functions that directly communicate with the GPIB board. The LabVIEW Instrument I/O VIs and functions directly call the driver software.

You configure an object listed in MAX by right-clicking on the item and making a selection from the shortcut menu. Figure A.8 shows the GPIB interface board in the MAX utility and the results of pressing the **Scan For Instruments** button at the top of the window.

The configuration utilities and hierarchy described in this example are specific to Windows platforms. If you are using a Macintosh or other operating system, refer to the manuals that came with your GPIB interface board for the appropriate information for configuring and testing that board.

◆

As discussed in Chapter 8, MAX is a configuration utility for your software and hardware. MAX executes system diagnostics, adds new channels, interfaces,

Select this option to enable the search for instruments

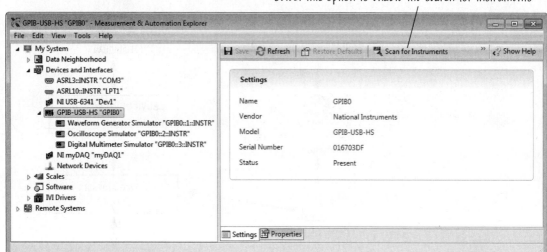

FIGURE A.8
Using MAX to search for instruments.

and tasks, and views devices and instruments connected to your system. Open MAX by double-clicking on its icon on the desktop or by selecting **Measurement & Automation Explorer** from the **Tools** menu of LabVIEW.

Four relevant selections in MAX are:

- **Data Neighborhood**—Use this selection to create virtual channels, aliases, and tags to your channels or measurements configured in Devices and Interfaces as you did in the DAQ chapter.

- **Devices and Interfaces**—Use this selection to configure resources and other physical properties of your devices and interfaces. Using this selection, you can view attributes of one or more devices, such as serial numbers.

- **Scales**—Use this selection to set up simple operations to perform on your data, such as scaling.

- **Software**—Use this selection to determine which drivers and application software are installed and their version numbers.

If you do not have an instrument attached to your computer, MAX will obviously not find any instruments. In the example in Figure A.9, the attached instrument is an NI-Instrument Simulator v2.0 that simulates three instruments denoted by **Waveform Generator Simulator "GPIB0::1::INSTR", Oscilloscope Simulator "GPIB0::2::INSTR,"** and **Digital Multimeter Simulator "GPIB0::3::INSTR"** in MAX.

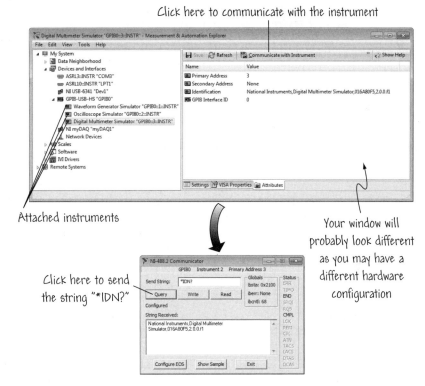

FIGURE A.9
A National Instruments GPIB device simulator is detected.

Click **Communicate with Instrument** to communicate with the instrument, as shown in Figure A.9. By default, the string *IDN? appears in the command to send to this GPIB device. The string *IDN? is an IEEE 488.2 standard identification request to the instrument. Click **Query** to send the string. In this example the string is sent to the Instrument Simulator v2.0. The instrument replies with the string "National Instruments, Digital Multimeter Simulator, 016A80F5, 2.0.0.f1." The next time you launch MAX, it will automatically contain your stored instrument configuration.

You can also examine the properties of your GPIB board. In MAX, highlight the GPIB hardware, as illustrated in Figure A.10. The properties provide all the information about your GPIB board.

A.3   USING THE INSTRUMENT I/O ASSISTANT

The Instrument I/O Assistant can be employed to establish communication with message-based instruments. The Instrument I/O Assistant organizes instrument communication into ordered steps using a dialog box associated with the I/O

Highlight the GPIB device

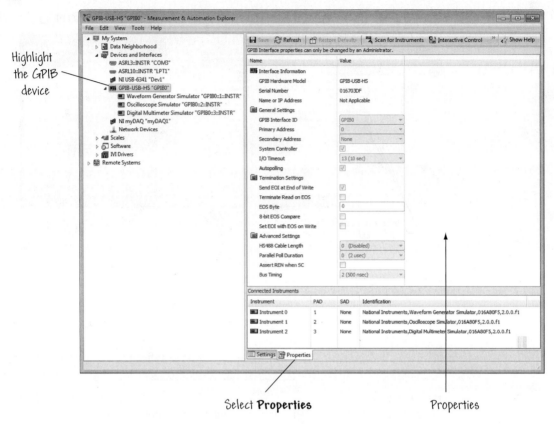

Select **Properties**

Properties

FIGURE A.10
Examining the properties of the GPIB devices.

Assistant. Like the DAQ Assistant (see Chapter 8), the Instrument I/O Assistant is readily configured to provide a communication link between the instrument and the computer. The Instrument I/O Assistant can be used whenever an instrument driver is not available.

To launch the Instrument I/O Assistant, place the Instrument I/O Assistant on the block diagram, as illustrated in Figure A.11. The Instrument I/O Assistant is located on the **Functions≫Instrument I/O** and **Functions≫Express≫ Input** palettes. Once the Instrument I/O Assistant is placed on the block diagram, the Instrument I/O Assistant configuration dialog box will automatically appear. The dialog box is shown in Figure A.12.

If the Instrument I/O Assistant configuration dialog box does not appear, double-click the Instrument I/O Assistant icon.

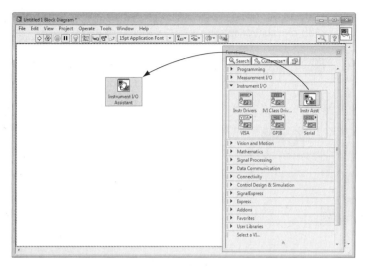

FIGURE A.11
Placing the Instrument I/O Assistant on the block diagram.

The desired communication steps in the instrument communication process are specified in the Instrument I/O Assistant configuration dialog box. Once the configuration dialog box appears, the general procedure for configuring the Instrument I/O Assistant is as follows:

1. **Select an instrument.**
    Instruments that have been configured in MAX appear in the **Select an instrument** pull-down menu.

2. **Choose a Code generation type.**
    VISA code generation allows for more flexibility and modularity than GPIB code generation.

3. **Specify the communication steps.**
    Select from the following communication steps using the **Add Step** button:

    (a) **Query and Parse**—Sends a query to the instrument and parses the returned string.

    (b) **Write**—Sends a command to the instrument.

    (c) **Read and Parse**—Reads and parses data from the instrument.

    To use the Instrument I/O Assistant, you place the communication steps into a sequence. As you add steps to the sequence, they appear in the **Step Sequence** window (see Figure A.12).

4. **Test the communication sequence.**
    After adding the desired number of steps, click the **Run** button to test the sequence of communication that you have configured for the Express VI.

Add communication steps

Refresh list of instruments that
have been configured in MAX

Select VISA or GPIB
code generation

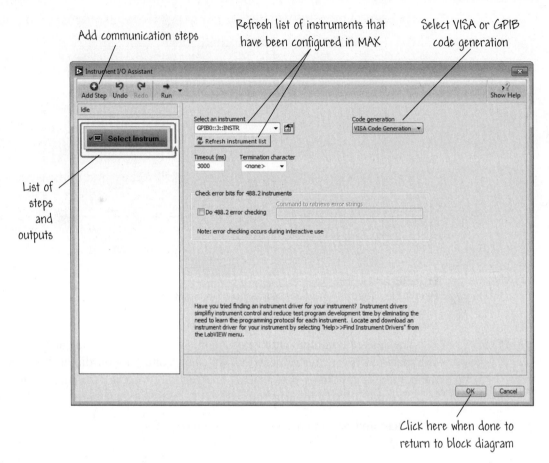

List of
steps
and
outputs

Click here when done to
return to block diagram

FIGURE A.12
The Instrument I/O Assistant dialog box.

5. **Return to the block diagram and complete the VI.**
    Click the **OK** button to exit the Instrument I/O Assistant configuration dialog box. LabVIEW adds input and output terminals to the Instrument I/O Assistant on the block diagram that correspond to the data you will receive from the instrument.

 **Practice
with the
Instrument I/O
Assistant**

In this example, we will construct a VI that uses the Instrument I/O Assistant to communicate with a NI Instrument Simulator. This device uses a GPIB interface.

To begin the VI development, open a blank VI and place the Instrument I/O Assistant (located on the **Functions≫Instrument I/O** palette) on the block diagram. The dialog box will appear to allow you to configure the Instrument I/O Assistant.

As previously discussed, the first step is to select an instrument. In the dialog box, make sure that the National Instruments Digital Multimeter Simulator is selected (denoted by GPIB0::3::INSTR in Figure A.12). You can click on the **Refresh instrument list** button to make sure all the available instruments are listed.

The second step is to choose a code generation type. Since VISA code generation allows for more flexibility and modularity than GPIB code generation, we will choose **VISA Code Generation** from the **Code generation type** pulldown menu.

We will add two communication steps using the **Query and Parse** sequence. To accomplish this, click the **Add Step** button, and in the dialog box that appears, select **Query and Parse** to write and read from the Digital Multimeter Simulator, as illustrated in Figure A.13. The Step Sequence window will update automatically to show the Query and Parse step. To test the communication link, type *IDN? as the command, and click the **Run this step** button. If no error warning appears in the lower half of the dialog box, this step has successfully completed. To parse the data received, click the **Auto parse** button. Notice that "**Token**"

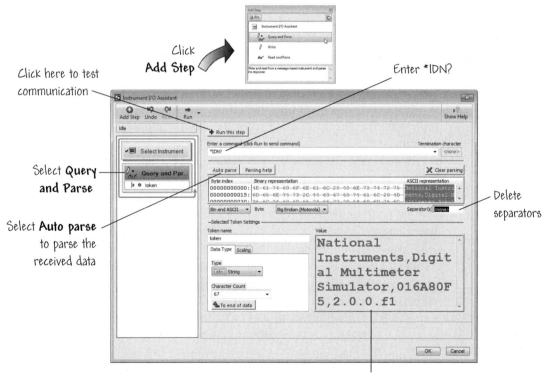

FIGURE A.13
Configuring the Instrument I/O Assistant.

Select **Run this step** to test the communication link

Add second communication step by entering MEAS?

Add step

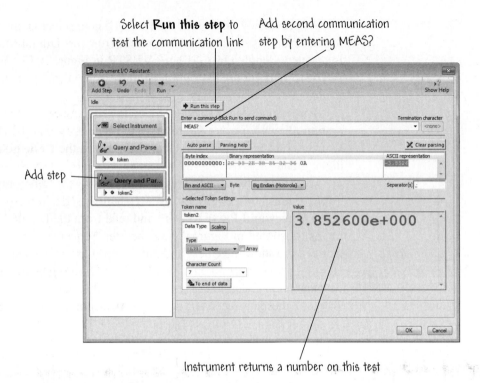

Instrument returns a number on this test

FIGURE A.14
Adding a second communication step.

now appears in the **Outputs** pane on the left side of the dialog box. This value represents the string returned from the identification query—the name of the instrument.

Now, repeat the previous configuration to include a second communication step, as illustrated in Figure A.14. Click the **Add Step** button and select **Query and Parse**. Enter the command MEAS? and click the **Run this step** button to test the communication. As before, to parse the data received, click the **Auto parse** button. The data returned is a random numeric value, which in this case is 3.8526 as shown in Figure A.14. This is a simulated voltage reading. The name "Token2" now appears in the **Outputs** pane on the left side of the dialog box.

Generally we should rename the outputs to more accurately represent the variables. Right-click on the output **Token** and select **Rename** to change the name of the output, as illustrated in Figure A.15. Rename **Token** to ID String and rename **Token2** to Voltage.

Click the **OK** button to exit the Instrument I/O Assistant and return to the block diagram. The configuration of the Instrument I/O Assistant is complete.

To view the code generated by the Instrument I/O Assistant, right-click the Instrument I/O Assistant icon and select **Open Front Panel** from the shortcut menu. When asked if you want to convert to a subVI select **Convert**. This

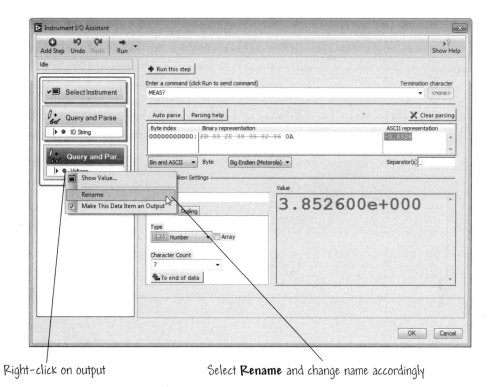

FIGURE A.15
Renaming the outputs.

converts the Express VI to a subVI. Switch to the block diagram to see the code generated. Figure A.16 shows the code generated by the Instrument I/O Assistant after completing the above steps.

*Once an Express VI has been converted to a subVI, it cannot be converted back. If you changed an Express VI to a subVI without making changes, you can **Undo** the conversion.*

The final step in the development of the VI is to wire appropriate controls and indicators on the block diagram. In our case, we need to wire an indicator to the **ID String** and another to the **Voltage** output. Right-click the **ID String** output and select Create≫Indicator from the shortcut menu. Similarly, right-click the **Voltage** output and select Create≫Indicator from the shortcut menu. The appropriate indicators will automatically wire to the outputs of the Instrument I/O Assistant, as illustrated in Figure A.17.

Now we finally test the VI. Display the front panel and select **Run**. You will most likely need to resize the string indicator to accommodate the instrument identification. As seen in Figure A.17, the Digital Multimeter Simulator returns

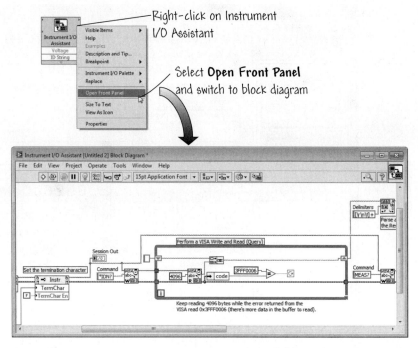

FIGURE A.16
Viewing the code generated by the Instrument I/O Assistant.

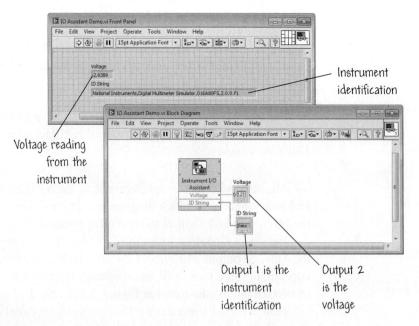

FIGURE A.17
Completing the instrument VI.

the identification National Instruments, Digital Multimeter, 016A80F5, 2.0.0.f1, and on this test it returned a voltage reading of 2.9389.

Save the VI as IO Assistant Demo.vi in the Users Stuff folder in the Learning directory.   ◆

A.4   INSTRUMENT DRIVERS

An **instrument driver** is software (that is, a VI) that controls a particular instrument. Instrument drivers eliminate the need to learn the complex, low-level programming commands for each instrument. LabVIEW is ideally suited for creating instrument drivers since the VI front panel can simulate the operation of an instrument front panel. The block diagram sends the necessary commands to the instrument to perform the various operations specified on the front panel. When using an instrument driver, you do not need to remember the commands necessary to control the instrument—this is specified via input on the front panel.

LabVIEW provides many VIs that can be used in the development of an instrument driver for your hardware. These VIs can be grouped into the following categories:

- Standard VISA I/O functions
- Traditional GPIB functions and added capability via the GPIB 488.2
- Serial port communication functions

VISA stands for Virtual Instrument Software Architecture. In essence, VISA is a VI library for controlling GPIB, serial, Ethernet, USB, or VXI instruments and making the appropriate calls depending on the type of instrument. VISA by itself does not provide instrumentation programming capability—it is a high-level application programming interface (API) that calls lower-level code to control the hardware. Each VISA instrument driver VI corresponds to a programmatic operation, such as configuring, reading from, writing to, and triggering an instrument. The GPIB and serial port functions provide similar capabilities.

Two questions come to mind:

1. When should you attempt to develop your own instrument driver from "scratch" using the VISA, GPIB, or serial port functions?

And if you must develop your own instrument driver,

2. When should you use the VISA functions, and when should you use the GPIB functions?

The answer to the first question is simple—students new to LabVIEW should not attempt to develop their own instrument drivers! They should use the instrument drivers developed by National Instruments, rather than attempt to develop drivers from "scratch." Instrument drivers can be downloaded from the National Instruments website using the Instrument Driver Network. To access the driver network, connect to

http://www.ni.com/idnet

directly or use Tools≫Instrumentation≫Visit Instrument Driver Network.

In LabVIEW, instrument drivers can be found and installed using the Find Instrument Drivers wizard found in Tools≫Instrumentation≫Find Instrument Drivers. This wizard allows you to automatically connect to the network, search for instrument drivers, view their rating, and install them, as illustrated in Figure A.18. If an instrument driver is not available, use the Instrument I/O Assistant.

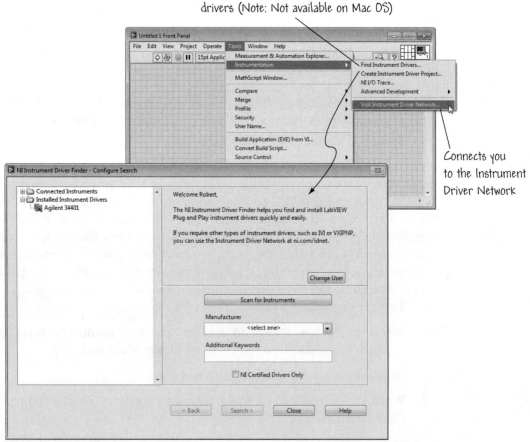

FIGURE A.18
Link automatically to the National Instruments, Instrument Driver Network.

The LabVIEW instrument driver library contains instrument drivers for a variety of programmable instruments that use the GPIB, serial, Ethernet, USB, or VXI interfaces. You can use an instrument driver from the library to control your instrument, or you can customize the instrument drivers, since they are distributed with their block diagram source code. Once you have properly installed the appropriate instrument drivers using MAX, you can access the instrument drivers on the **Functions** palette, as shown in Figure A.19.

You can link automatically to the National Instruments Instrument Driver Network, as shown in Figure A.18. The Instrument Driver Network provides the complete library of available instrument drivers for LabVIEW. On this web page, you can search for your instrument driver among over 9000 drivers with free source code available from more than 345 vendors. You then can download the instrument driver you need and install it into the instr.lib folder in the LabVIEW root directory.

If you decide to build your own instrument driver, VISA is the standard API throughout the instrumentation industry. In addition, the VISA functions

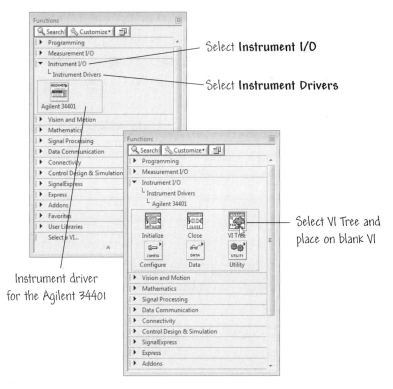

FIGURE A.19
Locating the installed instrument drivers on the **Functions** palette.

can control a suite of instruments of different types, including GPIB, serial, Ethernet, USB, or VXI. In other words, VISA provides interface independence. Students that need to program instruments for different interfaces only need to learn one API.

The Agilent 34401 Instrument Driver

In this example we will take a look at the Agilent 34401 instrument driver. Begin by first opening a new VI. Use Figure A.19 as a guide to locating the Agilent 34401 VI Tree. Drop the VI on the block diagram. Then double-click on the VI, which should open up the front panel. The value of this VI can be found on the block diagram, illustrated in Figure A.20. The VIs are organized in a Tree depicting the key VIs for performing various functions. A typical instrument driver application will include steps for initializing communication, configuring the instrument, passing data, and closing communication.

Open the NI Example Finder to see an example using these VIs. Navigate to the Example Finder by clicking Help≫Find Examples, then navigate through the folder structure to Hardware Input and Output≫Instrument Drivers≫ LabVIEW Plug and Play and select Agilent 34401 Read Single

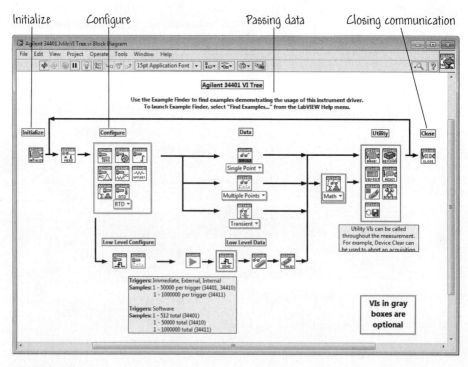

FIGURE A.20
The Agilent 34401 VI Tree.

Measurement.vi. The Instrument Driver Example VIs are used to verify communication with your instrument and test a typical programmatic instrument operation. The Agilent 34401 Read Single Measurement VI front panel is shown in Figure A.21.

With the exception of the address field, the defaults for most controls on the front panel will be sufficient for your first run. You will need to set the VISA resource name appropriately. Typically this is accomplished by clicking the down arrow on the control and selecting your device from the list. MAX populates the VISA resource name list box. If you do not know the VISA resource name, or it is not present in the list, refer to MAX for help. After running the VI, check to see that reasonable data was returned and an error was not reported in the error cluster (bottom right corner of the front panel).

If your Agilent 34401 Read Single Measurement VI does not work, you need to check that

- NI-VISA is installed. If you did not choose this as an option during your LabVIEW installation, you will need to download and install it from the NI website before rerunning your Agilent 34401 Read Single Measurement VI.

- The instrument VISA resource name is correct.

- The selected instrument driver supports the exact instrument model you are using.

Once you have verified basic communication with your instrument using the Agilent 34401 Read Single Measurement VI, you may want to customize instrument control for your particular needs. If your application needs are similar to

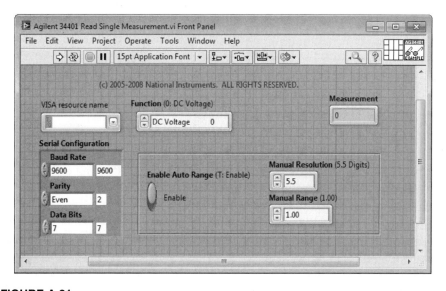

FIGURE A.21
The Agilent 34401 Read Single Measurement Vi front panel.

the Agilent 34401 Read Single Measurement VI, the simplest means of creating a customized VI is to save a copy of the Agilent 34401 Read Single Measurement VI by selecting **Save As** from the **File** menu. You can change the default values on the front panel by selecting **Make Current Values Default** from the **Operate** menu.

The block diagram of an instrument driver generally consists of three VIs: the Initialize VI, an application function VI, and the Close VI. A simple instrument driver can be assembled from the VIs provided by LabVIEW for the Agilent 34401. The three VIs we need are Agilent 34401 Initialize.vi, Agilent 34401 Read Multiple Points.vi, and Agilent 34401 Close.vi. A straightforward instrument driver comprised of these three basic VIs is shown in Figure A.22. Build the block diagram shown in Figure A.22, if you have the Agilent 34401 or do not have an instrument. If you are using another instrument driver, build a VI similar to Figure A.22 using the Initialize, Read Points, and Close VIs for that instrument.

A working version of **Figure A.22.vi** *can be found in the* **Appendix A** *folder of the* **Learning** *directory.* ◆

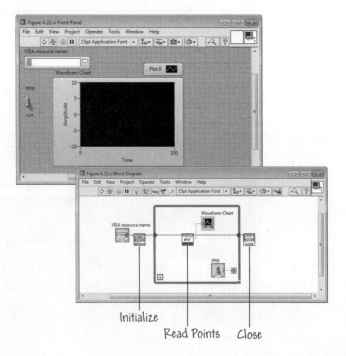

FIGURE A.22
A simple Agilent 34401 instrument driver.

A.4.1   Developing Your Own Instrument Driver

Make sure to check thoroughly for existing instrument drivers for your instrument before starting to build your own from scratch. You should check first at the National Instruments website as discussed in the previous sections. If that search fails, you should then search the website of the manufacturer of your equipment—they may have developed the instrument driver already. During your website searches, be on the lookout for instrument drivers that support a similar instrument, since instruments from the same model series often have similar command sets. If you find any such drivers, download them and assess the similarity of their command sets to that of your instrument—they may work directly or with minor modifications. For instruments from the same model series, you might need to contact the manufacturer and ask for details on the differences between the command sets.

If an instrument driver for your particular instrument does not exist, you can do one of several things (in rank order):

1. Use a driver for a similar instrument. Often similar instruments from the same manufacturer have similar if not identical command sets. The degree to which an instrument driver will need to be changed will depend on how similar the instruments and their command sets are. If the command sets are very different, you may be better off starting from scratch.

2. Use the Instrument I/O Assistant.

3. Develop a complete, fully functional instrument driver. To develop a National Instruments quality driver, you can read the *Developing LabVIEW Plug and Play Instrument Drivers* tutorial from the National Instruments website at

http://www.ni.com/idnet

This tutorial will help you to develop a complete instrument driver. To aid in the development of your instrument driver, National Instruments created an Instrument Driver Project Wizard to help you implement standards for instrument driver structure, device management, instrument I/O, and error reporting. The Instrument Driver Project Wizard can be accessed by navigating to Tools≫Instrumentation≫Create Instrument Driver Project in LabVIEW. The tutorial describes these standards, as well as the purpose of a LabVIEW instrument driver, its components, and the integration of these components. In addition, this tutorial suggests a process for developing useful instrument drivers.

As to the question of whether to use VISA, GPIB, or serial functions, the recommendation is to rely on VISA. There are four major reasons for using VISA VIs over GPIB VIs. VISA

- is the industry standard,

- provides interface independence,

- provides platform independence, and

- will be easily adaptable in future instrumentation control applications.

VISA uses the same operations to communicate with instruments regardless of the interface type. VISA is designed so that programs written using VISA function calls are easily portable from one interface type to another like GPIB, serial, Ethernet, USB, or VXI. To ensure platform independence, VISA strictly defines its own data types. The VISA function calls and their associated parameters are uniform across all platforms, so that software can be ported to other platforms and then recompiled. A LabVIEW program using VISA can be ported to any platform supporting LabVIEW. A final advantage of using VISA is that it is an object-oriented API that will easily adapt to new instrumentation interfaces as they are developed in the future, making application migration to the new interfaces easy.

If you must develop your own instrument driver, you should use VISA functions rather than GPIB functions because of the versatility of VISA.

A.5   FUTURE OF INSTRUMENT DRIVERS AND INSTRUMENT CONTROL

Current instrument drivers use VISA, which is a common software interface for controlling GPIB, serial, Ethernet, USB, or VXI instruments. At the present time and into the future, there will be one instrument driver for all oscilloscopes, for instance, no matter the manufacturer, model, or hardware interface (such as GPIB). These new instrument drivers are called Interchangeable Virtual Instruments (IVI) drivers and are supported by the IVI Foundation. The IVI Foundation is comprised of end-user test engineers and system integrators with many years of experience building GPIB- and VXI-based test systems. By defining a standard instrument driver model that enables engineers to swap instruments without requiring software changes, the IVI Foundation members believe that significant savings in time and money will result because of:

- Software that does not change when instruments become obsolete.

- A single software application that can be used on a system with different instrument hardware, which maximizes existing resources.

- Portable code that can be developed in test labs and hosted on different instruments in the production environment.

The following instrument types or classes have been defined by the IVI Foundation: oscilloscope, DMM, arbitrary waveform generator, switch, and power supply. More instrument types will be defined in the future. For more information on the IVI drivers and the IVI Foundation, refer to the tutorial *Using IVI Drivers to Build Hardware-Independent Test Systems with LabVIEW and LabWindows/CVI* on the National Instruments website, zone.ni.com/devzone/cda/tut/p/id/4558. You may also want to visit the IVI Foundation website at

http://www.ivifoundation.org/

A.6   SUMMARY

As we already know from the chapter on data acquisition, LabVIEW can communicate with external devices. As discussed in Chapter 8, you can use DAQ devices in conjunction with LabVIEW to read and generate analog input, analog output, and digital signals. In this appendix we learned that LabVIEW can also control external instruments (such as digital voltmeters and oscilloscopes). MAX is used to detect instruments and to install instrument drivers. An instrument driver is a VI that controls a particular instrument. Students should use the instrument drivers developed by National Instruments rather than attempt to develop drivers from "scratch." The Instrument I/O Assistant provides a relatively straightforward way to build instrument communication VIs. Instrument drivers can be downloaded from the National Instruments website using the Instrument Driver Network.

KEY TERMS

Bus: The means by which computers and instruments transfer data.

Controller: A GPIB device that manages the flow of information on the GPIB by sending commands to all devices. A computer usually handles the controller function.

Controller-in-charge: The active controller in a GPIB system.

Device-dependent messages: Messages that contain device-specific information such as programming instructions, measurement results, machine status, and data files. These are often called **data messages**.

Ethernet: Computer networking technology for local area networks. Extensive Ethernet networks are connected to over one hundred million Ethernet-capable computers worldwide, making this technology attractive for instrument control applications.

GPIB: General Purpose Interface Bus is the common name for the communications interface system defined in ANSI/IEEE Standard 488-1975 and 488.2-1987.

IEEE: Institute for Electrical and Electronics Engineers.

IEEE 1394: A high-performance serial bus developed by Apple in the 1980s. Data transfer rates up to 400 Mb/s make this an appealing option for instrument control.

Instrument driver: A set of LabVIEW VIs that communicates with an instrument using standard VISA I/O functions.

Interface messages: Messages that manage the bus itself and perform such tasks as initializing the bus, addressing and unaddressing devices, and setting device modes for remote or local programming. These are often called **command messages**.

Listener: A GPIB device that receives data messages from the talker.

MAX: A utility that provides information about connected instruments and installs and manages instrument drivers.

Serial communication: A popular means of transmitting data between a computer and a peripheral device by sending data one bit at a time over a single communication line.

Talker: A GPIB device that sends data messages to one or more listeners.

Universal Serial Bus (USB): A serial bus standard with plug-n-play capability designed to connect peripheral devices to PCs. USB offers the nice feature of ease of use in instrument control applications.

VISA: Virtual Instrument Software Architecture. VISA is a VI library for controlling GPIB, serial, Ethernet, USB, or VXI instruments.

APPENDIX B

LabVIEW Developer Certification

Certification provides a credible way to validate your LabVIEW expertise, and it can help you distinguish yourself, whether you are searching for a job or just looking to move up in your current organization. National Instruments offers several levels of LabVIEW certification, beginning with the Certified LabVIEW Associate Developer (CLAD), continuing with the Certified LabVIEW Developer (CLD), and culminating with the Certified LabVIEW Architect (CLA). We focus here on the CLAD level. To help you prepare for the CLAD introductory level certification examination, we provide a practice test with complete answers. We also provide information on additional resources to help you prepare for the examination. After completing *Learning with LabVIEW*, you should be well prepared for the CLAD exam.

GOALS

1.   Understand the NI LabVIEW Certification structure.
2.   Know the logistics for taking the CLAD exam.
3.   Understand the benefits of becoming NI Certified.
4.   Prepare for the CLAD exam using NI resources.

717

B.1   OVERVIEW OF THE NI LABVIEW CERTIFICATION STRUCTURE

National Instruments (NI) offers three certification designations to indicate different levels of expertise with LabVIEW. The Certified LabVIEW Associate Developer (CLAD), Certified LabVIEW Developer (CLD), and Certified LabVIEW Architect (CLA) designations represent basic, advanced, and expert knowledge, respectively, in using NI products for test and measurement applications. At each level of certification, you must demonstrate the following:

- **Certified LabVIEW Associate Developer:** You must demonstrate a basic understanding of the LabVIEW environment. You must show that you understand LabVIEW development best practices and that you can understand and interpret existing code.

- **Certified LabVIEW Developer:** You must demonstrate the ability to design and develop functional programs while minimizing development time and ensuring maintainability.

- **Certified LabVIEW Architect:** You must demonstrate the skill to develop a framework for an application to be executed by a team of developers, given a set of high-level requirements.

The three levels of certification are illustrated in Figure B.1. In order to take a higher-level certification examination, you must have successfully passed the lower-level examination(s). We focus here on the CLAD level.

B.2   LOGISTICS OF THE CLAD EXAMINATION

The CLAD examination is a one-hour multiple-choice test consisting of forty questions administered and proctored by Pearson VUE. Visit the website www.pearsonvue.com/ni/ to locate a test center near you and to schedule an examination. The CLAD examination is closed-book, and no reference materials or computers (other than the testing computer) are allowed at the test center during the examination. After you complete the test, the CLAD examination will be evaluated automatically and results will be made available to you. The minimum score to pass is 70%. You are allowed two attempts per certification level. If you are unsuccessful at both attempts, you must wait six months before reattempting the certification examination at that level. If you pass the examination, your certification will be valid for two years before renewal testing is required.

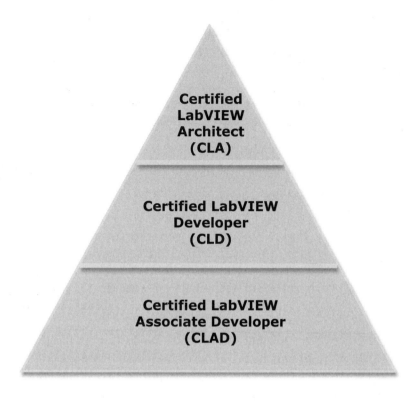

FIGURE B.1
NI LabVIEW certifications: CLAD, CLD, and CLA.

B.3 BENEFITS OF CLAD

Passing the CLAD examination can help you distinguish yourself for accelerated career development and better employment opportunities. Of course, being a Certified LabVIEW Associate Developer does not guarantee employment, but a demonstrated knowledge of LabVIEW may be especially useful in the job search. Job opportunities may be found at the website www.LabVIEWJobs.com or at one of the many NI Alliance Partners. Even with years of experience, the journey to achieve LabVIEW certified status may challenge you to hone coding and documentation skills that will make you a more efficient and effective developer.

Certified developers and architects also receive certification branded merchandise and can use the NI certification logo on business cards, websites, and resumes. You can also distinguish yourself within the NI Developer Community through listing on the National Instruments website, and you can add the NI certification logo to your www.ni.com/community profile. Certified users often receive recognition at developer events, such as NIWeek.

NI provides certified developers many opportunities to network with fellow technology leaders and engage in peer-to-peer discussions online and at events. Specifically, developers can connect at biquarterly LabVIEW Virtual User Group meetings segmented by certification level and during an annual roundtable lunch at NIWeek.

B.4 SAMPLE CLAD EXAMINATION

The CLAD examination is given in both a print and an online format. The sample CLAD test question format presented here is designed to prepare you for the type of questions asked. You can make best use of it by taking the practice test and then reviewing your answers (solutions provided in Section B.5) and the detailed descriptions to better understand the rationale for each solution.

Reading this textbook should be in itself sufficient preparation for the CLAD examination. We have covered all of the general question topics, but not every possible LabVIEW concept that may come up in a CLAD examination. The CLAD is designed to test your ability to use the LabVIEW software. Therefore, *solving homework and doing projects with the software* provide opportunities for important additional learning, which will help you pass the examination.

The Sample Exam and solutions were provided by National Instruments and are reprinted here with NI permission. Table B.1 is a blank answer sheet for you to use as you work through the examination.

Certified LabVIEW Associate Developer Exam Test Booklet

Note: The use of the computer or any reference materials is NOT allowed during the exam.

Instructions:
If you did not receive this exam in a sealed envelope stamped "NI Certification," **DO NOT ACCEPT** this exam. Return it to the proctor immediately. You will be provided with a replacement exam.

- **Please do not detach the binding staple of any section. If any part of the exam paper is missing or detached when returned to National Instruments, you will be deemed to have failed the exam.**

- Please do not ask the proctor for help. If you believe the intent of a question is not clear, you may note that question and your reasons for choosing the answer you believe best fits the question.

- This examination may not be taken from the examination area or reproduced in any way. You may not keep any portion of this exam after you have completed it.

Exam Details:

- Time allocated: 1 hour
- Type of exam items: Multiple choice
- Number of exam items: 40 questions
- Passing Grade: 70%

IMPORTANT: When you have completed this exam, place it in the provided envelope with your answer sheet and SEAL the envelope. Give the sealed envelope to your proctor.

TABLE B.1   Answer Sheet

> To check your answers against the solutions on the Solutions Page, record your answers here. This page is not included in the actual CLAD exam; it is included here for practice purposes only. The solutions are provided in Section B.5.
>
> | 1. _____ | 2. _____ | 3. _____ | 4. _____ | 5. _____ |
> | 6. _____ | 7. _____ | 8. _____ | 9. _____ | 10. _____ |
> | 11. _____ | 12. _____ | 13. _____ | 14. _____ | 15. _____ |
> | 16. _____ | 17. _____ | 18. _____ | 19. _____ | 20. _____ |
> | 21. _____ | 22. _____ | 23. _____ | 24. _____ | 25. _____ |
> | 26. _____ | 27. _____ | 28. _____ | 29. _____ | 30. _____ |
> | 31. _____ | 32. _____ | 33. _____ | 34. _____ | 35. _____ |
> | 36. _____ | 37. _____ | 38. _____ | 39. _____ | 40. _____ |

Sample Exam Items:

1. For implementing state diagrams that allow future application scalability, the best choice for a base structure is?

    a. Sequence structure

    b. Case structure

    c. Formula node

    d. Object-Oriented structure

2. Clicking on the _____ button allows you to bypass a node in the block diagram without single-stepping through the node.

   a. Step Into

   b. Step Over

   c. Step Out

   d. Step Through

3. Which of the following apply to Property Nodes? **(More than one answer may apply.)**

   a. Property Nodes allow attributes of front panel objects to be programmatically manipulated

   b. Property Nodes can be used to update the values contained in a front panel object

   c. More than one Property Node can be used for a single front panel object

   d. Property Nodes can be used to programmatically generate a Value Change event

4. You have a front panel control on a top-level VI that you must control from within a subVI. What must you pass to the subVI?

   a. The control's properties

   b. The control's methods

   c. The control's reference

   d. The control's data type

5. What value does the Value Out indicator display after the VI executes?

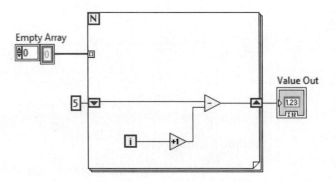

   a. 0

   b. 4

   c. 5

   d. 6

6.   You set a Breakpoint in the code contained within the False case of a Case structure. Will the VI execution pause if the code in True case of the Case structure is being executed?

　　a.   Yes

　　b.   No

7.   Which Property Node changes the text color attributes of any digital numeric control from a subVI?

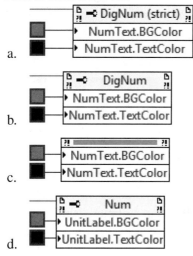

8.   Which chart update mode should be used to show running data continuously scrolling from left to right across the chart?

　　a.   Strip Chart

　　b.   Scope Chart

　　c.   Sweep Chart

　　d.   Step Chart

9.   Which of the following does not conform to the dataflow programming paradigm?

　　a.   Shift Registers

　　b.   Tunnels

　　c.   SubVIs

　　d.   Local variables

10. If possible, a Sequence structure should be replaced with a(n):

　　a.   Event structure

　　b.   For loop

　　c.   State Machine

　　d.   While loop

11. Which of the following statements is true about the following code segment?

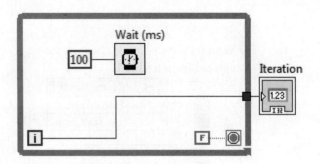

   a. The loop will execute once and the indicator Iteration will have a value of one

   b. The loop will execute once and the indicator Iteration will have a value of zero

   c. The loop will execute infinitely and the program will have to be aborted

   d. The loop will not execute and the indicator Iteration will have a value of zero

12. Which equation is equivalent to the code?

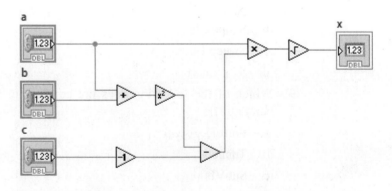

   a. $x = \sqrt{a \times \left((a + b)^2 - (c - 1)\right)}$

   b. $x = \sqrt{a \times \left((c - 1) - (a + b)^2\right)}$

   c. $x = \sqrt{a \times \left((a + b)^2 - (c - 1)\right)}$

   d. $x = \sqrt{a^2 \times \left((a + b)^2 - (c - 1)\right)}$

13. A VI has **Automatic Error Handling** enabled. It calls a SubVI that has **Automatic Error Handling** disabled. The Error Out terminal of the SubVI is unwired in the calling VI. The SubVI returns an error to the calling VI. Will an error dialog be posted because of automatic error handling?

 a. No, since the SubVI has **Automatic Error Handling** disabled

 b. Yes, since the SubVI has **Automatic Error Handling** disabled

 c. No, since the Error Out terminal from the subVI is not wired in the calling VI

 d. Yes, since the Error Out terminal from the subVI is not wired in the calling VI

14. Which of the following statements is false?

 a. A SubVI connector pane defines where to wire inputs and outputs

 b. The color of a SubVI connector pane terminal matches the data type it is connected to

 c. You must have an icon/connector pane to use a SubVI

 d. A SubVI icon can be edited from the functions palette

15. Which of the following cannot be used to transfer data?

 a. Semaphores

 b. Queues

 c. Notifiers

 d. Local variables

16. The Error list shows all of the following except:

 a. Items with errors

 b. Errors and warnings

 c. Details about the warnings

 d. Error Codes

17. Auto-indexing at loop boundary is a feature that allows: **(More than one answer may apply.)**

 a. Arrays to be automatically traversed from one iteration to the next

 b. Array functions to automatically index array elements

 c. Functions that use arrays to automatically index them

 d. Arrays to be automatically accumulated at loop boundaries

18. What is the output of the Initialize Array function after the following code has executed?

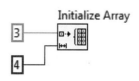

    a.   1-D Array of {3, 3, 3, 3}

    b.   1-D Array of {4, 4, 4}

    c.   1-D Array of {3, 4}

    d.   1-D Array of {4, 3}

19. Which of the following will cause an event to be captured by the LabVIEW Event Structure?

    a.   Changing a front panel control's value via a mouse click on the control

    b.   Changing a front panel control's value property via a Property Node

    c.   Changing a front panel control's value via a Control Reference

    d.   Changing a front panel control's value via a Local Variable

20. Which variable is commonly used to eliminate race conditions by preventing simultaneous access to code or data?

    a.   Functional global variable

    b.   Local variable

    c.   Global variable

    d.   Shared variable

21. What is the value in Shift Register Answer after the following code has executed?

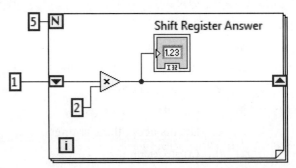

    a.   16

    b.   24

    c.   32

    d.   10

22. What is the result in subarray after the following code has executed?

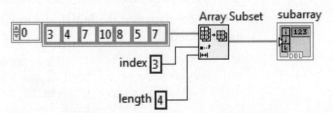

    a.   1-D Array of {8, 5, 7}

    b.   1-D Array of {10, 8, 5}

    c.   1-D Array of {10, 8, 5, 7}

    d.   1-D Array of {7, 10, 8, 5}

23. How do you document a VI so that the description appears in the **Show Context Help** popup window?

    a.   Use the **VI Properties Documentation** window

    b.   Type in the **Show Context Help** window

    c.   Create a free label on the front panel

    d.   Edit the LabVIEW help files

24. Which of the following terminals controls how many times a For loop executes?

    a.   [i]

    b.   [↻]

    c.   [◉]

    d.   [N]

25. Does this code snippet demonstrate an appropriate use of a Property Node?

    a.   Yes, because updating the value via a Property Node uses less memory than writing to the terminal

    b.   Yes, because updating the value via a Property Node allows for error checking

    c.   No, because you should always wire data directly into indicators when possible

    d.   No, because Property Nodes should only be used with more complex data types

26. The Wait Until Next ms Multiple function:

    a.   Begins timing at program start and completes upon a multiple of the PC timer

    b.   Begins timing after the code in the loop has completed and expires when the multiple of the (user) input time expires

   c. Begins timing after the code in the loop has completed and expires when the (user) input time completes

   d. Begins timing at program start and waits until the code in the loop has completed

27. Which data synchronization mechanism ensures that no data is lost when an application temporarily provides data faster than it is able to process it?

   a. Notifier

   b. Queue

   c. Semaphore

   d. Local Variable

28. Which of the following statements is not valid:

   a. You can make a Cluster of Clusters

   b. You can make an Array of Arrays

   c. You can make a Cluster of Arrays

   d. You can make an Array of Clusters

29. Which data type is not accepted by the case selector terminal on a case structure?

   a. Arrays

   b. Enumerated type values

   c. Strings

   d. Integers

30. How many bytes does LabVIEW write to the file when the code executes?

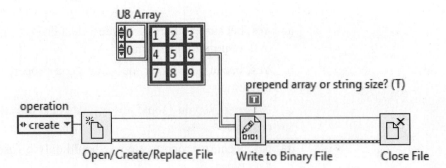

   a. 9 Bytes

   b. 11 Bytes

   c. 13 Bytes

   d. 17 Bytes

31. Why is the sequence local terminal displayed as unassigned in Frame 0 of the Stacked Sequence structure?

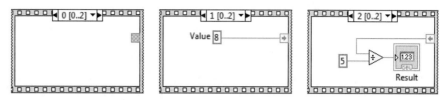

    a.   The developer chose not to wire the value to any terminal in this frame

    b.   The value is available only to frames after frame 1

    c.   The data type of the terminal is incompatible with the data type of Value

    d.   The developer disabled the terminal

32. The Wait function can be added to While Loops:

    a.   To free up available memory

    b.   To allocate memory used by the CPU

    c.   To allow the processor time to complete other tasks

    d.   To reserve which processor the code is running on

33. The data type of a Ring can be any of the allowable data types, whereas the numerical representation of the data type of an Enum can only be of unsigned data type.

    a.   True

    b.   False

34. When the user clicks the **Button** control, how many times is the Increment function called?

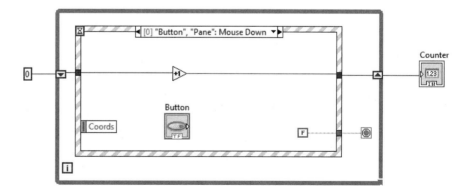

    a.   0

    b.   1

    c.   2

    d.   3

35. Why is LabVIEW a data flow programming language?

    a. It uses icons instead of text to create applications

    b. The flow of data through nodes on the block diagram determines execution

    c. It does not have a Goto function to control execution

    d. The memory location in which data is stored changes each iteration

36. If the value in a is 2 and b is 128, what value does the Result indicator display after the VI executes?

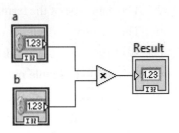

    a. -128

    b. 0

    c. 255

    d. 256

37. What VI is typically used to terminate an Error Cluster wire and to display any error message?

    a. Merge Errors

    b. One Button Dialog/Two Button Dialog

    c. Generate Front Panel Activity

    d. Simple Error Handler

38. Which of the following methods will create an XY Graph with two plots?

    a. Create a single cluster from two arrays of X values and two arrays of Y values in an X, Y, X, Y pattern

    b. Bundle the X and Y array pairs together into two clusters and then create an array of these two clusters

    c. Bundle the X and Y array pairs together into two clusters and create a cluster of these two clusters

    d. Create a single array of each of the X arrays and Y arrays in an X, Y, X, Y pattern

39. While the VI executes, the user presses and then releases Boolean Control with **Switch Until Released** mechanical action. Assuming that the starting value of Boolean Control is False, what value does the Result indicator display after execution?

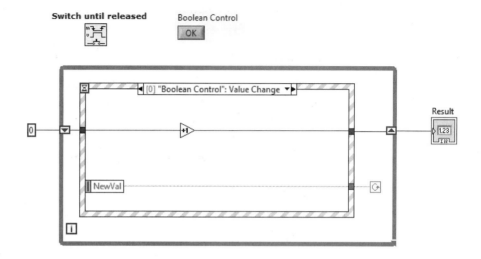

    a. 0

    b. 1

    c. 2

    d. 3

40. Which of the following is the best method to update an indicator on the front panel?

    a. Use a Value Property Node

    b. Wire directly to the indicator terminal

    c. Use a local variable

    d. Use a functional global variable

B.5 DETAILED SAMPLE CLAD TEST SOLUTIONS

1. **B** In LabVIEW, simple State Machines consist of a While Loop containing a Case structure. The Case structure allows the decision of which case to run to be determined programmatically. It is also scalable since it is easy to add new cases to the Case structure.

2. **B** Clicking the **Step Into** button causes LabVIEW to open up the node or subVI. The **Step Out** button is used to return from a subVI to a main VI

while single-stepping. There is no **Step Through** button. **Step Over** provides the functionality stated by the question; therefore this is the correct answer.

3. **A, B, C, D** All of the statements are true for Property Nodes.

4. **C** The only answer option that can be passed from calling VI to subVI is the control reference. The reference can then be used with Property Nodes and Invoke Nodes to call properties and methods, respectively. The data type is a property of the control.

5. **C** An empty array is wired to the For Loop using an auto-indexing tunnel. This causes the For Loop to iterate once for every element in the array, which, in this case, is zero. However, the value 5 is written to the shift register before loop execution, and since the loop iterates zero times, the same value of 5 is present at the output shift register.

6. **B** Code containing breakpoints executes like normal until data is passed on a wire that contains a breakpoint. Then, LabVIEW suspends execution. In order for the breakpoint to activate, the section of code containing it must be called. With a Case structure, only one case executes each time it is called. Since, in this case, the case containing the breakpoint was not called, the execution was not paused.

7. **B** Strict Property Nodes require precise knowledge of the numeric representation. Since the question asks for the means to change the text color of any control, a strict Property Node does not suffice. An implicit Property Node (C) can only be used locally. Answer D shows a property that is not even relevant to the question. Thus, option B is the best answer.

8. **A** Strip charts start plotting from left to right and continue to scroll while plotting. Scope charts start plotting from left to right and continue until the chart is full. Then the chart is cleared, and plotting resumes at the left. Sweep charts behave similarly to scope charts, except that once the chart is full, sweep charts start plotting at the left and progressively overwrite previously plotted data. There is no such thing as a Step chart in LabVIEW.

9. **D** Local variables do not conform to the dataflow paradigm because they communicate by reference, not by value. The basic premise of local variables is to allow transfer of data where it is impossible to use wires. This circumvents the dataflow paradigm.

10. **C** A State Machine can be used to achieve the same functionality as a Sequence structure. However, State Machines allow the developer to programmatically determine the sequence at run-time. This makes the State Machine implementation more scalable.

11. **C** The While Loop has a False Boolean wired to its conditional terminal, which is set to **Stop if True**. Thus, there is no condition that causes this

While Loop to stop. So, the While Loop could run infinitely and the VI must be aborted.

12. **A** Because LabVIEW is a dataflow language, we can trace the flow of data in the block diagram to see which operations execute first, second, and so forth.

13. **D** Since automatic error handling is enabled in the main VI, and the error terminals of the subVI are not wired, LabVIEW automatically handles the error by displaying a dialog.

14. **D** Icons cannot be edited from the **Functions** palette. SubVI icons can be edited by right-clicking the icon in the upper right of the VI and selecting **Edit Icon**.

15. **A** Queues, notifiers, and local variables are all designed to transfer data. Semaphores do not pass data. Instead, their sole purpose is to prevent certain sections of code from running while other critical sections are running.

16. **D** You must look at either an output error cluster or an error dialog to find the error code.

17. **A, D** Auto-indexing is a feature for loops interacting with arrays. Array functions themselves do not have iterative auto-indexing features.

18. **A** The Initialize Array function creates an array with a length specified by dimension size. The value of each element is specified by the element input. To initialize multidimensional arrays, you can simply expand the Initialize Array function to display more dimension size inputs.

19. **A** Each possible answer refers to a Value Change event. Value Change events are generated either by user interaction on the front panel or by calling a Value (Signaling) Property Node. Calling a Value Property Node does not generate an event.

20. **A** You can place critical data or sections of code in functional global variables. Since functional global variables are non-reentrant VIs, the possibility of race conditions is eliminated.

21. **C** The For Loop executes 5 times. Starting with the value of 1, the result of the previous iteration is multiplied by 2. Thus, the value in the indicator after 5 iterations is equivalent to $1 \times 2 \times 2 \times 2 \times 2 \times 2$ (25), or 32.

22. **C** The Array Subset function takes an input array and returns a section of that array as specified, starting at the given index and continuing for a number of elements equal to length. Here, the index value 3 specifies the fourth element of the array, or 10. Since the specified length is value 4, an array of length 4 is returned as follows: {10, 8, 5, 7}.

23. **A** The documentation window in **VI Properties** is the only place to edit the information about the VI that appears in **Context Help**.

24. **D** For Loops refer to the input to the Count terminal to determine how many iterations to execute.

25. **C** Since the indicator terminal is not isolated from the data being written by any structures, there is no reason not to wire the data directly to the indicator instead of using a Property Node. It is always best practice to wire directly when possible, because using variables or Property Nodes to update values can cause race conditions if not used carefully.

26. **A** Answers B and D are incorrect because the Wait Until Next ms Multiple function executes before loop iterations terminate, not after. Answer C is incorrect, because it describes the functionality of the Wait (ms) function.

27. **B** Answer C is incorrect because semaphores cannot pass data. Answer A is incorrect because notifiers pass data, but they can only pass one element at a time. Data is overwritten and lost if the program writes to the notifier twice before data is read. Answer D is incorrect because local variables have no mechanism for determining when data is updated, so there is no way to tell if data is newly-acquired or not. Queues support multiple elements and operate using a FIFO principle, guaranteeing that no data is lost or overwritten.

28. **B** There is no need to make an array of arrays since you can simply add dimensions to an existing array.

29. **A** Arrays are not accepted by the case selector terminal because the case selector terminal requires a scalar value.

30. **D** LabVIEW represents arrays as a list of bytes containing a header and array data. The header contains a 4-byte integer for each dimension that specifies the length of that dimension. Following the header is the actual array data. In the question, the array has two dimensions. There are 4 bytes in the header for each dimension for a total of 8 bytes in the header. Since the array is composed of 8-bit, or 1-byte integers, there is a total of 9 bytes of actual array data. This makes a total of 17 bytes being written to file.

31. **B** The sequence local is first written to in frame 1. Thus, in frame zero, no data is available to read since no data has yet been written.

32. **C** The Wait (ms) function does nothing to release or allocate memory or specify processor core. All it does is cause the execution of a VI to pause for a short time to allow the processor time to complete other tasks.

33. **A** Rings can use any numeric representation while enums can use only unsigned integers.

34. **C** The event case shown handles two events: the Mouse Down event on the **Button** control, and the Mouse Down Event on the front panel. These two events occur simultaneously, so the event structures queue them up, and handle one and then the other. Thus two iterations occur, and the value in the shift register is incremented twice.

35. **B** Functions and VIs execute as soon as all of their inputs have data. Thus, it is the flow of data through the program that determines the order of execution.

36. **B** The integers shown are unsigned 8-bit integers. The range for 8-bit integers is 0-255. The product of 2 times 128 is 256, one more than the maximum allowable value for an unsigned 8-bit integer. Thus, the value wraps around to 0.

37. **D** The Merge Errors function does not display any dialog. The One and Two Button Dialog functions are for general purpose and are not best applied for error handling applications. There is no error input on these functions. There is no VI with the name Generate Front Panel Activity. The Simple Error Handler is the best choice because it accepts an error cluster as an input and displays a dialog to the user in the event of an error.

38. **B** XY graphs accept a cluster of two arrays, an X array, and a Y array to generate a single plot. To generate multiple plots, XY graphs accept an array of these clusters.

39. **C** Since the mechanical action is set to **Switch Until Released**, two events are generated when a user clicks and releases. The first event is the False to True transition, and the second is the True to False transition. Latch actions are specifically designed to reset the value of the button after the change is read without generating a second event.

40. **B** When possible, it is always best to wire data directly to indicators. This helps maintain the dataflow paradigm.

B.6   ADDITIONAL STUDY RESOURCES

As you may have discovered when taking the sample examination, you need to master many concepts in order to become recognized as a Certified LabVIEW Associate Developer. While this textbook may not explicitly prepare you for every concept, it does prepare you to learn through discovery. If you are studying this book on your own and are looking for a quick, formal introduction to LabVIEW, consider the *LabVIEW Core I* and *LabVIEW Core II* courses offered by National Instruments. All of the required concepts are covered using both classroom instruction and interactive computer-based exercises with LabVIEW and data-acquisition hardware.

The links to the following resources are found at www.ni.com/training:

1. **CLAD Preparation Course:** National Instruments CLAD Preparation Course (Online) webcast at www.ni.com/webcast/566/en

2. **Complete List of CLAD Exam Topics:** CLAD Exam Preparation Guide found at ftp://ftp.ni.com/pub/devzone/tut/clad_exam_preparation_guide.pdf

3. **Training / Tutorials:**

   a. Three Hours LabVIEW Introduction Course

   b. Six Hours National Instruments

   c. LabVIEW Development Guidelines

   d. Free practice LabVIEW Fundamentals Exam

   e. LabVIEW Manuals Online (current manuals)

B.7   SUMMARY

The Certified LabVIEW Associate Developer (CLAD), Certified LabVIEW Developer (CLD), and Certified LabVIEW Architect (CLA) were discussed. We found that this textbook teaches topics specifically covered in the CLAD examination. A sample CLAD practice test with the complete answer key was provided to help familiarize you with the format and types of questions that will be asked. Certification provides a credible way to validate your LabVIEW expertise and can help you distinguish yourself, whether you are searching for a job or just looking to move up in your current organization. We also provided some additional resources to prepare for taking the CLAD introductory level certification exam.

KEY TERMS

Certified LabVIEW Associate Developer: A CLAD demonstrates a basic understanding of the LabVIEW environment and of LabVIEW development best practices.

Certified LabVIEW Developer: A CLD demonstrates the ability to design and develop functional programs while minimizing development time and ensuring maintainability.

Certified LabVIEW Architect: A CLA demonstrates the skill to develop a framework for an application to be executed by a team of developers, given a set of high-level requirements.

LabVIEW Virtual User Group: Technical support group for LabVIEW enabling virtual networking with programmers and developers worldwide.

NI Developer Community: A worldwide community of Lab-VIEW users seeking to discover and collaborate on the latest example code, tutorials, and textbooks, to share development techniques, learn about cutting-edge technologies, and connect with LabVIEW and other NI product experts working on similar applications.

Pearson VUE: Administers and proctors the certification examinations, including the CLAD.

INDEX